Cambridge IGCSE™

Mathematics
Core and Extended
Fifth Edition

Ric Pimentel
Frankie Pimentel
Terry Wall

Endorsement indicates that a resource has passed Cambridge International's rigorous quality-assurance process and is suitable to support the delivery of a Cambridge International syllabus. However, endorsed resources are not the only suitable materials available to support teaching and learning, and are not essential to be used to achieve the qualification. Resource lists found on the Cambridge International website will include this resource and other endorsed resources. Any example answers to questions taken from past question papers, practice questions, accompanying marks and mark schemes included in this resource have been written by the authors and are for guidance only. They do not replicate examination papers. In examinations the way marks are awarded may be different. Any references to assessment and/or assessment preparation are the publisher's interpretation of the syllabus requirements. Examiners will not use endorsed resources as a source of material for any assessment set by Cambridge International. While the publishers have made every attempt to ensure that advice on the qualification and its assessment is accurate, the official syllabus specimen assessment materials and any associated assessment guidance materials produced by the awarding body are the only authoritative source of information and should always be referred to for definitive guidance. Cambridge International recommends that teachers consider using a range of teaching and learning resources based on their own professional judgement of their students' needs. Cambridge International has not paid for the production of this resource, nor does Cambridge International receive any royalties from its sale. For more information about the endorsement process, please visit www.cambridgeinternational.org/endorsed-resources.

Cambridge International copyright material in this publication is reproduced under licence and remains the intellectual property of Cambridge Assessment International Education.

Photo credits: pp.2–3 © Chaoss/stock.adobe.com; **p.3** © Dinodia Photos/Alamy Stock Photo; **pp.106–7** © Denisismagilov/stock.adobe.com; **p.107** © Eduard Kim/Shutterstock; **pp.256–7** © Microgen/stock.adobe.com; **p.257** © Georgios Kollidas/123RF; **pp.288–9** © Tampatra/stock.adobe.com; **p.289** © Classic Image/Alamy Stock Photo; **pp.348–9** © Ardely/stock.adobe.com; **p.349** © Hirarchivum Press/Alamy Stock Photo; **pp.390–1** © Marina Sun/stock.adobe.com; **p.391** © Georgios Kollidas/stock.adobe.com; **pp.442–3** © Ohenze/stock.adobe.com; **p.443** © Photo Researchers/Science History Images/Alamy Stock Photo; **pp.478–9** © Princess Anmitsu/stock.adobe.com; **p.479** © Bernard 63/stock.adobe.com; **p.504–5** © J BOY/stock.adobe.com; **p.505** © Caifas/stock.adobe.com

Every effort has been made to trace all copyright holders, but if any have been inadvertently overlooked, the Publishers will be pleased to make the necessary arrangements at the first opportunity.

Although every effort has been made to ensure that website addresses are correct at time of going to press, Hodder Education cannot be held responsible for the content of any website mentioned in this book. It is sometimes possible to find a relocated web page by typing in the address of the home page for a website in the URL window of your browser.

Hachette UK's policy is to use papers that are natural, renewable and recyclable products and made from wood grown in well-managed forests and other controlled sources. The logging and manufacturing processes are expected to conform to the environmental regulations of the country of origin.

Orders: please contact Hachette UK Distribution, Hely Hutchinson Centre, Milton Road, Didcot, Oxfordshire, OX11 7HH. Telephone: +44 (0)1235 827827. Email education@hachette.co.uk Lines are open from 9 a.m. to 5 p.m., Monday to Friday. You can also order through our website: www.hoddereducation.com

© Ric Pimentel, Terry Wall and Frankie Pimentel 2023

First published in 1997
Second edition published in 2006
Third edition published in 2013
Fourth edition published in 2018

This edition published in 2023 by
Hodder Education, An Hachette UK Company, Carmelite House, 50 Victoria Embankment, London EC4Y 0DZ

www.hoddereducation.com

The authorised representative in the EEA is Hachette Ireland, 8 Castlecourt Centre, Dublin 15, D15 XTP3, Ireland (email: info@hbgi.ie)

Impression number   10 9 8 7 6

Year   2027 2026 2025

All rights reserved. Apart from any use permitted under UK copyright law, no part of this publication may be reproduced or transmitted in any form or by any means, electronic or mechanical, including photocopying and recording, or held within any information storage and retrieval system, without permission in writing from the publisher or under licence from the Copyright Licensing Agency Limited. Further details of such licences (for reprographic reproduction) may be obtained from the Copyright Licensing Agency Limited, www.cla.co.uk

Cover photo © Lev - stock.adobe.com

Print in India

Typeset by Integra Software Services Pvt. Ltd., Pondicherry, India

A catalogue record for this title is available from the British Library.

ISBN: 978 1 3983 7391 4

Contents

	Introduction		v
TOPIC 1	**Number**		**2**
	Chapter 1	Number and language	4
	Chapter 2	Accuracy	13
	Chapter 3	Calculations and order	25
	Chapter 4	Integers, fractions, decimals and percentages	31
	Chapter 5	Further percentages	45
	Chapter 6	Ratio and proportion	53
	Chapter 7	Indices, standard form and surds	63
	Chapter 8	Money and finance	79
	Chapter 9	Time	89
	Chapter 10	Set notation and Venn diagrams	92
		Mathematical investigations and ICT 1	102
TOPIC 2	**Algebra and graphs**		**106**
	Chapter 11	Algebraic representation and manipulation	108
	Chapter 12	Algebraic indices	128
	Chapter 13	Equations and inequalities	132
	Chapter 14	Graphing inequalities and regions	153
	Chapter 15	Sequences	157
	Chapter 16	Proportion	170
	Chapter 17	Graphs in practical situations	176
	Chapter 18	Graphs of functions	197
	Chapter 19	Differentiation and the gradient function	224
	Chapter 20	Functions	245
		Mathematical investigations and ICT 2	254
TOPIC 3	**Coordinate geometry**		**256**
	Chapter 21	Straight-line graphs	258
		Mathematical investigations and ICT 3	285
TOPIC 4	**Geometry**		**288**
	Chapter 22	Geometrical vocabulary and construction	290
	Chapter 23	Similarity and congruence	301
	Chapter 24	Symmetry	314
	Chapter 25	Angle properties	322
		Mathematical investigations and ICT 4	345
TOPIC 5	**Mensuration**		**348**
	Chapter 26	Measures	350
	Chapter 27	Perimeter, area and volume	355
		Mathematical investigations and ICT 5	388

CONTENTS

TOPIC 6	**Trigonometry**		**390**
	Chapter 28 Bearings		392
	Chapter 29 Trigonometry		394
	Chapter 30 Further trigonometry		422
	Mathematical investigations and ICT 6		440
TOPIC 7	**Vectors and transformations**		**442**
	Chapter 31 Vectors		444
	Chapter 32 Transformations		456
	Mathematical investigations and ICT 7		475
TOPIC 8	**Probability**		**478**
	Chapter 33 Probability		480
	Chapter 34 Further probability		491
	Mathematical investigations and ICT 8		501
TOPIC 9	**Statistics**		**504**
	Chapter 35 Mean, median, mode and range		506
	Chapter 36 Collecting, displaying and interpreting data		512
	Chapter 37 Cumulative frequency		535
	Mathematical investigations and ICT 9		544
	Glossary		547
	Index		560

Answers can be found at www.hoddereducation.com/CambridgeExtras

Introduction

This book has been written for all students of Cambridge IGCSE™ and IGCSE (9–1) Mathematics (0580/0980) for examination from 2025. It carefully and precisely follows the syllabus from Cambridge Assessment International Education. It provides the detail and guidance that are needed to support you throughout your course and help you to prepare for your examinations. The content is aimed particularly at students studying the Extended syllabus: our *Cambridge IGCSE™ Core Mathematics* Student's Book provides support for students focusing on the Core syllabus only.

How to use this book

To make your study of mathematics as rewarding and successful as possible, this book, endorsed by Cambridge International, offers the following important features:

Learning objectives

Each topic starts with an outline of the subject material and syllabus content to be covered. These opening pages show the learning objectives on the Extended syllabus, prefixed with an 'E'. All Core syllabus learning objectives are covered in the *Cambridge IGCSE Core Mathematics* Student's Book.

Organisation

Topics follow the order of the syllabus and are divided into chapters. In some cases, the order of the chapters is determined by continuity of the mathematics they cover, rather than the order of the syllabus. All instances where students should refer to other chapters are clearly explained in the text.

Within each chapter there is a blend of teaching, worked examples and exercises to help you build confidence and develop the skills and knowledge you need. In particular, there is an increased emphasis on non-calculator methods as well as suggestions for the use of scientific calculators.

At the end of each chapter there are comprehensive Student Assessment questions. You will also find sets of questions linked to the **Boost eBook** (boost-learning.com), which offer practice in topic areas that students often find difficult.

Although the authors have identified what material belongs exclusively to the Extended syllabus by use of a purple line, students studying the Extended syllabus are expected to be familiar with all the content of the Core syllabus as well. Material which is also part of the Core syllabus has been identified with a blue line.

ICT, mathematical modelling and problem-solving

Problem-solving is key to mathematical thinking and ICT can play a crucial role in this. Therefore, each topic ends with a section involving investigations and the use of ICT. The ICT investigations are, however, beyond the requirements of the syllabus and are identified with a yellow line as explained below.

Diagrams and working

Students are encouraged to draw diagrams when tackling questions where appropriate, and to show their full worked solutions. This is helpful for checking your own work, and also applies to any questions where use of a calculator is allowed.

Calculator and non-calculator questions

All exercise questions that should be attempted without a calculator are indicated by . Students should do as many calculations as possible without using a calculator. This will help to build understanding and confidence.

Some areas of mathematics, such as those using powers and roots, π, trigonometry and calculations with decimals, are more likely to require a calculator.

Core, Extended and Additional material

As this book covers the syllabus for both the Core and Extended content, we have used vertical coloured lines to distinguish between the two: a blue line and a purple line. Furthermore, there are a few instances where we have judged it to be appropriate to include some additional content that lies beyond the scope of the syllabus – where we consider it to be useful in supporting the syllabus content, and helpful in deepening understanding. This is indicated with a yellow line.

Key terms and glossary

It is important to understand and use mathematical terms; therefore, all key terms are highlighted in bold and explained in the glossary.

Answers and worked solutions

Answers to all questions are available free on hoddereducation.com/cambridgeextras

Worked solutions for the Student Assessment questions are available in *Cambridge IGCSE Core and Extended Mathematics Teacher's Guide with Boost Subscription*.

Callouts and Notes
These commentaries provide additional explanations and encourage full understanding of mathematical principles. Notes give additional clarifications and tips.

Worked examples
The worked examples cover important techniques and question styles. They are designed to reinforce the explanations, and give you step-by-step help for solving problems.

Exercises
These appear throughout the text and allow you to apply what you have learned. There are plenty of routine questions covering important mathematical techniques.

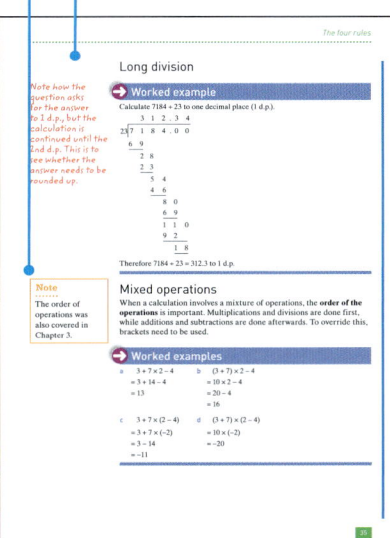

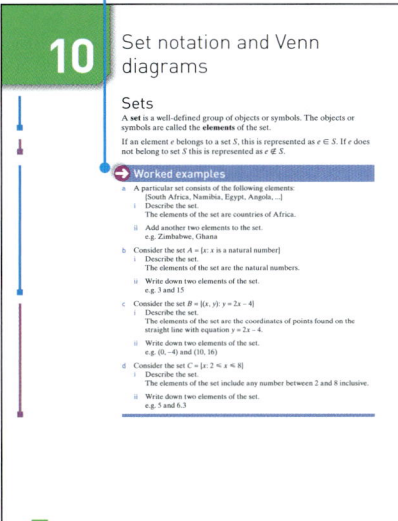

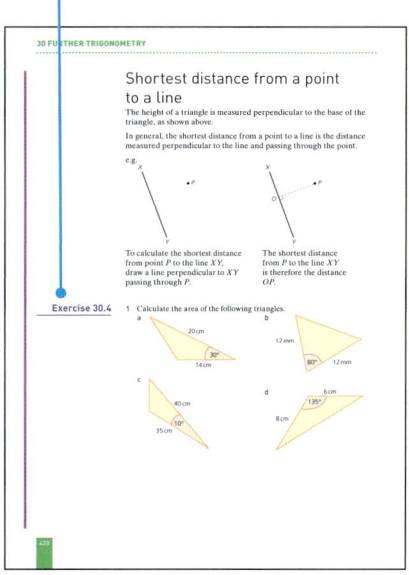

A blue line identifies content that is relevant for all students, regardless of whether they are studying the Core or Extended syllabus. A purple line identifies Extended content only. A yellow line identifies any material that lies beyond the scope of the syllabus (occasionally included because it can be helpful for students).

Mathematical investigations and ICT
More problem-solving activities are provided at the end of each section to put what you've learned into practice.

Student assessments
End-of-chapter questions to test your understanding of the key topics and help to prepare you for your examinations.

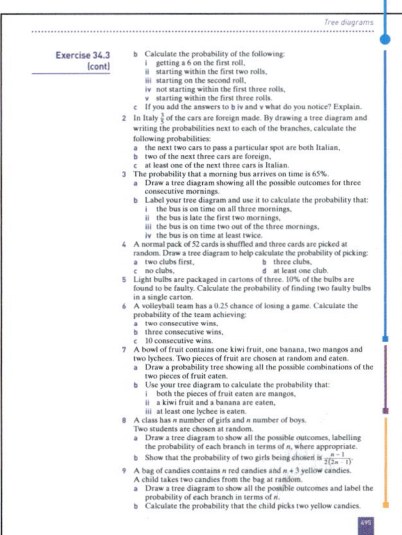

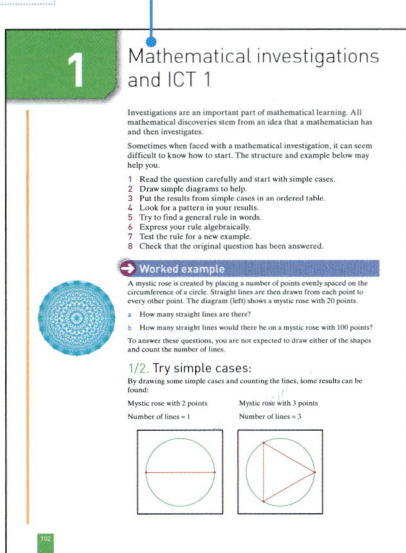

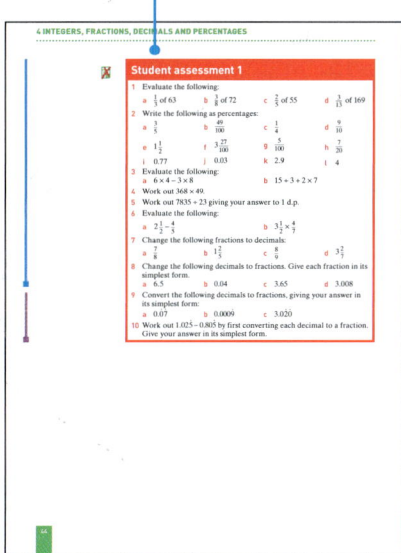

Assessment

The information in this section is based on the Cambridge International syllabus. You should always refer to the appropriate syllabus document for the year of examination to confirm the details and for more information. The syllabus document is available on the Cambridge International website at www.cambridgeinternational.org

For Cambridge IGCSE™ Mathematics you will take two papers. If you are studying the Core syllabus, you will take Paper 1 and Paper 3. If you are studying the Extended syllabus, you will take Paper 2 and Paper 4. You may use a scientific calculator only for Papers 3 and 4, Paper 1 and Paper 2 are non-calculator papers.

Paper	Length	Type of questions
Paper 1 Non-calculator (Core)	1 hour 30 minutes	Structured and unstructured questions
Paper 2 Non-calculator (Extended)	2 hours	Structured and unstructured questions
Paper 3 Calculator (Core)	1 hour 30 minutes	Structured and unstructured questions
Paper 4 Calculator (Extended)	2 hours	Structured and unstructured questions

Command words

The command words that may appear in examinations are listed below. The command word used will relate to the context of the question.

Command word	What it means
Calculate	Work out from given facts, figures or information
Construct	Make an accurate drawing
Describe	State the points of a topic / give characteristics and main features
Determine	Establish with certainty
Explain	Set out purposes or reasons / make the relationships between things clear / say why and/or how and support with relevant evidence
Give	Produce an answer from a given source or recall/memory
Plot	Mark point(s) on a graph
Show (that)	Provide structured evidence that leads to a given result
Sketch	Make a simple freehand drawing showing the key features
State	Express in clear terms
Work out	Calculate from given facts, figures or information with or without the use of a calculator
Write	Give an answer in a specific form
Write down	Give an answer without significant working

Examination techniques

Make sure you check the instructions on the question paper, the length of the paper and the number of questions you have to answer.

Allocate your time sensibly between each question. Be sure not to spend too long on some questions because this might mean you don't have enough time to complete all of them. Make sure you show your working to show how you've reached your answer.

From the authors

Mathematics comes from the Greek word meaning *knowledge* or *learning*. Galileo Galilei (1564–1642) wrote 'the universe cannot be read until we learn the language in which it is written. It is written in mathematical language.' Mathematics is used in science, engineering, medicine, art, finance and so on, but mathematicians have always studied the subject for pleasure. They look for patterns in nature, for fun, as a game or a puzzle.

A mathematician may find that their puzzle solving helps to solve 'real life' problems. But trigonometry was developed without a 'real life' application in mind, before it was then applied to navigation and many other things. The algebra of curves was not 'invented' to send a rocket to Jupiter.

The study of mathematics is across all lands and cultures. A mathematician in Africa may be working with another in Japan to extend work done by a Brazilian in the USA.

People in all cultures have tried to understand the world around them, and mathematics has been a common way of furthering that understanding, even in cultures which have left no written records.

Each topic in this textbook has an introduction which tries to show how, over thousands of years, mathematical ideas have been passed from one culture to another. So when you are studying from this textbook, remember that you are following in the footsteps of earlier mathematicians who were excited by the discoveries they had made. These discoveries changed our world.

You may find some of the questions in this book difficult. It is easy when this happens to ask the teacher for help. Remember though that mathematics is intended to stretch the mind. If you are trying to get physically fit you do not stop as soon as things get hard. It is the same with mental fitness. Think logically. Try harder. In the end you are responsible for your own learning. Teachers and textbooks can only guide you. Be confident that you can solve that difficult problem.

Ric Pimentel

Terry Wall

Frankie Pimentel

TOPIC 1

Number

Contents

Chapter 1   Number and language (E1.1, E1.3)
Chapter 2   Accuracy (E1.9, E1.10)
Chapter 3   Calculations and order (E1.5, E1.6)
Chapter 4   Integers, fractions, decimals and percentages (E1.4, E1.6)
Chapter 5   Further percentages (E1.13)
Chapter 6   Ratio and proportion (E1.11, E1.12)
Chapter 7   Indices, standard form and surds (E1.7, E1.8, E1.18, E2.4)
Chapter 8   Money and finance (E1.13, E1.14, E1.16, E1.17)
Chapter 9   Time (E1.14, E1.15)
Chapter 10 Set notation and Venn diagrams (E1.2)

Learning objectives

E1.1 Types of number
Identify and use:
- natural numbers
- integers (positive, zero and negative)
- prime numbers
- square numbers
- cube numbers
- common factors
- common multiples
- rational and irrational numbers
- reciprocals.

E1.2 Sets
Understand and use set language, notation and Venn diagrams to describe sets and represent relationships between sets.

E1.3 Powers and roots
Calculate with the following:
- squares
- square roots
- cubes
- cube roots
- other powers and roots of numbers.

E1.4 Fractions, decimals and percentages
1   Use the language and notation of the following in appropriate contexts:
- proper fractions
- improper fractions
- mixed numbers
- decimals
- percentages.
2   Recognise equivalence and convert between these forms.

E1.5 Ordering
Order quantities by magnitude and demonstrate familiarity with the symbols $=$, $\neq$, $>$, $<$, $\geq$ and $\leq$.

E1.6 The four operations
Use the four operations for calculations with integers, fractions and decimals, including correct ordering of operations and use of brackets.

E1.7 Indices I
1   Understand and use indices (positive, zero, negative and fractional).
2   Understand and use the rules of indices.

E1.8 Standard form
1. Use the standard form $A \times 10^n$ where n is a positive or negative integer, and $1 \leq A < 10$.
2. Convert numbers into and out of standard form.
3. Calculate with values in standard form.

E1.9 Estimation
1. Round values to a specified degree of accuracy.
2. Make estimates for calculations involving numbers, quantities and measurements.
3. Round answers to a reasonable degree of accuracy in the context of a given problem.

E1.10 Limits of accuracy
1. Give upper and lower bounds for data rounded to a specified accuracy.
2. Find upper and lower bounds of the results of calculations which have used data rounded to a specified accuracy.

E1.11 Ratio and proportion
Understand and use ratio and proportion, including:
- giving ratios in simplest form
- dividing a quantity in a given ratio
- using proportional reasoning and ratios in context.

E1.12 Rates
1. Use common measures of rate.
2. Apply other measures of rate.
3. Solve problems involving average speed.

E1.13 Percentages
1. Calculate a given percentage of a quantity.
2. Express one quantity as a percentage of another.
3. Calculate percentage increase or decrease.
4. Calculate with simple and compound interest.
5. Calculate using reverse percentages.

E1.14 Using a calculator
1. Use a calculator efficiently.
2. Enter values appropriately on a calculator.
3. Interpret the calculator display appropriately.

E1.15 Time
1. Calculations involving time: seconds (s), minutes (min), hours (h), days, weeks, months, years, including the relationship between units.
2. Calculate times in terms of the 24-hour and 12-hour clock.
3. Read clocks and timetables.

E1.16 Money
1. Calculate with money.
2. Convert from one currency to another.

E1.17 Exponential growth and decay
Use exponential growth and decay.

E1.18 Surds
1. Understand and use surds, including simplifying expressions.
2. Rationalise the denominator.

Hindu mathematicians

In 1300BCE a Hindu teacher named Laghada used geometry and trigonometry for his astronomical calculations. At around this time, other Indian mathematicians solved quadratic and simultaneous equations.

Much later in about 500CE, another Indian teacher, Aryabhata, worked on approximations for π (pi), and worked on the trigonometry of the sphere. He realised that not only did the planets go around the Sun but also that their paths were elliptical.

Brahmagupta, a Hindu, was the first to treat zero as a number in its own right. This helped to develop the decimal system of numbers.

One of the greatest mathematicians of all time was Bhascara, who, in the twelfth century, worked in algebra and trigonometry. He discovered that $\sin(A + B) = \sin A \cos B + \cos A \sin B$. His work was taken to Arabia and later to Europe.

Still alive today is the Indian woman mathematician Raman Parimala (born in 1948). Her work is famous in the fields of algebra and its connections with algebraic geometry and number theory.

A statue of Aryabhata (476–550)

1 Number and language

Vocabulary for sets of numbers

A **square** can be classified in many different ways. It is a quadrilateral but it is also a polygon and a two-dimensional shape. Just as shapes can be classified in many ways, so can numbers. Below is a description of some of the more common types of numbers.

Natural numbers

A child learns to count 'one, two, three, four, …'. These are sometimes called the counting numbers or whole numbers.

The child will say 'I am three', or 'I live at number 73'.

If we include the number 0, then we have the set of numbers called the **natural numbers**.

The set of natural numbers is {0, 1, 2, 3, 4, …}.

Integers

On a cold day, the temperature may be 4 °C at 10 p.m. If the temperature drops by a further 6 °C, then the temperature is 'below zero'; it is −2 °C.

If you are overdrawn at the bank by $200, this might be shown on a bank statement as −$200.

The set of **integers** is {…, −3, −2, −1, 0, 1, 2, 3, …}.

Integers therefore are an extension of natural numbers. Every natural number is an integer.

Reciprocal

The **reciprocal** of a number is obtained when 1 is divided by that number. The reciprocal of 5 is $\frac{1}{5}$, the reciprocal of $\frac{2}{5}$ is $\frac{1}{\frac{2}{5}}$ which simplifies to $\frac{5}{2}$.

> *You should learn how to convert between values expressed in numbers and values expressed in words. For example, 12 014 is twelve thousand and fourteen; 1 745 233 is one million, seven hundred and forty-five thousand, two hundred and thirty-three.*

 Exercise 1.1

1 Write the reciprocal of each of the following:
   a $\frac{1}{8}$
   b $\frac{7}{12}$
   c $\frac{3}{5}$
   d $1\frac{1}{2}$
   e $3\frac{3}{4}$
   f 6

Square numbers

The number 1 can be written as 1×1 or 1^2. This can be read as '1 squared' or '1 raised to the **power** of 2'.

The number 4 can be written as 2×2 or 2^2.

9 can be written as 3×3 or 3^2.

16 can be written as 4×4 or 4^2.

When an integer (whole number) is multiplied by itself, the result is a **square number**. In the examples above, 1, 4, 9 and 16 are all square numbers.

Cube numbers

The number 1 can be written as $1 \times 1 \times 1$ or 1^3. This can be read as '1 cubed' or '1 raised to the power of 3'.

The number 8 can be written as $2 \times 2 \times 2$ or 2^3.

27 can be written as $3 \times 3 \times 3$ or 3^3.

64 can be written as $4 \times 4 \times 4$ or 4^3.

When an integer is multiplied by itself and then by itself again, the result is a **cube number**. In the examples above 1, 8, 27 and 64 are all cube numbers.

Factors

The factors of 12 are all the numbers which will divide exactly into 12, i.e. 1, 2, 3, 4, 6 and 12.

 Exercise 1.2

1 List all the factors of the following numbers:
   a 6   b 9   c 7   d 15   e 24
   f 36   g 35   h 25   i 42   j 100

Prime numbers

A **prime number** is one whose only **factors** are 1 and itself. (Note: 1 is not a prime number.)

Prime factors

The factors of 12 are 1, 2, 3, 4, 6 and 12.

Of these, 2 and 3 are prime numbers, so 2 and 3 are the **prime factors** of 12.

1 NUMBER AND LANGUAGE

 Exercise 1.3

1 In a 10 by 10 square, write the numbers 1 to 100.
   Cross out number 1.
   Cross out all the even numbers after 2 (these have 2 as a factor).
   Cross out every third number after 3 (these have 3 as a factor).
   Continue with 5, 7, 11 and 13, then list all the prime numbers less than 100.

2 List the prime factors of the following numbers:
   a 15   b 18   c 24   d 16   e 20
   f 13   g 33   h 35   i 70   j 56

An easy way to find prime factors is to divide by the prime numbers in order, smallest first.

 Worked examples

a Find the prime factors of 18 and express it as a product of prime numbers:

	18
2	9
3	3
3	1

$18 = 2 \times 3 \times 3$ or 2×3^2

b Find the prime factors of 24 and express it as a product of prime numbers:

	24
2	12
2	6
2	3
3	1

$24 = 2 \times 2 \times 2 \times 3$ or $2^3 \times 3$

c Find the prime factors of 75 and express it as a product of prime numbers:

	75
3	25
5	5
5	1

$75 = 3 \times 5 \times 5$ or 3×5^2

Multiples

Exercise 1.4

1 Find the prime factors of the following numbers and express them as a product of prime numbers:
   a 12   b 32   c 36   d 40   e 44
   f 56   g 45   h 39   i 231   j 63

Highest common factor

The factors of 12 can be listed as 1, 2, 3, 4, 6, 12.

The factors of 18 can be listed as 1, 2, 3, 6, 9, 18.

As can be seen, the factors 1, 2, 3 and 6 are common to both numbers. They are known as **common factors**. As 6 is the largest of the common factors, it is called the **highest common factor (HCF)** of 12 and 18.

The prime factors of 12 are $2 \times 2 \times 3$.

The prime factors of 18 are $2 \times 3 \times 3$.

So the highest common factor can be seen by inspection to be 2×3, i.e. 6.

Multiples

Multiples of 2 are 2, 4, 6, 8, 10, etc.
Multiples of 3 are 3, 6, 9, 12, 15, etc.
The numbers 6, 12, 18, 24, etc., are **common multiples** as these appear in both lists.
The **lowest common multiple (LCM)** of 2 and 3 is 6, since 6 is the smallest number divisible by 2 and 3.

The LCM of 3 and 5 is 15.

The LCM of 6 and 10 is 30.

Exercise 1.5

1 Find the HCF of the following numbers:
   a 8, 12   b 10, 25   c 12, 18, 24
   d 15, 21, 27   e 36, 63, 108   f 22, 110
   g 32, 56, 72   h 39, 52   i 34, 51, 68
   j 60, 144

2 Find the LCM of the following:
   a 6, 14   b 4, 15   c 2, 7, 10   d 3, 9, 10
   e 6, 8, 20   f 3, 5, 7   g 4, 5, 10   h 3, 7, 11
   i 6, 10, 16   j 25, 40, 100

7

1 NUMBER AND LANGUAGE

Rational and irrational numbers

A **rational number** is any number which can be expressed as a fraction. Examples of some rational numbers and how they can be expressed as a fraction are shown below:

$0.2 = \frac{1}{5}$ $0.3 = \frac{3}{10}$ $7 = \frac{7}{1}$ $1.53 = \frac{153}{100}$ $0.\dot{2} = \frac{2}{9}$

An **irrational number** cannot be expressed as a fraction. Examples of irrational numbers include:

$\sqrt{2}, \quad \sqrt{5}, \quad 6 - \sqrt{3}, \quad \pi$

In summary:

Rational numbers include:
- whole numbers,
- fractions,
- recurring decimals,
- terminating decimals.

Irrational numbers include:
- the **square root** of any number other than square numbers,
- a decimal which does not repeat or terminate (e.g. π).

> *A recurring decimal is one which repeats itself and has no end, e.g. 1.33333333...*
>
> *A terminating decimal is one which has an end point, e.g. 5.2 or 0.45*

Real numbers

The set of rational and irrational numbers together form the set of **real numbers**.

Exercise 1.6

1 State whether each number below is rational or irrational:
   a 1.3
   b $0.\dot{6}$
   c $\sqrt{3}$
   d $-2\frac{3}{5}$
   e $\sqrt{25}$
   f $\sqrt[3]{8}$
   g $\sqrt{7}$
   h 0.625
   i $0.\dot{1}\dot{1}$

2 State whether each number below is rational or irrational:
   a $\sqrt{2} \times \sqrt{3}$
   b $\sqrt{2} + \sqrt{3}$
   c $(\sqrt{2} \times \sqrt{3})^2$
   d $\frac{\sqrt{8}}{\sqrt{2}}$
   e $\frac{2\sqrt{5}}{2\sqrt{20}}$
   f $4 + (\sqrt{9} - 4)$

3 In each of the following, decide whether the quantity required is rational or irrational. Give reasons for your answer.

   a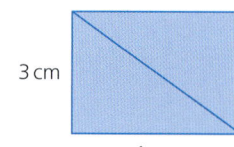
   3 cm
   4 cm
   The length of the diagonal

   b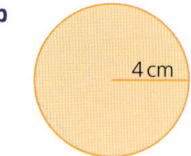
   4 cm
   The circumference of the circle

8

Using a graph

c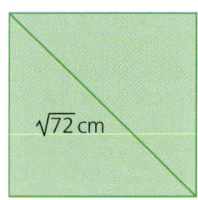

The side length of the square

d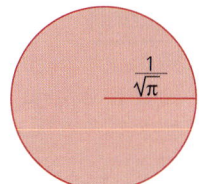

The area of the circle

Square roots

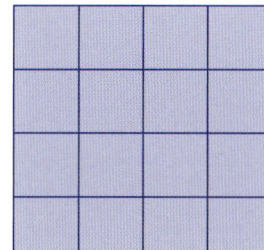

The square shown contains 16 squares. It has sides of length 4 units.

So the square root of 16 is 4.

This can be written as $\sqrt{16} = 4$.

Note that 4×4 is 16 so 4 is the square root of 16.

However, -4×-4 is also 16 so -4 is also the square root of 16.

By convention, $\sqrt{16}$ means 'the positive square root of 16' so

$\sqrt{16} = 4$ but the square root of 16 is ±4, i.e. +4 or −4.

Note: −16 has no square root since any integer squared is positive.

Exercise 1.7

1 Find the following:
   a $\sqrt{25}$ b $\sqrt{9}$ c $\sqrt{49}$ d $\sqrt{100}$
   e $\sqrt{121}$ f $\sqrt{169}$ g $\sqrt{0.01}$ h $\sqrt{0.04}$
   i $\sqrt{0.09}$ j $\sqrt{0.25}$

2 Use the $\sqrt{}$ key on your calculator to check your answers to Question 1.

3 Calculate the following:
   a $\sqrt{\frac{1}{9}}$ b $\sqrt{\frac{1}{16}}$ c $\sqrt{\frac{1}{25}}$ d $\sqrt{\frac{1}{49}}$
   e $\sqrt{\frac{1}{100}}$ f $\sqrt{\frac{4}{9}}$ g $\sqrt{\frac{9}{100}}$ h $\sqrt{\frac{49}{81}}$
   i $\sqrt{2\frac{7}{9}}$ j $\sqrt{6\frac{1}{4}}$

Using a graph

Exercise 1.8

1 Copy and complete the table below for the equation $y = \sqrt{x}$.

x	0	1	4	9	16	25	36	49	64	81	100
y											

Plot the graph of $y = \sqrt{x}$.

Use your graph to find the approximate values of the following:
   a $\sqrt{35}$ b $\sqrt{45}$ c $\sqrt{55}$ d $\sqrt{60}$ e $\sqrt{2}$

1 NUMBER AND LANGUAGE

Exercise 1.8 (cont)

2 Check your answers to Question 1 above by using the $\sqrt{}$ key on your calculator.

Cube roots

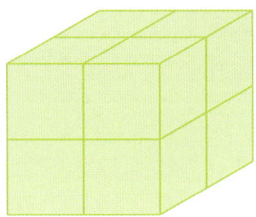

The cube shown has sides of 2 units and occupies 8 cubic units of space. (That is, $2 \times 2 \times 2$.)

So the **cube root** of 8 is 2.

This can be written as $\sqrt[3]{8} = 2$.

$\sqrt[3]{}$ is read as 'the cube root of …'.

$\sqrt[3]{64}$ is 4, since $4 \times 4 \times 4 = 64$.

Note that $\sqrt[3]{64}$ is not -4

since $-4 \times -4 \times -4 = -64$

but $\sqrt[3]{-64}$ is -4.

Exercise 1.9

1 Find the following cube roots:

a $\sqrt[3]{8}$
b $\sqrt[3]{125}$
c $\sqrt[3]{27}$
d $\sqrt[3]{0.001}$
e $\sqrt[3]{0.027}$
f $\sqrt[3]{216}$
g $\sqrt[3]{1000}$
h $\sqrt[3]{1\,000\,000}$
i $\sqrt[3]{-8}$
j $\sqrt[3]{-27}$
k $\sqrt[3]{-1000}$
l $\sqrt[3]{-1}$

Further powers and roots

We have seen that the square of a number is the same as raising that number to the power of 2. For example, the square of 5 is written as 5^2 and means 5×5. Similarly, the cube of a number is the same as raising that number to the power of 3. For example, the cube of 5 is written as 5^3 and means $5 \times 5 \times 5$.

Numbers can be raised by other powers too. Therefore, 5 raised to the power of 6 can be written as 5^6 and means $5 \times 5 \times 5 \times 5 \times 5 \times 5$.

You will find a button on your calculator to help you to do this. On most calculators, it will look like y^x.

We have also seen that the square root of a number can be written using the $\sqrt{}$ symbol. Therefore, the square root of 16 is written as $\sqrt{16}$ and is 4, because $4 \times 4 = 16$.

The cube root of a number can be written using the $\sqrt[3]{}$ symbol. Therefore, the cube root of 27 is written as $\sqrt[3]{27}$ and is 3, because $3 \times 3 \times 3 = 27$.

Numbers can be rooted by other values as well. The fourth root of a number can be written using the symbol $\sqrt[4]{\ }$. Therefore the fourth root of 625 can be expressed as $\sqrt[4]{625}$ and is 5, because $5 \times 5 \times 5 \times 5 = 625$.

You will find a button on your calculator to help you to calculate with roots. On most calculators, it will look like $\sqrt[x]{y}$.

Exercise 1.10

1 Work out the following:
   a 6^4
   b $3^5 + 2^4$
   c $(3^4)^2$
   d $0.1^6 \div 0.01^4$
   e $\sqrt[4]{2401}$
   f $\sqrt[8]{256}$
   g $(\sqrt[5]{243})^3$
   h $(\sqrt[9]{36})^9$
   i $2^7 \times \sqrt{\dfrac{1}{4}}$
   j $\sqrt[6]{\dfrac{1}{64}} \times 2^7$
   k $\sqrt[4]{5^4}$
   l $(\sqrt[10]{59049})^2$

Directed numbers

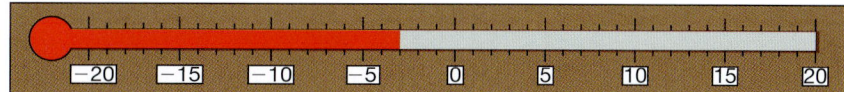

→ Worked example

The diagram above shows the scale of a thermometer. The temperature at 04 00 was −3 °C. By 09 00 the temperature had risen by 8 °C. What was the temperature at 09 00?

$-3 + 8 = 5$, so the temperature is 5 °C.

Exercise 1.11

1 The highest temperature ever recorded was in Libya. It was 58 °C. The lowest temperature ever recorded was −88 °C in Antarctica. What is the temperature difference?

2 Ms Okoro's bank account shows a positive amount of $105. Describe the amount in her account as a positive or negative number after each of these transactions is made in sequence:
   a rent $140
   b car insurance $283
   c 1 week's salary $230
   d food bill $72
   e credit transfer $250

3 The roof of an apartment block is 130 m above ground level. The car park beneath the apartment is 35 m below ground level. How high is the roof above the floor of the car park?

4 A submarine is at a depth of 165 m. If the ocean floor is 860 m from the surface, how far is the submarine from the ocean floor?

1 NUMBER AND LANGUAGE

Student assessment 1

1. State whether the following numbers are rational or irrational:
   a 1.5
   b $\sqrt{7}$
   c $0.\dot{7}$
   d $0.\dot{7}\dot{3}$
   e $\sqrt{121}$
   f π

2. Show, by expressing them as fractions or whole numbers, that the following numbers are rational:
   a 0.625
   b $\sqrt[3]{27}$
   c 0.44

3. Find the value of:
   a 9^2
   b 15^2
   c 0.2^2
   d 0.7^2

4. Calculate:
   a 3.5^2
   b 4.1^2
   c 0.15^2

5. Without using a calculator, find:
   a $\sqrt{225}$
   b $\sqrt{0.01}$
   c $\sqrt{0.81}$
   d $\sqrt{\frac{9}{25}}$
   e $\sqrt{5\frac{4}{9}}$
   f $\sqrt{2\frac{23}{49}}$

6. Without using a calculator, find:
   a 4^3
   b $(0.1)^3$
   c $\left(\frac{2}{3}\right)^3$

7. Without using a calculator, find:
   a $\sqrt[3]{27}$
   b $\sqrt[3]{1\,000\,000}$
   c $\sqrt[3]{\frac{64}{125}}$

8. Toby's bank statement for seven days in October is shown below:

Date	Payments ($)	Receipts ($)	Balance ($)
01/10			200
02/10	284		(a)
03/10		175	(b)
04/10	(c)		46
05/10		(d)	120
06/10	163		(e)
07/10		28	(f)

Copy and complete the statement by entering the amounts (a) to (f).

9. Using a calculator if necessary, work out:
   a $2^6 \div 2^8$
   b $4^5 \times \sqrt[6]{64}$
   c $\sqrt[4]{81} \times 4^3$

2 Accuracy

Approximation

In many instances exact numbers are not necessary or even desirable. In those circumstances approximations are given. Approximations can take several forms; the most common forms are dealt with below.

Rounding

If 28 617 people attend a gymnastics competition, this figure can be reported to various levels of accuracy.

To the nearest 10 000 this figure would be **rounded** up to 30 000.

To the nearest 1000 the figure would be rounded up to 29 000.

To the nearest 100 the figure would be rounded down to 28 600.

In this type of situation, it is unlikely that the exact number would be reported.

Note: If a number falls exactly half-way, then it is rounded up.
For example, rounding 16 500 to the nearest thousand can be visualised as follows:

16 000 and 17 000 are the numbers in thousands either side of 16 500. As 16 500 falls exactly half-way, it gets rounded up to 17 000 if the answer is wanted to the nearest thousand.

Exercise 2.1

1 Round the following numbers to the nearest 1000:
   - a 68 786
   - b 74 245
   - c 89 000
   - d 4020
   - e 99 500
   - f 999 999

2 Round the following numbers to the nearest 100:
   - a 78 540
   - b 6858
   - c 14 099
   - d 8084
   - e 950
   - f 2984

3 Round the following numbers to the nearest 10:
   - a 485
   - b 692
   - c 8847
   - d 83
   - e 4
   - f 997

Decimal places

A number can also be approximated to a given number of **decimal places** (d.p.). This refers to the number of digits written after a decimal point.

2 ACCURACY

 Worked examples

a Write 7.864 to 1 d.p.

The answer needs to be written with one digit after the decimal point. However, to do this, the second digit after the decimal point also needs to be considered. If it is 5 or more, then the first digit is rounded up.

i.e. 7.864 is written as 7.9 to 1 d.p.

b Write 5.574 to 2 d.p.

The answer here is to be given with two digits after the decimal point. In this case, the third digit after the decimal point needs to be considered. As the third digit after the decimal point is less than 5, the second digit is not rounded up.

i.e. 5.574 is written as 5.57 to 2 d.p.

Exercise 2.2

1 Give the following to 1 d.p.
- a 5.58
- b 0.73
- c 11.86
- d 157.39
- e 4.04
- f 15.045
- g 2.95
- h 0.98
- i 12.049

2 Give the following to 2 d.p.
- a 6.473
- b 9.587
- c 16.476
- d 0.088
- e 0.014
- f 9.3048
- g 99.996
- h 0.0048
- i 3.0037

Significant figures

Numbers can also be approximated to a given number of **significant figures** (s.f.). In the number 43.25 the 4 is the most significant figure as it has a value of 40. In contrast, the 5 is the least significant as it only has a value of 5 hundredths.

 Worked examples

a Write 43.25 to 3 s.f.

Only the three most significant digits are written in the answer; however, the fourth digit needs to be considered to see whether the third digit is to be rounded up or not.

i.e. 43.25 is written as 43.3 to 3 s.f.

b Write 0.0043 to 1 s.f.

In this example only two digits have any significance, the 4 and the 3. The 4 is the most significant and therefore is the only one of the two digits to be written in the answer.

i.e. 0.0043 is written as 0.004 to 1 s.f.

Estimating answers to calculations

Exercise 2.3

1 Write the following to the number of significant figures written in brackets:
   a 48 599 (1 s.f.)   b 48 599 (3 s.f.)   c 6841 (1 s.f.)
   d 7538 (2 s.f.)   e 483.7 (1 s.f.)   f 2.5728 (3 s.f.)
   g 990 (1 s.f.)   h 2045 (2 s.f.)   i 14.952 (3 s.f.)

2 Write the following to the number of significant figures written in brackets:
   a 0.085 62 (1 s.f.) b 0.5932 (1 s.f.)   c 0.942 (2 s.f.)
   d 0.954 (1 s.f.)   e 0.954 (2 s.f.)   f 0.003 05 (1 s.f.)
   g 0.003 05 (2 s.f.) h 0.009 73 (2 s.f.) i 0.009 73 (1 s.f.)

Appropriate accuracy

In many instances calculations carried out using a calculator produce answers which are not whole numbers. A calculator will give the answer to as many decimal places as will fit on its screen. In most cases this degree of accuracy is neither desirable nor necessary. Unless specifically asked for, answers should not be given to more than two decimal places. Indeed, one decimal place is usually sufficient. Alternatively, giving an answer correct to three significant figures is also considered an appropriate degree of accuracy.

➡ Worked example

Calculate 4.64 ÷ 2.3, giving your answer to an appropriate degree of accuracy.

The calculator will give the answer to 4.64 ÷ 2.3 as 2.0173913.

However, the answer given to 1 d.p. is sufficient.

Therefore 4.64 ÷ 2.3 = 2.0 (1 d.p.).

Estimating answers to calculations

Even though many calculations can be done quickly and effectively on a calculator, often an **estimate** for an answer, done without using a calculator, can be a useful check. Estimating an answer is done by rounding each of the numbers in such a way that the mental calculation becomes relatively straightforward.

➡ Worked examples

a Estimate the answer to 57 × 246.
   Here are two possibilities:
   i   60 × 200 = 12 000,
   ii 50 × 250 = 12 500.

b Estimate the answer to 6386 ÷ 27.
   6000 ÷ 30 = 200.

2 ACCURACY

> **Note**
> ≈ means approximately equal to.

c Estimate the answer to $\sqrt[3]{120} \times 48$.
   As $\sqrt[3]{125} = 5$, $\sqrt[3]{120} \approx 5$
   Therefore $\sqrt[3]{120} \times 48 \approx 5 \times 50$
   ≈ 250.

d Estimate the answer to $\frac{2^5 \times \sqrt[4]{600}}{8}$

   An approximate answer can be calculated using the knowledge that $2^5 = 32$ and $\sqrt[4]{625} = 5$

   Therefore $\frac{2^5 \times \sqrt[4]{600}}{8} \approx \frac{30 \times 5}{8} \approx \frac{150}{8}$
   ≈ 20.

Exercise 2.4

1 Calculate the following, giving your answer to an appropriate degree of accuracy:
   a 23.456×17.89
   b 0.4×12.62
   c 18×9.24
   d $76.24 \div 3.2$
   e 7.6^2
   f 16.42^3
   g $\frac{2.3 \times 3.37}{4}$
   h $\frac{8.31}{2.02}$
   i $9.2 \div 4^2$

2 Without using a calculator, estimate the answers to the following:
   a 62×19
   b 270×12
   c 55×60
   d 4950×28
   e 0.8×0.95
   f 0.184×475

3 Without using a calculator, estimate the answers to the following:
   a $3946 \div 18$
   b $8287 \div 42$
   c $906 \div 27$
   d $5520 \div 13$
   e $48 \div 0.12$
   f $610 \div 0.22$

4 Without using a calculator, estimate the answers to the following:
   a $78.45 + 51.02$
   b $168.3 - 87.09$
   c 2.93×3.14
   d $84.2 \div 19.5$
   e $\frac{4.3 \times 752}{15.6}$
   f $\frac{(9.8)^3}{(2.2)^2}$
   g $\frac{\sqrt[3]{78} \times 6}{5^3}$
   h $\frac{38 \times 6^3}{\sqrt[4]{9900}}$
   i $\sqrt[4]{24} \times \sqrt[4]{26}$

5 Using estimation, identify which of the following are definitely incorrect. Explain your reasoning clearly.
   a $95 \times 212 = 20140$
   b $44 \times 17 = 748$
   c $689 \times 413 = 28457$
   d $142656 \div 8 = 17832$
   e $77.9 \times 22.6 = 2512.54$
   f $\frac{8.42 \times 46}{0.2} = 19366$

6 Estimate the area of the shaded areas of the following shapes. Do *not* work out an exact answer.
   a

Estimating answers to calculations

b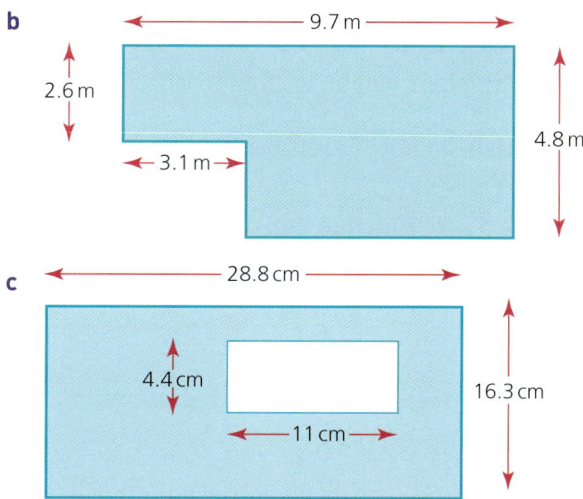

c

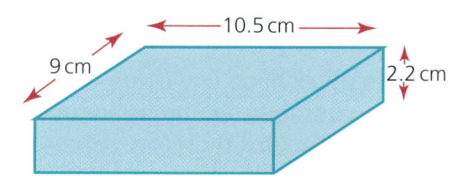

7 Estimate the volume of each of the solids below. Do *not* work out an exact answer.

a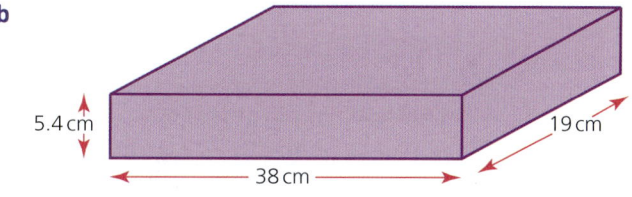

b

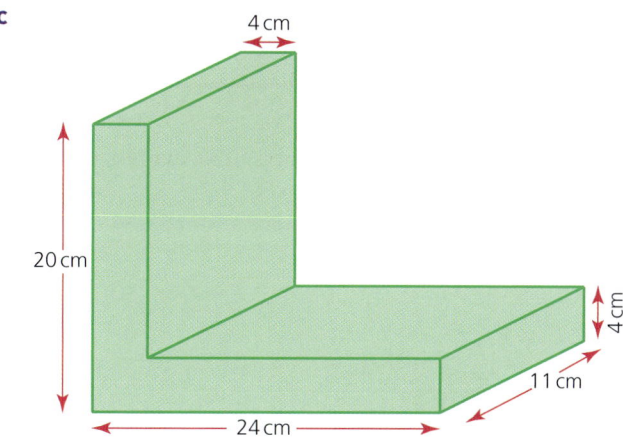

c

2 ACCURACY

Upper and lower bounds

Numbers can be written to different degrees of accuracy. For example, although 4.5, 4.50 and 4.500 appear to represent the same number, they do not. This is because they are written to different degrees of accuracy.

4.5 is rounded to one decimal place and therefore any number from 4.45 up to but not including 4.55 would be rounded to 4.5. On a number line, this would be represented as:

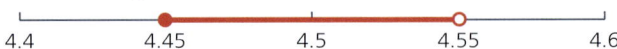

As an inequality where x represents the number it would be expressed as:

$4.45 \leqslant x < 4.55$

4.45 is known as the **lower bound** of 4.5, while 4.55 is known as the **upper bound**.
Note that ○——▶ implies that the number is not included in the solution while ●——▶ implies that the number is included in the solution.

4.50 on the other hand is written to two decimal places and therefore only numbers from 4.495 up to but not including 4.505 would be rounded to 4.50. This therefore represents a much smaller range of numbers than when it is rounded to 4.5. Similarly, the range of numbers being rounded to 4.500 would be even smaller.

➡ Worked example

A girl's **height** is given as 162 cm to the nearest centimetre.

a Work out the lower and upper bounds within which her height can lie.
Lower bound = 161.5 cm
Upper bound = 162.5 cm

b Represent this range of numbers on a number line.

c If the girl's height is h cm, express this range as an inequality.
$161.5 \leqslant h < 162.5$

Calculating with upper and lower bounds

Exercise 2.5

1. Each of the following numbers is expressed to the nearest whole number.
   i Give the upper and lower bounds of each.
   ii Using x as the number, express the range in which the number lies as an inequality.
   a 6
   b 83
   c 152
   d 1000
   e 100

2. Each of the following numbers is correct to one decimal place.
   i Give the upper and lower bounds of each.
   ii Using x as the number, express the range in which the number lies as an inequality.
   a 3.8
   b 15.6
   c 1.0
   d 10.0
   e 0.3

3. Each of the following numbers is correct to two significant figures.
   i Give the upper and lower bounds of each.
   ii Using x as the number, express the range in which the number lies as an inequality.
   a 4.2
   b 0.84
   c 420
   d 5000
   e 0.045
   f 25 000

4. The mass of a sack of vegetables is given as 5.4 kg.
   a Illustrate the lower and upper bounds of the mass on a number line.
   b Using M kg for the mass, express the range of values in which M must lie as an inequality.

5. At a school sports day, the winning time for the 100 m race was given as 11.8 seconds.
   a Illustrate the lower and upper bounds of the winning time on a number line.
   b Using T seconds for the time, express the range of values in which T must lie as an inequality.

6. The capacity of a swimming pool is given as 620 m^3 correct to two significant figures.
   a Calculate the lower and upper bounds of the pool's capacity.
   b Using x cubic metres for the capacity, express the range of values in which x must lie as an inequality.

7. Hadiza is a surveyor. She measures the dimensions of a rectangular field to the nearest 10 m. The length is recorded as 630 m and the width is recorded as 400 m.
   a Calculate the lower and upper bounds of the length.
   b Using W metres for the width, express the range of values in which W must lie as an inequality.

Calculating with upper and lower bounds

When numbers are written to a specific degree of accuracy, calculations involving those numbers also give a range of possible answers.

2 ACCURACY

> **Worked examples**

a Calculate the upper and lower bounds for the following calculation, given that each number is given to the nearest whole number.

34×65

34 lies in the range $33.5 \leqslant x < 34.5$.

65 lies in the range $64.5 \leqslant x < 65.5$.

The lower bound of the calculation is obtained by multiplying together the two lower bounds. Therefore the minimum product is 33.5×64.5, i.e. 2160.75.

The upper bound of the calculation is obtained by multiplying together the two upper bounds. Therefore the maximum product is 34.5×65.5, i.e. 2259.75.

b Calculate the upper and lower bounds to $\frac{33.5}{22.0}$ given that each of the numbers is accurate to 1 d.p.

33.5 lies in the range $33.45 \leqslant x < 33.55$.

22.0 lies in the range $21.95 \leqslant x < 22.05$.

The lower bound of the calculation is obtained by dividing the lower bound of the numerator by the *upper* bound of the denominator. So the minimum value is $33.45 \div 22.05$, i.e. 1.52 (2 d.p.).

The upper bound of the calculation is obtained by dividing the upper bound of the numerator by the *lower* bound of the denominator. So the maximum value is $33.55 \div 21.95$, i.e. 1.53 (2 d.p.).

Exercise 2.6

1 Calculate the lower and upper bounds for the following calculations, if each of the numbers is given to the nearest whole number.
- **a** 14×20
- **b** 135×25
- **c** 100×50
- **d** $\frac{40}{10}$
- **e** $\frac{33}{11}$
- **f** $\frac{125}{15}$
- **g** $\frac{12 \times 65}{16}$
- **h** $\frac{101 \times 28}{69}$
- **i** $\frac{250 \times 7}{100}$
- **j** $\frac{44}{3^2}$
- **k** $\frac{578}{17 \times 22}$
- **l** $\frac{1000}{4 \times (3+8)}$

2 Calculate the lower and upper bounds for the following calculations, if each of the numbers is given to 1 d.p.
- **a** $2.1 + 4.7$
- **b** 6.3×4.8
- **c** 10.0×14.9
- **d** $17.6 - 4.2$
- **e** $\frac{8.5 + 3.6}{6.8}$
- **f** $\frac{7.7 - 6.2}{3.5}$
- **g** $\frac{16.4^2}{(3.0 - 0.3)^2}$
- **h** $\frac{100.0}{(50.0 - 40.0)^2}$
- **i** $(0.1 - 0.2)^2$

3 Calculate the lower and upper bounds for the following calculations, if each of the numbers is given to 2 s.f.
- **a** 64×320
- **b** 1.7×0.65
- **c** 4800×240
- **d** $\frac{54\,000}{600}$
- **e** $\frac{4.2}{0.031}$
- **f** $\frac{200}{5.2}$
- **g** $\frac{6.8 \times 42}{120}$
- **h** $\frac{200}{(4.5 \times 6.0)}$
- **i** $\frac{180}{(7.3 - 4.5)}$

Calculating with upper and lower bounds

Exercise 2.7

1. The masses to the nearest 0.5 kg of two parcels are 1.5 kg and 2.5 kg. Calculate the lower and upper bounds of their combined mass.

2. Calculate the upper and lower bounds for the perimeter of the **rectangle** shown (below), if its dimensions are correct to 1 d.p.

3. Calculate the upper and lower bounds for the perimeter of the rectangle shown (below), whose dimensions are accurate to 2 d.p.

4. Calculate the upper and lower bounds for the **area** of the rectangle shown (below), if its dimensions are accurate to 1 d.p.

   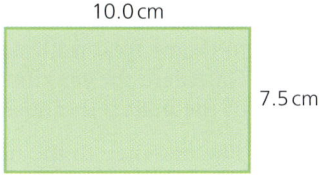

5. Calculate the upper and lower bounds for the area of the rectangle shown (below), whose dimensions are correct to 2 s.f.

6. Calculate the upper and lower bounds for the length marked x cm in the rectangle (below). The area and length are both given to 1 d.p.

   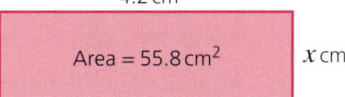

7. Calculate the upper and lower bounds for the length marked x cm in the rectangle (below). The area and length are both accurate to 2 s.f.

   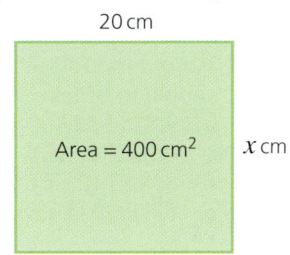

2 ACCURACY

Exercise 2.7 (cont)

8 The radius of the circle shown (below) is given to 1 d.p. Calculate the upper and lower bounds of:
   a the **circumference**,
   b the area.

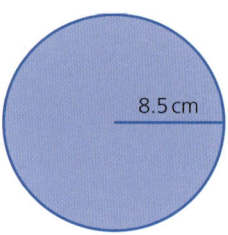

9 The area of the circle shown (below) is given to 2 s.f. Calculate the upper and lower bounds of:
   a the radius,
   b the circumference.

10 The mass of a cube of side 2 cm is given as 100 g. The side is accurate to the nearest millimetre and the mass accurate to the nearest gram. Calculate the maximum and minimum possible values for the density of the material (density = mass ÷ volume).

11 The distance to the nearest 100 000 km from Earth to the Moon is given as 400 000 km. The **average speed** to the nearest 500 km/h of a rocket to the Moon is given as 3500 km/h. Calculate the greatest and least time it could take the rocket to reach the Moon.

Student assessment 1

1 Round the following numbers to the degree of accuracy shown in brackets:
   a 2841 (nearest 100)   b 7096 (nearest 10)
   c 48 756 (nearest 1000)   d 951 (nearest 100)

2 Round the following numbers to the number of decimal places shown in brackets:
   a 3.84 (1 d.p.)   b 6.792 (1 d.p.)
   c 0.8506 (2 d.p.)   d 1.5849 (2 d.p.)
   e 9.954 (1 d.p.)   f 0.0077 (3 d.p.)

3 Round the following numbers to the number of significant figures shown in brackets:
   a 3.84 (1 s.f.)   b 6.792 (2 s.f.)
   c 0.7065 (1 s.f.)   d 9.624 (1 s.f.)
   e 834.97 (2 s.f.)   f 0.00451 (1 s.f.)
   g 62.4899 (5 s.f.)   h 0.9997 (3 s.f.)

Calculating with upper and lower bounds

4 1 mile is 1760 yards. Estimate the number of yards in 11.5 miles.

5 Estimate the shaded area of the figure below:

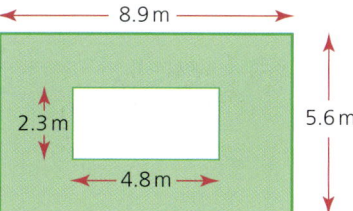

6 Estimate the answers to the following.
Do *not* work out an exact answer.

   a $\dfrac{5.3 \times 11.2}{2.1}$ **b** $\dfrac{(9.8)^2}{(4.7)^2}$ **c** $\dfrac{18.8 \times (7.1)^2}{(3.1)^2 \times (4.9)^2}$

7 A cuboid's dimensions are given as 12.32 cm by 1.8 cm by 4.16 cm. Calculate its volume, giving your answer to an appropriate degree of accuracy.

Student assessment 2

1 The following numbers are expressed to the nearest whole number. Illustrate on a number line the range in which each must lie.
   a 7   **b** 40   **c** 300   **d** 2000

2 The following numbers are expressed correct to two significant figures. Representing each number by the letter x, express the range in which each must lie, using an inequality.
   a 210   **b** 64   **c** 3.0   **d** 0.88

3 Some students measure the dimensions of their school's rectangular playing field to the nearest metre. The length was recorded as 350 m and the width as 200 m. Express the range in which the length and width lie using inequalities.

4 A boy's mass was measured to the nearest 0.1 kg. If his mass was recorded as 58.9 kg, illustrate on a number line the range within which it must lie.

5 An electronic clock is accurate to $\dfrac{1}{1000}$ of a second. The duration of a flash from a camera is timed at 0.004 seconds. Express the upper and lower bounds of the duration of the flash using inequalities.

6 The following numbers are rounded to the degree of accuracy shown in brackets. Express the lower and upper bounds of these numbers as an inequality.
   a $x = 4.83$ (2 d.p.)
   b $y = 5.05$ (2 d.p.)
   c $z = 10.0$ (1 d.p.)
   d $p = 100.00$ (2 d.p.)

2 ACCURACY

Student assessment 3

1. Five animals have a mass, given to the nearest 10 kg, of: 40 kg, 50 kg, 50 kg, 60 kg and 80 kg. Calculate the least possible total mass.

2. A water tank measures 30 cm by 50 cm by 20 cm. If each of these measurements is given to the nearest centimetre, calculate the largest possible volume of the tank.

3. The volume of a cube is given as 125 cm^3 to the nearest whole number.
   a. Express as an inequality the upper and lower bounds of the cube's volume.
   b. Express as an inequality the upper and lower bounds of the length of each of the cube's edges.

4. The radius of a circle is given as 4.00 cm to 2 d.p. Express as an inequality the upper and lower bounds for:
   a. the circumference of the circle,
   b. the area of the circle.

5. A cylindrical water tank has a volume of 6000 cm^3 correct to 1 s.f. A full cup of water from the tank has a volume of 300 cm^3 correct to 2 s.f. Calculate the maximum number of full cups of water that can be drawn from the tank.

6. A match measures 5 cm to the nearest centimetre. 100 matches end to end measure 5.43 m correct to 3 s.f.
   a. Calculate the upper and lower limits of the length of one match.
   b. How can the limits of the length of a match be found to 2 d.p.?

3 Calculations and order

Ordering

The following symbols have a specific meaning in mathematics:

= is equal to
≠ is not equal to
> is greater than
⩾ is greater than or equal to
< is less than
⩽ is less than or equal to

$x \geqslant 3$ implies that x is greater than or equal to 3, i.e. x can be 3, 4, 4.2, 5, 5.6, etc.

$3 \leqslant x$ implies that 3 is less than or equal to x, i.e. x is still 3, 4, 4.2, 5, 5.6, etc.

Therefore:

$5 > x$ can be rewritten as $x < 5$, i.e. x can be 4, 3.2, 3, 2.8, 2, 1, etc.

$-7 \leqslant x$ can be rewritten as $x \geqslant -7$, i.e. x can be $-7, -6, -5$, etc.

These **inequalities** can also be represented on a number line:

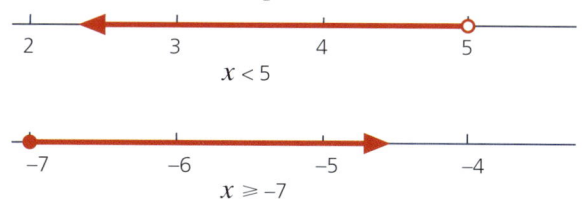

Note that ○⟶ implies that the number is not included in the solution while ●⟶ implies that the number is included in the solution.

➔ Worked examples

a The maximum number of players from one football team allowed on the pitch at any one time is eleven. Represent this information:
   i   as an inequality,
   ii on a number line.

   i   Let the number of players be represented by the letter n. n must be less than or equal to 11. Therefore $n \leqslant 11$.
   ii

3 CALCULATIONS AND ORDER

b The maximum number of players from one football team allowed on the pitch at any one time is eleven. The minimum allowed is seven players. Represent this information:
  i as an inequality,
  ii on a number line.
  i Let the number of players be represented by the letter n. n must be greater than or equal to 7, but less than or equal to 11. Therefore $7 \leq n \leq 11$.
  ii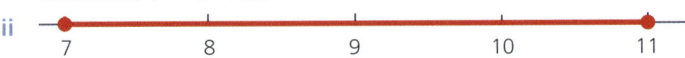

Exercise 3.1

1 Copy each of the following statements, and insert one of the symbols =, >, < into the space to make the statement correct:
   a $7 \times 2 \ldots 8 + 7$ b $6^2 \ldots 9 \times 4$
   c $5 \times 10 \ldots 7^2$ d $80\,cm \ldots 1\,m$
   e $1000\,litres \ldots 1\,m^3$ f $48 \div 6 \ldots 54 \div 9$

2 Represent each of the following inequalities on a number line, where x is a real number:
   a $x < 2$ b $x \geq 3$
   c $x \leq -4$ d $x \geq -2$
   e $2 < x < 5$ f $-3 < x < 0$
   g $-2 \leq x < 2$ h $2 \geq x \geq -1$

3 Write down the inequalities which correspond to the following number lines:

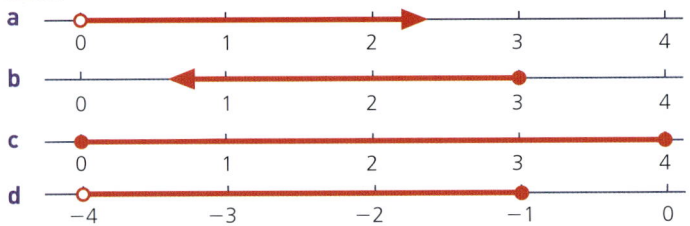

4 Write the following sentences using inequality signs:
   a The maximum capacity of an athletics stadium is 20 000 people.
   b In a class, the tallest student is 180 cm and the shortest is 135 cm.
   c Five times a number plus 3 is less than 20.
   d The maximum temperature in May was 25 °C.
   e A farmer has between 350 and 400 apples on each tree in her orchard.
   f In December, temperatures in Kenya were between 11 °C and 28 °C.

Exercise 3.2

1 Write the following decimals in order of magnitude, starting with the smallest:
   6.0   0.6   0.66   0.606   0.06   6.6   6.606

2 Write the following fractions in order of magnitude, starting with the largest:
   $\dfrac{1}{2}$ $\dfrac{1}{3}$ $\dfrac{6}{13}$ $\dfrac{4}{5}$ $\dfrac{7}{18}$ $\dfrac{2}{19}$

3 Write the following lengths in order of magnitude, starting with the smallest:
   2 m   60 cm   800 mm   180 cm   0.75 m

4 Write the following masses in order of magnitude, starting with the largest:
   4 kg   3500 g $\frac{3}{4}$ kg   700 g   1 kg

5 Write the following volumes in order of magnitude, starting with the smallest:
   1 l   430 ml   800 cm^3 120 cl   150 cm^3

The order of operations

When carrying out calculations, care must be taken to ensure that they are carried out in the correct order.

➡ Worked examples

a Use a scientific calculator to work out the answer to the following:
   $2 + 3 \times 4 =$
   [2] [+] [3] [×] [4] [=] 14

b Use a scientific calculator to work out the answer to the following:
   $(2 + 3) \times 4 =$
   [(] [2] [+] [3] [)] [×] [4] [=] 20

The reason why different answers are obtained is because, by convention, the operations have different priorities. These are as follows:

(1) brackets

(2) indices

(3) multiplication/division

(4) addition/subtraction.

Therefore in **Worked example a** 3×4 is **evaluated** first, and then the 2 is added, while in **Worked example b** $(2 + 3)$ is evaluated first, followed by multiplication by 4.

c Use a scientific calculator to work out why the answer to the following is −20:
   $-4 \times (8 + -3) = -20$

The $(8 + -3)$ is evaluated first as it is in the brackets, the answer 5 is then multiplied by −4.

d Use a scientific calculator to work out why the answer to the following is −35:
   $-4 \times 8 + -3 = -35$

The -4×8 is evaluated first as it is a multiplication, the answer −32 then has −3 added to it.

3 CALCULATIONS AND ORDER

Exercise 3.3 In the following questions, evaluate the answers:
  i in your head,
  ii using a scientific calculator.

1 a $8 \times 3 + 2$ b $4 \div 2 + 8$
   c $12 \times 4 - 6$ d $4 + 6 \times 2$
   e $10 - 6 \div 3$ f $6 - 3 \times 4$

2 a $7 \times 2 + 3 \times 2$ b $12 \div 3 + 6 \times 5$
   c $9 + 3 \times 8 - 1$ d $36 - 9 \div 3 - 2$
   e $-14 \times 2 - 16 \div 2$ f $4 + 3 \times 7 - 6 \div 3$

3 a $(4 + 5) \times 3$ b $8 \times (12 - 4)$
   c $3 \times (-8 + -3) - 3$ d $(4 + 11) \div (7 - 2)$
   e $4 \times 3 \times (7 + 5)$ f $24 \div 3 \div (10 - 5)$

Exercise 3.4 In each of the following questions:
  i Copy the calculation and put in any brackets which are needed to make it correct.
  ii Check your answer using a scientific calculator.

1 a $6 \times 2 + 1 = 18$ b $1 + 3 \times 5 = 16$
   c $8 + 6 \div 2 = 7$ d $9 + 2 \times 4 = 44$
   e $9 \div 3 \times 4 + 1 = 13$ f $3 + 2 \times 4 - 1 = 15$

2 a $12 \div 4 - 2 + 6 = 7$ b $12 \div 4 - 2 + 6 = 12$
   c $12 \div 4 - 2 + 6 = -5$ d $12 \div 4 - 2 + 6 = 1.5$
   e $4 + 5 \times 6 - 1 = 33$ f $4 + 5 \times 6 - 1 = 29$
   g $4 + 5 \times 6 - 1 = 53$ h $4 + 5 \times 6 - 1 = 45$

It is important to use brackets when dealing with more complex calculations.

→ Worked examples

a Evaluate the following using a scientific calculator:

$$\frac{12 + 9}{10 - 3} =$$

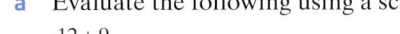 3

b Evaluate the following using a scientific calculator:

$$\frac{20 + 12}{4^2} =$$

 2

c Evaluate the following using a scientific calculator:

$$\frac{90 + 38}{4^3} =$$

[9 0 + 3 8] ÷ 4 x^y 3 = 2

Note: Different types of calculator have different 'to the power of' and 'fraction' buttons. It is therefore important that you get to know your calculator.

Your calculator may have a fraction button. It may look like this:

The order of operations

Exercise 3.5 Using a scientific calculator, evaluate the following:

1. a $\dfrac{9+3}{6}$ b $\dfrac{30-6}{5+3}$
   c $\dfrac{40+9}{12-5}$ d $\dfrac{15 \times 2}{7+8} + 2$
   e $\dfrac{100+21}{11} + 4 \times 3$ f $\dfrac{7+2 \times 4}{7-2} - 3$

2. a $\dfrac{4^2-6}{2+8}$ b $\dfrac{3^2+4^2}{5}$
   c $\dfrac{6^3-4^2}{4 \times 25}$ d $\dfrac{3^3 \times 4^4}{12^2} + 2$
   e $\dfrac{3+3^3}{5} + \dfrac{4^2-2^3}{8}$ f $\dfrac{(6+3) \times 4}{2^3} - 2 \times 3$

Exercise 3.6 In each of the following questions:
a Write the calculation represented by each problem.
b Work out the answer to each calculation.

1. The temperature of water in a beaker is initially 48°C. It is allowed to cool by 16°C before being heated up again. When heated up, the temperature of the water is trebled.
   What is the temperature (*T*°C) of the water once it has been heated up?

2. A submarine is initially at a depth of 400 m below the water's surface. It then dives a further distance so that it is at a depth double what it was initially. The submarine later climbs 620 m.
   Calculate the depth *D* (m) the submarine climbs to.

3. Luis arranges five counters in a line. He squares the number of counters, then adds a further 11 counters. Finally, he divides the number of counters equally between himself and Dari, so they each receive *N* counters.
   Calculate the number of counters (*N*) they each receive.

Student assessment 1

1. Write the information on the following number lines as inequalities:
   a
   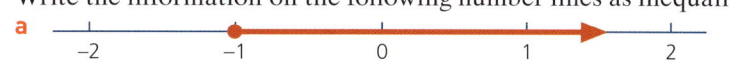
   b
   c
   d

2. a Illustrate each of the following inequalities on a number line:
   i $x \geqslant 3$ ii $x < 4$
   iii $0 < x < 4$ iv $-3 \leqslant x < 1$
   b Write down the smallest integer value which satisfies each inequality in part **a** above.

3. Write the following fractions in order of magnitude, starting with the smallest:
   $\dfrac{4}{7} \quad \dfrac{3}{14} \quad \dfrac{9}{10} \quad \dfrac{1}{2} \quad \dfrac{2}{5}$

3 CALCULATIONS AND ORDER

Student assessment 2

1. Evaluate the following:
   a $6 \times 8 - 4$
   b $3 + 5 \times 2$
   c $3 \times 3 + 4 \times 4$
   d $3 + 3 \times 4 + 4$
   e $(5 + 2) \times 7$
   f $18 \div 2 \div (5 - 2)$

2. Copy the following, if necessary inserting brackets to make the statement correct:
   a $7 - 4 \times 2 = 6$
   b $12 + 3 \times 3 + 4 = 33$
   c $5 + 5 \times 6 - 4 = 20$
   d $5 + 5 \times 6 - 4 = 56$

3. Evaluate the following using a calculator:
   a $\dfrac{2^4 - 3^2}{2}$
   b $\dfrac{(8 - 3) \times 3}{5} + 7$

Student assessment 3

1. Evaluate the following:
   a $3 \times 9 - 7$
   b $12 + 6 \div 2$
   c $3 + 4 \div 2 \times 4$
   d $6 + 3 \times 4 - 5$
   e $(5 + 2) \div 7$
   f $14 \times 2 \div (9 - 2)$

2. Copy the following, if necessary inserting brackets to make the statement correct:
   a $7 - 5 \times 3 = 6$
   b $16 + 4 \times 2 + 4 = 40$
   c $4 + 5 \times 6 - 1 = 45$
   d $1 + 5 \times 6 - 6 = 30$

3. Using a calculator, evaluate the following:
   a $\dfrac{3^3 - 4^2}{2}$
   b $\dfrac{(15 - 3) \div 3}{2} + 7$

4 Integers, fractions, decimals and percentages

Fractions

A single unit can be broken into equal parts called fractions, e.g. $\frac{1}{2}, \frac{1}{3}, \frac{1}{6}$. If, for example, the unit is broken into ten equal parts and three parts are then taken, the fraction is written as $\frac{3}{10}$. That is, three parts out of ten parts.

In the fraction $\frac{3}{10}$:

The three is called the **numerator**.

The ten is called the **denominator**.

A **proper fraction** has its numerator less than its denominator, e.g. $\frac{3}{4}$.

An **improper fraction** has its numerator more than its denominator, e.g. $\frac{9}{2}$.

A **mixed number** is made up of a whole number and a proper fraction, e.g. $4\frac{1}{5}$.

➜ Worked examples

a Find $\frac{1}{5}$ of 35.

This means 'divide 35 into 5 equal parts'.

$\frac{1}{5}$ of 35 is $35 \div 5 = 7$.

b Find $\frac{3}{5}$ of 35.

Since $\frac{1}{5}$ of 35 is 7, $\frac{3}{5}$ of 35 is 7×3.

That is, 21.

 Exercise 4.1

1 Evaluate the following:

a $\frac{3}{4}$ of 20 b $\frac{4}{5}$ of 20 c $\frac{4}{9}$ of 45 d $\frac{5}{8}$ of 64

e $\frac{3}{11}$ of 66 f $\frac{9}{10}$ of 80 g $\frac{5}{7}$ of 42 h $\frac{8}{9}$ of 54

i $\frac{7}{8}$ of 240 j $\frac{4}{5}$ of 65

4 INTEGERS, FRACTIONS, DECIMALS AND PERCENTAGES

Changing a mixed number to an improper fraction

> **Worked examples**

a Change $3\frac{5}{8}$ into an improper fraction.

$$3\frac{5}{8} = \frac{24}{8} + \frac{5}{8}$$
$$= \frac{24 + 5}{8}$$
$$= \frac{29}{8}$$

b Change the improper fraction $\frac{27}{4}$ into a mixed number.

$$\frac{27}{4} = \frac{24 + 3}{4}$$
$$= \frac{24}{4} + \frac{3}{4}$$
$$= 6\frac{3}{4}$$

Exercise 4.2

1 Change the following mixed numbers into improper fractions:

a $4\frac{2}{3}$ **b** $3\frac{3}{5}$ **c** $5\frac{7}{8}$ **d** $2\frac{5}{6}$

e $8\frac{1}{2}$ **f** $9\frac{5}{7}$ **g** $6\frac{4}{9}$ **h** $4\frac{1}{4}$

i $5\frac{4}{11}$ **j** $7\frac{6}{7}$ **k** $4\frac{3}{10}$ **l** $11\frac{3}{13}$

2 Change the following improper fractions into mixed numbers:

a $\frac{29}{4}$ **b** $\frac{33}{5}$ **c** $\frac{41}{6}$ **d** $\frac{53}{8}$

e $\frac{49}{9}$ **f** $\frac{17}{12}$ **g** $\frac{66}{7}$ **h** $\frac{33}{10}$

i $\frac{19}{2}$ **j** $\frac{73}{12}$

Decimals

H	T	U.	$\frac{1}{10}$	$\frac{1}{100}$	$\frac{1}{1000}$
		3.	2	7	
		0.	0	3	8

Percentages

3.27 is 3 units, 2 tenths and 7 hundredths.

i.e. $3.27 = 3 + \frac{2}{10} + \frac{7}{100}$

0.038 is 3 hundredths and 8 thousandths.

i.e. $0.038 = \frac{3}{100} + \frac{8}{1000}$

Note that 2 tenths and 7 hundredths is equivalent to 27 hundredths

i.e. $\frac{2}{10} + \frac{7}{100} = \frac{27}{100}$

and that 3 hundredths and 8 thousandths is equivalent to 38 thousandths.

i.e. $\frac{3}{100} + \frac{8}{1000} = \frac{38}{1000}$

 Exercise 4.3

1 Write the following fractions as decimals:

   a $4\frac{5}{10}$ b $6\frac{3}{10}$ c $17\frac{8}{10}$ d $3\frac{7}{100}$

   e $9\frac{27}{100}$ f $11\frac{36}{100}$ g $4\frac{6}{1000}$ h $5\frac{27}{1000}$

   i $4\frac{356}{1000}$ j $9\frac{204}{1000}$

2 Evaluate the following:
   a 2.7 + 0.35 + 16.09   b 1.44 + 0.072 + 82.3
   c 23.8 − 17.2   d 16.9 − 5.74
   e 121.3 − 85.49   f 6.03 + 0.5 − 1.21
   g 72.5 − 9.08 + 3.72   h 100 − 32.74 − 61.2
   i 16.0 − 9.24 − 5.36   j 1.1 − 0.92 − 0.005

Percentages

A fraction whose denominator is 100 can be expressed as a **percentage**.

$\frac{29}{100}$ can be written as 29%

$\frac{45}{100}$ can be written as 45%

By using **equivalent fractions** to change the denominator to 100, other fractions can be written as percentages.

 Worked example

Change $\frac{3}{5}$ to a percentage.

$\frac{3}{5} = \frac{3}{5} \times \frac{20}{20} = \frac{60}{100}$

$\frac{60}{100}$ can be written as 60%

4 INTEGERS, FRACTIONS, DECIMALS AND PERCENTAGES

Exercise 4.4

1 Express each of the following as a fraction with denominator 100, then write them as percentages:

a $\frac{29}{50}$ b $\frac{17}{25}$ c $\frac{11}{20}$ d $\frac{3}{10}$

e $\frac{23}{25}$ f $\frac{19}{50}$ g $\frac{3}{4}$ h $\frac{2}{5}$

2 Copy and complete the table of equivalents.

Fraction	$\frac{1}{10}$		$\frac{4}{10}$			$\frac{4}{5}$	$\frac{1}{4}$	
Decimal		0.2		0.5	0.7	0.9		
Percentage			30%		60%			75%

The four rules

Addition, subtraction, multiplication and division are mathematical operations.

Long multiplication

When carrying out long multiplication, it is important to remember place value.

Worked example

184 × 37 =

```
      1 8 4
   ×    3 7
    1 2 8 8   (184 × 7)
    5 5 2 0   (184 × 30)
    6 8 0 8   (184 × 37)
```

Short division

Worked example

453 ÷ 6 =

```
      7  5 r3
   6)4 5 ³3
```

It is usual, however, to give the final answer in decimal form rather than with a remainder. The division should therefore be continued:

453 ÷ 6

```
      7  5 . 5
   6)4 5 ³3 . ³0
```

The four rules

Long division

→ Worked example

Calculate 7184 ÷ 23 to one decimal place (1 d.p.).

```
        3 1 2 . 3 4
    23 ) 7 1 8 4 . 0 0
         6 9
         ───
           2 8
           2 3
           ───
             5 4
             4 6
             ───
               8 0
               6 9
               ───
               1 1 0
                 9 2
                 ───
                   1 8
```

Therefore 7184 ÷ 23 = 312.3 to 1 d.p.

Note how the question asks for the answer to 1 d.p., but the calculation is continued until the 2nd d.p. This is to see whether the answer needs to be rounded up.

Mixed operations

When a calculation involves a mixture of operations, the **order of the operations** is important. Multiplications and divisions are done first, while additions and subtractions are done afterwards. To override this, brackets need to be used.

> **Note**
> The order of operations was also covered in Chapter 3.

→ Worked examples

a $3 + 7 \times 2 - 4$
     $= 3 + 14 - 4$
     $= 13$

b $(3 + 7) \times 2 - 4$
     $= 10 \times 2 - 4$
     $= 20 - 4$
     $= 16$

c $3 + 7 \times (2 - 4)$
     $= 3 + 7 \times (-2)$
     $= 3 - 14$
     $= -11$

d $(3 + 7) \times (2 - 4)$
     $= 10 \times (-2)$
     $= -20$

4 INTEGERS, FRACTIONS, DECIMALS AND PERCENTAGES

Exercise 4.5

1. Evaluate the answer to each of the following:
   a. $3 + 5 \times 2 - 4$
   b. $12 \div 8 + 6 \div 4$

2. Copy these equations and put brackets in the correct places to make them correct:
   a. $6 \times 4 + 6 \div 3 = 20$
   b. $9 - 3 \times 7 + 2 = 54$

3. i. Without using a calculator, work out the solutions to the following multiplications:
   a. 785×38
   b. 164×253
   ii. Use the answers to the above questions to deduce the answer to the following:
   a. 7.85×3.8
   b. 1.64×2530

4. Work out the remainders in the following divisions:
   a. $72 \div 7$
   b. $430 \div 9$

5. a. A length of rail track is 9 m long. How many complete lengths will be needed to lay 1 km of track?
   b. How many 35-cent stamps can be bought for 10 dollars?

6. Work out the following long divisions to 1 d.p.:
   a. $7892 \div 7$
   b. $7892 \div 15$

Calculations with fractions

Equivalent fractions

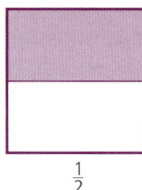
$\frac{1}{2}$

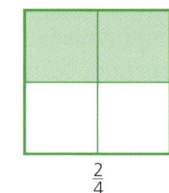
$\frac{2}{4}$

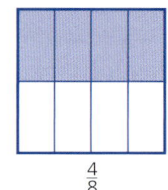
$\frac{4}{8}$

It should be apparent that $\frac{1}{2}$, $\frac{1}{4}$ and $\frac{4}{8}$ are equivalent fractions.

Similarly, $\frac{1}{3}$, $\frac{2}{6}$, $\frac{3}{9}$ and $\frac{4}{12}$ are equivalent, as are $\frac{1}{5}$, $\frac{10}{50}$ and $\frac{20}{100}$. Equivalent fractions are mathematically the same as each other. In the example diagrams, $\frac{1}{2}$ is mathematically the same as $\frac{4}{8}$. However, $\frac{1}{2}$ is a simplified form of $\frac{4}{8}$.

When carrying out calculations involving fractions it is usual to give your answer in its **simplest form**. Another way of saying 'simplest form' is '**lowest terms**'.

Writing a fraction in its 'simplest form' or in its 'lowest terms' means the same thing.

Calculations with fractions

Worked examples

a Write $\frac{4}{22}$ in its simplest form.
   Divide both the numerator and the denominator by their highest common factor.
   The highest common factor of both 4 and 22 is 2.
   Dividing both 4 and 22 by 2 gives $\frac{2}{11}$.
   Therefore $\frac{2}{11}$ is $\frac{4}{22}$ written in its simplest form.

b Write $\frac{12}{40}$ in its lowest terms.
   Divide both the numerator and the denominator by their highest common factor.
   The highest common factor of both 12 and 40 is 4.
   Dividing both 12 and 40 by 4 gives $\frac{3}{10}$.
   Therefore $\frac{3}{10}$ is $\frac{12}{40}$ written in its lowest terms.

 Exercise 4.6

1 Express the following fractions in their lowest terms. e.g. $\frac{12}{16} = \frac{3}{4}$

   a $\frac{5}{10}$ b $\frac{7}{21}$ c $\frac{8}{12}$

   d $\frac{16}{36}$ e $\frac{75}{100}$ f $\frac{81}{90}$

Addition and subtraction of fractions

For fractions to be either added or subtracted, the denominators need to be the same.

Worked examples

a $\frac{3}{11} + \frac{5}{11} = \frac{8}{11}$ b $\frac{7}{8} + \frac{5}{8} = \frac{12}{8} = 1\frac{1}{2}$

c $\frac{1}{2} + \frac{1}{3}$ d $\frac{4}{5} - \frac{1}{3}$

   $= \frac{3}{6} + \frac{2}{6} = \frac{5}{6}$ $= \frac{12}{15} - \frac{5}{15} = \frac{7}{15}$

When dealing with calculations involving mixed numbers, it is sometimes easier to change them to improper fractions first.

4 INTEGERS, FRACTIONS, DECIMALS AND PERCENTAGES

Worked examples

a $\quad 5\frac{3}{4} - 2\frac{5}{8}$ $\qquad$ b $\quad 1\frac{4}{7} + 3\frac{3}{4}$

$= \frac{23}{4} - \frac{21}{8}$ $\qquad\qquad = \frac{11}{7} + \frac{15}{4}$

$= \frac{46}{8} - \frac{21}{8}$ $\qquad\qquad = \frac{44}{28} + \frac{105}{28}$

$= \frac{25}{8} = 3\frac{1}{8}$ $\qquad\qquad = \frac{149}{28} = 5\frac{9}{28}$

Exercise 4.7

Evaluate each of the following and write the answer as a fraction in its simplest form:

1. a $\frac{3}{5} + \frac{4}{5}$ b $\frac{3}{11} + \frac{7}{11}$ c $\frac{2}{3} + \frac{1}{4}$
   d $\frac{3}{5} + \frac{4}{9}$ e $\frac{8}{13} + \frac{2}{5}$ f $\frac{1}{2} + \frac{2}{3} + \frac{3}{4}$

2. a $\frac{1}{8} + \frac{3}{8} + \frac{5}{8}$ b $\frac{3}{7} + \frac{5}{7} + \frac{4}{7}$ c $\frac{1}{3} + \frac{1}{2} + \frac{1}{4}$
   d $\frac{1}{5} + \frac{1}{3} + \frac{1}{4}$ e $\frac{3}{8} + \frac{3}{5} + \frac{3}{4}$ f $\frac{3}{13} + \frac{1}{4} + \frac{1}{2}$

3. a $\frac{3}{7} - \frac{2}{7}$ b $\frac{4}{5} - \frac{7}{10}$ c $\frac{8}{9} - \frac{1}{3}$
   d $\frac{7}{12} - \frac{1}{2}$ e $\frac{5}{8} - \frac{2}{5}$ f $\frac{3}{4} - \frac{2}{5} + \frac{7}{10}$

4. a $\frac{3}{4} + \frac{1}{5} - \frac{2}{3}$ b $\frac{3}{8} + \frac{7}{11} - \frac{1}{2}$ c $\frac{4}{5} - \frac{3}{10} + \frac{7}{20}$
   d $\frac{9}{13} + \frac{1}{3} - \frac{4}{5}$ e $\frac{9}{10} - \frac{1}{5} - \frac{1}{4}$ f $\frac{8}{9} - \frac{1}{3} - \frac{1}{2}$

5. a $2\frac{1}{2} + 3\frac{1}{4}$ b $3\frac{3}{5} + 1\frac{7}{10}$ c $6\frac{1}{2} - 3\frac{2}{5}$
   d $8\frac{5}{8} - 2\frac{1}{3}$ e $5\frac{7}{8} - 4\frac{3}{4}$ f $3\frac{1}{4} - 2\frac{5}{9}$

6. a $2\frac{1}{2} + 1\frac{1}{4} + 1\frac{3}{8}$ b $2\frac{4}{5} + 3\frac{1}{8} + 1\frac{3}{10}$ c $4\frac{1}{2} - 1\frac{1}{4} - 3\frac{5}{8}$
   d $6\frac{1}{2} - 2\frac{3}{4} - 3\frac{2}{5}$ e $2\frac{4}{7} - 3\frac{1}{4} - 1\frac{3}{5}$ f $4\frac{7}{20} - 5\frac{1}{2} + 2\frac{2}{5}$

Multiplication and division of fractions

Worked examples

a $\quad \frac{3}{4} \times \frac{2}{3}$ $\qquad$ b $\quad 3\frac{1}{2} \times 4\frac{4}{7}$

$= \frac{6}{12}$ $\qquad\qquad = \frac{7}{2} \times \frac{32}{7}$

$= \frac{1}{2}$ $\qquad\qquad = \frac{224}{14}$

$\qquad\qquad\qquad = 16$

Calculations with fractions

As already defined in Chapter 1, the reciprocal of a number is obtained when 1 is divided by that number. Therefore, the reciprocal of 5 is $\frac{1}{5}$ and the reciprocal of $\frac{2}{5}$ is $\frac{5}{2}$, and so on.

Dividing one fraction by another gives the same result as multiplying by the reciprocal.

➡ Worked examples

a $\quad \frac{3}{8} \div \frac{3}{4}$ 	b $\quad 5\frac{1}{2} \div 3\frac{2}{3}$

$= \frac{3}{8} \times \frac{4}{3}$ 	$= \frac{11}{2} \div \frac{11}{3}$

$= \frac{12}{24}$ 	$= \frac{11}{2} \times \frac{3}{11}$

$= \frac{1}{2}$ 	$= \frac{3}{2}$

Exercise 4.8

1. Write the reciprocal of each of the following:
   a $\quad \frac{1}{8}$ 	b $\quad \frac{7}{12}$ 	c $\quad \frac{3}{5}$
   d $\quad 1\frac{1}{2}$ 	e $\quad 3\frac{3}{4}$ 	f $\quad 6$

2. Evaluate the following:
   a $\quad \frac{3}{8} \times \frac{4}{9}$ 	b $\quad \frac{2}{3} \times \frac{9}{10}$ 	c $\quad \frac{5}{7} \times \frac{4}{15}$
   d $\quad \frac{3}{4}$ of $\frac{8}{9}$ 	e $\quad \frac{5}{6}$ of $\frac{3}{10}$ 	f $\quad \frac{7}{8}$ of $\frac{2}{5}$

3. Evaluate the following:
   a $\quad \frac{5}{8} \div \frac{3}{4}$ 	b $\quad \frac{5}{6} \div \frac{1}{3}$ 	c $\quad \frac{4}{5} \div \frac{7}{10}$
   d $\quad 1\frac{2}{3} \div \frac{2}{5}$ 	e $\quad \frac{3}{7} \div 2\frac{1}{7}$ 	f $\quad 1\frac{1}{4} \div 1\frac{7}{8}$

4. Evaluate the following:
   a $\quad \frac{3}{4} \times \frac{4}{5}$ 	b $\quad \frac{7}{8} \times \frac{2}{3}$ 	c $\quad \frac{3}{4} \times \frac{4}{7} \times \frac{3}{10}$
   d $\quad \frac{4}{5} \div \frac{2}{3} \times \frac{7}{10}$ 	e $\quad \frac{1}{2}$ of $\frac{3}{4}$ 	f $\quad 4\frac{1}{2} \div 3\frac{1}{9}$

5. Evaluate the following:
   a $\quad \left(\frac{3}{8} \times \frac{4}{5}\right) + \left(\frac{1}{2} \text{ of } \frac{3}{5}\right)$ 	b $\quad \left(1\frac{1}{2} \times 3\frac{3}{4}\right) - \left(2\frac{3}{5} \div 1\frac{1}{2}\right)$
   c $\quad \left(\frac{3}{5} \text{ of } \frac{4}{9}\right) + \left(\frac{4}{9} \text{ of } \frac{3}{5}\right)$ 	d $\quad \left(1\frac{1}{3} \times 2\frac{5}{8}\right)^2$

6. Using the correct order of operations, evaluate the following:
   a $\quad \frac{1}{4} + \frac{2}{7} \times \frac{3}{4}$ 	b $\quad \frac{1}{2} + \frac{3}{8} \div \frac{1}{4} - 2\frac{1}{2}$
   c $\quad \left(\frac{1}{6}\right)^2 - \frac{1}{8} \times \frac{5}{9}$ 	d $\quad \left(5\frac{2}{3} - \frac{14}{3}\right)^2 \div \frac{2}{7} + 3\frac{1}{2}$

4 INTEGERS, FRACTIONS, DECIMALS AND PERCENTAGES

Changing a fraction to a decimal

To change a fraction to a decimal, divide the numerator by the denominator.

→ Worked examples

a Change $\frac{5}{8}$ to a decimal.

$$\begin{array}{r} 0.\,6\;2\;5 \\ 8\overline{)5.0\;^20\;^40} \end{array}$$

b Change $2\frac{3}{5}$ to a decimal.

This can be represented as $2 + \frac{3}{5}$.

$$\begin{array}{r} 0.6 \\ 5\overline{)3.0} \end{array}$$

Therefore $2\frac{3}{5} = 2.6$

Exercise 4.9

1 Change the following fractions to decimals:

a $\frac{3}{4}$ b $\frac{4}{5}$ c $\frac{9}{20}$

d $\frac{17}{50}$ e $\frac{1}{3}$ f $\frac{3}{8}$

g $\frac{7}{16}$ h $\frac{2}{9}$ i $\frac{7}{11}$

2 Change the following mixed numbers to decimals:

a $2\frac{3}{4}$ b $3\frac{3}{5}$ c $4\frac{7}{20}$

d $6\frac{11}{50}$ e $5\frac{2}{3}$ f $6\frac{7}{8}$

g $5\frac{9}{16}$ h $4\frac{2}{9}$ i $5\frac{3}{7}$

Changing a decimal to a fraction

Changing a decimal to a fraction is done by knowing the 'value' of each of the digits in any decimal.

→ Worked examples

a Change 0.45 from a decimal to a fraction.

units . tenths hundredths
  0   .   4   5

Changing a recurring decimal to a fraction

0.45 is therefore equivalent to 4 tenths and 5 hundredths, which in turn is the same as 45 hundredths.

Therefore $0.45 = \frac{45}{100} = \frac{9}{20}$

b Change 2.325 from a decimal to a fraction.

units	.	tenths	hundredths	thousandths
2	.	3	2	5

Therefore $2.325 = 2\frac{325}{1000} = 2\frac{13}{40}$

 Exercise 4.10

1 Change the following decimals to fractions. Give your fraction in its simplest form:
- **a** 0.5
- **b** 0.7
- **c** 0.6
- **d** 0.75
- **e** 0.825
- **f** 0.05
- **g** 0.050
- **h** 0.402
- **i** 0.0002

2 Change the following decimals to mixed numbers in their simplest form:
- **a** 2.4
- **b** 6.5
- **c** 8.2
- **d** 3.75
- **e** 10.55
- **f** 9.204
- **g** 15.455
- **h** 30.001
- **i** 1.0205

Recurring decimals

In Chapter 1, the definition of a rational number was given as any number that can be written as a fraction. These include integers, terminating decimals and recurring decimals. The examples given were:

$$0.2 = \frac{1}{5} \quad 0.3 = \frac{3}{10} \quad 7 = \frac{7}{1} \quad 1.53 = \frac{153}{100} \quad \text{and} \quad 0.\dot{2} = \frac{2}{9}$$

The first four examples are more straightforward to understand in terms of how the number can be expressed as a fraction. The fifth example shows a recurring decimal also written as a fraction. Any recurring decimal can be written as a fraction as any recurring decimal is also a rational number.

Changing a recurring decimal to a fraction

A recurring decimal is one in which the numbers after the decimal point repeat themselves infinitely. Which numbers are repeated is indicated by a dot above them.

$0.\dot{2}$ implies $0.222\,222\,222\,222\,222\ldots$

$0.\dot{4}\dot{5}$ implies $0.454\,545\,454\,545\ldots$

$0.\dot{6}0\dot{2}\dot{4}$ implies $0.602\,460\,246\,024\ldots$

4 INTEGERS, FRACTIONS, DECIMALS AND PERCENTAGES

Note: The last example is usually written in one of two other ways: $0.\dot{6}02\dot{4}$, where the dot only appears above the first and last numbers to be repeated, or as $0.\overline{6024}$, where a horizontal line is drawn above the numbers being repeated.

Entering $\frac{4}{9}$ into a calculator will produce 0.444 444 444...

Therefore $0.\dot{4} = \frac{4}{9}$.

The example below will prove this.

➡ Worked examples

a Convert $0.\dot{4}$ to a fraction.

   Let $x = 0.\dot{4}$.   i.e. $x = 0.444\,444\,444\,444\ldots$

   $10x = 4.\dot{4}$.   i.e. $10x = 4.444\,444\,444\,444\ldots$

   Subtracting x from $10x$ gives:

   $10x = 4.444\,444\,444\,444\ldots$
   $-\ \ x = 0.444\,444\,444\,444\ldots$
   $\ \ 9x = 4$

   Rearranging gives $x = \frac{4}{9}$

   But $x = 0.\dot{4}$

   Therefore $0.\dot{4} = \frac{4}{9}$

b Convert $0.\dot{6}\dot{8}$ to a fraction.

   Let $x = 0.\dot{6}\dot{8}$ i.e. $x = 0.686\,868\,686\,868\,686\ldots$

   $100x = 68.\dot{6}\dot{8}$ i.e. $100x = 68.686\,868\,686\,868\,686\ldots$

   Subtracting x from $100x$ gives:

   $100x = 68.686\,868\,686\,868\,686\ldots$
   $-\ \ \ x =\ \ 0.686\,868\,686\,868\,686\ldots$
   $\ \ 99x = 68$

   Rearranging gives $x = \frac{68}{99}$

   But $x = 0.\dot{6}\dot{8}$

   Therefore $0.\dot{6}\dot{8} = \frac{68}{99}$

c Convert $0.0\dot{3}\dot{1}$ to a fraction.

   Let $x = 0.0\dot{3}\dot{1}$ i.e. $x = 0.031\,313\,131\,313\,131\ldots$

   $100x = 3.\dot{1}\dot{3}$ i.e. $100x = 3.131\,313\,131\,313\,131\ldots$

   Subtracting x from $100x$ gives:

   $100x = 3.131\,313\,131\,313\,131\ldots$
   $-\ \ \ x = 0.031\,313\,131\,313\,131\ldots$
   $\ \ 99x = 3.1$

Changing a recurring decimal to a fraction

Multiplying both sides of the equation by 10 eliminates the decimal to give:

$990x = 31$

Rearranging gives $x = \dfrac{31}{990}$

But $x = 0.0\dot{3}\dot{1}$

Therefore $0.0\dot{3}\dot{1} = \dfrac{31}{990}$

The method is therefore to let the recurring decimal equal x and then to multiply this by a multiple of 10 so that when one is subtracted from the other, either an integer (whole number) or terminating decimal (a decimal that has an end point) is left.

d Convert $2.0\overline{406}$ to a fraction.

Let $x = 2.0\overline{406}$ i.e. $x = 2.040\,640\,640\,640\ldots$

$1000x = 2040.\overline{640}$ i.e. $1000x = 2040.640\,640\,640\,640\ldots$

Subtracting x from $1000x$ gives:

$1000x = 2040.640\,640\,640\,640\ldots$

$-x = 2.040\,640\,640\,640\ldots$

$999x = 2038.6$

Multiplying both sides of the equation by 10 eliminates the decimal to give:

$9990x = 20\,386$

Rearranging gives $x = \dfrac{20\,386}{9990} = 2\dfrac{406}{9990}$, which simplifies further to $2\dfrac{203}{4995}$.

But $x = 2.0\overline{406}$.

Therefore $2.0\overline{406} = 2\dfrac{203}{4995}$

Exercise 4.11

1 Convert each of the following recurring decimals to fractions in their simplest form:
   a $0.\dot{3}$ b $0.\dot{7}$ c $0.\dot{4}\dot{2}$ d $0.\dot{6}\dot{5}$

2 Convert each of the following recurring decimals to fractions in their simplest form:
   a $0.0\dot{5}$ b $0.0\dot{6}\dot{2}$ c $1.0\dot{2}$ d $4.00\dot{3}\dot{8}$

3 Work out the sum $0.1\dot{5} + 0.0\dot{4}$ by converting each decimal to a fraction first. Give your answer as a fraction in its simplest form.

4 Evaluate $0.\dot{2}\dot{7} - 0.1\dot{0}\dot{6}$ by converting each decimal to a fraction first. Give your answer as a fraction in its simplest form.

4 INTEGERS, FRACTIONS, DECIMALS AND PERCENTAGES

Student assessment 1

1. Evaluate the following:
   a $\frac{1}{3}$ of 63
   b $\frac{3}{8}$ of 72
   c $\frac{2}{5}$ of 55
   d $\frac{3}{13}$ of 169

2. Write the following as percentages:
   a $\frac{3}{5}$
   b $\frac{49}{100}$
   c $\frac{1}{4}$
   d $\frac{9}{10}$
   e $1\frac{1}{2}$
   f $3\frac{27}{100}$
   g $\frac{5}{100}$
   h $\frac{7}{20}$
   i 0.77
   j 0.03
   k 2.9
   l 4

3. Evaluate the following:
   a $6 \times 4 - 3 \times 8$
   b $15 \div 3 + 2 \times 7$

4. Work out 368×49.

5. Work out $7835 \div 23$ giving your answer to 1 d.p.

6. Evaluate the following:
   a $2\frac{1}{2} - \frac{4}{5}$
   b $3\frac{1}{2} \times \frac{4}{7}$

7. Change the following fractions to decimals:
   a $\frac{7}{8}$
   b $1\frac{2}{5}$
   c $\frac{8}{9}$
   d $3\frac{2}{7}$

8. Change the following decimals to fractions. Give each fraction in its simplest form.
   a 6.5
   b 0.04
   c 3.65
   d 3.008

9. Convert the following decimals to fractions, giving your answer in its simplest form:
   a $0.0\dot{7}$
   b $0.000\dot{9}$
   c $3.0\dot{2}\dot{0}$

10. Work out $1.02\dot{5} - 0.80\dot{5}$ by first converting each decimal to a fraction. Give your answer in its simplest form.

5 Further percentages

You should already be familiar with the percentage equivalents of simple fractions and decimals as outlined in the table below.

Fraction	Decimal	Percentage
$\frac{1}{2}$	0.5	50%
$\frac{1}{4}$	0.25	25%
$\frac{3}{4}$	0.75	75%
$\frac{1}{8}$	0.125	12.5%
$\frac{3}{8}$	0.375	37.5%
$\frac{5}{8}$	0.625	62.5%
$\frac{7}{8}$	0.875	87.5%
$\frac{1}{10}$	0.1	10%
$\frac{2}{10}$ or $\frac{1}{5}$	0.2	20%
$\frac{3}{10}$	0.3	30%
$\frac{4}{10}$ or $\frac{2}{5}$	0.4	40%
$\frac{6}{10}$ or $\frac{3}{5}$	0.6	60%
$\frac{7}{10}$	0.7	70%
$\frac{8}{10}$ or $\frac{4}{10}$	0.8	80%
$\frac{9}{10}$	0.9	90%

5 FURTHER PERCENTAGES

Simple percentages

> **Worked examples**

a Of 100 sheep in a field, 88 are ewes.
   i What percentage of the sheep are ewes?

   88 out of 100 are ewes
   $= 88\%$

   ii What percentage are not ewes?

   12 out of 100
   $= 12\%$

b Convert the following percentages into fractions and decimals:
   i 27%   ii 5%

   $\frac{27}{100} = 0.27$ $\frac{5}{100} = \frac{1}{20} = 0.05$

c Convert $\frac{3}{16}$ to a percentage:

   This example is more complicated as 16 is not a factor of 100.

   Convert $\frac{3}{16}$ to a decimal first.

   $3 \div 16 = 0.1875$

   Convert the decimal to a percentage by multiplying by 100.

   $0.1875 \times 100 = 18.75$

   Therefore $\frac{3}{16} = 18.75\%$.

 Exercise 5.1

1 There are 200 birds in a flock. 120 of them are female. What percentage of the flock are:
   a female?   b male?

2 Write these fractions as percentages:
   a $\frac{7}{8}$ b $\frac{11}{15}$ c $\frac{7}{24}$ d $\frac{1}{7}$

3 Convert the following percentages to decimals:
   a 39%   b 47%   c 83%
   d 7%   e 2%   f 20%

4 Convert the following decimals to percentages:
   a 0.31   b 0.67   c 0.09
   d 0.05   e 0.2   f 0.75

Calculating a percentage of a quantity

> **Worked examples**
>
> a Find 25% of 300 m.
>
> 25% can be written as 0.25.
>
> $0.25 \times 300\,m = 75\,m$.
>
> b Find 35% of 280 m.
>
> 35% can be written as 0.35.
>
> $0.35 \times 280\,m = 98\,m$.

 Exercise 5.2

1. Write the percentage equivalent of the following fractions:
   a $\frac{1}{4}$
   b $\frac{2}{3}$
   c $\frac{5}{8}$
   d $1\frac{4}{5}$
   e $4\frac{9}{10}$
   f $3\frac{7}{8}$

2. Write the decimal equivalent of the following:
   a $\frac{3}{4}$
   b 80%
   c $\frac{1}{5}$
   d 7%
   e $1\frac{7}{8}$
   f $\frac{1}{6}$

3. Evaluate the following:
   a 25% of 80
   b 80% of 125
   c 62.5% of 80
   d 30% of 120
   e 90% of 5
   f 25% of 30

4. Evaluate the following:
   a 17% of 50
   b 50% of 17
   c 65% of 80
   d 80% of 65
   e 7% of 250
   f 250% of 7

5. In a class of 30 students, 20% travel to school by car, 10% walk and 70% travel by bus. Calculate the number of students who travel to school by:
   a car
   b walking
   c bus

6. A survey conducted among 120 school children looked at which type of fruit they preferred. 55% said they preferred apple, 20% said they preferred mango, 15% preferred pineapple and 10% grapes. Calculate the number of children in each category.

7. A survey was carried out in a school to see what nationality its students were. Of the 220 students in the school, 65% were Australian, 20% were Pakistani, 5% were Greek and 10% belonged to other nationalities. Calculate the number of students of each nationality and how many were of other nationalities.

8. A shopkeeper keeps a record of the number of items she sells in one day. Of the 150 items she sold, 46% were newspapers, 24% were pens, 12% were books while the remaining 18% were other items. Calculate the number of each item sold and how many were other items.

5 FURTHER PERCENTAGES

Expressing one quantity as a percentage of another

To express one quantity as a percentage of another, first write the first quantity as a fraction of the second and then multiply by 100.

➡ Worked example

In an examination, Yuji obtains 69 marks out of 75. Express this result as a percentage.

$\frac{69}{75} \times 100\% = 92\%$

Exercise 5.3

1 Express the first quantity as a percentage of the second.
   - a 24 out of 50
   - b 46 out of 125
   - c 7 out of 20
   - d 45 out of 90
   - e 9 out of 20
   - f 16 out of 40
   - g 13 out of 39
   - h 20 out of 35

2 A hockey team plays 42 matches. It wins 21, draws 14 and loses the rest. Express each of these results as a percentage of the total number of games played.

3 Four candidates stood in an election:
   A received 24 500 votes
   B received 18 200 votes
   C received 16 300 votes
   D received 12 000 votes
   Express each of these as a percentage of the total votes cast.

4 A car manufacturer produces 155 000 cars a year. The cars are available for sale in six different colours. In one year, the numbers sold of each colour were:
   Red 55 000
   Blue 48 000
   White 27 500
   Silver 10 200
   Green 9300
   Black 5000

   Express each of these as a percentage of the total number of cars produced. Give your answers to 1 d.p.

Percentage increases and decreases

Worked examples

a A shop assistant has a salary of $16000 per year. If his salary increases by 8%, calculate:
  i the amount extra he receives each year,
  ii his new annual salary.
  i Increase = 8% of $16000
         = 0.08 × $16000 = $1280
  ii New salary = old salary + increase
         = $16000 + $1280 per year
         = $17280 per year

b A garage increases the price of a truck by 12%. If the original price was $14500, calculate its new price.

The original price represents 100%, therefore the increased price can be represented as 112%.

New price = 112% of $14500
      = 1.12 × $14500
      = $16240

c A shop is having a sale. It sells a set of tools costing $130 at a 15% discount. Calculate the sale price of the tools.

The old price represents 100%, therefore the new price can be represented as (100 − 15)% = 85%.

85% of $130 = 0.85 × $130
       = $110.50

Exercise 5.4

1 Increase the following by the given percentage:
  a 150 by 25%   b 230 by 40%   c 7000 by 2%
  d 70 by 250%   e 80 by 12.5%   f 75 by 62%

2 Decrease the following by the given percentage:
  a 120 by 25%   b 40 by 5%   c 90 by 90%
  d 1000 by 10%   e 80 by 37.5%   f 75 by 42%

3 In the following questions the first number is increased to become the second number. Calculate the percentage increase in each case.
  a 50 → 60   b 75 → 135   c 40 → 84
  d 30 → 31.5   e 18 → 33.3   f 4 → 13

4 In the following questions the first number is decreased to become the second number. Calculate the percentage decrease in each case.
  a 50 → 25   b 80 → 56   c 150 → 142.5
  d 3 → 0   e 550 → 352   f 20 → 19

5 FURTHER PERCENTAGES

Exercise 5.4 (cont)

5 A farmer increases the yield on her farm by 15%. If her previous yield was 6500 tonnes, what is her present yield?

6 The cost of a computer in a store is reduced by 12.5% in a sale. If the computer was priced at $7800, what is its price in the sale?

7 A winter coat is priced at $100. In the sale, its price is reduced by 25%.
   a Calculate the sale price of the coat.
   b After the sale, its price is increased by 25%. Calculate the coat's price after the sale.

8 A farmer takes 250 chickens to be sold at a market. In the first hour, he sells 8% of his chickens. In the second hour, he sells 10% of those that were left.
   a How many chickens has he sold in total?
   b What percentage of the original number did he manage to sell in the two hours?

9 The number of fish on a fish farm increases by approximately 10% each month. If there were originally 350 fish, calculate to the nearest 100 how many fish there would be after 12 months.

Reverse percentages

> **Worked examples**
>
> a In a test, Ahmed answered 92% of the questions correctly. If he answered 23 questions correctly, how many did he get wrong?
>
> 92% of the marks is equivalent to 23 questions.
>
> 1% of the marks therefore is equivalent to $\frac{23}{92}$ questions.
>
> So 100% is equivalent to $\frac{23}{92} \times 100 = 25$ questions.
>
> Ahmed got 2 questions wrong.
>
> b A boat is sold for $15 360. This represents a profit of 28% to the seller. What did the boat originally cost the seller?
>
> The selling price is 128% of the original cost to the seller.
>
> 128% of the original cost is $15 360.
>
> 1% of the original cost is $\frac{\$15\,360}{128}$.
>
> 100% of the original cost is $\frac{\$15\,360}{128} \times 100$, i.e. $12 000.

Reverse percentages

Exercise 5.5

1 Calculate the value of *X* in each of the following:
   a 40% of *X* is 240
   b 24% of *X* is 84
   c 85% of *X* is 765
   d 4% of *X* is 10
   e 15% of *X* is 18.75
   f 7% of *X* is 0.105

2 Calculate the value of *Y* in each of the following:
   a 125% of *Y* is 70
   b 140% of *Y* is 91
   c 210% of *Y* is 189
   d 340% of *Y* is 68
   e 150% of *Y* is 0.375
   f 144% of *Y* is −54.72

3 In a geography textbook, 35% of the pages are coloured. If there are 98 coloured pages, how many pages are there in the whole book?

4 A town has 3500 families who own a car. If this represents 28% of the families in the town, how many families are there in total?

5 In a test, Isabel scored 88%. If she got three questions incorrect, how many did she get correct?

6 Water expands when it freezes. Ice is less dense than water so it floats. If the increase in volume is 4%, what volume of water will make an iceberg of 12 700 000 m^3? Give your answer to 3 s.f.

Student assessment 1

1 Find 40% of 1600 m.

2 A shop increases the price of a television by 8%. If the present price is $320, what is the new price?

3 A car loses 55% of its value after four years. If it cost $22 500 when new, what is its value after the four years?

4 Express the first quantity as a percentage of the second.
   a 40 cm, 2 m
   b 25 mins, 1 hour
   c 450 g, 2 kg
   d 3 m, 3.5 m
   e 70 kg, 1 tonne
   f 75 cl, 2.5 l

5 A house is bought for $75 000 and then resold for $87 000. Calculate the percentage profit.

6 A pair of shoes is priced at $45. During a sale the price is reduced by 20%.
   a Calculate the sale price of the shoes.
   b What is the percentage increase in the price if after the sale it is restored to $45?

5 FURTHER PERCENTAGES

Student assessment 2

1. Find 30% of 2500 m.

2. In a sale a shop reduces its prices by 12.5%. What is the sale price of a desk previously costing $600?

3. In the last six years the value of a house has increased by 35%. If it cost $72 000 six years ago, what is its value now?

4. Express the first quantity as a percentage of the second.
   a 35 mins, 2 hours
   b 650 g, 3 kg
   c 5 m, 4 m
   d 15 s, 3 mins
   e 600 kg, 3 tonnes
   f 35 cl, 3.5 l

5. Shares in a company are bought for $600. After a year, the same shares are sold for $550. Calculate the percentage loss.

6. In a sale, the price of a jacket originally costing $1700 is reduced by $400. Any item not sold by the last day of the sale is reduced by a further 50%. If the jacket is sold on the last day of the sale:
   a calculate the price it is finally sold for,
   b calculate the overall percentage reduction in price.

Student assessment 3

1. Calculate the original price for each of the following:

Selling price	Profit
$224	12%
$62.50	150%
$660.24	26%
$38.50	285%

2. Calculate the original price for each of the following:

Selling price	Loss
$392.70	15%
$2480	38%
$3937.50	12.5%
$4675	15%

3. In an examination, Aliya obtained 87.5% by gaining 105 marks. How many marks did she lose?

4. At the end of a year, a factory has produced 38 500 televisions. If this represents a 10% increase in productivity on last year, calculate the number of televisions that were made last year.

5. A computer manufacturer is expected to have produced 24 000 units by the end of this year. If this represents a 4% decrease on last year's output, calculate the number of units produced last year.

6. A company increased its productivity by 10% each year for the last two years. If it produced 56 265 units this year, how many units did it produce two years ago?

6 Ratio and proportion

Direct proportion

Workers in a pottery factory are paid according to how many plates they produce. The wage paid to them is said to be in **direct proportion** to the number of plates made. As the number of plates made increases so does their wage. Other workers are paid for the number of hours worked. For them the wage paid is in **direct proportion** to the number of hours worked. There are two main methods for solving problems involving direct proportion: the **ratio method** and the **unitary method**.

> ### Worked example
> A bottling machine fills 500 bottles in 15 minutes. How many bottles will it fill in $1\frac{1}{2}$ hours?
>
> Note: The time units must be the same, so for either method the $1\frac{1}{2}$ hours must be changed to 90 minutes.
>
> ### The ratio method
> Let x be the number of bottles filled. Then:
>
> $$\frac{x}{90} = \frac{500}{15}$$
>
> so $x = \frac{500 \times 90}{15} = 3000$
>
> 3000 bottles are filled in $1\frac{1}{2}$ hours.
>
> ### The unitary method
> In 15 minutes, 500 bottles are filled.
>
> Therefore in 1 minute, $\frac{500}{15}$ bottles are filled.
>
> So in 90 minutes, $90 \times \frac{500}{15}$ bottles are filled.
>
> In $1\frac{1}{2}$ hours, 3000 bottles are filled.

Exercise 6.1   Use either the ratio method or the unitary method to solve the problems below.

1. A machine prints four books in 10 minutes. How many will it print in 2 hours?

2. A gardener plants five apple trees in 25 minutes. If he continues to work at a constant rate, how long will it take him to plant 200 trees?

6 RATIO AND PROPORTION

Exercise 6.1 (cont)

3. A television uses 3 units of electricity in 2 hours. How many units will it use in 7 hours? Give your answer to the nearest unit.

4. A bricklayer lays 1500 bricks in an 8-hour day. Assuming she continues to work at the same rate, calculate:
   a. how many bricks she would expect to lay in a five-day week,
   b. how long to the nearest hour it would take her to lay 10000 bricks.

5. A machine used to paint white lines on a road uses 250 litres of paint for each 8 km of road marked. Calculate:
   a. how many litres of paint would be needed for 200 km of road,
   b. what length of road could be marked with 4000 litres of paint.

6. An aircraft is cruising at 720 km/h and covers 1000 km. How far would it travel in the same period of time if the speed increased to 800 km/h?

7. A production line travelling at 2 m/s labels 150 tins. In the same period of time how many will it label at:
   a. 6 m/s   b. 1 m/s   c. 1.6 m/s?

8. A car travels at an average speed of 80 km/h for 6 hours.
   a. How far will it travel in the 6 hours?
   b. What average speed will it need to travel at in order to cover the same distance in 5 hours?

If the information is given in the form of a ratio, the method of solution is the same.

Worked example

Tin and copper are mixed in the ratio 8 : 3. How much tin is needed to mix with 36 g of copper?

The ratio method

Let x grams be the mass of tin needed.

$$\frac{x}{36} = \frac{8}{3}$$

Therefore $x = \frac{8 \times 36}{3}$

$\qquad\quad\;\; = 96$

So 96 g of tin is needed.

The unitary method

3 g of copper mixes with 8 g of tin.

1 g of copper mixes with $\frac{8}{3}$ g of tin.

So 36 g of copper mixes with $36 \times \frac{8}{3}$ g of tin.

Therefore 36 g of copper mixes with 96 g of tin.

Divide a quantity in a given ratio

Exercise 6.2

1. Sand and gravel are mixed in the ratio 5 : 3 to form ballast.
   a. How much gravel is mixed with 750 kg of sand?
   b. How much sand is mixed with 750 kg of gravel?

2. A recipe uses 150 g butter, 500 g flour, 50 g sugar and 100 g currants to make 18 small cakes.
   a. Write the ratio of the amount of butter : flour : sugar : currants in its simplest form.
   b. How much of each ingredient will be needed to make 72 cakes?
   c. How many whole cakes could be made with 1 kg of butter?

3. A paint mix uses red and white paint in a ratio of 1 : 12.
   a. How much white paint will be needed to mix with 1.4 litres of red paint?
   b. If a total of 15.5 litres of paint is mixed, calculate the amount of white paint and the amount of red paint used. Give your answers to the nearest 0.1 litre.

4. Rebecca sells sacks of mixed bulbs to local people. The bulbs develop into two different colours of tulips, red and yellow. The colours are packaged in a ratio of 8 : 5 respectively.
   a. If a sack contains 200 red bulbs, calculate the number of yellow bulbs.
   b. If a sack contains 351 bulbs in total, how many of each colour would you expect to find?
   c. One sack is packaged with a bulb mixture in the ratio 7 : 5 by mistake. If the sack contains 624 bulbs, how many more yellow bulbs would you expect to have compared with a normal sack of 624 bulbs?

5. A pure fruit juice is made by mixing the juices of oranges and mangos in the ratio of 9 : 2.
   a. If 189 litres of orange juice are used, calculate the number of litres of mango juice needed.
   b. If 605 litres of the juice are made, calculate the number of litres of orange juice and mango juice used.

Divide a quantity in a given ratio

→ Worked examples

a. Divide 20 m in the ratio 3 : 2.

The ratio method

3 : 2 gives 5 parts.

$\frac{3}{5} \times 20$ m = 12 m

$\frac{2}{5} \times 20$ m = 8 m

20 m divided in the ratio 3 : 2 is 12 m : 8 m.

6 RATIO AND PROPORTION

The unitary method

3 : 2 gives 5 parts.
5 parts is equivalent to 20 m.
1 part is equivalent to $\frac{20}{5}$ m.
Therefore 3 parts is $3 \times \frac{20}{5}$ m; that is 12 m.
Therefore 2 parts is $2 \times \frac{20}{5}$ m; that is 8 m.

b A factory produces cars in red, blue, white and green in the ratio 7 : 5 : 3 : 1. Out of a production of 48 000 cars how many are white?

7 + 5 + 3 + 1 gives a total of 16 parts.
Therefore, the total number of white cars = $\frac{3}{16} \times 48\,000 = 9000$.

Exercise 6.3

1 Divide 150 in the ratio 2 : 3.
2 Divide 72 in the ratio 2 : 3 : 4.
3 Divide 5 kg in the ratio 13 : 7.
4 Divide 45 minutes in the ratio 2 : 3.
5 Divide 1 hour in the ratio 1 : 5.
6 $\frac{7}{8}$ of a can of drink is water, the rest is syrup. What is the ratio of water to syrup?
7 $\frac{5}{9}$ of a litre carton of orange is pure orange juice, the rest is water. How many millilitres of each are in the carton?
8 55% of students in a school are boys.
   a What is the ratio of boys to girls?
   b How many boys and how many girls are there if the school has 800 students?
9 A piece of wood is cut in the ratio 2 : 3.
   a What fraction of the length is the longer piece?
   b If the piece of wood is 80 cm long, how long is the shorter piece?
10 A gas pipe is 7 km long. A valve is positioned in such a way that it divides the length of the pipe in the ratio 4 : 3. Calculate the distance of the valve from each end of the pipe.
11 The size of the angles of a quadrilateral are in the ratio 1 : 2 : 3 : 3. Calculate the size of each angle.
12 The angles of a triangle are in the ratio 3 : 5 : 4. Calculate the size of each angle.
13 A millionaire leaves 1.4 million dollars in her will to be shared between her three children in the ratio of their ages. If they are 24, 28 and 32 years old, calculate to the nearest dollar the amount they will each receive.
14 A small company makes a **profit** of $8000. This is divided between the directors in the ratio of their initial investments. If Malik put $20 000 into the firm, Zahra $35 000 and Ahmet $25 000, calculate the amount of the profit they will each receive.

Inverse proportion

Sometimes an increase in one quantity causes a decrease in another quantity. For example, if fruit is to be picked by hand, the more people there are picking the fruit, the less time it will take.

> ### Worked examples

a If 8 people can pick the apples from the trees in 6 days, how long will it take 12 people?

8 people take 6 days.

1 person will take 6×8 days.

Therefore 12 people will take $\frac{6 \times 8}{12}$ days, i.e. 4 days.

b A cyclist averages a speed of 27 km/h for 4 hours. At what average speed would she need to cycle to cover the same distance in 3 hours?

Completing it in 1 hour would require cycling at 27×4 km/h.

Completing it in 3 hours requires cycling at $\frac{27 \times 4}{3}$ km/h; that is 36 km/h.

Exercise 6.4

1 A teacher shares sweets among 8 students so that they get 6 each. How many sweets would they each have had if there had been 12 students?

2 The table below represents the relationship between the speed and the time taken for a train to travel between two stations.

Speed (km/h)	60			120	90	50	10
Time (h)	2	3	4				

Copy and complete the table.

3 Six people can dig a trench in 8 hours.
   a How long would it take:
      i 4 people   **ii** 12 people   **iii** 1 person?
   b How many people would it take to dig the trench in:
      i 3 hours   **ii** 16 hours   **iii** 1 hour?

4 Chairs in a hall are arranged in 35 rows of 18.
   a How many rows would there be with 21 chairs to a row?
   b How many chairs would there be in each row if there were 15 rows?

5 A train travelling at 100 km/h takes 4 hours for a journey. How long would it take a train travelling at 60 km/h?

6 A worker in a sugar factory packs 24 cardboard boxes with 15 bags of sugar in each. If he had boxes which held 18 bags of sugar each, how many fewer boxes would be needed?

7 A swimming pool is filled in 30 hours by two identical pumps. How much quicker would it be filled if five similar pumps were used instead?

6 RATIO AND PROPORTION

Compound measures

A **compound measure** is one made up of two or more other measures. The most common ones, which we will consider here, are speed, density and population density.

Speed is a compound measure as it is measured using distance and time.

$$\text{Speed} = \frac{\text{Distance}}{\text{Time}}$$

Units of speed include metres per second (m/s) or kilometres per hour (km/h).

The relationship between speed, distance and time is often presented as shown in the diagram below:

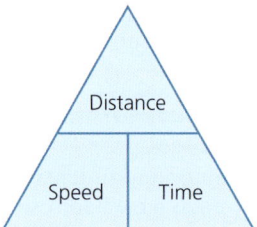

i.e. $\text{Speed} = \frac{\text{Distance}}{\text{Time}}$

$\text{Distance} = \text{Speed} \times \text{Time}$

$\text{Time} = \frac{\text{Distance}}{\text{Speed}}$

Similarly, $\text{Average speed} = \frac{\text{Total distance}}{\text{Total time}}$

Density, which is a measure of the mass of a substance per unit of its volume, is calculated using the following formula:

$$\text{Density} = \frac{\text{Mass}}{\text{Volume}}$$

Units of density include kilograms per cubic metre (kg/m³) or grams per millilitre (g/ml).

The relationship between density, mass and volume, like speed, can also be presented in a helpful diagram as shown:

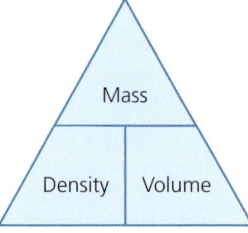

i.e. $\text{Density} = \frac{\text{Mass}}{\text{Volume}}$

$\text{Mass} = \text{Density} \times \text{Volume}$

$\text{Volume} = \frac{\text{Mass}}{\text{Density}}$

Population density is also a compound measure as it is a measure of a population per unit of area.

$$\text{Population density} = \frac{\text{Population}}{\text{Area}}$$

An example of its units include the number of people per square kilometre (people/km²).

Compound measures

Once again this can be represented in a triangular diagram as shown:

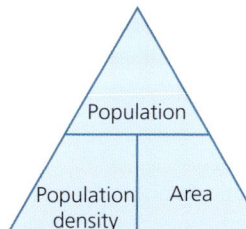

i.e. Population density = $\frac{\text{Population}}{\text{Area}}$

Population = Population density × Area

Area = $\frac{\text{Population}}{\text{Population density}}$

→ Worked examples

a A train travels a total distance of 140 km in $1\frac{1}{2}$ hours.

  i Calculate the average speed of the train during the journey.

  Average speed = $\frac{\text{Total distance}}{\text{Total time}}$

  $= \frac{140}{1\frac{1}{2}}$

  $= 93\frac{1}{3}$ km/h

  ii During the journey, the train spent 15 minutes stopped at stations. Calculate the average speed of the train while it was moving.

  Notice that the original time was given in hours, while the time spent stopped at stations is given in minutes. To proceed with the calculation, the units have to be consistent, i.e. either both in hours or both in minutes.
  The time spent travelling is $1\frac{1}{2} - \frac{1}{4} = 1\frac{1}{4}$ hours.
  Therefore:

  Average speed = $\frac{140}{1\frac{1}{4}}$

  $= 112$ km/h

  iii If the average speed was 120 km/h, calculate how long the journey took.

  Total time = $\frac{\text{Total distance}}{\text{Average speed}}$

  $= \frac{140}{120} = 1.1\dot{6}$ hours

  Note: It may be necessary to convert a decimal answer to hours and minutes.
  To convert a decimal time to minutes, multiply by 60.
  $0.1\dot{6} \times 60 = 10$
  Therefore total time is 1 hr 10 mins or 70 mins.

b A village has a population of 540. Its total area is 8 km².

  i Calculate the population density of the village.

  Population density = $\frac{\text{Population}}{\text{Area}}$

  $= \frac{540}{8} = 67.5$ people/km².

59

6 RATIO AND PROPORTION

ii A building company wants to build some new houses in the existing area of the village. It is decided that the maximum desirable population density of the village should not exceed 110 people/km². Calculate the extra number of people the village can have.

Population = Population density × Area
= 110 × 8
= 880 people

Therefore the maximum number of extra people who will need housing is 880 − 540 = 340.

Exercise 6.5

1 Aluminium has a density of 2900 kg/m³. A construction company needs four cubic metres of aluminium.
Calculate the mass of the aluminium needed.

2 A marathon race is 42 195 m in length. The world record in 2022 was 2 hrs, 1 min and 9 seconds held by Eliud Kipchoge of Kenya.
   a How many seconds in total did Eliud take to complete the race?
   b Calculate his average speed in m/s for the race, giving your answer to 2 decimal places.
   c What average speed would the runner need to maintain to complete the marathon in under two hours?

3 The approximate densities of four metals in g/cm³ are given below:
   Aluminium   2.9 g/cm³
   Brass   8.8 g/cm³
   Copper   9.3 g/cm³
   Steel   8.2 g/cm³
   A cube of an unknown metal has side lengths of 5 cm. The mass of the cube is 1.1 kg.
   a By calculating the cube's density, determine which metal the cube is likely to be made from.
   Another cube made of steel has a mass of 4.0 kg.
   b Calculate the length of each of the sides of the steel cube, giving your answer to 1 d.p.

4 Singapore is the country with the highest population density in the world. Its population is 5 954 000 and it has a total area of 719 km².
   a Calculate Singapore's population density.
   China is the country with the largest population.
   b Explain why China has not got the world's highest population density.
   c Find the area and population of your own country. Calculate your country's population density.

5 Kwabena has a rectangular field measuring 600 m × 800 m. He uses the field for grazing his sheep.
   a Calculate the area of the field in km².
   b 40 sheep graze in the field. Calculate the population density of sheep in the field, giving your answer in sheep/km².
   Guidelines for keeping sheep state that the maximum population density for grazing sheep is 180/km².
   c Calculate the number of sheep Kwabena is allowed to graze in his field.

6 The formula linking pressure (P N/m^2), force (F N) and surface area (A m^2) is given as $P = \dfrac{F}{A}$. A square-based box exerts a force of 160 N on a floor. If the pressure on the floor is 1000 N/m^2, calculate the length, in cm, of each side of the base of the box.

Student assessment 1

1 A boat travels at an average speed of 15 km/h for 1 hour.
  a Calculate the distance it travels in one hour.
  b At what average speed will the boat have to travel to cover the same distance in $2\frac{1}{2}$ hours?

2 A ruler 30 cm long is broken into two parts in the ratio 8 : 7. How long are the two parts?

3 A recipe needs 400 g of flour to make 8 cakes. How much flour would be needed in order to make 24 cakes?

4 To make 6 jam tarts, 120 g of jam is needed. How much jam is needed to make 10 tarts?

5 The scale of a map is 1 : 25 000.
  a Two villages are 8 cm apart on the map. How far apart are they in real life? Give your answer in kilometres.
  b The distance from a village to the edge of a lake is 12 km in real life. How far apart would they be on the map? Give your answer in centimetres.

6 A motorbike uses petrol and oil mixed in the ratio 13 : 2.
  a How much of each is there in 30 litres of mixture?
  b How much petrol would be mixed with 500 ml of oil?

7 **a** A model car is a $\frac{1}{40}$ scale model. Express this as a ratio.
  b If the length of the real car is 5.5 m, what is the length of the model car?

8 An aunt gives a brother and sister $2000 to be divided in the ratio of their ages. If the girl is 13 years old and the boy 12 years old, how much will each get?

9 The angles of a triangle are in the ratio 2 : 5 : 8. Find the size of each of the angles.

10 A photocopying machine is capable of making 50 copies each minute.
  a If four identical copiers are used simultaneously, how long would it take to make a total of 50 copies?
  b How many copiers would be needed to make 6000 copies in 15 minutes?

11 It takes 16 hours for three bricklayers to build a wall. Calculate how long it would take for eight bricklayers to build a similar wall.

6 RATIO AND PROPORTION

Student assessment 2

1. A cyclist travels at an average speed of 20 km/h for 1.5 hours.
   a. Calculate the distance the cyclist travels in 1.5 hours.
   b. What average speed will the cyclist need to travel in order to cover the same distance in 1 hour?

2. A piece of wood is cut in the ratio 3 : 7.
   a. What fraction of the whole is the longer piece?
   b. If the wood is 1.5 m long, how long is the shorter piece?

3. A recipe for two people requires $\frac{1}{4}$ kg of rice to 150 g of meat.
   a. How much meat would be needed for five people?
   b. How much rice would there be in 1 kg of the final dish?

4. The scale of a map is 1 : 10 000.
   a. Two rivers are 4.5 cm apart on the map. How far apart are they in real life? Give your answer in metres.
   b. Two towns are 8 km apart in real life. How far apart are they on the map? Give your answer in centimetres.

5. a. A model train is a $\frac{1}{25}$ scale model. Express this as a ratio.
   b. If the length of the model engine is 7 cm, what is the true length of the engine?

6. Divide 3 tonnes in the ratio 2 : 5 : 13.

7. The ratio of the angles of a quadrilateral is 2 : 3 : 3 : 4. Calculate the size of each of the angles.

8. The sum of the interior angles of a pentagon is 540°. The ratio of the interior angles is 2 : 3 : 4 : 4 : 5. Calculate the size of the largest angle.

9. A large swimming pool takes 36 hours to fill using three identical pumps.
   a. How long would it take to fill using eight identical pumps?
   b. If the pool needs to be filled in 9 hours, how many pumps will be needed?

10. The first triangle is an enlargement of the second. Calculate the size of the missing sides and angles.

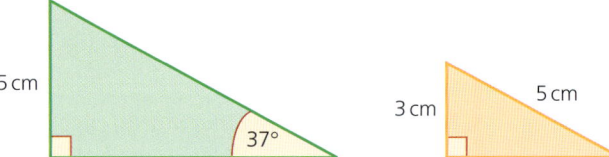

11. A tap issuing water at a rate of 1.2 litres per minute fills a container in 4 minutes.
    a. How long would it take to fill the same container if the rate was decreased to 1 litre per minute? Give your answer in minutes and seconds.
    b. If the container is to be filled in 3 minutes, calculate the rate at which the water should flow.

7 Indices, standard form and surds

The index refers to the power to which a number is raised. In the example 5^3, the number 5 is raised to the power 3. The 3 is known as the **index**. Indices is the plural of index.

> **Worked examples**
>
> a $5^3 = 5 \times 5 \times 5$
> $= 125$
>
> b $7^4 = 7 \times 7 \times 7 \times 7$
> $= 2401$
>
> c $3^1 = 3$

Laws of indices

When working with numbers involving indices there are three basic laws which can be applied. These are:

1 $a^m \times a^n = a^{m+n}$
   e.g. $2^3 \times 2^4 = 2^{3+4} = 2^7$

2 $a^m \div a^n$ or $\dfrac{a^m}{a^n} = a^{m-n}$
   e.g. $5^6 \div 5^4 = 5^{6-4} = 5^2$

3 $(a^m)^n = a^{mn}$
   e.g. $(4^5)^3 = 4^{5 \times 3} = 4^{15}$

Positive indices

> **Worked examples**
>
> a Simplify $4^3 \times 4^2$.
> $4^3 \times 4^2 = 4^{(3+2)}$
> $= 4^5$
>
> b Simplify $2^5 \div 2^3$.
> $2^5 \div 2^3 = 2^{(5-3)}$
> $= 2^2$
>
> c Evaluate $3^3 \times 3^4$.
> $3^3 \times 3^4 = 3^{(3+4)}$
> $= 3^7$
> $= 2187$
>
> d Evaluate $(4^2)^3$.
> $(4^2)^3 = 4^{(2 \times 3)}$
> $= 4^6$
> $= 4096$

Exercise 7.1

1 Using indices, simplify the following expressions:
   a $3 \times 3 \times 3$
   b $2 \times 2 \times 2 \times 2 \times 2$
   c 4×4
   d $6 \times 6 \times 6 \times 6$
   e $8 \times 8 \times 8 \times 8 \times 8 \times 8$
   f 5

2 Simplify the following using indices:
   a $2 \times 2 \times 2 \times 3 \times 3$
   b $4 \times 4 \times 4 \times 4 \times 4 \times 5 \times 5$

7 INDICES, STANDARD FORM AND SURDS

Exercise 7.1 (cont)

  c $3 \times 3 \times 4 \times 4 \times 4 \times 5 \times 5$
  d $2 \times 7 \times 7 \times 7 \times 7$
  e $1 \times 1 \times 6 \times 6$
  f $3 \times 3 \times 3 \times 4 \times 4 \times 6 \times 6 \times 6 \times 6 \times 6$

3 Write out the following in full:
  a 4^2
  b 5^7
  c 3^5
  d $4^3 \times 6^3$
  e $7^2 \times 2^7$
  f $3^2 \times 4^3 \times 2^4$

4 Work out the value of the following:
  a 2^5
  b 3^4
  c 8^2
  d 6^3
  e 10^6
  f 4^4
  g $2^3 \times 3^2$
  h $10^3 \times 5^3$

Exercise 7.2

1 Simplify the following using indices:
  a $3^2 \times 3^4$
  b $8^5 \times 8^2$
  c $5^2 \times 5^4 \times 5^3$
  d $4^3 \times 4^5 \times 4^2$
  e $2^1 \times 2^3$
  f $6^2 \times 3^2 \times 3^3 \times 6^4$
  g $4^5 \times 4^3 \times 5^5 \times 5^4 \times 6^2$
  h $2^4 \times 5^7 \times 5^3 \times 6^2 \times 6^6$

2 Simplify the following:
  a $4^6 \div 4^2$
  b $5^7 \div 5^4$
  c $2^5 \div 2^4$
  d $6^5 \div 6^2$
  e $\dfrac{6^5}{6^2}$
  f $\dfrac{8^6}{8^5}$
  g $\dfrac{4^8}{4^5}$
  h $\dfrac{3^9}{3^2}$

3 Simplify the following:
  a $(5^2)^2$
  b $(4^3)^4$
  c $(10^2)^5$
  d $(3^3)^5$
  e $(6^2)^4$
  f $(8^2)^3$

4 Simplify the following:
  a $\dfrac{2^2 \times 2^4}{2^3}$
  b $\dfrac{3^4 \times 3^2}{3^5}$
  c $\dfrac{5^6 \times 5^7}{5^2 \times 5^8}$
  d $\dfrac{(4^2)^5 \times 4^2}{4^7}$
  e $\dfrac{4^4 \times 2^5 \times 4^2}{4^3 \times 2^3}$
  f $\dfrac{6^3 \times 6^3 \times 8^5 \times 8^6}{8^6 \times 6^2}$
  g $\dfrac{(5^5)^2 \times (4^4)^3}{5^8 \times 4^9}$
  h $\dfrac{(6^3)^4 \times 6^3 \times 4^9}{6^8 \times 4^8}$

The zero index

The zero index indicates that a number is raised to the power 0. A number raised to the power 0 is equal to 1. This can be explained by applying the laws of indices.

$$\dfrac{a^m}{a^n} = a^{m-n} \text{ therefore } \dfrac{a^m}{a^m} = a^{m-m}$$
$$= a^0$$

However, $\dfrac{a^m}{a^m} = 1$

therefore $a^0 = 1$

Negative indices

A negative index indicates that a number is being raised to a negative power: e.g. 4^{-3}.

Another **law of indices** states that $a^{-m} = \frac{1}{a^m}$. This can be proved as follows.

$$a^{-m} = a^{0-m}$$
$$= \frac{a^0}{a^m} \text{ (from the second law of indices)}$$
$$= \frac{1}{a^m}$$

therefore $a^{-m} = \frac{1}{a^m}$

Exercise 7.3

Evaluate the following:

1. **a** $2^3 \times 2^0$ **b** $5^2 \div 6^0$ **c** $5^2 \times 5^{-2}$
   d $6^3 \times 6^{-3}$ **e** $(4^0)^2$ **f** $4^0 \div 2^2$

2. **a** 4^{-1} **b** 3^{-2} **c** 6×10^{-2}
   d 5×10^{-3} **e** 100×10^{-2} **f** 10^{-3}

3. **a** 9×3^{-2} **b** 16×2^{-3} **c** 64×2^{-4}
   d 4×2^{-3} **e** 36×6^{-3} **f** 100×10^{-1}

4. **a** $\frac{3}{2^{-2}}$ **b** $\frac{4}{2^{-3}}$ **c** $\frac{9}{5^{-2}}$
   d $\frac{5}{4^{-2}}$ **e** $\frac{7^{-3}}{7^{-4}}$ **f** $\frac{8^{-6}}{8^{-8}}$

Exponential equations

Equations that involve indices as unknowns are known as **exponential equations**.

> ### Worked examples
>
> **a** Find the value of x if $2^x = 32$.
>
> 32 can be expressed as a power of 2, $32 = 2^5$.
>
> Therefore $2^x = 2^5$
> $x = 5$
>
> **b** Find the value of m if $3^{(m-1)} = 81$.
>
> 81 can be expressed as a power of 3, $81 = 3^4$.
>
> Therefore $3^{(m-1)} = 3^4$
> $m - 1 = 4$
> $m = 5$

7 INDICES, STANDARD FORM AND SURDS

Exercise 7.4

1. Find the value of x in each of the following:
   a $2^x = 4$
   b $2^x = 16$
   c $4^x = 64$
   d $10^x = 1000$
   e $5^x = 625$
   f $3^x = 1$

2. Find the value of z in each of the following:
   a $2^{(z-1)} = 8$
   b $3^{(z+2)} = 27$
   c $4^{2z} = 64$
   d $10^{(z+1)} = 1$
   e $3^z = 9^{(z-1)}$
   f $5^z = 125^z$

3. Find the value of n in each of the following:
   a $\left(\frac{1}{2}\right)^n = 8$
   b $\left(\frac{1}{3}\right)^n = 81$
   c $\left(\frac{1}{2}\right)^n = 32$
   d $\left(\frac{1}{2}\right)^n = 4^{(n+1)}$
   e $\left(\frac{1}{2}\right)^{(n+1)} = 2$
   f $\left(\frac{1}{16}\right)^n = 4$

4. Find the value of x in each of the following:
   a $3^{-x} = 27$
   b $2^{-x} = 128$
   c $2^{(-x+3)} = 64$
   d $4^{-x} = \frac{1}{16}$
   e $2^{-x} = \frac{1}{256}$
   f $3^{(-x+1)} = \frac{1}{81}$

Standard form

Standard form is also known as standard index form or sometimes as scientific notation. It involves writing large numbers or very small numbers in terms of powers of 10.

Positive indices and large numbers

$100 = 1 \times 10^2$
$1000 = 1 \times 10^3$
$10\,000 = 1 \times 10^4$
$3000 = 3 \times 10^3$

For a number to be in standard form it must take the form $A \times 10^n$ where the index n is a positive or negative integer and A must lie in the range $1 \leqslant A < 10$.

e.g. 3100 can be written in many different ways:

3.1×10^3 31×10^2 0.31×10^4 etc.

However, only 3.1×10^3 satisfies the above conditions and therefore is the only one which is written in standard form.

Positive indices and large numbers

> **Worked examples**

a Write 72 000 in standard form.

   7.2×10^4

b Write 4×10^4 as an ordinary number.

   $4 \times 10^4 = 4 \times 10 000$
   $= 40 000$

c Multiply the following and write your answer in standard form:

   600×4000
   $= 2 400 000$
   $= 2.4 \times 10^6$

d Multiply the following and write your answer in standard form:

   $(2.4 \times 10^4) \times (5 \times 10^7)$
   $= 12 \times 10^{11}$
   $= 1.2 \times 10^{12}$ when written in standard form

e Divide the following and write your answer in standard form:

   $(6.4 \times 10^7) \div (1.6 \times 10^3)$
   $= 4 \times 10^4$

f Add the following and write your answer in standard form:
   $(3.8 \times 10^6) + (8.7 \times 10^4)$

   Changing the indices to the same value gives the sum:
   $(380 \times 10^4) + (8.7 \times 10^4)$
   $= 388.7 \times 10^4$
   $= 3.887 \times 10^6$ when written in standard form

g Subtract the following and write your answer in standard form:
   $(6.5 \times 10^7) - (9.2 \times 10^5)$
   Changing the indices to the same value gives
   $(650 \times 10^5) - (9.2 \times 10^5)$
   $= 640.8 \times 10^5$
   $= 6.408 \times 10^7$ when written in standard form

Exercise 7.5

1 Which of the following are not in standard form?
   a 6.2×10^5 b 7.834×10^{16}
   c 8.0×10^5 d 0.46×10^7
   e 82.3×10^6 f 6.75×10^1

2 Write the following numbers in standard form:
   a 600 000   b 48 000 000
   c 784 000 000 000   d 534 000
   e 7 million   f 8.5 million

67

7 INDICES, STANDARD FORM AND SURDS

Exercise 7.5 (cont)

3 Write the following in standard form:
   a 68×10^5 b 720×10^6
   c 8×10^5 d 0.75×10^8
   e 0.4×10^{10} f 50×10^6

4 Write the following as ordinary numbers:
   a 3.8×10^3 b 4.25×10^6
   c 9.003×10^7 d 1.01×10^5

5 Multiply the following and write your answers in standard form:
   a 200×3000 b 6000×4000
   c 7 million $\times 20$ d 500×6 million
   e 3 million $\times 4$ million   f 4500×4000

6 Light from the Sun takes approximately 8 minutes to reach Earth. If light travels at a speed of 3×10^8 m/s, calculate to three significant figures (s.f.) the distance from the Sun to the Earth.

> **Note**
>
> Core students may use a calculator for Questions 7–9.

7 Find the value of the following and write your answers in standard form:
   a $(4.4 \times 10^3) \times (2 \times 10^5)$ b $(6.8 \times 10^7) \times (3 \times 10^3)$
   c $(4 \times 10^5) \times (8.3 \times 10^5)$ d $(5 \times 10^9) \times (8.4 \times 10^{12})$
   e $(8.5 \times 10^6) \times (6 \times 10^{15})$ f $(5.0 \times 10^{12})^2$

8 Find the value of the following and write your answers in standard form:
   a $(3.8 \times 10^8) \div (1.9 \times 10^6)$ b $(6.75 \times 10^9) \div (2.25 \times 10^4)$
   c $(9.6 \times 10^{11}) \div (2.4 \times 10^5)$ d $(1.8 \times 10^{12}) \div (9.0 \times 10^7)$
   e $(2.3 \times 10^{11}) \div (9.2 \times 10^4)$ f $(2.4 \times 10^8) \div (6.0 \times 10^3)$

9 Find the value of the following and write your answers in standard form:
   a $(3.8 \times 10^5) + (4.6 \times 10^4)$ b $(7.9 \times 10^9) + (5.8 \times 10^8)$
   c $(6.3 \times 10^7) + (8.8 \times 10^5)$ d $(3.15 \times 10^9) + (7.0 \times 10^6)$
   e $(5.3 \times 10^8) - (8.0 \times 10^7)$ f $(6.5 \times 10^7) - (4.9 \times 10^6)$
   g $(8.93 \times 10^{10}) - (7.8 \times 10^9)$ h $(4.07 \times 10^7) - (5.1 \times 10^6)$

Negative indices and small numbers

A negative index is used when writing a number between 0 and 1 in standard form.

e.g.
$$\begin{aligned}
100 &= 1 \times 10^2 \\
10 &= 1 \times 10^1 \\
1 &= 1 \times 10^0 \\
0.1 &= 1 \times 10^{-1} \\
0.01 &= 1 \times 10^{-2} \\
0.001 &= 1 \times 10^{-3} \\
0.0001 &= 1 \times 10^{-4}
\end{aligned}$$

Note that A must still lie within the range $1 \leqslant A < 10$.

Fractional indices

> ## Worked examples
>
> a Write 0.0032 in standard form.
> 3.2×10^{-3}
>
> b Write 1.8×10^{-4} as an ordinary number.
> $1.8 \times 10^{-4} = 1.8 \div 10^4$
> $\phantom{1.8 \times 10^{-4}} = 1.8 \div 10\,000$
> $\phantom{1.8 \times 10^{-4}} = 0.000\,18$
>
> c Write the following numbers in order of magnitude, starting with the largest:
> $3.6 \times 10^{-3}\quad 5.2 \times 10^{-5}\quad 1 \times 10^{-2}\quad 8.35 \times 10^{-2}\quad 6.08 \times 10^{-8}$
> $8.35 \times 10^{-2}\quad 1 \times 10^{-2}\quad 3.6 \times 10^{-3}\quad 5.2 \times 10^{-5}\quad 6.08 \times 10^{-8}$

Exercise 7.6

1 Write the following numbers in standard form:
   a 0.0006
   b 0.000053
   c 0.000864
   d 0.000000088
   e 0.0000007
   f 0.0004145

2 Write the following numbers in standard form:
   a 68×10^{-5}
   b 750×10^{-9}
   c 42×10^{-11}
   d 0.08×10^{-7}
   e 0.057×10^{-9}
   f 0.4×10^{-10}

3 Write the following as ordinary numbers:
   a 8×10^{-3}
   b 4.2×10^{-4}
   c 9.03×10^{-2}
   d 1.01×10^{-5}

4 Deduce the value of n in each of the following cases:
   a $0.000\,25 = 2.5 \times 10^n$
   b $0.003\,57 = 3.57 \times 10^n$
   c $0.00000006 = 6 \times 10^n$
   d $0.004^2 = 1.6 \times 10^n$
   e $0.00065^2 = 4.225 \times 10^n$
   f $0.0002^n = 8 \times 10^{-12}$

5 Write these numbers in order of magnitude, starting with the largest:
   $3.2 \times 10^{-4}\quad 6.8 \times 10^5\quad 5.57 \times 10^{-9}\quad 6.2 \times 10^3$
   $5.8 \times 10^{-7}\quad 6.741 \times 10^{-4}\quad 8.414 \times 10^2$

Fractional indices

$16^{\frac{1}{2}}$ can be written as $(4^2)^{\frac{1}{2}}$.

$(4^2)^{\frac{1}{2}} = 4^{\left(2 \times \frac{1}{2}\right)}$
$\phantom{(4^2)^{\frac{1}{2}}} = 4^1$
$\phantom{(4^2)^{\frac{1}{2}}} = 4$

Therefore $16^{\frac{1}{2}} = 4$
but $\sqrt{16} = 4$
Therefore $16^{\frac{1}{2}} = \sqrt{16}$

7 INDICES, STANDARD FORM AND SURDS

Similarly:

$125^{\frac{1}{3}}$ can be written as $(5^3)^{\frac{1}{3}}$

$$(5^3)^{\frac{1}{3}} = 5^{(3 \times \frac{1}{3})}$$
$$= 5^1$$
$$= 5$$

Therefore $125^{\frac{1}{3}} = 5$

But $\sqrt[3]{125} = 5$

Therefore $125^{\frac{1}{3}} = \sqrt[3]{125}$

In general:

$a^{\frac{1}{n}} = \sqrt[n]{a}$

$a^{\frac{m}{n}} = \sqrt[n]{(a^m)}$ or $\left(\sqrt[n]{a}\right)^m$

→ Worked examples

a Evaluate $16^{\frac{1}{4}}$ without the use of a calculator.

$16^{\frac{1}{4}} = \sqrt[4]{16}$ Alternatively: $16^{\frac{1}{4}} = (2^4)^{\frac{1}{4}}$
$\phantom{16^{\frac{1}{4}}} = \sqrt[4]{2^4}$ $= 2^1$
$\phantom{16^{\frac{1}{4}}} = 2$ $= 2$

b Evaluate $25^{\frac{3}{2}}$ without the use of a calculator.

$25^{\frac{3}{2}} = (25^{\frac{1}{2}})^3$ Alternatively: $25^{\frac{3}{2}} = (5^2)^{\frac{3}{2}}$
$\phantom{25^{\frac{3}{2}}} = (\sqrt{25})^3$ $= 5^3$
$\phantom{25^{\frac{3}{2}}} = 5^3$ $= 125$
$\phantom{25^{\frac{3}{2}}} = 125$

c Solve $32^x = 2$

$32^x = 2$
$(2^5)^x = 2^1$
$2^{5x} = 2^1$
$5x = 1$ Therefore $x = \frac{1}{5}$

d Solve $125^x = 5$

$125^x = 5$
$(5^3)^x = 5^1$
$5^{3x} = 5^1$
$3x = 1$ Therefore $x = \frac{1}{3}$

Surds

Exercise 7.7 Evaluate the following:

1. a $16^{\frac{1}{2}}$ b $25^{\frac{1}{2}}$ c $100^{\frac{1}{2}}$
   d $27^{\frac{1}{3}}$ e $81^{\frac{1}{2}}$ f $1000^{\frac{1}{3}}$

2. a $16^{\frac{1}{4}}$ b $81^{\frac{1}{4}}$ c $32^{\frac{1}{5}}$
   d $64^{\frac{1}{6}}$ e $216^{\frac{1}{3}}$ f $256^{\frac{1}{4}}$

3. a $4^{\frac{3}{2}}$ b $4^{\frac{5}{2}}$ c $9^{\frac{3}{2}}$
   d $16^{\frac{3}{2}}$ e $1^{\frac{5}{2}}$ f $27^{\frac{2}{3}}$

4. a $125^{\frac{2}{3}}$ b $32^{\frac{3}{5}}$ c $64^{\frac{5}{6}}$
   d $1000^{\frac{2}{3}}$ e $16^{\frac{5}{4}}$ f $81^{\frac{3}{4}}$

Solve the following:

5. a $16^x = 4$ b $8^x = 2$ c $9^x = 3$
   d $27^x = 3$ e $100^x = 10$ f $64^x = 2$

6. a $1000^x = 10$ b $49^x = 7$ c $81^x = 3$
   d $343^x = 7$ e $1000000^x = 10$ f $216^x = 6$

Exercise 7.8 Evaluate the following:

1. a $\dfrac{27^{\frac{2}{3}}}{3^2}$ b $\dfrac{7^{\frac{3}{2}}}{\sqrt{7}}$ c $\dfrac{4^{\frac{5}{2}}}{4^2}$
   d $\dfrac{16^{\frac{3}{2}}}{2^6}$ e $\dfrac{27^{\frac{5}{3}}}{\sqrt{9}}$ f $\dfrac{6^{\frac{4}{3}}}{6^{\frac{1}{3}}}$

2. a $5^{\frac{2}{3}} \times 5^{\frac{4}{3}}$ b $4^{\frac{1}{4}} \times 4^{\frac{1}{4}}$ c 8×2^{-2}
   d $3^{\frac{4}{3}} \times 3^{\frac{5}{3}}$ e $2^{-2} \times 16$ f $8^{\frac{5}{3}} \times 8^{-\frac{4}{3}}$

3. a $\dfrac{2^{\frac{1}{2}} \times 2^{\frac{5}{2}}}{2}$ b $\dfrac{4^{\frac{5}{6}} \times 4^{\frac{1}{6}}}{4^{\frac{1}{2}}}$ c $\dfrac{2^3 \times 8^{\frac{3}{2}}}{\sqrt{8}}$
   d $\dfrac{(3^2)^{\frac{3}{2}} \times 3^{-\frac{1}{2}}}{3^{\frac{1}{2}}}$ e $\dfrac{8^{\frac{1}{3}} + 7}{27^{\frac{1}{3}}}$ f $\dfrac{9^{\frac{1}{2}} \times 3^{\frac{5}{2}}}{3^{\frac{2}{3}} \times 3^{-\frac{1}{6}}}$

Surds

You have already encountered the various types of roots of numbers and have also worked with indices too.

So you will know that $\sqrt{9} = 3$, $\sqrt{\frac{4}{81}} = \frac{2}{9}$ and $\sqrt[4]{625} = 5$.

In each of these cases the answer was a rational number. If the root cannot be expressed as a rational number, then it is called a **surd**. Surds, therefore, are irrational.

$\sqrt{3}$, $\sqrt{7}$ and $\sqrt[3]{20}$ are examples of surds.

Mathematicians tend to prefer writing their answers using surds as surds are more accurate than providing a decimal approximation. For example, if an answer is given as $\sqrt{5}$ this is preferable to using a calculator to work out the answer as 2.236 067 977 (9 d.p.).

7 INDICES, STANDARD FORM AND SURDS

Rules of surds

As surds involve roots, they have similar rules to those of indices.

Multiplication rule:

e.g. $\sqrt{5} \times \sqrt{7} = \sqrt{35}$

Therefore, in general $\sqrt{a} \times \sqrt{b} = \sqrt{a \times b}$

Division rule:

e.g. $\frac{\sqrt{5}}{\sqrt{8}} = \sqrt{\frac{5}{8}}$

Therefore, in general $\frac{\sqrt{a}}{\sqrt{b}} = \sqrt{\frac{a}{b}}$

Using the two rules above, calculations involving surds can often be simplified.

→ Worked examples

a Simplify the following calculation:
$\sqrt{3} \times \sqrt{12}$
$\sqrt{36} = 6$

Note that in this case, the result of multiplying the two numbers produced a square number, so it could be simplified further.

b Simplify $\frac{\sqrt{8}}{\sqrt{12}}$

$\sqrt{\frac{8}{12}} = \sqrt{\frac{2}{3}}$

A single surd can often be simplified, too. In general, a surd can be simplified if it has a square number as a factor.

To simplify $\sqrt{48}$, identify the largest square number factor of 48. In this case, 16 is the largest factor that is a square number. Therefore, using the multiplication rule, $\sqrt{48}$ can be written as $\sqrt{16} \times \sqrt{3}$. However, $\sqrt{16} = 4$; therefore, $\sqrt{48}$ can be simplified to $4\sqrt{3}$.

→ Worked examples

Simplify each of the following surds:

a $\sqrt{72}$
$\sqrt{72} = \sqrt{36} \times \sqrt{2} = 6\sqrt{2}$

b $\sqrt{45}$
$\sqrt{45} = \sqrt{9} \times \sqrt{5} = 3\sqrt{5}$

Rationalising the denominator

By checking with your calculator, you will be able to see that $\frac{1}{\sqrt{3}}$ is the same as $\frac{\sqrt{3}}{3}$.

In mathematics, if a fraction involves a surd, it is more usual to write it without the surd in the denominator. In the two fractions above, it is more usual to write $\frac{\sqrt{3}}{3}$ rather than $\frac{1}{\sqrt{3}}$. The process of removing the surd from the denominator is known as **rationalising the denominator**.

➡ Worked examples

Rationalise the denominator of the following fractions:

a $\frac{1}{\sqrt{3}}$

Multiplying both the numerator and denominator by $\sqrt{3}$ produces an equivalent fraction because $\frac{\sqrt{3}}{\sqrt{3}} = 1$

Therefore, $\frac{1}{\sqrt{3}} \times \frac{\sqrt{3}}{\sqrt{3}} = \frac{\sqrt{3}}{\sqrt{3} \times \sqrt{3}} = \frac{\sqrt{3}}{3}$

b $\frac{2}{3\sqrt{5}}$

To eliminate the surd from the denominator, multiply both the numerator and denominator by $\sqrt{5}$. Once again this produces an equivalent fraction to the original one as $\frac{\sqrt{5}}{\sqrt{5}} = 1$

$\frac{2}{3\sqrt{5}} \times \frac{\sqrt{5}}{\sqrt{5}} = \frac{2\sqrt{5}}{3\sqrt{5} \times \sqrt{5}} = \frac{2\sqrt{5}}{3 \times 5} = \frac{2\sqrt{5}}{15}$

More complicated expressions include ones where the denominator has more than one term, e.g. $\frac{1}{1+\sqrt{3}}$.

In this case, simply multiplying the numerator and denominator by $\sqrt{3}$ will not eliminate all the surds from the denominator, as it will produce $\frac{\sqrt{3}}{\sqrt{3}+3}$.

To rationalise $\frac{1}{1+\sqrt{3}}$, multiply both the numerator and the denominator by $1 - \sqrt{3}$ (note that the sign has changed).

$\frac{1}{1+\sqrt{3}} \times \frac{1-\sqrt{3}}{1-\sqrt{3}} = \frac{1-\sqrt{3}}{(1+\sqrt{3})(1-\sqrt{3})}$

$= \frac{1-\sqrt{3}}{1-\sqrt{3}+\sqrt{3}-3} = \frac{1-\sqrt{3}}{-2} = \frac{-1+\sqrt{3}}{2}$

> **Note**
>
> To understand the maths behind the following method you may find it helpful to know what is meant by 'the difference of two squares', which is covered in Chapter 11.

7 INDICES, STANDARD FORM AND SURDS

> **Worked examples**

a Rationalise the expression $\dfrac{4}{\sqrt{3}-2}$

Multiplying both the numerator and the denominator by $\sqrt{3}+2$ gives:

$$\dfrac{4}{\sqrt{3}-2} \times \dfrac{\sqrt{3}+2}{\sqrt{3}+2} = \dfrac{4(\sqrt{3}+2)}{(\sqrt{3}-2)(\sqrt{3}+2)}$$

$$= \dfrac{4\sqrt{3}+8}{3+2\sqrt{3}-2\sqrt{3}-4} = \dfrac{4\sqrt{3}+8}{-1} = -4\sqrt{3}-8$$

b Rationalise the expression $\dfrac{\sqrt{2}}{2\sqrt{3}+\sqrt{2}}$

Multiply both the numerator and the denominator by $2\sqrt{3}-\sqrt{2}$

$$\dfrac{\sqrt{2}}{2\sqrt{3}+\sqrt{2}} \times \dfrac{2\sqrt{3}-\sqrt{2}}{2\sqrt{3}-\sqrt{2}} = \dfrac{\sqrt{2}(2\sqrt{3}-\sqrt{2})}{(2\sqrt{3}+\sqrt{2})(2\sqrt{3}-\sqrt{2})}$$

$$= \dfrac{2\sqrt{6}-2}{12-2\sqrt{6}+2\sqrt{6}-2} = \dfrac{2\sqrt{6}-2}{10} = \dfrac{\sqrt{6}-1}{5}$$

Exercise 7.9

1 Simplify the following surds:
   a $\sqrt{3} \times \sqrt{27}$ **b** $\sqrt{5} \times \sqrt{20}$ **c** $\sqrt{24} \times \sqrt{6}$
   d $(2\sqrt{3})^2$ **e** $(5\sqrt{2})^2$ **f** $(3\sqrt{6})^2$

2 Simplify the following surds:
   a $\dfrac{\sqrt{50}}{\sqrt{5}}$ **b** $\dfrac{\sqrt{18}}{\sqrt{2}}$ **c** $\left(\dfrac{\sqrt{10}}{\sqrt{5}}\right)^2$
   d $\dfrac{\sqrt{2}}{\sqrt{5}\sqrt{6}}$ **e** $\left(\dfrac{3\sqrt{8}}{2\sqrt{7}}\right)^2$ **f** $\left(\dfrac{(6\sqrt{5})^2+(2\sqrt{5})^2}{20}\right)^2$

3 Express the following surds in their simplest form:
   a $\sqrt{8}$ **b** $\sqrt{50}$ **c** $\sqrt{18}$ **d** $\sqrt{45}$
   e $\sqrt{75}$ **f** $\sqrt{72}$ **g** $\sqrt{700}$ **h** $\sqrt{162}$
   i $\sqrt{98}$ **j** $\sqrt{242}$ **k** $\sqrt{192}$ **l** $\sqrt{450}$

4 Rationalise the following fractions, simplifying where possible:
   a $\dfrac{1}{\sqrt{6}}$ **b** $\dfrac{3}{\sqrt{5}}$ **c** $\dfrac{2}{\sqrt{8}}$ **d** $\dfrac{6}{\sqrt{6}}$
   e $\dfrac{3}{5\sqrt{2}}$ **f** $-\dfrac{1}{2\sqrt{5}}$ **g** $\dfrac{5}{4\sqrt{10}}$ **h** $\dfrac{3}{8\sqrt{6}}$

5 Rationalise the denominators of the following fractions, simplifying where possible:
   a $\dfrac{2}{5+\sqrt{3}}$ **b** $\dfrac{2}{\sqrt{2}-1}$ **c** $\dfrac{5}{\sqrt{2}+\sqrt{5}}$
   d $\dfrac{7\sqrt{3}}{2\sqrt{2}+5}$ **e** $\dfrac{\sqrt{6}+\sqrt{3}}{\sqrt{6}-3\sqrt{3}}$ **f** $\dfrac{2+\sqrt{5}}{2-\sqrt{5}}$

6 A rectangle is shown below.

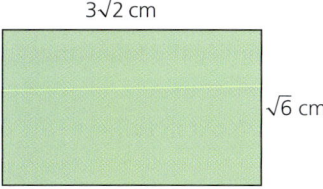

  a Calculate its area.
  b Calculate its perimeter.

7 A right-angled triangle has dimensions as shown below.

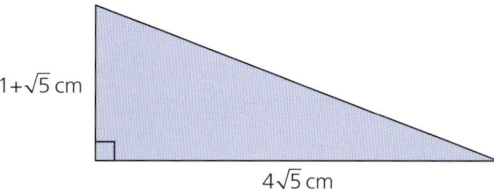

Calculate the area of the triangle.

8 A rectangle of area $6\,cm^2$ is shown below.

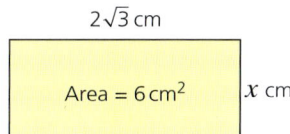

Calculate the length of the side marked x, giving your answer in its simplest form.

9 A square has an area of $18\,cm^2$. Calculate its side length, giving your answer in its simplest form.

10 The shape below consists of two squares joined together.
The length of each side of square A is $4\sqrt{3}\,cm$.
The total area of the whole shape is $60\,cm^2$.

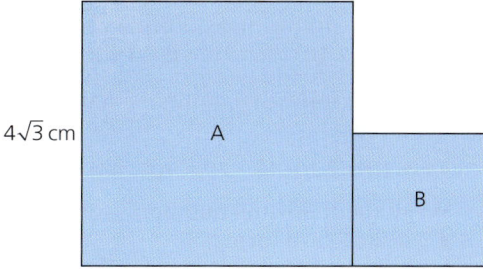

Calculate the following:
  a the area of square A
  b the area of square B
  c the length of each side of square B
  d the perimeter of the whole shape.

7 INDICES, STANDARD FORM AND SURDS

Student assessment 1

1. Using indices, simplify the following:
   a. $3 \times 2 \times 2 \times 3 \times 27$
   b. $2 \times 2 \times 4 \times 4 \times 4 \times 2 \times 32$

2. Write the following out in full:
   a. 6^5
   b. 2^{-5}

3. Work out the value of the following without using a calculator:
   a. $3^3 \times 10^3$
   b. $1^{-4} \times 5^3$

4. Simplify the following using indices:
   a. $2^4 \times 2^3$
   b. $7^5 \times 7^2 \times 3^4 \times 3^8$
   c. $\dfrac{4^8}{2^{10}}$
   d. $\dfrac{(3^3)^4}{27^3}$
   e. $\dfrac{7^6 \times 4^2}{4^3 \times 7^6}$
   f. $\dfrac{8^{-2} \times 2^6}{2^{-2}}$

5. Without using a calculator, evaluate the following:
   a. $5^2 \times 5^{-1}$
   b. $\dfrac{4^5}{4^3}$
   c. $\dfrac{7^{-5}}{7^{-7}}$
   d. $\dfrac{3^{-5} \times 4^2}{3^{-6}}$

6. Find the value of x in each of the following:
   a. $2^{(2x+2)} = 128$
   b. $\dfrac{1}{4^{-x}} = \dfrac{1}{2}$
   c. $3^{(-x+4)} = 81$
   d. $8^{-3x} = \dfrac{1}{4}$

Student assessment 2

1. Write the following numbers in standard form:
   a. 6 million
   b. 0.0045
   c. 3 800 000 000
   d. 0.000 000 361
   e. 460 million
   f. 3

2. Write the following as ordinary numbers:
   a. 8.112×10^6
   b. 4.4×10^5
   c. 3.05×10^{-4}

3. Write the following numbers in order of magnitude, starting with the largest:

   $3.6 \times 10^2 \quad 2.1 \times 10^{-3} \quad 9 \times 10^1 \quad 4.05 \times 10^8 \quad 1.5 \times 10^{-2} \quad 7.2 \times 10^{-3}$

4. Write the following numbers:
   a. in standard form,
   b. in order of magnitude, starting with the smallest.

   15 million   430 000   0.000 435   4.8   0.0085

Surds

5 Deduce the value of n in each of the following:
  a $4750 = 4.75 \times 10^n$
  b $6\,440\,000\,000 = 6.44 \times 10^n$
  c $0.0040 = 4.0 \times 10^n$
  d $1000^2 = 1 \times 10^n$
  e $0.9^3 = 7.29 \times 10^n$
  f $800^3 = 5.12 \times 10^n$

6 Write the answers to the following calculations in standard form:
  a $50\,000 \times 2400$
  b $(3.7 \times 10^6) \times (4.0 \times 10^4)$
  c $(5.8 \times 10^7) + (9.3 \times 10^6)$
  d $(4.7 \times 10^6) - (8.2 \times 10^5)$

7 The speed of light is 3×10^8 m/s. Jupiter is 778 million km from the Sun. Calculate the number of minutes it takes for sunlight to reach Jupiter.

8 A star is 300 light years away from Earth. The speed of light is 3×10^5 km/s. Calculate the distance from the star to Earth. Give your answer in kilometres and written in standard form.

Student assessment 3

1 Evaluate the following:
  a $81^{\frac{1}{2}}$
  b $27^{\frac{1}{3}}$
  c $9^{\frac{1}{2}}$
  d $625^{\frac{3}{4}}$
  e $343^{\frac{2}{3}}$
  f $16^{-\frac{1}{4}}$
  g $\dfrac{1}{25^{-\frac{1}{2}}}$
  h $\dfrac{2}{16^{-\frac{3}{4}}}$

2 Evaluate the following:
  a $\dfrac{16^{\frac{1}{2}}}{2^2}$
  b $\dfrac{9^{\frac{5}{2}}}{3^3}$
  c $\dfrac{8^{\frac{4}{3}}}{8^{\frac{2}{3}}}$
  d $5^{\frac{6}{5}} \times 5^{\frac{4}{5}}$
  e $4^{\frac{3}{2}} \times 2^{-2}$
  f $\dfrac{27^{\frac{2}{3}} \times 3^{-2}}{4^{-\frac{3}{2}}}$
  g $\dfrac{(4^3)^{-\frac{1}{2}} \times 2^{\frac{3}{2}}}{2^{-\frac{3}{2}}}$
  h $\dfrac{(5^{\frac{2}{3}})^{\frac{1}{2}} \times 5^{\frac{2}{3}}}{3^{-2}}$

3 Draw a pair of axes with x from -4 to 4 and y from 0 to 10.
  a Plot a graph of $y = 3^{\frac{x}{2}}$.
  b Use your graph to estimate when $3^{\frac{x}{2}} = 5$.

4 Express the following surds in their simplest form.
  a $\sqrt{125}$
  b $\sqrt{80}$
  c $\sqrt{12} + \sqrt{48}$

5 Rationalise the following fractions, simplifying where possible.
  a $\dfrac{5}{2\sqrt{2}}$
  b $\dfrac{1}{2 - \sqrt{7}}$

6 The area (A) of a circle is calculated using the formula $A = \pi r^2$, where r is the radius of the circle.

Calculate the area of a circle with a radius of $3\sqrt{5}$ cm. Leave π in your answer.

7 INDICES, STANDARD FORM AND SURDS

Student assessment 4

1 Evaluate the following:
   a $64^{\frac{1}{6}}$
   b $27^{\frac{4}{3}}$
   c $9^{-\frac{1}{2}}$
   d $512^{\frac{2}{3}}$
   e $\sqrt[3]{27}$
   f $\sqrt[4]{16}$
   g $\dfrac{1}{36^{-\frac{1}{2}}}$
   h $\dfrac{2}{64^{-\frac{2}{3}}}$

2 Evaluate the following:
   a $\dfrac{25^{\frac{1}{2}}}{9^{-\frac{1}{2}}}$
   b $\dfrac{4^{\frac{5}{2}}}{2^3}$
   c $\dfrac{27^{\frac{4}{3}}}{3^3}$
   d $25^{\frac{3}{2}} \times 5^2$
   e $4^{\frac{6}{4}} \times 4^{-\frac{1}{2}}$
   f $\dfrac{27^{\frac{2}{3}} \times 3^{-3}}{9^{-\frac{1}{2}}}$
   g $\dfrac{(4^2)^{-\frac{1}{4}} \times 9^{\frac{3}{2}}}{\left(\frac{1}{4}\right)^{\frac{1}{2}}}$
   h $\dfrac{(5^{\frac{1}{3}})^{\frac{1}{2}} \times 5^{\frac{5}{6}}}{4^{-\frac{1}{2}}}$

3 Draw a pair of axes with x from -4 to 4 and y from 0 to 18.
   a Plot a graph of $y = 4^{-\frac{x}{2}}$.
   b Use your graph to estimate when $4^{-\frac{x}{2}} = 6$.

4 Simplify the following surds.
   a $(6\sqrt{3})^2$
   b $3\sqrt{2} \times 4\sqrt{6}$

5 Rationalise the following fractions, simplifying where possible.
   a $\dfrac{7}{3\sqrt{3}}$
   b $\dfrac{2}{\sqrt{7}+4}$

6 Calculate the area of a rectangle with a length of $(\sqrt{2} + 3\sqrt{6})$ cm and width $\sqrt{2}$ cm.

8 Money and finance

Currency conversions

In 2022, 1 euro (€) could be exchanged for 1.50 Australian dollars (A$).

➡ Worked examples

a How many Australian dollars can be bought for €400?

€1 buys A$1.50.

€400 buys $1.50 \times 400 = $ A$600.

b How much does it cost, in euros, to buy A$940?

A$1.50 costs €1.

A$940 costs $\frac{1 \times 940}{1.5} = $ €626.67.

Exercise 8.1 The table shows the exchange rate for €1 into various currencies.

Australia	1.50 Australian dollars (A$)
India	75 rupees
Zimbabwe	412.8 Zimbabwe dollars (ZIM$)
South Africa	15 rand
Turkey	4.0 Turkish lira (L)
Japan	130 yen
Kuwait	0.35 dinar
USA	1.15 US dollars (US$)

1 Convert the following:
   a €25 into Australian dollars
   b €50 into rupees
   c €20 into Zimbabwean dollars
   d €300 into rand
   e €130 into Turkish lira
   f €40 into yen
   g €400 into dinar
   h €150 into US dollars

2 How many euros does it cost to buy the following:
   a A$500
   b 200 rupees
   c ZIM$10 000
   d 500 rand
   e 750 Turkish lira
   f 1200 yen
   g 50 dinar
   h US$150?

Earnings

Net pay is what is left after deductions such as tax, insurance and pension contributions are taken from **gross earnings**.

That is, Net pay = Gross pay − Deductions.

8 MONEY AND FINANCE

A **bonus** is an extra payment sometimes added to an employee's **basic pay**.

In many companies there is a fixed number of hours that an employee is expected to work. Any work done in excess of this **basic week** is paid at a higher rate, referred to as **overtime**. Overtime may be 1.5 times basic pay, called **time and a half**, or twice basic pay, called **double time**.

Piece work is another method of payment. Employees are paid for the number of articles made, not for the time taken.

Exercise 8.2

1. Kamal's gross pay is $188.25. Deductions amount to $33.43. What is his net pay?

2. Keiko's basic pay is $128. She earns $36 for overtime and receives a bonus of $18. What is her gross pay?

3. Ella's gross pay is $203. She pays $54 in tax and $18 towards her pension. What is her net pay?

4. Nameen works 35 hours for an hourly rate of $8.30. What is his basic pay?

5. a. Leeza works 38 hours for an hourly rate of $4.15. In addition she works 6 hours of overtime at time and a half. What is her total gross pay?
   b. Deductions amount to 32% of her total gross pay. What is her net pay?

6. Pepe is paid $5.50 for each basket of grapes he picks. One week he picks 25 baskets. How much is he paid?

7. Farah is paid $5 for every 12 plates that she makes. This is her record for one week.

Mon	240
Tues	360
Wed	288
Thurs	192
Fri	180

How much is she paid?

8. Harry works at home, making clothes. The patterns and materials are provided by the company. The table shows the rates he is paid and the number of items he makes in one week:

Item	Rate	Number made
Jacket	$25	3
Trousers	$11	12
Shirt	$13	7
Dress	$12	0

a. What are his gross earnings?
b. Deductions amount to 15% of gross earnings. What is his net pay?

Profit and loss

Foodstuffs and manufactured goods are produced at a cost, known as the **cost price**, and sold at the **selling price**. If the selling price is greater than the cost price, a profit is made.

→ Worked example

A market trader buys oranges in boxes of 12 dozen for $14.40 per box. He buys three boxes and sells all the oranges for 12c each. What is his profit or **loss**?

Cost price: $3 \times \$14.40 = \$43.20$

Selling price: $3 \times 144 \times 12c = \$51.84$

In this case he makes a profit of $51.84 − $43.20

His profit is $8.64.

A second way of solving this problem would be:

$14.40 for a box of 144 oranges is 10c each.

So cost price of each orange is 10c, and selling price of each orange is 12c. The profit is 2c per orange.

So 3 boxes would give a profit of $3 \times 144 \times 2c$.

That is, $8.64.

Sometimes, particularly during sales or promotions, the selling price is reduced; this is known as a **discount**.

→ Worked example

In a sale, a skirt usually costing $35 is sold at a 15% discount. What is the discount?

15% of $35 = $0.15 \times \$35 = \$5.25$

The discount is $5.25.

Exercise 8.3

1 A market trader buys peaches in boxes of 120. She buys 4 boxes at a cost price of $13.20 per box. She sells 425 peaches at 12c each. The rest are ruined. How much profit or loss does she make?

2 A shopkeeper buys 72 bars of chocolate for $5.76. What is his profit if he sells them for 12c each?

3 A holiday company charters an aircraft to fly to Qatar at a cost of $22 000. It then sells 150 seats at $185 each and a further 35 seats at a 20% discount. Calculate the profit made per seat if the plane has 200 seats.

4 A car is priced at $7200. The car dealer allows a customer to pay a one-third deposit and 12 payments of $420 per month. How much extra does it cost the customer?

5 At an auction, a company sells 150 televisions for an **average** of $65 each. The production cost was $10 000. How much loss did the company make?

Percentage profit and loss

Most profits or losses are expressed as a percentage.

Profit or loss, divided by cost price, multiplied by 100 = % profit or loss.

> ### Worked example
>
> Abi buys a car for $7500 and sells it two years later for $4500. Calculate her loss over two years as a percentage of the cost price.
>
> cost price = $7500   selling price = $4500   loss = $3000
>
> % Loss = $\frac{3000}{7500} \times 100 = 40$
>
> Her loss is 40%.

When something becomes worth less over a period of time, it is said to **depreciate**.

Exercise 8.4

1. Find the depreciation of the following cars as a percentage of the cost price. (C.P. = cost price, S.P. = selling price)
   a Car A   C.P. $4500   S.P. $4005
   b Car B   C.P. $9200   S.P. $6900

2. A company manufactures electrical items for the kitchen. Find the percentage profit on each of the following:
   a Fridge   C.P. $50   S.P. $65
   b Freezer   C.P. $80   S.P. $96

3. A developer builds a number of different types of house. Which type gives the developer the largest percentage profit?
   Type A   C.P. $40 000   S.P. $52 000
   Type B   C.P. $65 000   S.P. $75 000
   Type C   C.P. $81 000   S.P. $108 000

4. Students in a school organise a disco. The disco company charges $350 hire charge. The students sell 280 tickets at $2.25. What is the percentage profit?

Interest

Interest can be defined as money added by a bank to sums deposited by customers. The money deposited is called the **principal**. The **percentage interest** is the given rate and the money is left for a fixed period of time.

A formula can be obtained for **simple interest**:

$$SI = \frac{Ptr}{100}$$

where SI = simple interest, i.e. the interest paid
P = the principal
t = time in years
r = rate percent

Interest

→ Worked examples

a Find the simple interest earned on $250 deposited for 6 years at 8% p.a.

$$SI = \frac{Ptr}{100}$$

$$SI = \frac{250 \times 6 \times 8}{100}$$

$$SI = 120$$

p.a. stands for per annum which means 'each year'

So the interest paid is $120.

b How long will it take for a sum of $250 invested at 8% to earn interest of $80?

$$SI = \frac{Ptr}{100}$$

$$80 = \frac{250 \times t \times 8}{100}$$

$$80 = 20t$$

$$4 = t$$

It will take 4 years.

c What rate per year must be paid for a principal of $750 to earn interest of $180 in 4 years?

$$SI = \frac{Ptr}{100}$$

$$180 = \frac{750 \times 4 \times r}{100}$$

$$180 = 30r$$

$$6 = r$$

The rate must be 6% per year.

d Find the principal which will earn interest of $120 in 6 years at 4%.

$$SI = \frac{Ptr}{100}$$

$$120 = \frac{P \times 6 \times 4}{100}$$

$$120 = \frac{24P}{100}$$

$$12\,000 = 24P$$

$$500 = P$$

So the principal is $500.

8 MONEY AND FINANCE

Exercise 8.5

All rates of interest given here are annual rates.

1 Find the simple interest paid in the following cases:
   a Principal $300   rate 6%   time 4 years
   b Principal $750   rate 8%   time 7 years

2 Calculate how long it will take for the following amounts of interest to be earned at the given rate.
   a $P = \$500$ $r = 6\%$ $SI = \$150$
   b $P = \$400$ $r = 9\%$ $SI = \$252$

3 Calculate the rate of interest per year which will earn the given amount of interest:
   a Principal $400   time 4 years   interest $112
   b Principal $800   time 7 years   interest $224

4 Calculate the principal which will earn the interest below in the given number of years at the given rate:
   a $SI = \$36$ time = 3 years   rate = 6%
   b $SI = \$340$ time = 5 years   rate = 8%

5 What rate of interest is paid on a deposit of $2000 which earns $400 interest in 5 years?

6 How long will it take a principal of $350 to earn $56 interest at 8% per year?

7 A principal of $480 earns $108 interest in 5 years. What rate of interest was being paid?

8 A principal of $750 becomes a total of $1320 in 8 years. What rate of interest was being paid?

9 $1500 is invested for 6 years at 3.5% per year. What is the interest earned?

10 $500 is invested for 11 years and becomes $830 in total. What rate of interest was being paid?

Compound interest

Compound interest means that interest is paid not only on the principal amount, but also on the interest itself: it is compounded (or added to). This sounds complicated but the example below will make it clearer.

For example, a builder is going to build six houses on a plot of land. He borrows $500 000 at 10% compound interest per annum and will pay the loan off in full after three years.

10% of $500 000 is $50 000, therefore at the end of the first year he will owe a total of $550 000 as shown:

> **Note**
> A principal amount can be money deposited or an amount loaned.

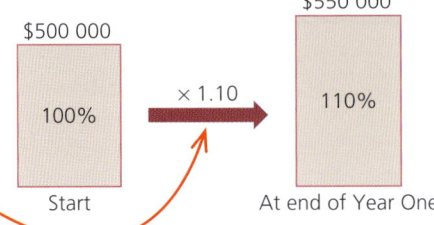

An increase of 10% is the same as multiplying by 1.10.

For the second year, the amount he owes increases again by 10%, but this is calculated by adding 10% to the amount he owed at the end of the first year, i.e. 10% of $550 000. This can be represented using this diagram:

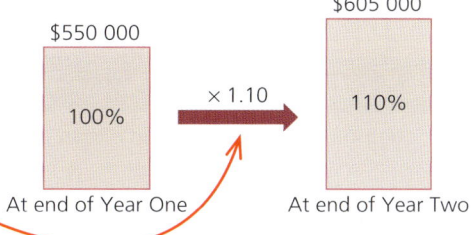

Once again an increase of 10% is the same as multiplying by 1.10.

For the third year, the amount he owes increases again by 10%. This is calculated by adding 10% to the amount he owed at the end of the second year, i.e. 10% of $605 000 as shown:

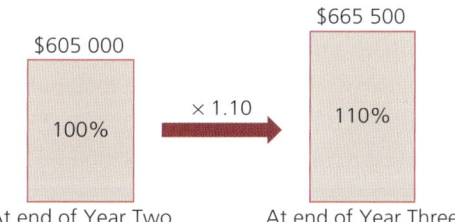

Therefore, the compound interest he has to pay at the end of three years is $665 500 − $500 000 = $165 500.

By looking at the diagrams above it can be seen that the principal amount has in effect been multiplied by 1.10 three times (this is the same as multiplying by 1.10^3), i.e. $500\,000 \times 1.10^3 = \$665\,500$.

The time taken for a debt to grow at compound interest can be calculated as shown in the next example.

➡ Worked example

How long will it take for a debt to double at a compound interest rate of 27% per annum?

An interest rate of 27% implies a multiplier of 1.27.

The effect of applying this multiplier to a principal amount, P, is shown in the table:

Time (years)	0	1	2	3
Debt	P	$1.27P$	$1.27^2 P = 1.61P$	$1.27^3 P = 2.05P$

$\times 1.27 \quad \times 1.27 \quad \times 1.27$

The debt will have more than doubled after 3 years.

8 MONEY AND FINANCE

Note

The opposite of exponential growth is known as **exponential decay**.

Compound interest is an example of a geometric sequence and therefore of **exponential** growth.

The interest is usually calculated annually, but there can be other time periods. Compound interest can be charged yearly, half-yearly, quarterly, monthly or daily. (In theory any time period can be chosen.)

→ Worked examples

a Find the compound interest paid on a loan of $600 at a rate of 5% for 3 years.

An increase of 5% is equivalent to a multiplier of 1.05.
Therefore 3 years at 5% is calculated as $600 \times 1.05^3 = 694.58$ (2 d.p.).
The total payment is $694.58, so the interest paid is $694.58 − $600 = $94.58.

b Find the compound interest when $3000 is invested for 18 months at a rate of 8.5% per year if the interest is calculated every 6 months.

Note: The interest for each time period of 6 months is $\frac{8.5}{2}\% = 4.25\%$.
There will be 3 time periods of 6 months each.
An increase of 4.25% is equivalent to a multiplier of 1.0425.
Therefore the total amount is $3000 \times 1.0425^3 = 3398.99$.
The interest earned is $3398.99 − $3000 = $398.99.

There is a formula for calculating the compound interest. It is written as:

$$I = P\left(1 + \frac{r}{100}\right)^n - P$$

Where I = compound interest
P = the principal (amount originally borrowed)
r = interest rate
n = number of years

Note

You should learn how to use the formula for calculating compound interest.

For the example of the builder earlier in this section, $P = 500\,000$ dollars, $r = 10\%$ and $n = 3$.

Therefore $I = 500\,000\left(1 + \frac{10}{100}\right)^3 - 500\,000 = 165\,500$ dollars.

Exercise 8.6

Using the formula for compound interest or otherwise, calculate the following:

1. A shipping company borrows $70 million at 5% p.a. compound interest to build a new cruise ship. If it repays the debt after 3 years, how much interest will the company pay?

2. A woman borrows $100 000 for home improvements. The compound interest rate is 15% p.a. and she repays the loan in full after 3 years. How much interest did she pay?

3. A man owes $5000 on his credit cards. The rate of interest is 20% per year. If he doesn't repay any of the debt, how much will he owe after 4 years?

Compound interest

4 A school increases its intake by 10% each year. If it starts with 1000 students, how many will it have at the beginning of the fourth year of expansion?

5 8 million tonnes of fish were caught in the North Sea in 2019. If the catch is reduced by 20% each year for 4 years, what mass is caught at the end of this time?

6 How many years will it take for a debt to double at 42% p.a. compound interest?

7 How many years will it take for a debt to double at 15% p.a. compound interest?

8 A car loses value at a rate of 27% each year. How long will it take for its value to halve?

Student assessment 1

1 A visitor from Hong Kong receives 13.5 Pakistani rupees for each Hong Kong dollar.
   a How many Pakistani rupees would he get for HK$240?
   b How many Hong Kong dollars does it cost for 1 thousand rupees?

2 Below is a currency conversion table showing the amount of foreign currency received for 1 euro.

New Zealand	1.60 dollars (NZ$)
Brazil	3.70 reals

   a How many euros does it cost for NZ$1000?
   b How many euros does it cost for 500 Brazilian reals?

3 Laila works in a shop on Saturdays for 8.5 hours. She is paid $3.60 per hour. What is her gross pay for 4 weeks' work?

4 Razik makes cups and saucers in a factory. He is paid $1.44 per batch of cups and $1.20 per batch of saucers. What is his gross pay if he makes 9 batches of cups and 11 batches of saucers in one day?

5 Calculate the missing numbers from the simple interest table below:

Principal ($)	Rate (%)	Time (years)	Interest ($)
300	6	4	a
250	b	3	60
480	5	c	96
650	d	8	390
e	3.75	4	187.50

6 A family house was bought for $48 000 twelve years ago. It is now valued at $120 000. What is the average annual increase in the value of the house?

7 An electrician bought five broken washing machines for $550. She repaired them and sold them for $143 each. What was her percentage profit?

8 MONEY AND FINANCE

Student assessment 2

1. Find the simple interest paid on the following principal sums P, deposited in a savings account for t years at a fixed rate of interest of $r\%$:
   a $P = \$550$ $t = 5$ years $r = 3\%$
   b $P = \$8000$ $t = 10$ years $r = 6\%$
   c $P = \$12\,500$ $t = 7$ years $r = 2.5\%$

2. A sum of $\$25\,000$ is deposited in a bank. After 8 years, the simple interest gained was $\$7000$. Calculate the annual rate of interest on the account assuming it remained constant over the 8 years.

3. A bank lends a business $\$250\,000$. The annual rate of interest is 8.4%. When paying back the loan, the business pays an amount of $\$105\,000$ in simple interest. Calculate the number of years the business took out the loan for.

4. A small business wishes to take out a $\$10\,000$ loan from a bank. The bank has two loan options.
   Option A: A loan with a **simple** interest rate of 4% per year.
   Option B: A loan with a **compound** interest rate of 3% per year.
   a If the loan is to be taken out for 10 years, calculate which loan is cheapest for the small business.
   b How much will the business save over the 10 years by choosing the cheaper of the two loans?

5. Find the compound interest paid on the following principal sums P, deposited in a savings account for n years at a fixed rate of interest of $r\%$:
   a $P = \$400$ $n = 2$ years $r = 3\%$
   b $P = \$5000$ $n = 8$ years $r = 6\%$
   c $P = \$18\,000$ $n = 10$ years $r = 4.5\%$

6. A car is bought for $\$12\,500$. Its value depreciates by 15% per year.
   a Calculate its value after:
      i 1 year   ii 2 years
   b After how many years will the car be worth less than $\$1000$?

9 Time

Times may be given in terms of the 12-hour clock. We tend to say, 'I get up at seven o'clock in the morning, play football at half past two in the afternoon, and go to bed before eleven o'clock'.

These times can be written as 7 a.m., 2.30 p.m. and 11 p.m.

In order to save confusion, most timetables are written using the 24-hour clock.

  7 a.m. is written as 07 00

  2.30 p.m. is written as 14 30

  11.00 p.m. is written as 23 00

→ Worked example

A train covers the 480 km journey from Paris to Lyon at an average speed of 100 km/h. If the train leaves Paris at 08 35, when does it arrive in Lyon?

Time taken = $\frac{\text{distance}}{\text{speed}}$

Paris to Lyon = $\frac{480}{100}$ hours, that is, 4.8 hours.

4.8 hours is 4 hours and (0.8 × 60 minutes), that is, 4 hours and 48 minutes.

Departure 08 35; arrival 08 35 + 04 48

Arrival time is 13 23.

Note that 4.80 hrs does not represent 4 hrs and 80 minutes. This is because time is not a decimal system which has 10 as its base number. Time is a **sexagesimal** number system, with 60 as its base number, i.e. there are 60 minutes in an hour and 60 seconds in a minute.

As shown above, converting 0.8 hrs into minutes can be done by multiplying by 60.

i.e. 0.8 hrs is equivalent to 0.8 × 60 = 48 minutes.

Your scientific calculator will have a sexagesimal button and it will look similar to $\boxed{°\,'\text{ and }''}$.

To convert 4.80 hrs to hours, minutes and seconds, enter the number into your calculator followed by the $\boxed{°\,'\text{ and }''}$ button. The calculator will show an answer of $\boxed{4}\,\boxed{°}\,\boxed{48}\,\boxed{'}\,\boxed{0}\,\boxed{''}$ which implies 4 hrs, 48 mins and 0 secs.

9 TIME

Exercise 9.1

1. Using your calculator, convert the following times written as decimals into time using hours, minutes and seconds.
   - **a** 0.25 hrs
   - **b** 3.765 hrs
   - **c** 0.22 hrs

2. A train leaves a station at 06 24. The journey has 4 stops. Calculate the time the train arrives at each stop if the time taken from one to stop to the next is as follows:
   Start to stop 1 takes 0.35 hrs.
   Stop 1 to stop 2 takes 1.30 hrs.
   Stop 2 to stop 3 takes 1.65 hrs.
   Stop 3 to final stop takes 2.91 hrs.

3. **a** A journey to work takes Sangita three quarters of an hour. If she catches the bus at 07 55, when does she arrive?
   b Sangita catches a bus home each evening. The journey takes 55 minutes. If she catches the bus at 17 50, when does she arrive?

4. Jake cycles to school each day. His journey takes 70 minutes. When will he arrive if he leaves home at 07 15?

5. Find the time in hours and minutes for the following journeys of the given distance at the average speed stated:
   - **a** 230 km at 100 km/h
   - **b** 70 km at 50 km/h

6. Grand Prix racing cars cover a 120 km race at the following average speeds. How long do the first five cars take to complete the race? Answer in minutes and seconds.
   First 240 km/h   Second 220 km/h   Third 210 km/h
   Fourth 205 km/h Fifth 200 km/h

7. A train covers the 1500 km distance from Amsterdam to Barcelona at an average speed of 90 km/h. If the train leaves Amsterdam at 9.30 a.m. on Tuesday, when does it arrive in Barcelona?

8. A plane takes off at 16 25 for the 3200 km journey from Moscow to Athens. If the plane flies at an average speed of 600 km/h, when will it land in Athens?

9. A plane leaves London for Boston, a distance of 5200 km, at 09 45. The plane travels at an average speed of 800 km/h. If Boston time is five hours behind British time, what is the time in Boston when the aircraft lands?

Student assessment 1

1. A journey to school takes Monica 0.4 hours. What time does she arrive if she leaves home at 08 38?
2. A car travels 295 km at 50 km/h. How long does the journey take? Give your answer in hours and minutes.
3. A bus leaves Deltaville at 11 32. It travels at an average speed of 42 km/h. If it arrives in Eastwich at 12 42, what is the distance between the two towns?
4. A plane leaves Betatown at 17 58 and arrives at Fleckley at 05 03 the following morning. How long does the journey take? Give your answer in hours and minutes.

Student assessment 2

1. A journey to school takes Fouad 22 minutes. What is the latest time he can leave home if he must be at school at 08 40?
2. A plane travels 270 km at 120 km/h. How long does the journey take? Give your answer in hours and minutes.
3. A train leaves Alphaville at 13 27. It travels at an average speed of 56 km/h. If it arrives in Westwich at 16 12, what is the distance between the two towns?
4. A car leaves Gramton at 16 39. It travels a distance of 315 km and arrives at Halfield at 20 09.
    a How long does the journey take?
    b What is the car's average speed?

10 Set notation and Venn diagrams

Sets

A **set** is a well-defined group of objects or symbols. The objects or symbols are called the **elements** of the set.

If an element e belongs to a set S, this is represented as $e \in S$. If e does not belong to set S this is represented as $e \notin S$.

➔ Worked examples

a A particular set consists of the following elements:
   {South Africa, Namibia, Egypt, Angola, ...}
   i   Describe the set.
       The elements of the set are countries of Africa.

   ii Add another two elements to the set.
       e.g. Zimbabwe, Ghana

b Consider the set $A = \{x : x \text{ is a natural number}\}$
   i   Describe the set.
       The elements of the set are the natural numbers.

   ii Write down two elements of the set.
       e.g. 3 and 15

c Consider the set $B = \{(x, y) : y = 2x - 4\}$
   i   Describe the set.
       The elements of the set are the coordinates of points found on the straight line with equation $y = 2x - 4$.

   ii Write down two elements of the set.
       e.g. $(0, -4)$ and $(10, 16)$

d Consider the set $C = \{x : 2 \leqslant x \leqslant 8\}$
   i   Describe the set.
       The elements of the set include any number between 2 and 8 inclusive.

   ii Write down two elements of the set.
       e.g. 5 and 6.3

Exercise 10.1

1 In the following questions:
   i describe the set in words,
   ii write down another two elements of the set.

   a {Asia, Africa, Europe, ...}
   b {2, 4, 6, 8, ...}
   c {Sunday, Monday, Tuesday, ...}
   d {January, March, July, ...}
   e {1, 3, 6, 10, ...}
   f {Mehmet, Michael, Mustapha, Matthew, ...}
   g {11, 13, 17, 19, ...}
   h {a, e, i, ...}
   i {Earth, Mars, Venus, ...}
   j $A = \{x: 3 \leq x \leq 12\}$
   k $S = \{y: -5 \leq y \leq 5\}$

2 The number of elements in a set A is written as n(A).
   Give the value of n(A) for the finite sets in Questions 1 a–k above.

Subsets

If all the elements of one set X are also elements of another set Y, then X is said to be a **subset** of Y.
This is written as $X \subseteq Y$.

If a set A is empty (i.e. it has no elements in it), then this is called the **empty set** and it is represented by the symbol $\emptyset$. Therefore $A = \emptyset$. The empty set is a subset of all sets.

e.g.   Three girls, Winnie, Natalie and Emma, form a set A
       $A = \{$Winnie, Natalie, Emma$\}$
       All the possible subsets of A are given below:
       $B = \{$Winnie, Natalie, Emma$\}$
       $C = \{$Winnie, Natalie$\}$
       $D = \{$Winnie, Emma$\}$
       $E = \{$Natalie, Emma$\}$
       $F = \{$Winnie$\}$
       $G = \{$Natalie$\}$
       $H = \{$Emma$\}$
       $I = \emptyset$

Note that the sets B and I above are considered as subsets of A.
i.e. $B \subseteq A$ and $I \subseteq A$

Similarly, $G \not\subseteq H$ implies that G is not a subset of H

➡ Worked examples

$A = \{1, 2, 3, 4, 5, 6, 7, 8, 9, 10\}$

a List subset B {even numbers}.
   $B = \{2, 4, 6, 8, 10\}$

b List subset C {prime numbers}.
   $C = \{2, 3, 5, 7\}$

10 SET NOTATION AND VENN DIAGRAMS

Exercise 10.2

1. P = {whole numbers less than 30}
   a. List the subset Q {even numbers}.
   b. List the subset R {odd numbers}.
   c. List the subset S {prime numbers}.
   d. List the subset T {square numbers}.
   e. List the subset U {triangle numbers}.

2. A = {whole numbers between 50 and 70}
   a. List the subset B {multiples of 5}.
   b. List the subset C {multiples of 3}.
   c. List the subset D {square numbers}.

3. State whether each of the following statements is true or false:
   a. {Algeria, Mozambique} $\subseteq$ {countries in Africa}
   b. {mango, banana} $\subseteq$ {fruit}
   c. {1, 2, 3, 4} $\subseteq$ {1, 2, 3, 4}
   d. {volleyball, basketball} $\not\subseteq$ {team sport}
   e. {potatoes, carrots} $\subseteq$ {vegetables}

The universal set

The **universal set** ($\mathscr{E}$) for any particular problem is the set which contains all the possible elements for that problem.

The **complement** of a set A is the set of elements which are in $\mathscr{E}$ but not in A. The complement of A is identified as A'.

Notice that $\mathscr{E}' = \varnothing$ and $\varnothing' = \mathscr{E}$.

➜ Worked examples

a. If $\mathscr{E}$ = {1, 2, 3, 4, 5, 6, 7, 8, 9, 10} and A = {1, 2, 3, 4, 5} what set is represented by A'?

   A' consists of those elements in $\mathscr{E}$ which are not in A.

   Therefore A' = {6, 7, 8, 9, 10}.

b. If $\mathscr{E}$ = {all 3D shapes} and P = {prisms} what set is represented by P'?

   P' = {all 3D shapes except prisms}.

Set notation and Venn diagrams

Venn diagrams are the principal way of showing sets diagrammatically. The method consists primarily of entering the elements of a set into a circle or circles.

Some examples of the uses of Venn diagrams are shown.
A = {2, 4, 6, 8, 10} can be represented as:

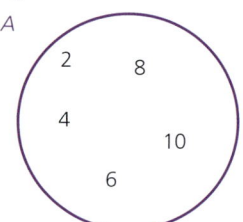

Elements which are in more than one set can also be represented using a Venn diagram.

$P = \{3, 6, 9, 12, 15, 18\}$ and $Q = \{2, 4, 6, 8, 10, 12\}$ can be represented as:

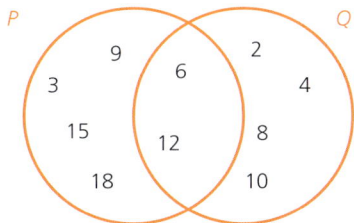

In the diagram above, it can be seen that those elements which belong to both sets are placed in the region of overlap of the two circles.

When two sets P and Q overlap as they do above, the notation $P \cap Q$ is used to denote the set of elements in the **intersection**, i.e. $P \cap Q = \{6, 12\}$.

6 belongs to the set of $P \cap Q$; whilst 8 does not belong to the set of $P \cap Q$.

$X = \{1, 3, 6, 7, 14\}$ and $Y = \{3, 9, 13, 14, 18\}$ are represented as:

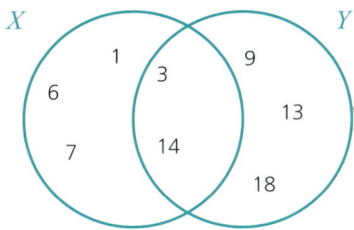

The **union** of two sets is everything which belongs to either or both sets and is represented by the symbol $\cup$.
Therefore in the example above $X \cup Y = \{1, 3, 6, 7, 9, 13, 14, 18\}$.

$J = \{10, 20, 30, 40, 50, 60, 70, 80, 90, 100\}$ and $K = \{60, 70, 80\}$; as discussed earlier, $K \subseteq J$ and can be represented as shown below:

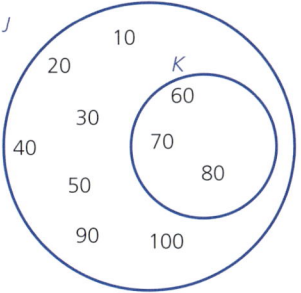

10 SET NOTATION AND VENN DIAGRAMS

Exercise 10.3

1 Complete the statement $A \cap B = \{...\}$ for each of the Venn diagrams below.

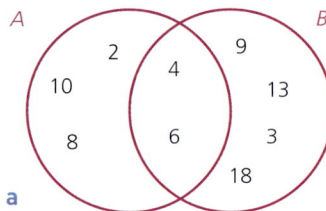
a

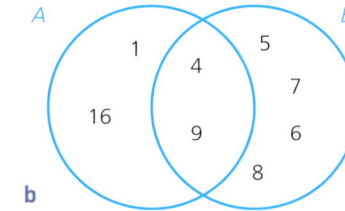
b

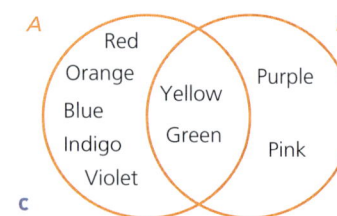
c

2 Complete the statement $A \cup B = \{...\}$ for each of the Venn diagrams in Question 1 above.

3

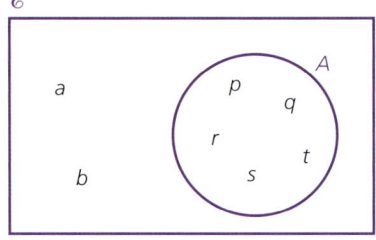

Copy and complete the following statements:
a $\mathcal{E} = \{...\}$ b $A' = \{...\}$

4 $\mathcal{E}$

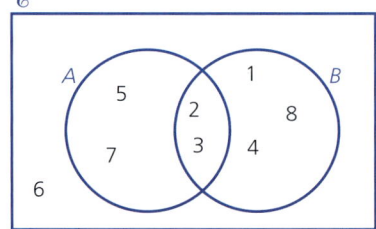

Copy and complete the following statements:
a $\mathcal{E} = \{...\}$ b $A' = \{...\}$ c $A \cap B = \{...\}$
d $A \cup B = \{...\}$ e $(A \cap B)' = \{...\}$ f $A \cap B' = \{...\}$

5 Using the Venn diagram (right), indicate whether the following statements are true or false. $\in$ means 'is an element of' and $\notin$ means 'is not an element of'.
a $5 \in A$ b $20 \in B$
c $20 \notin A$ d $50 \in A$
e $50 \notin B$ f $A \cap B = \{10, 20\}$

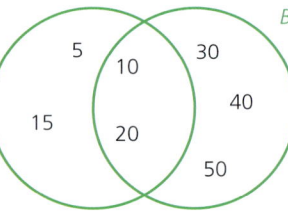

6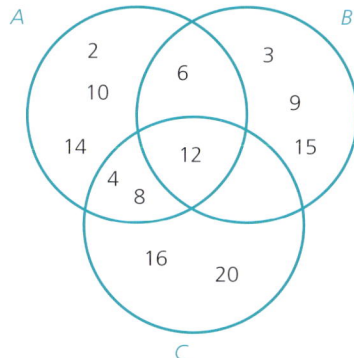

  a Describe in words the elements of:
   i set A **ii** set B **iii** set C
  b Copy and complete the following statements:
   i $A \cap B = \{...\}$ **ii** $A \cap C = \{...\}$ **iii** $B \cap C = \{...\}$
   iv $A \cap B \cap C = \{...\}$ **v** $A \cup B = \{...\}$ **vi** $C \cup B = \{...\}$

7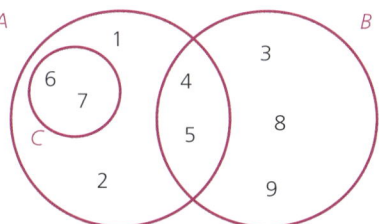

  a Copy and complete the following statements:
   i $A = \{...\}$ **ii** $B = \{...\}$ **iii** $C' = \{...\}$
   iv $A \cap B = \{...\}$ **v** $A \cup B = \{...\}$ **vi** $(A \cap B)' = \{...\}$
  b State, using set notation, the relationship between C and A.

8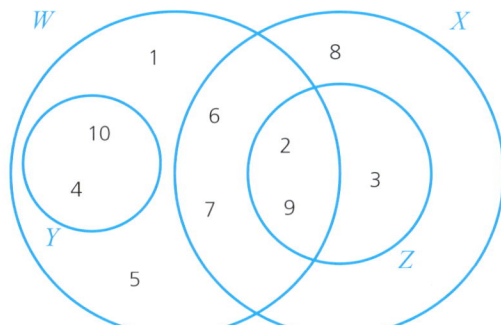

  a Copy and complete the following statements:
   i $W = \{...\}$ **ii** $X = \{...\}$ **iii** $Z' = \{...\}$
   iv $W \cap Z = \{...\}$ **v** $W \cap X = \{...\}$ **vi** $Y \cap Z = ...$
  b Which of the named sets is a subset of X?

10 SET NOTATION AND VENN DIAGRAMS

Exercise 10.4

1. $A = \{$Egypt, Libya, Morocco, Chad$\}$
   $B = \{$Iran, Iraq, Turkey, Egypt$\}$
   a. Draw a Venn diagram to illustrate the above information.
   b. Copy and complete the following statements:
      i. $A \cap B = \{...\}$
      ii. $A \cup B = \{...\}$

2. $P = \{2, 3, 5, 7, 11, 13, 17\}$
   $Q = \{11, 13, 15, 17, 19\}$
   a. Draw a Venn diagram to illustrate the above information.
   b. Copy and complete the following statements:
      i. $P \cap Q = \{...\}$
      ii. $P \cup Q = \{...\}$

3. $B = \{2, 4, 6, 8, 10\}$
   $A \cup B = \{1, 2, 3, 4, 6, 8, 10\}$
   $A \cap B = \{2, 4\}$
   Represent the above information on a Venn diagram.

4. $X = \{a, c, d, e, f, g, l\}$
   $Y = \{b, c, d, e, h, i, k, l, m\}$
   $Z = \{c, f, i, j, m\}$
   Represent the above information on a Venn diagram.

5. $P = \{1, 4, 7, 9, 11, 15\}$
   $Q = \{5, 10, 15\}$
   $R = \{1, 4, 9\}$
   Represent the above information on a Venn diagram.

Problems involving sets

→ Worked example

In a class of 31 students, some study Physics and some study Chemistry. If 22 study Physics, 20 study Chemistry and 5 study neither, calculate the number of students who take both subjects.

The information given above can be entered in a Venn diagram in stages.

The students taking neither Physics nor Chemistry can be put in first (as shown left).

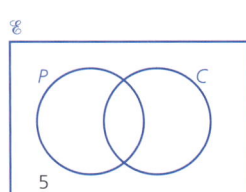

This leaves 26 students to be entered into the set circles.

If x students take both subjects then
$\quad$ n$(P) = 22 - x + x$
$\quad$ n$(C) = 20 - x + x$
$\quad P \cup C = 31 - 5 = 26$

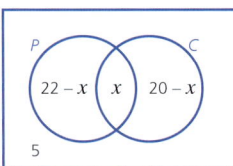

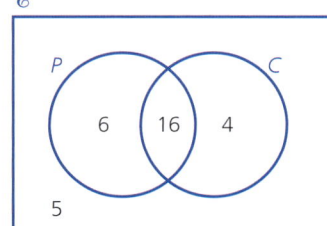

Therefore $22 - x + x + 20 - x = 26$
$\qquad 42 - x = 26$
$\qquad x = 16$

Substituting the value of x into the Venn diagram gives:

Therefore the number of students taking both Physics and Chemistry is 16.

Problems involving sets

 Exercise 10.5

1. In a class of 35 students, 19 take Spanish, 18 take Korean and 3 take neither. Calculate how many take:
   a both Korean and Spanish,
   b just Spanish,
   c just Korean.

2. In a year group of 108 students, 60 liked football, 53 liked tennis and 10 liked neither. Calculate the number of students who liked football but not tennis.

3. In a year group of 113 students, 60 liked hockey, 45 liked rugby and 18 liked neither. Calculate the number of students who:
   a liked both hockey and rugby,
   b liked only hockey.

4. One year, 37 students sat an examination in Physics, 48 sat Chemistry and 45 sat Biology. 15 students sat Physics and Chemistry, 13 sat Chemistry and Biology, 7 sat Physics and Biology and 5 students sat all three.
   a Draw a Venn diagram to represent this information.
   b Calculate n($P \cup C \cup B$).

Student assessment 1

1. Describe the following sets in words:
   a {2, 4, 6, 8}
   b {2, 4, 6, 8, ...}
   c {1, 4, 9, 16, 25, ...}
   d {Arctic, Atlantic, Indian, Pacific}

2. Calculate the value of n(A) for each of the sets shown below:
   a A = {days of the week}
   b A = {prime numbers between 50 and 60}
   c A = {x: x is an integer and $5 \leqslant x \leqslant 10$}
   d A = {days in a leap year}

3. Copy out the Venn diagram (below) twice.
   a On one copy, shade and label the region which represents $A \cap B$.
   b On the other copy, shade and label the region which represents $A \cup B$.

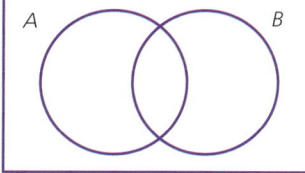

4. If $\mathscr{E}$ = {m, a, t, h, s} and A = {a, s}, what set is represented by A'?

5. If A = {a, b}, list all the subsets of A.

10 SET NOTATION AND VENN DIAGRAMS

Student assessment 2

1. J = {London, Paris, Rome, Washington, Canberra, Ankara, Cairo}
   K = {Cairo, Nairobi, Pretoria, Ankara}
   a Draw a Venn diagram to represent the above information.
   b Copy and complete the statement $J \cap K$ = {...}.
   c Copy and complete the statement $J' \cap K$ = {...}.

2. $M = \{x: x \text{ is an integer and } 2 \leqslant x \leqslant 20\}$
   N = {prime numbers less than 30}
   a Draw a Venn diagram to illustrate the information above.
   b Copy and complete the statement $M \cap N$ = {...}.
   c Copy and complete the statement $(M \cap N)'$ = {...}.

3. $\mathscr{E}$ = {natural numbers}, M = {even numbers} and N = {multiples of 5}.
   a Draw a Venn diagram and place the numbers 1, 2, 3, 4, 5, 6, 7, 8, 9, 10 in the appropriate places in it.
   b If $X = M \cap N$, describe set X in words.

4. If A = {2, 4, 6, 8}, write all the subsets of A with two or more elements.

5. In a region of mixed farming, farms keep goats, chickens or sheep. There are 77 farms altogether. 19 farms keep only goats, 8 keep only chickens and 13 keep only sheep. 13 keep both goats and chickens, 28 keep both chickens and sheep and 8 keep both goats and sheep.
   a Draw a Venn diagram to show the above information.
   b Calculate n($G \cap C \cap S$).

Student assessment 3

1. M = {a, e, i, o, u}
   a How many subsets are there of M?
   b List the subsets of M with four or more elements.

2. X = {lion, tiger, cheetah, leopard, puma, jaguar, cat}
   Y = {elephant, lion, zebra, cheetah, gazelle}
   Z = {anaconda, jaguar, tarantula, mosquito}
   a Draw a Venn diagram to represent the above information.
   b Copy and complete the statement $X \cap Y$ = {...}.
   c Copy and complete the statement $Y \cap Z$ = {...}.
   d Copy and complete the statement $X \cap Y \cap Z$ = {...}.

3. A group of 40 people were asked whether they like cricket (C) and football (F). The number liking both cricket and football was three times the number liking only cricket. Adding 3 to the number liking only cricket and doubling the answer equals the number of people liking only football. Four said they did not like sport at all.
   a Draw a Venn diagram to represent this information.
   b Calculate n($C \cap F$).
   c Calculate n($C \cap F'$).
   d Calculate n($C' \cap F$).

Problems involving sets

4 The Venn diagram below shows the number of elements in three sets P, Q and R.

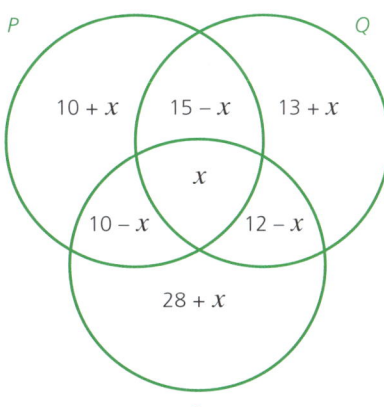

If $n(P \cup Q \cup R) = 93$ calculate:

a x
b $n(P)$
c $n(Q)$
d $n(R)$
e $n(P \cap Q)$
f $n(Q \cap R)$
g $n(P \cap R)$
h $n(R \cup Q)$
i $n(P \cap Q)'$

Mathematical investigations and ICT 1

Investigations are an important part of mathematical learning. All mathematical discoveries stem from an idea that a mathematician has and then investigates.

Sometimes when faced with a mathematical investigation, it can seem difficult to know how to start. The structure and example below may help you.

1 Read the question carefully and start with simple cases.
2 Draw simple diagrams to help.
3 Put the results from simple cases in an ordered table.
4 Look for a pattern in your results.
5 Try to find a general rule in words.
6 Express your rule algebraically.
7 Test the rule for a new example.
8 Check that the original question has been answered.

→ Worked example

A mystic rose is created by placing a number of points evenly spaced on the circumference of a circle. Straight lines are then drawn from each point to every other point. The diagram (left) shows a mystic rose with 20 points.

a How many straight lines are there?

b How many straight lines would there be on a mystic rose with 100 points?

To answer these questions, you are not expected to draw either of the shapes and count the number of lines.

1/2. Try simple cases:

By drawing some simple cases and counting the lines, some results can be found:

Mystic rose with 2 points   Mystic rose with 3 points

Number of lines = 1   Number of lines = 3

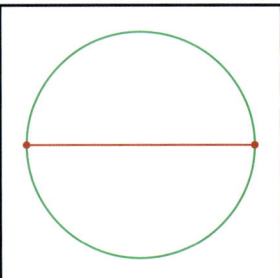
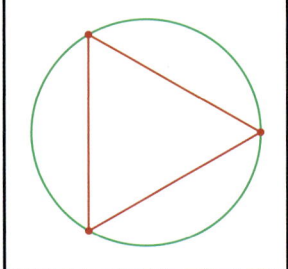

Mathematical investigations and ICT 1

Mystic rose with 4 points

Number of lines = 6

Mystic rose with 5 points

Number of lines = 10

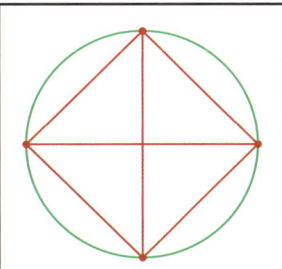
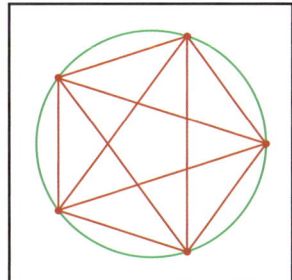

3. Enter the results in an ordered table:

Number of points	2	3	4	5
Number of lines	1	3	6	10

4/5. Look for a pattern in the results:

There are two patterns.

The first shows how the values change.

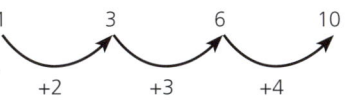

It can be seen that the difference between successive **terms** is increasing by one each time.

The problem with this pattern is that to find the 20th and 100th terms, it would be necessary to continue this pattern and find all the terms leading up to the 20th and 100th terms.

The second is the relationship between the number of points and the number of lines.

Number of points	2	3	4	5
Number of lines	1	3	6	10

It is important to find a relationship that works for all values. For example, subtracting 1 from the number of points gives the number of lines in the first example only, so is not useful. However, halving the number of points and multiplying this by 1 less than the number of points works each time, i.e. number of lines = (half the number of points) × (one less than the number of points).

6. Express the rule algebraically:

The rule expressed in words above can be written more elegantly using algebra. Let the number of lines be l and the number of points be p.

$$l = \frac{1}{2}p(p-1)$$

Note: Any letters can be used to represent the number of lines and the number of points, not just l and p.

MATHEMATICAL INVESTIGATIONS AND ICT 1

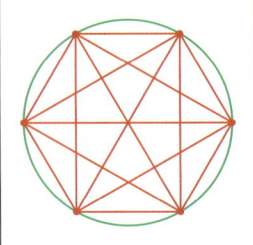

7. Test the rule:

The rule was derived from the original results. It can be tested by generating a further result.

If the number of points $p = 6$, then the number of lines l is:

$l = \frac{1}{2} \times 6(6 - 1)$
$= 3 \times 5$
$= 15$

From the diagram to the left, the number of lines can also be counted as 15.

8. Check that the original questions have been answered

Using the formula, the number of lines in a mystic rose with 20 points is:

$l = \frac{1}{2} \times 20(20 - 1)$
$= 10 \times 19$
$= 190$

The number of lines in a mystic rose with 100 points is:

$l = \frac{1}{2} \times 100(100 - 1)$
$= 50 \times 99$
$= 4950$

Primes and squares

13, 41 and 73 are prime numbers.

Two different square numbers can be added together to make these prime numbers, e.g. $3^2 + 8^2 = 73$.

1. Find the two square numbers that can be added to make 13 and 41.
2. List the prime numbers less than 100.
3. Which of the prime numbers less than 100 can be shown to be the sum of two different square numbers?
4. Is there a rule to the numbers in Question 3?
5. Your rule is a predictive rule not a formula. Discuss the difference.

Football leagues

There are 18 teams in a football league.
1. If each team plays the other teams twice, once at home and once away, then how many matches are played in a season?
2. If there are t teams in a league, how many matches are played in a season?

ICT activity 1

In this activity, you will be using a spreadsheet to track the price of a company's shares over a period of time.

1. **a** Using the internet or a newspaper as a resource, find the value of a particular company's shares.
   b Over a period of a month (or week), record the value of the company's shares. This should be carried out on a daily basis.
2. When you have collected all the values, enter them into a spreadsheet similar to the one shown on the left.
3. In column C, enter formulas that will calculate the value of the shares as a percentage of their value on day 1.
4. When the spreadsheet is complete, produce a graph showing how the percentage value of the share price changed over time.
5. Write a short report explaining the performance of the company's shares during that time.

ICT activity 2

The following activity requires the use of a graphing package.
The velocity of a student at different parts of a 100 m sprint will be analysed.
A running track is set out as shown below:

1. A student must stand at each of points A–F. The student at A runs the 100 m and is timed as they run past each of the points B–F by the students at these points, who each have a stopwatch.
2. Using the graphing package, plot a distance–time graph of the results by entering the data as pairs of coordinates, i.e. (time, distance).
3. Ensure that all the points are selected and draw a curve of best fit through them.
4. Select the curve and plot a coordinate of your choice on it. This point can now be moved along the curve using the cursor keys on the keyboard.
5. Draw a **tangent to the curve** through the point.
6. What does the gradient of the tangent represent?
7. At what point of the race was the student running fastest? How did you reach this answer?
8. Collect similar data for other students. Compare their graphs and running speeds.
9. Carefully analyse one of the graphs and write a brief report to the runner in which you should identify, giving reasons, the parts of the race they need to improve on.

TOPIC 2

Algebra and graphs

Contents

Chapter 11 Algebraic representation and manipulation (E2.1, E2.2, E2.3, E2.5)
Chapter 12 Algebraic indices (E2.4)
Chapter 13 Equations and inequalities (E2.5)
Chapter 14 Graphing inequalities and regions (E2.6)
Chapter 15 Sequences (E2.7)
Chapter 16 Proportion (E2.8)
Chapter 17 Graphs in practical situations (E2.9)
Chapter 18 Graphs of functions (E2.10, E2.11)
Chapter 19 Differentiation and the gradient function (E2.12)
Chapter 20 Functions (E2.13)

Learning objectives

E2.1 Introduction to algebra
1. Know that letters can be used to represent generalised numbers.
2. Substitute numbers into expressions and formulas.

E2.2 Algebraic manipulation
1. Simplify expressions by collecting like terms.
2. Expand products of algebraic expressions.
3. Factorise by extracting common factors.
4. Factorise expressions of the form:
   - $ax + bx + kay + kby$
   - $a^2x^2 - b^2y^2$
   - $a^2 + 2ab + b^2$
   - $ax^2 + bx + c$
   - $ax^3 + bx^2 + cx$.
5. Complete the square for expressions in the form $ax^2 + bx + c$.

E2.3 Algebraic fractions
1. Manipulate algebraic fractions.
2. Factorise and simplify rational expressions.

E2.4 Indices II
1. Understand and use indices (positive, zero, negative and fractional).
2. Understand and use the rules of indices.

E2.5 Equations
1. Construct expressions, equations and formulas.
2. Solve linear equations in one unknown.
3. Solve fractional equations with numerical and linear algebraic denominators.
4. Solve simultaneous linear equations in two unknowns.
5. Solve simultaneous equations, involving one linear and one non-linear.

6 Solve quadratic equations by factorisation, completing the square and by use of the quadratic formula.
7 Change the subject of formulas.

E2.6 Inequalities
1 Represent and interpret inequalities, including on a number line.
2 Construct, solve and interpret linear inequalities.
3 Represent and interpret linear inequalities in two variables graphically.
4 List inequalities that define a given region.

E2.7 Sequences
1 Continue a given number sequence or pattern.
2 Recognise patterns in sequences, including the term-to-term rule, and relationships between different sequences.
3 Find and use the nth term of sequences.

E2.8 Proportion
Express direct and inverse proportion in algebraic terms and use this form of expression to find unknown quantities.

E2.9 Graphs in practical situations
1 Use and interpret graphs in practical situations, including travel graphs and conversion graphs.
2 Draw graphs from given data.
3 Apply the idea of rate of change to simple kinematics involving distance–time and speed–time graphs, acceleration and deceleration.
4 Calculate distance travelled as area under a speed–time graph.

E2.10 Graphs of functions
1 Construct tables of values, and draw, recognise and interpret graphs for functions of the following forms:
- ax^n (includes sums of no more than three of these)
- $ab^x + c$

where $n = -2, -1, -\frac{1}{2}, 0, \frac{1}{2}, 1, 2, 3$; a and c are rational numbers; and b is a positive integer.
2 Solve associated equations graphically, including finding and interpreting roots by graphical methods.
3 Draw and interpret graphs representing exponential growth and decay problems.

E2.11 Sketching curves
Recognise, sketch and interpret graphs of the following functions:
a linear
b quadratic
c cubic
d reciprocal
e exponential.

E2.12 Differentiation
1 Estimate gradients of curves by drawing tangents.
2 Use the derivatives of functions of the form ax^n, where a is a rational constant and n is a positive integer or zero, and simple sums of not more than three of these.
3 Apply differentiation to gradients and stationary points (turning points).
4 Discriminate between maxima and minima by any method.

E2.13 Functions
1 Understand functions, domain and range and use function notation.
2 Understand and find inverse functions $f^{-1}(x)$.
3 Form composite functions as defined by $gf(x) = g(f(x))$.

The founders of algebra

Abū Ja'far Muḥammad ibn Mūsā al-Khwārizmī is called the 'father of algebra'. He was born in Baghdad in 790CE. He wrote the book *Hisab al-jabr w'al-muqabala* in 830CE when Baghdad had the greatest university in the world and the greatest mathematicians studied there. He gave us the word 'algebra' and worked on quadratic equations. He also introduced the decimal system from India.

Muhammad al-Karaji was born in North Africa in what is now Morocco. He lived in the eleventh century and worked on the theory of indices. He also worked on an algebraic method of calculating square and cube roots. He may also have travelled to the University of Granada (then part of the Moorish Empire) where works of his can be found in the University library.

The poet Omar Khayyam is known for his long poem *The Rubaiyat*. He was also a fine mathematician, working on the binomial theorem. He introduced the symbol 'shay', which became our 'x'.

Al-Khwārizmī (790–850)

11 Algebraic representation and manipulation

Algebra is a mathematical language and is at the heart of mathematics. As well as numbers, letters are also used. The letters are used to represent unknown quantities, or a variety of possible different values.

Using algebra may at first seem complicated, but as with any language, the more you use it and the more you understand its rules, the easier it becomes.

This topic deals with those rules.

Expanding a bracket

When removing brackets, every term inside the bracket must be multiplied by whatever is outside the bracket.

Worked examples

a $\quad 3(x + 4)$
$\quad = 3x + 12$

b $\quad 5x(2y + 3)$
$\quad = 10xy + 15x$

c $\quad 2a(3a + 2b - 3c)$
$\quad = 6a^2 + 4ab - 6ac$

d $\quad -4p(2p - q + r^2)$
$\quad = -8p^2 + 4pq - 4pr^2$

e $\quad -2x^2\left(1x + 3y - \frac{1}{x}\right)$
$\quad = -2x^3 - 6x^2y + 2x$

f $\quad \frac{-2}{x}\left(-2x + 4y + \frac{1}{x}\right)$
$\quad = 4 - \frac{8y}{x} - \frac{2}{x^2}$

Exercise 11.1

Expand the following:

1. a $\quad 4(x - 3)$
   b $\quad 5(2p - 4)$
   c $\quad -6(7x - 4y)$
   d $\quad 3(2a - 3b - 4c)$
   e $\quad -7(2m - 3n)$
   f $\quad -2(8x - 3y)$

2. a $\quad 3x(x - 3y)$
   b $\quad a(a + b + c)$
   c $\quad 4m(2m - n)$
   d $\quad -5a(3a - 4b)$
   e $\quad -4x(-x + y)$
   f $\quad -8p(-3p + q)$

3. a $\quad -(2x^2 - 3y^2)$
   b $\quad -(-a + b)$
   c $\quad -(-7p + 2q)$
   d $\quad \frac{1}{2}(6x - 8y + 4z)$
   e $\quad \frac{3}{4}(4x - 2y)$
   f $\quad \frac{1}{5}x(10x - 15y)$

4. a $\quad 3r(4r^2 - 5s + 2t)$
   b $\quad a^2(a + b + c)$
   c $\quad 3a^2(2a - 3b)$
   d $\quad pq(p + q - pq)$
   e $\quad m^2(m - n + nm)$
   f $\quad a^3(a^3 + a^2b)$

Expanding a pair of brackets

Exercise 11.2

Expand and simplify the following:

1. **a** $3a - 2(2a + 4)$ **b** $8x - 4(x + 5)$
   c $3(p - 4) - 4$ **d** $7(3m - 2n) + 8n$
   e $6x - 3(2x - 1)$ **f** $5p - 3p(p + 2)$

2. **a** $7m(m + 4) + m^2 + 2$ **b** $3(x - 4) + 2(4 - x)$
   c $6(p + 3) - 4(p - 1)$ **d** $5(m - 8) - 4(m - 7)$
   e $3a(a + 2) - 2(a^2 - 1)$ **f** $7a(b - 2c) - c(2a - 3)$

3. **a** $\frac{1}{2}(6x + 4) + \frac{1}{3}(3x + 6)$ **b** $\frac{1}{4}(2x + 6y) + \frac{3}{4}(6x - 4y)$
   c $\frac{1}{8}(6x - 12y) + \frac{1}{2}(3x - 2y)$ **d** $\frac{1}{5}(15x + 10y) + \frac{3}{10}(5x - 5y)$
   e $\frac{2}{3}(6x - 9y) + \frac{1}{3}(9x + 6y)$ **f** $\frac{x}{7}(14x - 21y) - \frac{x}{2}(4x - 6y)$

Expanding a pair of brackets

When multiplying together expressions in brackets, it is necessary to multiply all the terms in one bracket by all the terms in the other bracket.

➔ Worked examples

Expand the following:

a $(x + 3)(x + 5)$

	x	$+3$
x	x^2	$3x$
$+5$	$5x$	15

$= x^2 + 5x + 3x + 15$
$= x^2 + 8x + 15$

b $(x + 2)(2x - 1)$

	$2x$	-1
x	$2x^2$	$-x$
$+2$	$4x$	-2

$= 2x^2 - x + 4x - 2$
$= 2x^2 + 3x - 2$

Exercise 11.3

Expand the following and simplify your answer:

1. **a** $(x + 2)(x + 3)$ **b** $(x + 3)(x + 4)$
   c $(x + 5)(x + 2)$ **d** $(x + 6)(x + 1)$
   e $(x - 2)(x + 3)$ **f** $(x + 8)(x - 3)$

2. **a** $(x - 4)(x + 6)$ **b** $(x - 7)(x + 4)$
   c $(x + 5)(x - 7)$ **d** $(x + 3)(x - 5)$
   e $(x + 1)(x - 3)$ **f** $(x - 7)(x + 9)$

3. **a** $(2x - 3)(x - 3)$ **b** $(2x - 5)(x - 2)$
   c $(x - 4)(3x - 8)$ **d** $(x + 3)(5x - 3)$
   e $(2x - 3)^2$ **f** $(2x - 7)(3x - 5)$

4. **a** $(x + 3)(x - 3)$ **b** $(x + 7)(x - 7)$
   c $(x - 8)(x + 8)$ **d** $(x + y)(x - y)$
   e $(a + b)(a - b)$ **f** $(p - q)(p + q)$

11 ALGEBRAIC REPRESENTATION AND MANIPULATION

Simple factorising

When factorising, the largest possible factor is removed from each of the terms and placed outside the brackets.

Worked examples

Factorise the following expressions:

a $10x + 15$
  $= 5(2x + 3)$

b $8p - 6q + 10r$
  $= 2(4p - 3q + 5r)$

c $-2q - 6p + 12$
  $= 2(-q - 3p + 6)$

d $2a^2 + 3ab - 5ac$
  $= a(2a + 3b - 5c)$

e $6ax - 12ay - 18a^2$
  $= 6a(x - 2y - 3a)$

f $3b + 9ba - 6bd$
  $= 3b(1 + 3a - 2d)$

Exercise 11.4

Factorise the following:

1 a $4x - 6$
   b $18 - 12p$
   c $6y - 3$
   d $4a + 6b$
   e $3p - 3q$
   f $8m + 12n + 16r$

2 a $3ab + 4ac - 5ad$
   b $8pq + 6pr - 4ps$
   c $a^2 - ab$
   d $4x^2 - 6xy$
   e $abc + abd + fab$
   f $3m^2 + 9m$

3 a $3pqr - 9pqs$
   b $5m^2 - 10mn$
   c $8x^2y - 4xy^2$
   d $2a^2b^2 - 3b^2c^2$
   e $12p - 36$
   f $42x - 54$

4 a $18 + 12y$
   b $14a - 21b$
   c $11x + 11xy$
   d $4s - 16t + 20r$
   e $5pq - 10qr + 15qs$
   f $4xy + 8y^2$

5 a $m^2 + mn$
   b $3p^2 - 6pq$
   c $pqr + qrs$
   d $ab + a^2b + ab^2$
   e $3p^3 - 4p^4$
   f $7b^3c + b^2c^2$

6 a $m^3 - m^2n + mn^2$
   b $4r^3 - 6r^2 + 8r^2s$
   c $56x^2y - 28xy^2$
   d $72m^2n + 36mn^2 - 18m^2n^2$

Substitution

Worked examples

Evaluate the expressions below if $a = 3$, $b = 4$, $c = -5$:

a $2a + 3b - c$
  $= 6 + 12 + 5$
  $= 23$

b $3a - 4b + 2c$
  $= 9 - 16 - 10$
  $= -17$

c $-2a + 2b - 3c$
    $= -6 + 8 + 15$
    $= 17$

d $a^2 + b^2 + c^2$
    $= 9 + 16 + 25$
    $= 50$

e $3a(2b - 3c)$
    $= 9(8 + 15)$
    $= 9 \times 23$
    $= 207$

f $-2c(-a + 2b)$
    $= 10(-3 + 8)$
    $= 10 \times 5$
    $= 50$

 Exercise 11.5

Evaluate the following expressions if $p = 4$, $q = -2$, $r = 3$ and $s = -5$:

1   a $2p + 4q$
    b $5r - 3s$
    c $3q - 4s$
    d $6p - 8q + 4s$
    e $3r - 3p + 5q$
    f $-p - q + r + s$

2   a $2p - 3q - 4r + s$
    b $3s - 4p + r + q$
    c $p^2 + q^2$
    d $r^2 - s^2$
    e $p(q - r + s)$
    f $r(2p - 3q)$

3   a $2s(3p - 2q)$
    b $pq + rs$
    c $2pr - 3rq$
    d $q^3 - r^2$
    e $s^3 - p^3$
    f $r^4 - q^5$

4   a $-2pqr$
    b $-2p(q + r)$
    c $-2rq + r$
    d $(p + q)(r - s)$
    e $(p + s)(r - q)$
    f $(r + q)(p - s)$

5   a $(2p + 3q)(p - q)$
    b $(q + r)(q - r)$
    c $q^2 - r^2$
    d $p^2 - r^2$
    e $(p + r)(p - r)$
    f $(-s + p)q^2$

Rearrangement of formulas

In the formula $a = 2b + c$, 'a' is the subject. In order to make either b or c the subject, the formula has to be rearranged.

→ Worked examples

Rearrange the following formulas to make the red letter the subject:

a $a = 2b + c$
    $a - 2b = c$

b $2r + p = q$
    $p = q - 2r$

c $ab = cd$
    $\frac{ab}{d} = c$

d $\frac{a}{b} = \frac{c}{d}$
    $ad = cb$
    $d = \frac{cb}{a}$

11 ALGEBRAIC REPRESENTATION AND MANIPULATION

Exercise 11.6 In the following questions, make the letter in red the subject of the formula:

1.
   a $m + n = r$
   b $m + n = p$
   c $2m + n = 3p$
   d $3x = 2p + q$
   e $ab = cd$
   f $ab = cd$

2.
   a $3xy = 4m$
   b $7pq = 5r$
   c $3x = c$
   d $3x + 7 = y$
   e $5y - 9 = 3r$
   f $5y - 9 = 3x$

3.
   a $6b = 2a - 5$
   b $6b = 2a - 5$
   c $3x - 7y = 4z$
   d $3x - 7y = 4z$
   e $3x - 7y = 4z$
   f $2pr - q = 8$

4.
   a $\frac{p}{4} = r$
   b $\frac{4}{p} = 3r$
   c $\frac{1}{5}n = 2p$
   d $\frac{1}{5}n = 2p$
   e $p(q + r) = 2t$
   f $p(q + r) = 2t$

5.
   a $3m - n = rt(p + q)$
   b $3m - n = rt(p + q)$
   c $3m - n = rt(p + q)$
   d $3m - n = rt(p + q)$
   e $3m - n = rt(p + q)$
   f $3m - n = rt(p + q)$

6.
   a $\frac{ab}{c} = de$
   b $\frac{ab}{c} = de$
   c $\frac{ab}{c} = de$
   d $\frac{a+b}{c} = d$
   e $\frac{a}{c} + b = d$
   f $\frac{a}{c} + b = d$

Further expansion

You will have seen earlier in this chapter how to expand a pair of brackets of the form $(x - 3)(x + 4)$. A similar method can be used to expand a pair of brackets of the form $(2x - 3)(3x - 6)$.

Worked examples

a Expand $(2x - 3)(3x - 6)$.

	$2x$	-3
$3x$	$6x^2$	$-9x$
-6	$-12x$	18

$= 6x^2 - 9x - 12x + 18$
$= 6x^2 - 21x + 18$

Further expansion

b Expand $(x-1)(x+2)(2x-5)$.

The expansion can be shown using diagrams as in example **a** above. To do so, it is easier to carry out the multiplication in steps.

Step 1: Expand $(x-1)(x+2)$

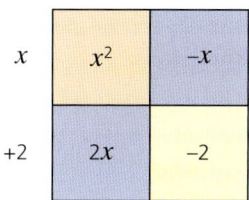

$= x^2 - x + 2x - 2$
$= x^2 + x - 2$

Step 2: Expand $(x^2 + x - 2)(2x - 5)$

	x^2	$+x$	-2
$2x$	$2x^3$	$2x^2$	$-4x$
-5	$-5x^2$	$-5x$	10

$= 2x^3 + 2x^2 - 4x - 5x^2 - 5x + 10$
$= 2x^3 - 3x^2 - 9x + 10$

Exercise 11.7

Expand the following brackets, giving your answer in its simplest form:

1. **a** $(y+2)(2y+3)$ **b** $(y+7)(3y+4)$
   c $(2y+1)(y+8)$ **d** $(2y+1)(2y+2)$
   e $(3y+4)(2y+5)$ **f** $(6y+3)(3y+1)$

2. **a** $(2p-3)(p+8)$ **b** $(4p-5)(p+7)$
   c $(3p-4)(2p+3)$ **d** $(4p-5)(3p+7)$
   e $(6p+2)(3p-1)$ **f** $(7p-3)(4p+8)$

3. **a** $(2x-1)(2x-1)$ **b** $(3x+1)^2$
   c $(4x-2)^2$ **d** $(5x-4)^2$
   e $(2x+6)^2$ **f** $(2x+3)(2x-3)$

4. **a** $(3+2x)(3-2x)$ **b** $(4x-3)(4x+3)$
   c $(3+4x)(3-4x)$ **d** $(7-5y)(7+5y)$
   e $(3+2y)(4y-6)$ **f** $(7-5y)^2$

5. **a** $(x+3)(3x+1)(x+2)$ **b** $(2x+4)(2x+1)(x-2)$
   c $(-x+1)(3x-1)(4x+3)$ **d** $(-2x-3)(-x+1)(-x+5)$
   e $(2x^2 - 3x + 1)(-x + 4)$ **f** $(4x-1)(-3x^2 - 3x - 2)$

11 ALGEBRAIC REPRESENTATION AND MANIPULATION

Further factorisation

Factorisation by grouping

> **Worked examples**

Factorise the following expressions:

a $6x + 3 + 2xy + y$
  $= 3(2x + 1) + y(2x + 1)$
  $= (3 + y)(2x + 1)$
  Note that $(2x + 1)$ was a common factor of both terms.

b $ax + ay - bx - by$
  $= a(x + y) - b(x + y)$
  $= (a - b)(x + y)$

c $2x^2 - 3x + 2xy - 3y$
  $= x(2x - 3) + y(2x - 3)$
  $= (x + y)(2x - 3)$

Exercise 11.8 Factorise the following by grouping:

1 a $ax + bx + ay + by$ b $ax + bx - ay - by$
   c $3m + 3n + mx + nx$ d $4m + mx + 4n + nx$
   e $3m + mx - 3n - nx$ f $6x + xy + 6z + zy$

2 a $pr - ps + qr - qs$ b $pq - 4p + 3q - 12$
   c $pq + 3q - 4p - 12$ d $rs + rt + 2ts + 2t^2$
   e $rs - 2ts + rt - 2t^2$ f $ab - 4cb + ac - 4c^2$

3 a $xy + 4y + x^2 + 4x$ b $x^2 - xy - 2x + 2y$
   c $ab + 3a - 7b - 21$ d $ab - b - a + 1$
   e $pq - 4p - 4q + 16$ f $mn - 5m - 5n + 25$

4 a $mn - 2m - 3n + 6$ b $mn - 2mr - 3rn + 6r^2$
   c $pr - 4p - 4qr + 16q$ d $ab - a - bc + c$
   e $x^2 - 2xz - 2xy + 4yz$ f $2a^2 + 2ab + b^2 + ab$

Difference of two squares

On expanding
  $(x + y)(x - y)$
  $= x^2 - xy + xy - y^2$
  $= x^2 - y^2$

The reverse is that $x^2 - y^2$ factorises to $(x + y)(x - y)$. x^2 and y^2 are both square and therefore $x^2 - y^2$ is known as the **difference of two squares**.

Worked examples

a $p^2 - q^2$
   $= (p+q)(p-q)$

b $4a^2 - 9b^2$
   $= (2a)^2 - (3b)^2$
   $= (2a+3b)(2a-3b)$

c $(mn)^2 - 25k^2$
   $= (mn)^2 - (5k)^2$
   $= (mn+5k)(mn-5k)$

d $4x^2 - (9y)^2$
   $= (2x)^2 - (9y)^2$
   $= (2x+9y)(2x-9y)$

Exercise 11.9

Factorise the following:

1. a $a^2 - b^2$ b $m^2 - n^2$ c $x^2 - 25$
   d $m^2 - 49$ e $81 - x^2$ f $100 - y^2$

2. a $144 - y^2$ b $q^2 - 169$ c $m^2 - 1$
   d $1 - t^2$ e $4x^2 - y^2$ f $25p^2 - 64q^2$

3. a $9x^2 - 4y^2$ b $16p^2 - 36q^2$ c $64x^2 - y^2$
   d $x^2 - 100y^2$ e $(qr)^2 - 4p^2$ f $(ab)^2 - (cd)^2$

4. a $m^2n^2 - 9y^2$ b $\frac{1}{4}x^2 - \frac{1}{9}y^2$ c $(2x)^2 - (3y)^4$
   d $p^4 - q^4$ e $4m^4 - 36y^4$ f $16x^4 - 81y^4$

Evaluation

Once factorised, numerical expressions can be evaluated.

Worked examples

Evaluate the following expressions:

a $13^2 - 7^2$
   $= (13+7)(13-7)$
   $= 20 \times 6$
   $= 120$

b $6.25^2 - 3.75^2$
   $= (6.25+3.75)(6.25-3.75)$
   $= 10 \times 2.5$
   $= 25$

Exercise 11.10

By factorising, evaluate the following:

1. a $8^2 - 2^2$ b $16^2 - 4^2$ c $49^2 - 1$
   d $17^2 - 3^2$ e $88^2 - 12^2$ f $96^2 - 4^2$

2. a $45^2 - 25$ b $99^2 - 1$ c $27^2 - 23^2$
   d $66^2 - 34^2$ e $999^2 - 1$ f $225 - 8^2$

11 ALGEBRAIC REPRESENTATION AND MANIPULATION

Exercise 11.10 (cont)

3 a $8.4^2 - 1.6^2$ b $9.3^2 - 0.7^2$ c $42.8^2 - 7.2^2$

   d $\left(8\tfrac{1}{2}\right)^2 - \left(1\tfrac{1}{2}\right)^2$ e $\left(7\tfrac{3}{4}\right)^2 - \left(2\tfrac{1}{4}\right)^2$ f $5.25^2 - 4.75^2$

4 a $8.62^2 - 1.38^2$ b $0.9^2 - 0.1^2$ c $3^4 - 2^4$

   d $2^4 - 1$ e $1111^2 - 111^2$ f $2^8 - 2^5$

Factorising quadratic expressions

$x^2 + 5x + 6$ is known as a quadratic expression as the highest power of any of its terms is squared – in this case x^2.

It can be factorised by writing it as a product of two brackets.

➡ Worked examples

a Factorise $x^2 + 5x + 6$.

On setting up a 2 × 2 grid, some of the information can immediately be entered.

As there is only one term in x^2, this can be entered, as can the constant +6. The only two values which multiply to give x^2 are x and x. These too can be entered.

We now need to find two values which multiply to give +6 and which add to give +5x.

The only two values which satisfy both these conditions are +3 and +2.

Therefore $x^2 + 5x + 6 = (x + 3)(x + 2)$

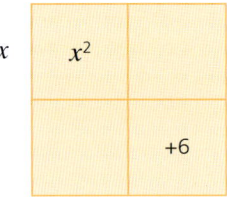

Factorising quadratic expressions

b Factorise $x^2 + 2x - 24$.

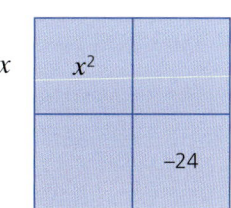

Therefore $x^2 + 2x - 24 = (x + 6)(x - 4)$

c Factorise $2x^2 + 11x + 12$.

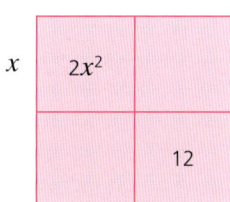 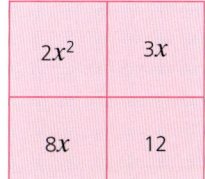

Therefore $2x^2 + 11x + 12 = (2x + 3)(x + 4)$

d Factorise $3x^2 + 7x - 6$.

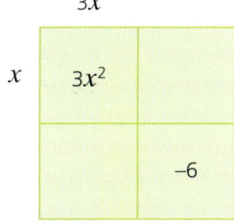 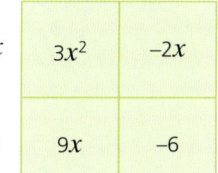

Therefore $3x^2 + 7x - 6 = (3x - 2)(x + 3)$

Exercise 11.11

Factorise the following quadratic expressions:

1.
   a $x^2 + 7x + 12$
   b $x^2 + 8x + 12$
   c $x^2 + 13x + 12$
   d $x^2 - 7x + 12$
   e $x^2 - 8x + 12$
   f $x^2 - 13x + 12$

2.
   a $x^2 + 6x + 5$
   b $x^2 + 6x + 8$
   c $x^2 + 6x + 9$
   d $x^2 + 10x + 25$
   e $x^2 + 22x + 121$
   f $x^2 - 13x + 42$

3.
   a $x^2 + 14x + 24$
   b $x^2 + 11x + 24$
   c $x^2 - 10x + 24$
   d $x^2 + 15x + 36$
   e $x^2 + 20x + 36$
   f $x^2 - 12x + 36$

4.
   a $x^2 + 2x - 15$
   b $x^2 - 2x - 15$
   c $x^2 + x - 12$
   d $x^2 - x - 12$
   e $x^2 + 4x - 12$
   f $x^2 - 15x + 36$

5.
   a $x^2 - 2x - 8$
   b $x^2 - x - 20$
   c $x^2 + x - 30$
   d $x^2 - x - 42$
   e $x^2 - 2x - 63$
   f $x^2 + 3x - 54$

6.
   a $2x^2 + 3x + 1$
   b $2x^2 + 7x + 6$
   c $2x^2 + x - 6$
   d $2x^2 - 7x + 6$
   e $3x^2 + 8x + 4$
   f $3x^2 + 11x - 4$
   g $4x^2 + 12x + 9$
   h $9x^2 - 6x + 1$
   i $6x^2 - x - 1$

11 ALGEBRAIC REPRESENTATION AND MANIPULATION

Rearrangement of complex formulas

→ Worked examples

Make the letters in red the subject of each formula:

a $\quad C = 2\pi r$

$\quad \dfrac{C}{2\pi} = r$

b $\quad A = \dfrac{h}{2}(a+b)$

$\quad 2A = h(a+b)$

$\quad \dfrac{2A}{h} = a+b$

$\quad \dfrac{2A}{h} - a = b$

c $\quad x^2 + y^2 = h^2$

$\quad y^2 = h^2 - x^2$

Note: not $y = h - x \longrightarrow$ $\quad y = \pm\sqrt{h^2 - x^2}$

d $\quad f = \sqrt{\dfrac{x}{k}}$

$\quad f^2 = \dfrac{x}{k}$

$\quad f^2 k = x$

e $\quad m = 3a\sqrt{\dfrac{p}{x}}$

Square both sides $\longrightarrow$ $\quad m^2 = \dfrac{9a^2 p}{x}$

$\quad m^2 x = 9a^2 p$

$\quad x = \dfrac{9a^2 p}{m^2}$

f $\quad A = \dfrac{y+x}{p+q^2}$

$\quad A(p+q^2) = y+x$

$\quad p+q^2 = \dfrac{y+x}{A}$

$\quad q^2 = \dfrac{y+x}{A} - p$

$\quad q = \pm\sqrt{\dfrac{y+x}{A} - p}$

g $\quad \dfrac{x}{4} = \dfrac{a-b}{3x}$

$\quad 3x^2 = 4(a-b)$

$\quad x^2 = \dfrac{4(a-b)}{3}$

$\quad x = \pm\sqrt{\dfrac{4(a-b)}{3}}$

h $\quad \dfrac{a}{bx+1} = \dfrac{b}{x}$

$\quad ax = b(bx+1)$

$\quad ax = b^2 x + b$

$\quad ax - b^2 x = b$

$\quad x(a - b^2) = b$

$\quad x = \dfrac{b}{a - b^2}$

Rearrangement of complex formulas

Exercise 11.12

1. In the formulas below, make x the subject:
   a. $P = 2mx$
   b. $\dfrac{P}{Q} = rx$

2. In the following questions, make the letter in red the subject of the formula:
   a. $v = u + at$
   b. $v^2 = u^2 + 2as$
   c. $s = ut + \frac{1}{2}at^2$
   d. $s = ut + \frac{1}{2}at^2$

Exercise 11.13

In the formulas below, make x the subject:

1. a. $T = 3x^2$
   b. $mx^2 = y^2$
   c. $x^2 + y^2 = p^2 - q^2$
   d. $m^2 + x^2 = y^2 - n^2$
   e. $p^2 - q^2 = 4x^2 - y^2$

2. a. $\dfrac{P}{Q} = rx^2$
   b. $\dfrac{P}{Q} = \dfrac{x^2}{r}$
   c. $\dfrac{m}{n} = \dfrac{1}{x^2}$
   d. $\dfrac{r}{st} = \dfrac{w}{x^2}$
   e. $\dfrac{p+q}{r} = \dfrac{w}{x^2}$

3. a. $\sqrt{x} = rp$
   b. $\dfrac{mn}{p} = \sqrt{x}$
   c. $g = \sqrt{\dfrac{k}{x}}$
   d. $r = 2\pi\sqrt{\dfrac{x}{g}}$
   e. $p^2 = \dfrac{4m^2 r}{x}$
   f. $p = 2m\sqrt{\dfrac{r}{x}}$

In the following questions, make the letter in red the subject of the formula:

4. a. $v^2 = u^2 + 2as$
   b. $s = ut + \frac{1}{2}at^2$

5. a. $A = \pi r \sqrt{s^2 + t^2}$
   b. $A = \pi r \sqrt{h^2 + r^2}$
   c. $\dfrac{1}{f} = \dfrac{1}{u} + \dfrac{1}{v}$
   d. $\dfrac{1}{f} = \dfrac{1}{u} + \dfrac{1}{v}$
   e. $t = 2\pi\sqrt{\dfrac{l}{g}}$
   f. $t = 2\pi\sqrt{\dfrac{l}{g}}$

6. a. $\dfrac{xt}{7} = \dfrac{p+2}{3x}$
   b. $\sqrt{a+2} = \dfrac{b-3}{\sqrt{a-2}}$

Exercise 11.14

1. The cost x of printing n newspapers is given by the formula $x = 1.50 + 0.05n$.
   a. Calculate the cost of printing 5000 newspapers.
   b. Make n the subject of the formula.
   c. How many newspapers can be printed for $25?

2. The formula $C = \frac{5}{9}(F - 32)$ can be used to convert temperatures in degrees Fahrenheit (°F) into degrees Celsius (°C).
   a. What temperature in °C is equivalent to 150°F?
   b. What temperature in °C is equivalent to 12°F?
   c. Make F the subject of the formula.
   d. Use your rearranged formula to find what temperature in °F is equivalent to 160°C.

11 ALGEBRAIC REPRESENTATION AND MANIPULATION

Exercise 11.14 (cont)

3 The height of Mount Kilimanjaro is given as 5900 m. The formula for the time taken, T hours, to climb to a height H metres is:

$$T = \frac{H}{1200} + k$$

where k is a constant.
   a Calculate the time taken, to the nearest hour, to climb to the top of the mountain if $k = 9.8$.
   b Make H the subject of the formula.
   c How far up the mountain, to the nearest 100 m, could you expect to be after 14 hours?

4 The **volume of a cylinder** is given by the formula $V = \pi r^2 h$, where h is the height of the cylinder and r is the radius.
   a Find the volume of a cylindrical post of length 7.5 m and a diameter of 30 cm.
   b Make r the subject of the formula.
   c A cylinder of height 75 cm has a volume of 6000 cm³, find its radius correct to 3 s.f.

5 The formula for the **volume V of a sphere** is given as $V = \frac{4}{3}\pi r^3$.
   a Find V if $r = 5$ cm.
   b Make r the subject of the formula.
   c Find the radius of a sphere of volume 2500 m³.

Algebraic fractions

Simplifying algebraic fractions

The rules for fractions involving algebraic terms are the same as those for numeric fractions. However, the actual calculations are often easier when using algebra.

> ### Worked examples
>
> a $\frac{3}{4} \times \frac{5}{7} = \frac{15}{28}$
>
> b $\frac{a}{c} \times \frac{b}{d} = \frac{ab}{cd}$
>
> c $\frac{\cancel{2}}{4} \times \frac{5}{\cancel{6}_2} = \frac{5}{8}$
>
> d $\frac{\cancel{a}}{c} \times \frac{b}{2\cancel{a}} = \frac{b}{2c}$
>
> e $\frac{\cancel{a}b}{e\cancel{c}} \times \frac{\cancel{c}d}{f\cancel{a}} = \frac{bd}{ef}$
>
> f $\frac{x^5}{x^3} = \frac{\cancel{x} \times \cancel{x} \times \cancel{x} \times x \times x}{\cancel{x} \times \cancel{x} \times \cancel{x}} = x^2$
>
> g $\frac{2b}{5} \div \frac{b}{7} = \frac{2b}{5} \times \frac{7}{b} = \frac{14\cancel{b}}{5\cancel{b}} = \frac{14}{5} = 2.8$

Exercise 11.15

Simplify the following **algebraic fractions**:

1 a $\frac{x}{y} \times \frac{p}{q}$
   b $\frac{x}{y} \times \frac{q}{x}$
   c $\frac{p}{q} \times \frac{q}{r}$

   d $\frac{ab}{c} \times \frac{d}{ab}$
   e $\frac{ab}{c} \times \frac{d}{ac}$
   f $\frac{p^2}{q^2} \times \frac{q^2}{p}$

Addition and subtraction of fractions

2 a $\dfrac{m^3}{m}$ b $\dfrac{r^7}{r^2}$ c $\dfrac{x^9}{x^3}$

   d $\dfrac{x^2 y^4}{xy^2}$ e $\dfrac{a^2 b^3 c^4}{ab^2 c}$ f $\dfrac{pq^2 r^4}{p^2 q^3 r}$

3 a $\dfrac{4ax}{2ay}$ b $\dfrac{12pq^2}{3p}$ c $\dfrac{15mn^2}{3mn}$

   d $\dfrac{24x^5 y^3}{8x^2 y^2}$ e $\dfrac{36p^2 qr}{12pqr}$ f $\dfrac{16m^2 n}{24m^3 n^2}$

4 a $\dfrac{2}{b} \times \dfrac{a}{3}$ b $\dfrac{4}{x} \times \dfrac{y}{2}$ c $\dfrac{8}{x} \times \dfrac{x}{4}$

   d $\dfrac{9y}{2} \times \dfrac{2x}{3}$ e $\dfrac{12x}{7} \times \dfrac{7}{4x}$ f $\dfrac{4x^3}{3y} \times \dfrac{9y^2}{2x^2}$

5 a $\dfrac{2ax}{3bx} \times \dfrac{4by}{a}$ b $\dfrac{3p^2}{2q} \times \dfrac{5q}{3p}$

   c $\dfrac{p^2 q}{rs} \times \dfrac{pr}{q}$ d $\dfrac{a^2 b}{fc^2} \times \dfrac{cd}{bd} \times \dfrac{ef^2}{ca^2}$

6 a $\dfrac{8x^2}{3} \div \dfrac{2x}{5}$ b $\dfrac{3b^3}{2} \div \dfrac{4b^2}{3}$

Addition and subtraction of fractions

In arithmetic it is easy to add or subtract fractions with the same denominator. It is the same process when dealing with algebraic fractions.

➜ Worked examples

a $\dfrac{4}{11} + \dfrac{3}{11}$ b $\dfrac{a}{11} + \dfrac{b}{11}$ c $\dfrac{4}{x} + \dfrac{3}{x}$

   $= \dfrac{7}{11}$ $= \dfrac{a+b}{11}$ $= \dfrac{7}{x}$

If the denominators are different, the fractions need to be changed to form fractions with the same denominator.

➜ Worked examples

a $\dfrac{2}{9} + \dfrac{1}{3}$ b $\dfrac{a}{9} + \dfrac{b}{3}$ c $\dfrac{4}{5a} + \dfrac{7}{10a}$

   $= \dfrac{2}{9} + \dfrac{3}{9}$ $= \dfrac{a}{9} + \dfrac{3b}{9}$ $= \dfrac{8}{10a} + \dfrac{7}{10a}$

   $= \dfrac{5}{9}$ $= \dfrac{a+3b}{9}$ $= \dfrac{15}{10a}$

                              $= \dfrac{3}{2a}$

11 ALGEBRAIC REPRESENTATION AND MANIPULATION

Similarly, with subtraction, the denominators need to be the same.

Worked examples

a $\dfrac{7}{a} - \dfrac{1}{2a}$
$= \dfrac{14}{2a} - \dfrac{1}{2a}$
$= \dfrac{13}{2a}$

b $\dfrac{p}{3} - \dfrac{q}{15}$
$= \dfrac{5p}{15} - \dfrac{q}{15}$
$= \dfrac{5p - q}{15}$

c $\dfrac{5}{3b} - \dfrac{8}{9b}$
$= \dfrac{15}{9b} - \dfrac{8}{9b}$
$= \dfrac{7}{9b}$

Exercise 11.16

Simplify the following fractions:

1. a $\dfrac{1}{7} + \dfrac{3}{7}$ b $\dfrac{a}{7} + \dfrac{b}{7}$ c $\dfrac{5}{13} + \dfrac{6}{13}$
   d $\dfrac{c}{13} + \dfrac{d}{13}$ e $\dfrac{x}{3} + \dfrac{y}{3} + \dfrac{z}{3}$ f $\dfrac{p^2}{5} + \dfrac{q^2}{5}$

2. a $\dfrac{5}{11} - \dfrac{2}{11}$ b $\dfrac{c}{11} - \dfrac{d}{11}$ c $\dfrac{6}{a} - \dfrac{2}{a}$
   d $\dfrac{2a}{3} - \dfrac{5b}{3}$ e $\dfrac{2x}{7} - \dfrac{3y}{7}$ f $\dfrac{3}{4x} - \dfrac{5}{4x}$

3. a $\dfrac{5}{6} - \dfrac{1}{3}$ b $\dfrac{5}{2a} - \dfrac{1}{a}$ c $\dfrac{2}{3c} + \dfrac{1}{c}$
   d $\dfrac{2}{x} + \dfrac{3}{2x}$ e $\dfrac{5}{2p} - \dfrac{1}{p}$ f $\dfrac{1}{w} - \dfrac{3}{2w}$

4. a $\dfrac{p}{4} - \dfrac{q}{12}$ b $\dfrac{x}{4} - \dfrac{y}{2}$ c $\dfrac{m}{3} - \dfrac{n}{9}$
   d $\dfrac{x}{12} - \dfrac{y}{6}$ e $\dfrac{r}{2} + \dfrac{m}{10}$ f $\dfrac{s}{3} - \dfrac{t}{15}$

5. a $\dfrac{3x}{4} - \dfrac{2x}{12}$ b $\dfrac{3x}{5} - \dfrac{2y}{15}$ c $\dfrac{3m}{7} + \dfrac{m}{14}$
   d $\dfrac{4m}{3p} - \dfrac{3m}{9p}$ e $\dfrac{4x}{3y} - \dfrac{5x}{6y}$ f $\dfrac{3r}{7s} + \dfrac{2r}{14s}$

Often one denominator is not a multiple of the other. In these cases the **lowest common multiple** of both denominators has to be found.

Worked examples

a $\dfrac{1}{4} + \dfrac{1}{3}$
$= \dfrac{3}{12} + \dfrac{4}{12}$
$= \dfrac{7}{12}$

b $\dfrac{1}{5} + \dfrac{2}{3}$
$= \dfrac{3}{15} + \dfrac{10}{15}$
$= \dfrac{13}{15}$

c $\dfrac{a}{3} + \dfrac{b}{4}$
$= \dfrac{4a}{12} + \dfrac{3b}{12}$
$= \dfrac{4a + 3b}{12}$

d $\dfrac{2a}{3} + \dfrac{3b}{5}$
$= \dfrac{10a}{15} + \dfrac{9b}{15}$
$= \dfrac{10a + 9b}{15}$

Simplifying complex algebraic fractions

Exercise 11.17 Simplify the following fractions:

1. a) $\frac{a}{2}+\frac{b}{3}$ b) $\frac{a}{3}+\frac{b}{5}$ c) $\frac{p}{4}+\frac{q}{7}$
   d) $\frac{2a}{5}+\frac{b}{3}$ e) $\frac{x}{4}+\frac{5y}{9}$ f) $\frac{2x}{7}+\frac{2y}{5}$

2. a) $\frac{a}{2}-\frac{a}{3}$ b) $\frac{a}{3}-\frac{a}{5}$ c) $\frac{p}{4}+\frac{p}{7}$
   d) $\frac{2a}{5}+\frac{a}{3}$ e) $\frac{x}{4}+\frac{5x}{9}$ f) $\frac{2x}{7}+\frac{2x}{5}$

3. a) $\frac{3m}{5}-\frac{m}{2}$ b) $\frac{3r}{5}-\frac{r}{2}$ c) $\frac{5x}{4}-\frac{3x}{2}$
   d) $\frac{2x}{7}+\frac{3x}{4}$ e) $\frac{11x}{2}-\frac{5x}{3}$ f) $\frac{2p}{3}-\frac{p}{2}$

4. a) $p-\frac{p}{2}$ b) $c-\frac{c}{3}$ c) $x-\frac{x}{5}$
   d) $m-\frac{2m}{3}$ e) $q-\frac{4q}{5}$ f) $w-\frac{3w}{4}$

5. a) $2m-\frac{m}{2}$ b) $3m-\frac{2m}{3}$ c) $2m-\frac{5m}{2}$
   d) $4m-\frac{3m}{2}$ e) $2p-\frac{5p}{3}$ f) $6q-\frac{6q}{7}$

6. a) $p-\frac{p}{r}$ b) $\frac{x}{y}+x$ c) $m+\frac{m}{n}$
   d) $\frac{a}{b}+a$ e) $2x-\frac{x}{y}$ f) $2p-\frac{3p}{q}$

7. a) $\frac{a}{3}+\frac{a+4}{2}$ b) $\frac{2b}{5}+\frac{b-4}{3}$
   c) $\frac{c+2}{4}-\frac{2-c}{2}$ d) $\frac{2(d-3)}{7}-\frac{3(2-d)}{2}$

Simplifying complex algebraic fractions

With more complex algebraic fractions, the method of getting a common denominator is still required.

> **Worked examples**

a) $\frac{2}{x+1}+\frac{3}{x+2}$

$=\frac{2(x+2)}{(x+1)(x+2)}+\frac{3(x+1)}{(x+1)(x+2)}$

$=\frac{2(x+2)+3(x+1)}{(x+1)(x+2)}$

$=\frac{2x+4+3x+3}{(x+1)(x+2)}$

$=\frac{5x+7}{(x+1)(x+2)}$

b) $\frac{5}{p+3}-\frac{3}{p-5}$

$=\frac{5(p-5)}{(p+3)(p-5)}-\frac{3(p+3)}{(p+3)(p-5)}$

$=\frac{5(p-5)-3(p+3)}{(p+3)(p-5)}$

$=\frac{5p-25-3p-9}{(p+3)(p-5)}$

$=\frac{2p-34}{(p+3)(p-5)}$

11 ALGEBRAIC REPRESENTATION AND MANIPULATION

c $\dfrac{x^2 - 2x}{x^2 + x - 6}$

$= \dfrac{x\cancel{(x-2)}}{(x+3)\cancel{(x-2)}}$

$= \dfrac{x}{x+3}$

d $\dfrac{x^2 - 3x}{x^2 + 2x - 15}$

$= \dfrac{x\cancel{(x-3)}}{\cancel{(x-3)}(x+5)}$

$= \dfrac{x}{x+5}$

Exercise 11.18

Simplify the following algebraic fractions:

1 a $\dfrac{1}{x+1} + \dfrac{2}{x+2}$ b $\dfrac{3}{m+2} - \dfrac{2}{m-1}$ c $\dfrac{2}{p-3} + \dfrac{1}{p-2}$

   d $\dfrac{3}{w-1} - \dfrac{2}{w+3}$ e $\dfrac{4}{y+4} - \dfrac{4}{y+1}$ f $\dfrac{2}{m-2} - \dfrac{3}{m+3}$

2 a $\dfrac{x(x-4)}{(x-4)(x+2)}$ b $\dfrac{y(y-3)}{(y+3)(y-3)}$ c $\dfrac{(m+2)(m-2)}{(m-2)(m-3)}$

   d $\dfrac{p(p+5)}{(p-5)(p+5)}$ e $\dfrac{m(2m+3)}{(m+4)(2m+3)}$ f $\dfrac{(m+1)(m-1)}{(m+2)(m-1)}$

3 a $\dfrac{x^2 - 5x}{(x+3)(x-5)}$ b $\dfrac{x^2 - 3x}{(x+4)(x-3)}$ c $\dfrac{y^2 - 7y}{(y-7)(y-3)}$

   d $\dfrac{x(x-1)}{x^2 + 2x - 3}$ e $\dfrac{x(x+2)}{x^2 + 4x + 4}$ f $\dfrac{x(x+4)}{x^2 + 5x + 4}$

4 a $\dfrac{x^2 - x}{x^2 - 1}$ b $\dfrac{x^2 + 2x}{x^2 + 5x + 6}$ c $\dfrac{x^2 + 4x}{x^2 + x - 12}$

   d $\dfrac{x^2 - 5x}{x^2 - 3x - 10}$ e $\dfrac{x^2 + 3x}{x^2 - 9}$ f $\dfrac{x^2 - 7x}{x^2 - 49}$

Student assessment 1

1 Expand the following and simplify where possible:
   a $5(2a - 6b + 3c)$
   b $3x(5x - 9)$
   c $-5y(3xy + y^2)$
   d $3x^2(5xy + 3y^2 - x^3)$
   e $5p - 3(2p - 4)$
   f $4m(2m - 3) + 2(3m^2 - m)$
   g $\dfrac{1}{3}(6x - 9) + \dfrac{1}{4}(8x + 24)$
   h $\dfrac{m}{4}(6m - 8) + \dfrac{m}{2}(10m - 2)$

2 Factorise the following:
   a $12a - 4b$
   b $x^2 - 4xy$
   c $8p^3 - 4p^2q$
   d $24xy - 16x^2y + 8xy^2$

Simplifying complex algebraic fractions

3 If $x = 2$, $y = -3$ and $z = 4$, evaluate the following:
  a $2x + 3y - 4z$
  b $10x + 2y^2 - 3z$
  c $z^2 - y^3$
  d $(x + y)(y - z)$
  e $z^2 - x^2$
  f $(z + x)(z - x)$

4 Rearrange the following formulas to make the green letter the subject:
  a $x = 3p + q$
  b $3m - 5n = 8r$
  c $2m = \frac{3y}{t}$
  d $x(w + y) = 2y$
  e $\frac{xy}{2p} = \frac{rs}{t}$
  f $\frac{x+y}{w} = m + n$

Student assessment 2

1 Expand the following and simplify where possible:
  a $3(2x - 3y + 5z)$
  b $4p(2m - 7)$
  c $-4m(2mn - n^2)$
  d $4p^2(5pq - 2q^2 - 2p)$
  e $4x - 2(3x + 1)$
  f $4x(3x - 2) + 2(5x^2 - 3x)$
  g $\frac{1}{5}(15x - 10) - \frac{1}{3}(9x - 12)$
  h $\frac{x}{2}(4x - 6) + \frac{x}{4}(2x + 8)$

2 Factorise the following:
  a $16p - 8q$
  b $p^2 - 6pq$
  c $5p^2q - 10pq^2$
  d $9pq - 6p^2q + 12q^2p$

3 If $a = 4$, $b = 3$ and $c = -2$, evaluate the following:
  a $3a - 2b + 3c$
  b $5a - 3b^2$
  c $a^2 + b^2 + c^2$
  d $(a + b)(a - b)$
  e $a^2 - b^2$
  f $b^3 - c^3$

4 Rearrange the following formulas to make the green letter the subject:
  a $p = 4m + n$
  b $4x - 3y = 5z$
  c $2x = \frac{3y}{5p}$
  d $m(x + y) = 3w$
  e $\frac{pq}{4r} = \frac{mn}{t}$
  f $\frac{p+q}{r} = m - n$

11 ALGEBRAIC REPRESENTATION AND MANIPULATION

Student assessment 3

1. Expand the following and simplify where possible:
   a. $(x-4)(x+2)$
   b. $(x-8)^2$
   c. $(x+y)^2$
   d. $(x-11)(x+11)$
   e. $(3x-2)(2x-3)$
   f. $(5-3x)^2$

2. a. Factorise the following fully:
      i. $pq - 3rq + pr - 3r^2$
      ii. $1 - t^4$
   b. By factorising, evaluate the following:
      i. $875^2 - 125^2$
      ii. $7.5^2 - 2.5^2$

3. Factorise the following:
   a. $x^2 - 4x - 77$
   b. $x^2 - 6x + 9$
   c. $x^2 - 144$
   d. $3x^2 + 3x - 18$
   e. $2x^2 + 5x - 12$
   f. $4x^2 - 20x + 25$

4. Make the letter in green the subject of the formula:
   a. $mf^2 = p$
   b. $m = 5t^2$
   c. $A = \pi r \sqrt{p + q}$
   d. $\frac{1}{x} + \frac{1}{y} = \frac{1}{t}$

5. Simplify the following algebraic fractions:
   a. $\frac{x^7}{x^3}$
   b. $\frac{mn}{p} \times \frac{pq}{m}$
   c. $\frac{(y^3)^3}{(y^2)^3}$
   d. $\frac{28pq^2}{7pq^3}$
   e. $\frac{m^2n}{2} \div \frac{m^2}{n^2}$
   f. $\frac{7b^3}{c} \div \frac{4b^2}{3c^3}$

6. Simplify the following algebraic fractions:
   a. $\frac{m}{11} + \frac{3m}{11} - \frac{2m}{11}$
   b. $\frac{3p}{8} - \frac{9p}{16}$
   c. $\frac{4x}{3y} - \frac{7x}{12y}$
   d. $\frac{3m}{15p} + \frac{4n}{5p} - \frac{11n}{30p}$
   e. $\frac{2(y+4)}{3} - (y-2)$
   f. $3(y+2) - \frac{2y+3}{2}$

7. Simplify the following:
   a. $\frac{4}{(x-5)} + \frac{3}{(x-2)}$
   b. $\frac{a^2 - b^2}{(a+b)^2}$
   c. $\frac{x-2}{x^2+x-6}$

Simplifying complex algebraic fractions

Student assessment 4

1. The volume V of a cylinder is given by the formula $V = \pi r^2 h$, where h is the height of the cylinder and r is the radius.
   a. Find the volume of a cylindrical post 6.5 m long and with a diameter of 20 cm.
   b. Make r the subject of the formula.
   c. A cylinder of height 60 cm has a volume of 5500 cm³, find its radius correct to 3 s.f.

2. The formula for the surface area of a closed cylinder is $A = 2\pi r(r + h)$, where r is the radius of the cylinder and h is its height.
   a. Find the surface area of a cylinder of radius 12 cm and height 20 cm, giving your answer to 3 s.f.
   b. Rearrange the formula to make h the subject.
   c. What is the height of a cylinder of surface area 500 cm² and radius 5 cm? Give your answer to 3 s.f.

3. The formula for finding the length d of the body diagonal of a cuboid whose dimensions are x, y and z is:
   $$d = \sqrt{x^2 + y^2 + z^2}$$
   a. Find d when $x = 2$, $y = 3$ and $z = 4$.
   b. How long is the body diagonal of a block of concrete in the shape of a rectangular prism of dimensions 2 m, 3 m and 75 cm?
   c. Rearrange the formula to make x the subject.
   d. Find x when $d = 0.86$, $y = 0.25$ and $z = 0.41$.

4. A pendulum of length l metres takes T seconds to complete one full oscillation. The formula for T is:
   $$T = 2\pi\sqrt{\frac{l}{g}}$$
   where g m/s² is the acceleration due to gravity.
   a. Find T if $l = 5$ and $g = 10$.
   b. Rearrange the formula to make l the subject.
   c. How long is a pendulum which takes 3 seconds for one oscillation, if $g = 10$?

> **Note**
>
> The body diagonal of a cuboid is the straight line connecting any of its two non-adjacent vertices.
>
>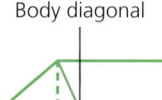
> Body diagonal
>
>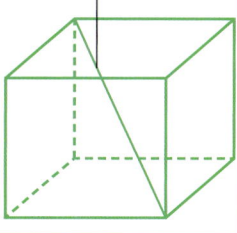

12 Algebraic indices

In Chapter 7, you saw how numbers can be expressed using indices. For example, $5 \times 5 \times 5 = 125$, therefore $125 = 5^3$. The 3 is called the **index**. **Indices** is the plural of index.

Three laws of indices were introduced:

1. $a^m \times a^n = a^{m+n}$
2. $a^m \div a^n$ or $\dfrac{a^m}{a^n} = a^{m-n}$
3. $(a^m)^n = a^{mn}$

Positive indices

➤ Worked examples

a Simplify $d^3 \times d^4$

$d^3 \times d^4 = d^{(3+4)}$
$= d^7$

b Simplify $\dfrac{(p^2)^4}{p^2 \times p^4}$

$\dfrac{(p^2)^4}{p^2 \times p^4} = \dfrac{p^{2 \times 4}}{p^{2+4}}$
$= \dfrac{p^8}{p^6}$
$= p^{8-6}$
$= p^2$

Exercise 12.1

1 Simplify the following:
   a $c^5 \times c^3$
   b $m^4 \div m^2$
   c $(b^3)^5 \div b^6$
   d $\dfrac{m^4 n^9}{mn^3}$
   e $\dfrac{6a^6 b^4}{3a^2 b^3}$
   f $\dfrac{12x^5 y^7}{4x^2 y^5}$
   g $\dfrac{4u^3 v^6}{8u^2 y^3}$
   h $\dfrac{3x^6 y^5 z^3}{9x^4 y^2 z}$

2 Simplify the following:
   a $4a^2 \times 3a^3$
   b $2a^2 b \times 4a^3 b^2$
   c $(2p^2)^3$
   d $(4m^2 n^3)^2$
   e $(5p^2)^2 \times (2p^3)^3$
   f $(4m^2 n^2) \times (2mn^3)^3$
   g $\dfrac{(6x^2 y^4)^2 \times (2xy)^3}{12xy^6 y^8}$
   h $(ab)^d \times (ab)^e$

128

The zero index

As shown in Chapter 7, the zero index indicates that a number or algebraic term is raised to the power of zero. A term raised to the power of zero is always equal to 1. This is shown below.

$$a^m \div a^n = a^{m-n} \qquad \text{therefore } \frac{a^m}{a^m} = a^{m-m}$$
$$= a^0$$

However, $\frac{a^m}{a^m} = 1$

therefore $a^0 = 1$

Negative indices

A negative index indicates that a number or an algebraic term is being raised to a negative power, e.g. a^{-4}.

As shown in Chapter 7, one law of indices states that:
$a^{-m} = \frac{1}{a^m}$. This is proved as follows.

$$a^{-m} = a^{0-m}$$
$$= \frac{a^0}{a^m} \text{ (from the second law of indices)}$$
$$= \frac{1}{a^m}$$

therefore $a^{-m} = \frac{1}{a^m}$

Exercise 12.2

1 Simplify the following:
   a $c^3 \times c^0$
   b $g^{-2} \times g^3 \div g^0$
   c $(p^0)^3 (q^2)^{-1}$
   d $(m^3)^3 (m^{-2})^5$

2 Simplify the following:
   a $\frac{a^{-3} \times a^5}{(a^2)^0}$
   b $\frac{(r^3)^{-2}}{(p^{-2})^3}$
   c $(t^3 \div t^{-5})^2$
   d $\frac{m^0 \div m^{-6}}{(m^{-1})^3}$

Fractional indices

It was shown in Chapter 7 that $16^{\frac{1}{2}} = \sqrt{16}$ and that $27^{\frac{1}{3}} = \sqrt[3]{27}$. This can be applied to algebraic indices too.

In general:

$a^{\frac{1}{n}} = \sqrt[n]{a}$

$a^{\frac{m}{n}} = \sqrt[n]{a^m}$ or $(\sqrt[n]{a})^m$

12 ALGEBRAIC INDICES

The last rule can be proved as shown below.

Using the laws of indices:

$a^{\frac{m}{n}}$ can be written as $(a^m)^{\frac{1}{n}}$ which in turn can be written as $\sqrt[n]{a^m}$.

Similarly:

$a^{\frac{m}{n}}$ can be written as $(a^{\frac{1}{n}})^m$ which in turn can be written as $(\sqrt[n]{a})^m$.

➡ Worked examples

a Express $(\sqrt[3]{a})^4$ in the form $a^{\frac{m}{n}}$

$(\sqrt[3]{a}) = a^{\frac{1}{3}}$

Therefore $(\sqrt[3]{a})^4 = (a^{\frac{1}{3}})^4 = a^{\frac{4}{3}}$

b Express $b^{\frac{2}{5}}$ in the form $(\sqrt[n]{b})^m$

$b^{\frac{2}{5}}$ can be expressed as $(b^{\frac{1}{5}})^2$

$b^{\frac{1}{5}} = \sqrt[5]{b}$

Therefore $b^{\frac{2}{5}} = (b^{\frac{1}{5}})^2 = (\sqrt[5]{b})^2$

c Simplify $\dfrac{p^{\frac{1}{2}} \times p^{\frac{1}{3}}}{p}$

Using the laws of indices, the numerator $p^{\frac{1}{2}} \times p^{\frac{1}{3}}$ can be simplified to $p^{(\frac{1}{2}+\frac{1}{3})} = p^{\frac{5}{6}}$.

Therefore $\dfrac{p^{\frac{5}{6}}}{p}$ can now be written as $p^{\frac{5}{6}} \times p^{-1}$

Using the laws of indices again, this can be simplified as $p^{(\frac{5}{6}-1)} = p^{-\frac{1}{6}}$

Therefore $\dfrac{p^{\frac{1}{2}} \times p^{\frac{1}{3}}}{p} = p^{-\frac{1}{6}}$

Other possible simplifications are $(\sqrt[6]{p})^{-1}$ or $\dfrac{1}{(\sqrt[6]{p})}$

Exercise 12.3

1 Rewrite the following in the form $a^{\frac{m}{n}}$:
   a $(\sqrt[5]{a})^3$ **b** $(\sqrt[6]{a})^2$ **c** $(\sqrt[4]{a})^4$ **d** $(\sqrt[7]{a})^3$

2 Rewrite the following in the form $(\sqrt[n]{b})^m$:
   a $b^{\frac{2}{7}}$ **b** $b^{\frac{8}{3}}$ **c** $b^{-\frac{2}{5}}$ **d** $b^{-\frac{4}{3}}$

3 Simplify the following algebraic expressions, giving your answer in the form $a^{\frac{m}{n}}$:
   a $a^{\frac{1}{2}} \times a^{\frac{1}{4}}$ **b** $a^{\frac{2}{5}} \times a^{-\frac{1}{4}}$ **c** $\dfrac{\sqrt{a}}{a^{-2}}$ **d** $\dfrac{\sqrt[3]{a}}{a}$

Fractional indices

4 Simplify the following algebraic expressions, giving your answer in the form $(\sqrt[n]{b})^m$:

a $\dfrac{\sqrt{b} \times b^{\frac{1}{4}}}{b^{-\frac{1}{5}}}$
b $\dfrac{b^{-\frac{1}{3}} \times \sqrt[3]{b}}{b^{\frac{2}{5}} \times b}$
c $\dfrac{b^3 \times b^{-\frac{1}{3}}}{b^{-2}}$
d $\dfrac{b^{-2} \times \sqrt[3]{b}}{\sqrt{b} \times (\sqrt[3]{b})^{-1}}$

5 Simplify the following:

a $\dfrac{1}{3} x^{\frac{1}{2}} \div 4 x^{-2}$
b $\dfrac{2}{5} y^{\frac{1}{3}} \times 5 y^{-\frac{2}{3}}$
c $\left(2 p^{-\frac{1}{4}}\right)^2 \div \dfrac{1}{2} p^2$
d $3 x^{-\frac{2}{3}} \div \dfrac{2}{3} x^{-\frac{1}{3}}$

Student assessment 1

1 Simplify the following using indices:
a $a \times a \times a \times b \times b$
b $d \times d \times e \times e \times e \times e \times e$

2 Write the following out in full:
a m^3
b r^4

3 Simplify the following using indices:
a $a^4 \times a^3$
b $p^3 \times p^2 \times q^4 \times q^5$
c $\dfrac{b^7}{b^4}$
d $\dfrac{(e^4)^5}{e^{14}}$

4 Simplify the following:
a $r^4 \times t^0$
b $\dfrac{(a^3)^0}{b^2}$
c $\dfrac{(m^0)^5}{n^{-3}}$

5 Simplify the following:
a $\dfrac{(p^2 \times p^{-5})^2}{p^3}$
b $\dfrac{(h^{-2} \times h^{-5})^{-1}}{h^0}$

Student assessment 2

1 Rewrite the following in the form $a^{\frac{m}{n}}$:
a $(\sqrt[8]{a})$
b $(\sqrt[5]{a})^{-2}$

2 Rewrite the following in the form $(\sqrt[n]{b^m})$:
a $b^{\frac{4}{9}}$
b $b^{-\frac{2}{3}}$

3 Simplify the following algebraic expressions, giving your answer in the form $a^{\frac{m}{n}}$:
a $a^{\frac{1}{3}} \times a^{\frac{3}{2}}$
b $\dfrac{\sqrt[3]{a}}{a^{-\frac{5}{6}}} \times a^2$

4 Simplify the following algebraic expressions, giving your answer in the form $(\sqrt[n]{t})^m$:
a $\dfrac{\sqrt{t} \times t^{\frac{2}{3}}}{t^{-\frac{1}{3}}}$
b $\dfrac{\sqrt[3]{t}}{t^2 \times t^{-\frac{2}{5}}}$

13 Equations and inequalities

An equation is formed when the value of an unknown quantity is needed.

Derive and solve linear equations with one unknown

Worked examples

Solve the following **linear equations**:

a) $3x + 8 = 14$
$3x = 6$
$x = 2$

b) $12 = 20 + 2x$
$-8 = 2x$
$-4 = x$

c) $3(p + 4) = 21$
$3p + 12 = 21$
$3p = 9$
$p = 3$

d) $4(x - 5) = 7(2x - 5)$
$4x - 20 = 14x - 35$
$4x + 15 = 14x$
$15 = 10x$
$1.5 = x$

e) $6 = \frac{2x}{x - 4}$
$6(x - 4) = 2x$
$6x - 24 = 2x$
$4x - 24 = 0$
$4x = 24$
$x = 6$

f) $\frac{-x}{2(x - 4)} = -3$
$\frac{-x}{2x - 8} = -3$
$-x = -3(2x - 8)$
$-x = -6x + 24$
$5x = 24$
$x = \frac{24}{5}$

Exercise 13.1

Solve the following linear equations:

1.
   a) $3x = 2x - 4$
   b) $5y = 3y + 10$
   c) $2y - 5 = 3y$
   d) $p - 8 = 3p$
   e) $3y - 8 = 2y$
   f) $7x + 11 = 5x$

2.
   a) $3x - 9 = 4$
   b) $4 = 3x - 11$
   c) $6x - 15 = 3x + 3$
   d) $4y + 5 = 3y - 3$
   e) $8y - 31 = 13 - 3y$
   f) $4m + 2 = 5m - 8$

3.
   a) $7m - 1 = 5m + 1$
   b) $5p - 3 = 3 + 3p$
   c) $12 - 2k = 16 + 2k$
   d) $6x + 9 = 3x - 54$
   e) $8 - 3x = 18 - 8x$
   f) $2 - y = y - 4$

4.
   a) $\frac{x}{2} = 3$
   b) $\frac{1}{2}y = 7$
   c) $\frac{x}{4} = 1$
   d) $\frac{1}{4}m = 3$
   e) $7 = \frac{x}{5}$
   f) $4 = \frac{1}{5}p$

5 a $\frac{x}{3} - 1 = 4$ b $\frac{x}{5} + 2 = 1$

c $\frac{2}{3}x = 5$ d $\frac{3}{4}x = 6$

e $\frac{1}{5}x = \frac{1}{2}$ f $\frac{2x}{5} = 4$

6 a $\frac{x+1}{2} = 3$ b $4 = \frac{x-2}{3}$

c $\frac{x-10}{3} = 4$ d $8 = \frac{5x-1}{3}$

e $\frac{2(x-5)}{3} = 2$ f $\frac{3(x-2)}{4} = 4x - 8$

7 a $6 = \frac{2(y-1)}{3}$ b $2(x+1) = 3(x-5)$

c $5(x-4) = 3(x+2)$ d $\frac{3+y}{2} = \frac{y+1}{4}$

e $\frac{7+2x}{3} = \frac{9x-1}{7}$ f $\frac{2x+3}{4} = \frac{4x-2}{6}$

8 a $\frac{14}{2x-3} = 2$ b $\frac{13}{3(x+5)} = \frac{1}{3}$

c $\frac{1}{x-1} - \frac{2}{x+4} = 0$ d $\frac{1}{x+2} - \frac{4}{x+5} = 0$

e $\frac{1}{x+3} - \frac{2}{3x} = 0$ f $\frac{5}{2x+6} = \frac{15}{44x-1}$

Constructing expressions and equations

In many cases, when dealing with the practical applications of mathematics, equations need to be constructed first before they can be solved. Often the information is either given within the context of a problem or in a diagram.

➡ Worked examples

> **Note**
>
> All diagrams are not drawn to scale.

a i Write an expression for the sum of the angles in the triangle (left).
   $(x + 30) + (x - 30) + 90$

   ii Find the size of each of the angles in the triangle by constructing an equation and solving it to find the value of x.
   The sum of the angles of a triangle is 180°.

   $(x + 30) + (x - 30) + 90 = 180$
   $2x + 90 = 180$
   $2x = 90$
   $x = 45$

The three angles are therefore: 90°, $x + 30 = 75°$, $x - 30 = 15°$.
Check: $90° + 75° + 15° = 180°$.

13 EQUATIONS AND INEQUALITIES

b **i** Write an expression for the sum of the angles in the quadrilateral (below).

$$4x + 30 + 3x + 10 + 3x + 2x + 20$$

ii Find the size of each of the angles in the quadrilateral by constructing an equation and solving it to find the value of x.
The sum of the angles of a quadrilateral is $360°$.

$$4x + 30 + 3x + 10 + 3x + 2x + 20 = 360$$
$$12x + 60 = 360$$
$$12x = 300$$
$$x = 25$$

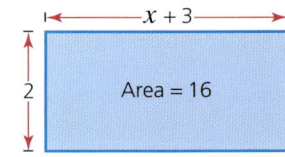

The angles are:

$$4x + 30 = (4 \times 25) + 30 = 130°$$
$$3x + 10 = (3 \times 25) + 10 = 85°$$
$$3x = 3 \times 25 = 75°$$
$$2x + 20 = (2 \times 25) + 20 = 70°$$
$$\text{Total} = 360°$$

c Construct an equation and solve it to find the value of x in the diagram (right).

Area of rectangle = base × height
$$2(x + 3) = 16$$
$$2x + 6 = 16$$
$$2x = 10$$
$$x = 5$$

Exercise 13.2

In Questions 1–3:
i write an expression for the sum of the angles in each case, giving your answer in its simplest form,
ii construct an equation in terms of x,
iii solve the equation,
iv calculate the size of each of the angles,
v check your answers.

1

a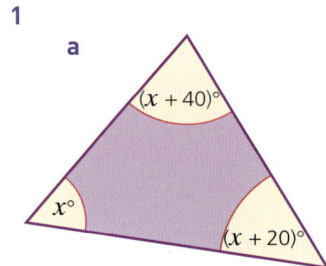

b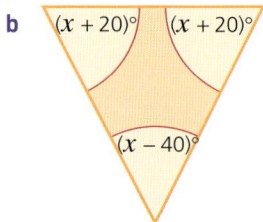

c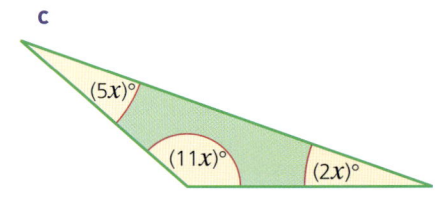

Constructing expressions and equations

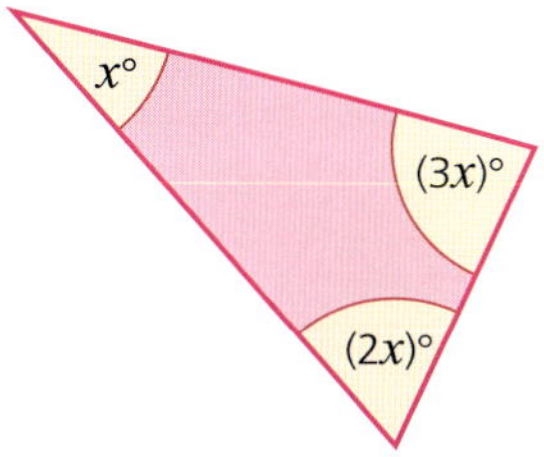

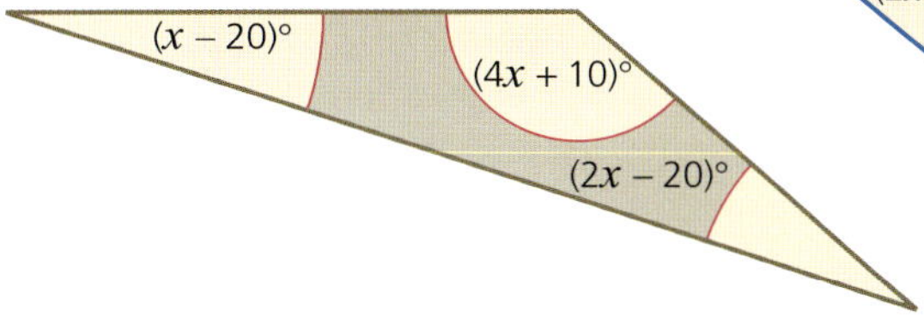

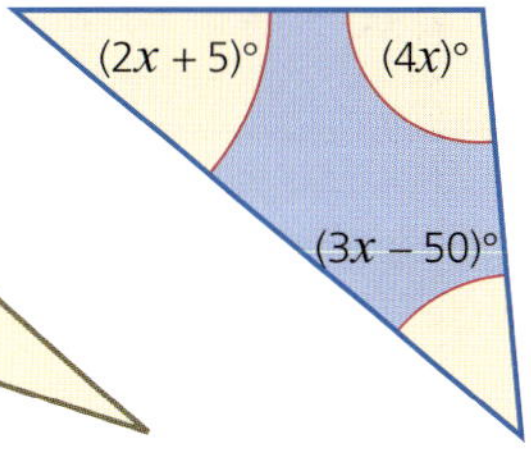

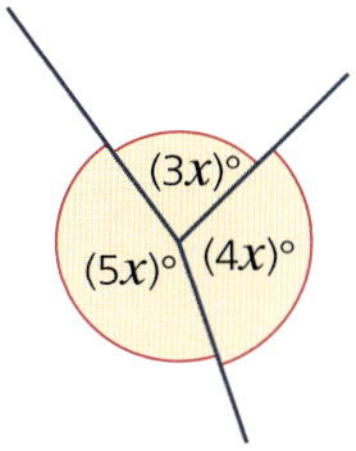

d Triangle with angles $x°$, $(3x)°$, $(2x)°$

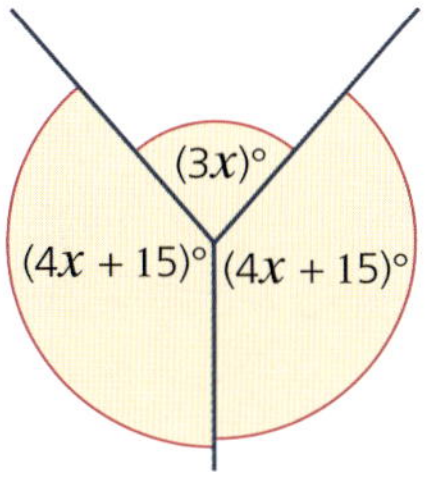

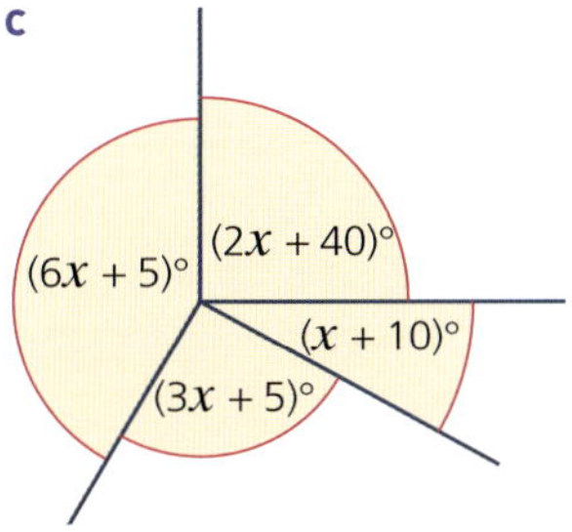

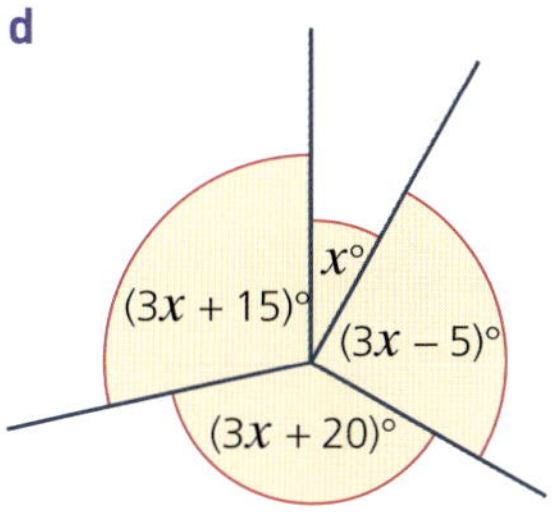

e Triangle with angles $(x-20)°$, $(4x+10)°$, $(2x-20)°$

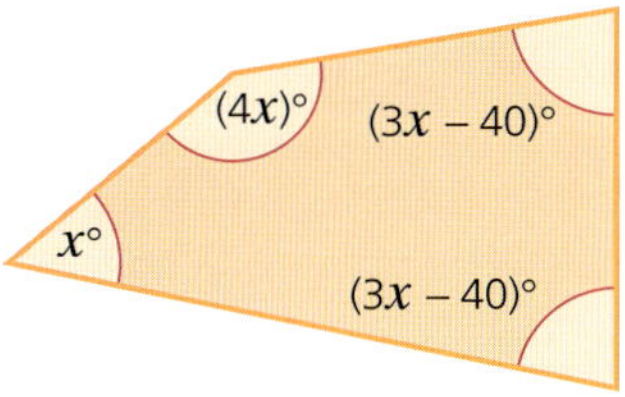

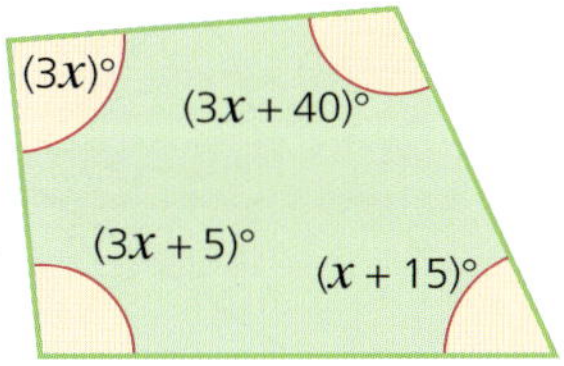

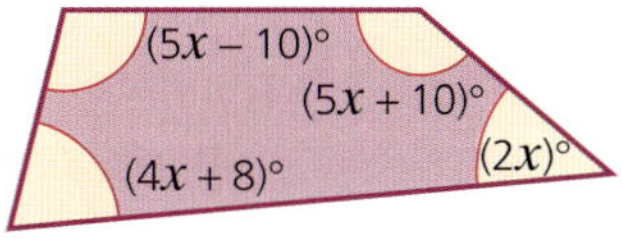

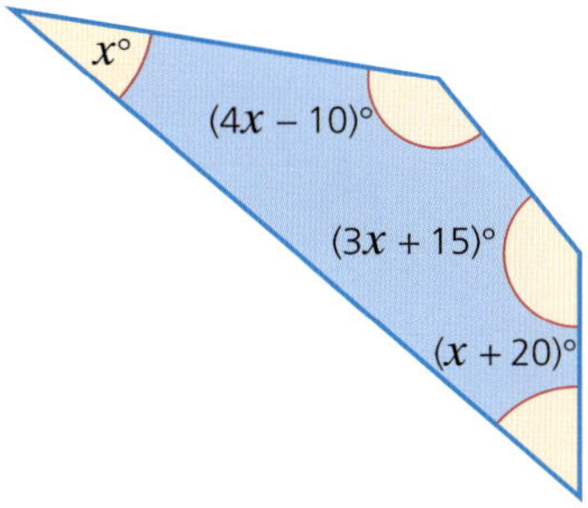

f Triangle with angles $(2x+5)°$, $(4x)°$, $(3x-50)°$

2 a Angles around a point: $(3x)°$, $(5x)°$, $(4x)°$

b Angles around a point: $(3x)°$, $(4x+15)°$, $(4x+15)°$

c Angles around a point: $(2x+40)°$, $(x+10)°$, $(3x+5)°$, $(6x+5)°$

d Angles around a point: $x°$, $(3x-5)°$, $(3x+20)°$, $(3x+15)°$

3 a Quadrilateral with angles $x°$, $(4x)°$, $(3x-40)°$, $(3x-40)°$

b Quadrilateral with angles $(3x)°$, $(3x+40)°$, $(3x+5)°$, $(x+15)°$

c Quadrilateral with angles $(5x-10)°$, $(5x+10)°$, $(2x)°$, $(4x+8)°$

d Quadrilateral with angles $x°$, $(4x-10)°$, $(3x+15)°$, $(x+20)°$

e Parallelogram with angles $(x+23)°$ and $(5x-5)°$

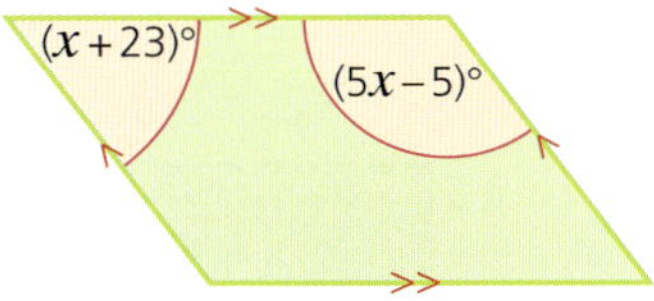

4 By constructing an equation and solving it, find the value of x in each of these isosceles triangles:

a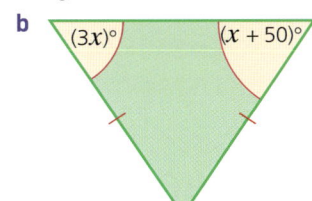
Isosceles triangle with apex angle $x°$ and base angle $(4x)°$

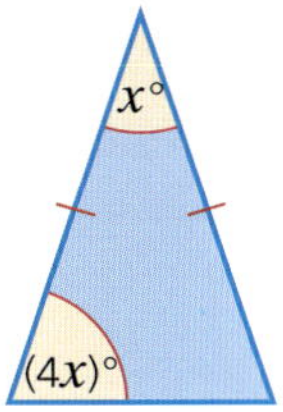

b
Isosceles triangle with base angles $(3x)°$ and $(x+50)°$

c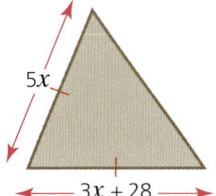
Isosceles triangle with equal side $5x$ and base $3x+28$

13 EQUATIONS AND INEQUALITIES

Exercise 13.2 (cont)

d 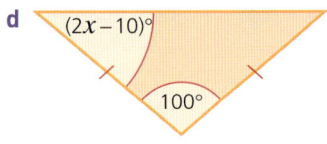   e 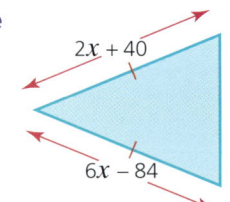   f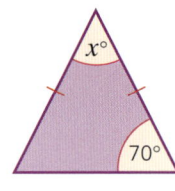

5 Using angle properties, calculate the value of x in each of these questions:

a 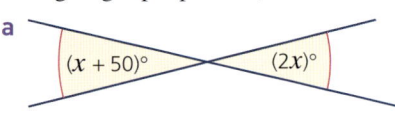   b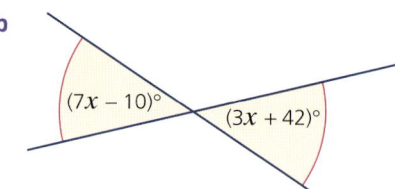

c 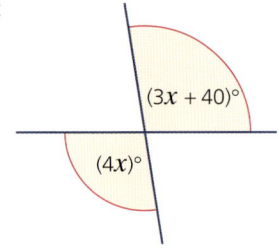   d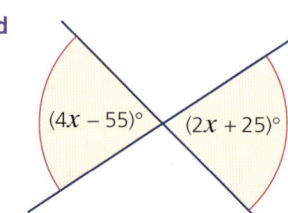

6 Calculate the value of x:

a 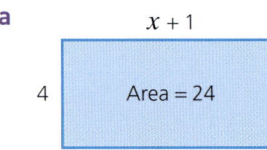   b 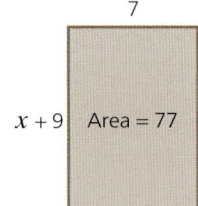   c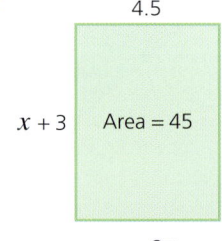

d 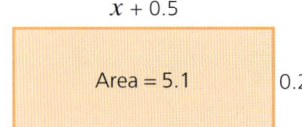   e (see figure)   f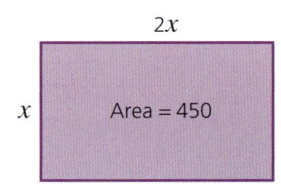

Simultaneous equations

When the values of two unknowns are needed, two equations need to be formed and solved. The process of solving two equations and finding a common solution is known as solving equations simultaneously.

Simultaneous equations

The two most common ways of solving **simultaneous equations** algebraically are by **elimination** and by **substitution**.

By elimination

The aim of this method is to eliminate one of the unknowns by either adding or subtracting the two equations.

➜ Worked examples

Solve the following simultaneous equations by finding the values of x and y which satisfy both equations.

a $3x + y = 9$ (1)
   $5x - y = 7$ (2)

By adding equations (1) + (2) we eliminate the variable y:

$8x = 16$
$x = 2$

To find the value of y, we substitute $x = 2$ into either equation (1) or (2).
Substituting $x = 2$ into equation (1):

$3x + y = 9$
$6 + y = 9$
$y = 3$

To check that the solution is correct, the values of x and y are substituted into equation (2). If it is correct, then the left-hand side of the equation will equal the right-hand side.

$5x - y = 7$
LHS $= 10 - 3 = 7$
$ = $ RHS ✓

b $4x + y = 23$ (1)
   $x + y = 8$ (2)

By subtracting the equations, i.e. (1) − (2), we eliminate the variable y:

$3x = 15$
$x = 5$

By substituting $x = 5$ into equation (2), y can be calculated:

$x + y = 8$
$5 + y = 8$
$y = 3$

Check by substituting both values into equation (1):

$4x + y = 23$
$20 + 3 = 23$
$23 = 23$

13 EQUATIONS AND INEQUALITIES

By substitution

The same equations can also be solved by the method known as **substitution**.

→ Worked examples

a $3x + y = 9$ (1)
   $5x - y = 7$ (2)

Equation (2) can be rearranged to give: $y = 5x - 7$

This can now be substituted into equation (1):

$3x + (5x - 7) = 9$
$3x + 5x - 7 = 9$
$8x - 7 = 9$
$8x = 16$
$x = 2$

To find the value of y, $x = 2$ is substituted into either equation (1) or (2) as before, giving $y = 3$.

b $4x + y = 23$ (1)
   $x + y = 8$ (2)

Equation (2) can be rearranged to give $y = 8 - x$.

This can be substituted into equation (1):

$4x + (8 - x) = 23$
$4x + 8 - x = 23$
$3x + 8 = 23$
$3x = 15$
$x = 5$

y can be found as before, giving a result of $y = 3$.

 Exercise 13.3

Solve the following simultaneous equations either by elimination or by substitution:

1 a $x + y = 6$ b $x + y = 11$ c $x + y = 5$
     $x - y = 2$ $x - y - 1 = 0$ $x - y = 7$
  d $2x + y = 12$ e $3x + y = 17$ f $5x + y = 29$
     $2x - y = 8$ $3x - y = 13$ $5x - y = 11$

2 a $3x + 2y = 13$ b $6x + 5y = 62$ c $x + 2y = 3$
     $4x = 2y + 8$ $4x - 5y = 8$ $8x - 2y = 6$
  d $9x + 3y = 24$ e $7x - y = -3$ f $3x = 5y + 14$
     $x - 3y = -14$ $4x + y = 14$ $6x + 5y = 58$

3 a $2x + y = 14$ b $5x + 3y = 29$ c $4x + 2y = 50$
     $x + y = 9$ $x + 3y = 13$ $x + 2y = 20$
  d $x + y = 10$ e $2x + 5y = 28$ f $x + 6y = -2$
     $3x = -y + 22$ $4x + 5y = 36$ $3x + 6y = 18$

4 a $x - y = 1$ b $3x - 2y = 8$ c $7x - 3y = 26$
     $2x - y = 6$ $2x - 2y = 4$ $2x - 3y = 1$
  d $x = y + 7$ e $8x - 2y = -2$ f $4x - y = -9$
     $3x - y = 17$ $3x - 2y = -7$ $7x - y = -18$

5
a $x + y = -7$
$x - y = -3$
b $2x + 3y = -18$
$2x = 3y + 6$
c $5x - 3y = 9$
$2x + 3y = 19$
d $7x + 4y = 42$
$9x - 4y = -10$
e $4x - 4y = 0$
$8x + 4y = 12$
f $x - 3y = -25$
$5x - 3y = -17$

6
a $2x + 3y = 13$
$2x - 4y + 8 = 0$
b $2x + 4y = 50$
$2x + y = 20$
c $x + y = 10$
$3y = 22 - x$
d $5x + 2y = 28$
$5x + 4y = 36$
e $2x - 8y = 2$
$2x - 3y = 7$
f $x - 4y = 9$
$x - 7y = 18$

7
a $-4x = 4y$
$4x - 8y = 12$
b $3x = 19 + 2y$
$-3x + 5y = 5$
c $3x + 2y = 12$
$-3x + 9y = -12$
d $3x + 5y = 29$
$3x + y = 13$
e $-5x + 3y = 14$
$5x + 6y = 58$
f $-2x + 8y = 6$
$2x = 3 - y$

Further simultaneous equations

If neither x nor y can be eliminated by simply adding or subtracting the two equations then it is necessary to multiply one or both of the equations. The equations are multiplied by a number in order to make the coefficients of x (or y) numerically equal.

➜ Worked examples

a $3x + 2y = 22$ (1)
$x + y = 9$ (2)

To eliminate y, equation (2) is multiplied by 2:

$3x + 2y = 22$ (1)
$2x + 2y = 18$ (3)

By subtracting (3) from (1), the variable y is eliminated:
$x = 4$

Substituting $x = 4$ into equation (2), we have:
$x + y = 9$
$4 + y = 9$
$y = 5$

Check by substituting both values into equation (1):
$3x + 2y = 22$
LHS $= 12 + 10 = 22$
$=$ RHS ✓

b $5x - 3y = 1$ (1)
$3x + 4y = 18$ (2)

To eliminate the variable y, equation (1) is multiplied by 4, and equation (2) is multiplied by 3.

$20x - 12y = 4$ (3)
$9x + 12y = 54$ (4)

By adding equations (3) and (4) the variable y is eliminated:

$29x = 58$
$x = 2$

Substituting $x = 2$ into equation (2) gives:

$3x + 4y = 18$
$6 + 4y = 18$
$4y = 12$
$y = 3$

Check by substituting both values into equation (1):

$5x - 3y = 1$
LHS $= 10 - 9 = 1$
$=$ RHS ✓

 Exercise 13.4

Solve the following:

1 a $2x + y = 7$
      $3x + 2y = 12$
   b $5x + 4y = 21$
      $x + 2y = 9$
   c $x + y = 7$
      $3x + 4y = 23$
   d $2x - 3y = -3$
      $3x + 2y = 15$
   e $4x = 4y + 8$
      $x + 3y = 10$
   f $x + 5y = 11$
      $2x - 2y = 10$

2 a $x + y = 5$
      $3x - 2y + 5 = 0$
   b $2x - 2y = 6$
      $x - 5y = -5$
   c $2x + 3y = 15$
      $2y = 15 - 3x$
   d $x - 6y = 0$
      $3x - 3y = 15$
   e $2x - 5y = -11$
      $3x + 4y = 18$
   f $x + y = 5$
      $2x - 2y = -2$

3 a $3y = 9 + 2x$
      $3x + 2y = 6$
   b $x + 4y = 13$
      $3x - 3y = 9$
   c $2x = 3y - 19$
      $3x + 2y = 17$
   d $2x - 5y = -8$
      $-3x - 2y = -26$
   e $5x - 2y = 0$
      $2x + 5y = 29$
   f $8y = 3 - x$
      $3x - 2y = 9$

4 a $4x + 2y = 5$
      $3x + 6y = 6$
   b $4x + y = 14$
      $6x - 3y = 3$
   c $10x - y = -2$
      $-15x + 3y = 9$
   d $-2y = 0.5 - 2x$
      $6x + 3y = 6$
   e $x + 3y = 6$
      $2x - 9y = 7$
   f $5x - 3y = -0.5$
      $3x + 2y = 3.5$

 **Exercise 13.5**

1 The sum of two numbers is 17 and their difference is 3. Find the two numbers by forming two equations and solving them simultaneously.
2 The difference between two numbers is 7. If their sum is 25, find the two numbers by forming two equations and solving them simultaneously.
3 Find the values of x and y.

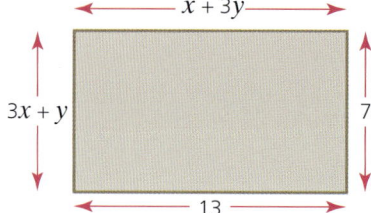

Constructing further equations

4 Find the values of x and y.

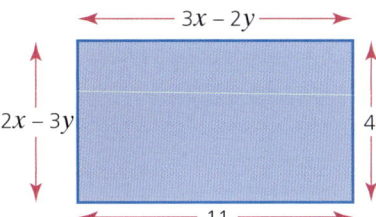

5 A man's age is three times his son's age. Ten years ago he was five times his son's age. By forming two equations and solving them simultaneously, find both of their ages.

6 A grandfather is ten times older than his granddaughter. He is also 54 years older than her. How old is each of them?

Constructing further equations

Earlier in this chapter we looked at some simple examples of constructing and solving equations when we were given geometrical diagrams. This section extends this work with more complicated formulas and equations.

→ Worked examples

Construct and solve the equations below.

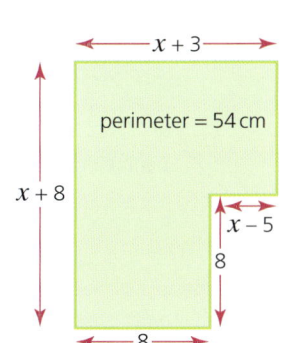

a Using the shape (left), construct an equation for the perimeter in terms of x. Find the value of x by solving the equation.

$$x + 3 + x + x - 5 + 8 + 8 + x + 8 = 54$$
$$4x + 22 = 54$$
$$4x = 32$$
$$x = 8$$

b A number is doubled, 5 is subtracted from it, and the total is 17. Find the number.

Let x be the unknown number.
$$2x - 5 = 17$$
$$2x = 22$$
$$x = 11$$

c 3 is added to a number. The result is multiplied by 8. If the answer is 64, calculate the value of the original number.

Let x be the unknown number.
$$8(x + 3) = 64$$
$$8x + 24 = 64$$
$$8x = 40$$
$$x = 5$$

or $8(x + 3) = 64$
$$x + 3 = 8$$
$$x = 5$$

The original number = 5

13 EQUATIONS AND INEQUALITIES

Exercise 13.6

1 Calculate the value of x:

a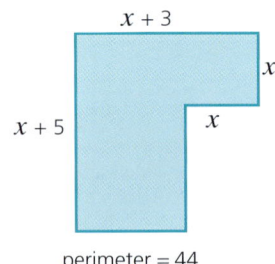
perimeter = 44

b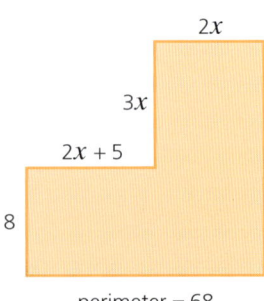
perimeter = 68

c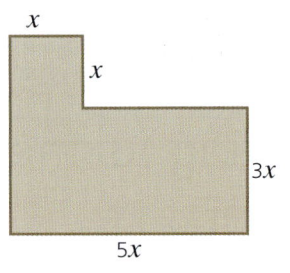
perimeter = 108

d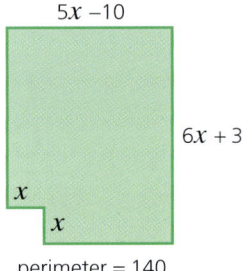
perimeter = 140

e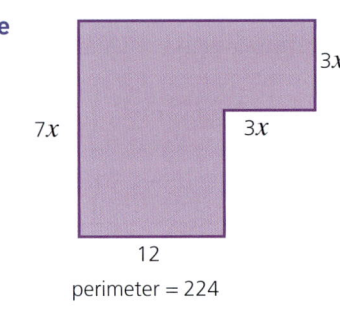
perimeter = 224

f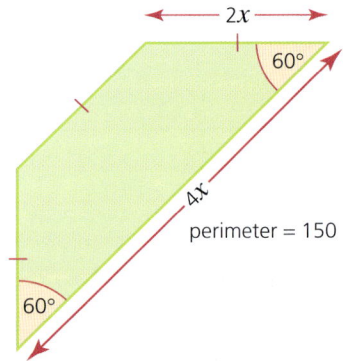
perimeter = 150

2 a A number is trebled and then 7 is added to it. If the total is 28, find the number.
   b Multiply a number by 4 and then add 5 to it. If the total is 29, find the number.
   c If 31 is the result of adding 1 to 5 times a number, find the number.
   d Double a number and then subtract 9. If the answer is 11, what is the number?
   e If 9 is the result of subtracting 12 from 7 times a number, find the number.

3 a Add 3 to a number and then double the result. If the total is 22, find the number.
   b 27 is the answer when you add 4 to a number and then treble it. What is the number?
   c Subtract 1 from a number and multiply the result by 5. If the answer is 35, what is the number?

d Add 3 to a number. If the result of multiplying this total by 7 is 63, find the number.
e Add 3 to a number. Quadruple the result. If the answer is 36, what is the number?

4 a Gabriella is x years old. Her brother is 8 years older and her sister is 12 years younger than she is. If their total age is 50 years, how old are they?
b A series of mathematics textbooks consists of four volumes. The first volume has x pages, the second has 54 pages more. The third and fourth volume each have 32 pages more than the second. If the total number of pages in all four volumes is 866, calculate the number of pages in each of the volumes.
c The five interior angles (in °) of a pentagon are x, $x + 30$, $2x$, $2x + 40$ and $3x + 20$. The sum of the interior angles of a pentagon is 540°. Calculate the size of each of the angles.
d A hexagon consists of three interior angles of equal size and a further three which are double this size. The sum of all six angles is 720°. Calculate the size of each of the angles.
e Four of the exterior angles of an octagon are the same size. The other four are twice as big. If the sum of the exterior angles is 360°, calculate the size of the interior angles.

Solving quadratic equations by factorising

You will need to be familiar with the work covered in Chapter 11 on the factorising of quadratics.

$x^2 - 3x - 10 = 0$ is a **quadratic equation**, which when factorised can be written as $(x - 5)(x + 2) = 0$.

Therefore either $x - 5 = 0$ or $x + 2 = 0$ since, if two things multiply to make zero, then one or both of them must be zero.

$x - 5 = 0$ or $x + 2 = 0$
$x = 5$ or $x = -2$

➜ Worked examples

Solve the following equations to give two solutions for x:

a
$x^2 - x - 12 = 0$
$(x - 4)(x + 3) = 0$
so either $x - 4 = 0$ or $x + 3 = 0$
$x = 4$ or $x = -3$

b
$x^2 + 2x = 24$
This becomes $x^2 + 2x - 24 = 0$
$(x + 6)(x - 4) = 0$
so either $x + 6 = 0$ or $x - 4 = 0$
$x = -6$ or $x = 4$

13 EQUATIONS AND INEQUALITIES

    c $\qquad x^2 - 6x = 0$
$\qquad\qquad\qquad\qquad x(x - 6) = 0$
    so either $\qquad x = 0 \qquad$ or $\quad x - 6 = 0$
$\qquad\qquad\qquad\qquad\qquad\qquad\qquad$ or $\quad x = 6$

    d $\qquad x^2 - 4 = 0$
$\qquad\qquad\qquad (x - 2)(x + 2) = 0$
    so either $\qquad x - 2 = 0 \quad$ or $\quad x + 2 = 0$
$\qquad\qquad\qquad\qquad x = 2 \qquad$ or $\quad x = -2$

Exercise 13.7

Solve the following quadratic equations by factorising:

1. **a** $x^2 + 7x + 12 = 0$ **b** $x^2 + 8x + 12 = 0$
   c $x^2 + 13x + 12 = 0$ **d** $x^2 - 7x + 10 = 0$
   e $x^2 - 5x + 6 = 0$ **f** $x^2 - 6x + 8 = 0$

2. **a** $x^2 + 3x - 10 = 0$ **b** $x^2 - 3x - 10 = 0$
   c $x^2 + 5x - 14 = 0$ **d** $x^2 - 5x - 14 = 0$
   e $x^2 + 2x - 15 = 0$ **f** $x^2 - 2x - 15 = 0$

3. **a** $x^2 + 5x = -6$ **b** $x^2 + 6x = -9$
   c $x^2 + 11x = -24$ **d** $x^2 - 10x = -24$
   e $x^2 + x = 12$ **f** $x^2 - 4x = 12$

4. **a** $x^2 - 2x = 8$ **b** $x^2 - x = 20$
   c $x^2 + x = 30$ **d** $x^2 - x = 42$
   e $x^2 - 2x = 63$ **f** $x^2 + 3x = 54$

Exercise 13.8

Solve the following quadratic equations:

1. **a** $x^2 - 9 = 0$ **b** $x^2 - 16 = 0$
   c $x^2 = 25$ **d** $x^2 = 121$
   e $x^2 - 144 = 0$ **f** $x^2 - 220 = 5$

2. **a** $4x^2 - 25 = 0$ **b** $9x^2 - 36 = 0$
   c $25x^2 = 64$ **d** $x^2 = \frac{1}{4}$
   e $x^2 - \frac{1}{9} = 0$ **f** $16x^2 - \frac{1}{25} = 0$

3. **a** $x^2 + 5x + 4 = 0$ **b** $x^2 + 7x + 10 = 0$
   c $x^2 + 6x + 8 = 0$ **d** $x^2 - 6x + 8 = 0$
   e $x^2 - 7x + 10 = 0$ **f** $x^2 + 2x - 8 = 0$

4. **a** $x^2 - 3x - 10 = 0$ **b** $x^2 + 3x - 10 = 0$
   c $x^2 - 3x - 18 = 0$ **d** $x^2 + 3x - 18 = 0$
   e $x^2 - 2x - 24 = 0$ **f** $x^2 - 2x - 48 = 0$

5. **a** $x^2 + x - 12 = 0$ **b** $x^2 + 8x = -12$
   c $x^2 + 5x = 36$ **d** $x^2 + 2x = -1$
   e $x^2 + 4x = -4$ **f** $x^2 + 17x = -72$

6. **a** $x^2 - 8x = 0$ **b** $x^2 - 7x = 0$
   c $x^2 + 3x = 0$ **d** $x^2 + 4x = 0$
   e $x^2 - 9x = 0$ **f** $4x^2 - 16x = 0$

7. **a** $2x^2 + 5x + 3 = 0$ **b** $2x^2 - 3x - 5 = 0$
   c $3x^2 + 2x - 1 = 0$ **d** $2x^2 + 11x + 5 = 0$
   e $2x^2 - 13x + 15 = 0$ **f** $12x^2 + 10x - 8 = 0$

Solving quadratic equations by factorising

8 a $x^2 + 12x = 0$ b $x^2 + 12x + 27 = 0$
  c $x^2 + 4x = 32$ d $x^2 + 5x = 14$
  e $2x^2 = 72$ f $3x^2 - 12 = 288$

Exercise 13.9

In the following questions, construct equations from the information given and then solve to find the unknown.

1 When a number x is added to its square, the total is 12. Find two possible values for x.

2 A number x is equal to its own square minus 42. Find two possible values for x.

Although a solution to a quadratic will give two solutions, sometimes one solution can be ignored due to the context of the question.

3 If the area of the rectangle (below) is $10\,\text{cm}^2$, calculate the only possible value for x.

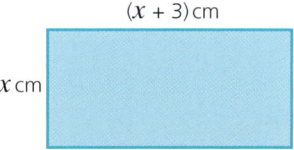

4 If the area of the rectangle (below) is $52\,\text{cm}^2$, calculate the only possible value for x.

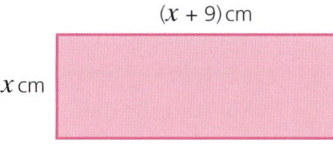

5 A triangle has a base length of $2x\,\text{cm}$ and a height of $(x - 3)\,\text{cm}$. If its area is $18\,\text{cm}^2$, calculate its height and base length.

6 A triangle has a base length of $(x - 8)\,\text{cm}$ and a height of $2x\,\text{cm}$. If its area is $20\,\text{cm}^2$, calculate its height and base length.

7 A right-angled triangle has a base length of $x\,\text{cm}$ and a height of $(x - 1)$ cm. If its area is $15\,\text{cm}^2$, calculate the base length and height.

8 A rectangular garden has a square flower bed of side length $x\,\text{m}$ in one of its corners. The remainder of the garden consists of lawn and has dimensions as shown (below). If the total area of the lawn is $50\,\text{m}^2$:
  a form an equation in terms of x,
  b solve the equation,
  c calculate the length and width of the whole garden.

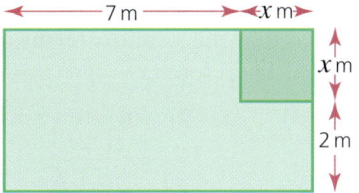

13 EQUATIONS AND INEQUALITIES

The quadratic formula

In general a quadratic equation takes the form $ax^2 + bx + c = 0$ where a, b and c are integers. Quadratic equations can be solved by the use of the **quadratic formula** which states that:

$$x = \frac{-b \pm \sqrt{b^2 - 4ac}}{2a}$$

→ Worked examples

a Solve the quadratic equation $x^2 + 7x + 3 = 0$.

   $a = 1$, $b = 7$ and $c = 3$.

   Substituting these values into the quadratic formula gives:

   $x = \dfrac{-7 \pm \sqrt{7^2 - 4 \times 1 \times 3}}{2 \times 1}$

   $x = \dfrac{-7 \pm \sqrt{49 - 12}}{2}$

   $x = \dfrac{-7 \pm \sqrt{37}}{2}$

   Therefore $x = \dfrac{-7 + 6.083}{2}$ or $x = \dfrac{-7 - 6.083}{2}$

   $x = -0.459$ (3 s.f.) or $x = -6.54$ (3 s.f.)

When answers are left in this form, they are left in 'surd form', meaning that the answer is left with a surd and not written as an approximate decimal.

This answer is given as a decimal. It is an approximate answer.

b Solve the quadratic equation $x^2 - 4x - 2 = 0$.

   $a = 1$, $b = -4$ and $c = -2$.

   Substituting these values into the quadratic formula gives:

   $x = \dfrac{-(-4) \pm \sqrt{(-4)^2 - (4 \times 1 \times -2)}}{2 \times 1}$

   $x = \dfrac{4 \pm \sqrt{16 + 8}}{2}$

   $x = \dfrac{4 \pm \sqrt{24}}{2}$

   $x = 2 \pm \sqrt{6}$

   Therefore $x = 2 + 2.449$ or $x = 2 - 2.449$

   $x = 4.45$ (3 s.f.) or $x = -0.449$ (3 s.f.)

This answer is in surd form.

This answer is given as a decimal. It is an approximate answer.

Completing the square

Quadratics can also be solved by expressing them in terms of a **perfect square**. We look once again at the quadratic $x^2 - 4x - 2 = 0$.
The perfect square $(x - 2)^2$ can be expanded to give $x^2 - 4x + 4$. Notice that the x^2 and x terms are the same as those in the original quadratic. Therefore $(x - 2)^2 - 6 = x^2 - 4x - 2$ and can be used to solve the quadratic.

$(x - 2)^2 - 6 = 0$
$(x - 2)^2 = 6$
$x - 2 = \pm\sqrt{6}$
$x = 2 \pm \sqrt{6}$

$x = 4.45$ (3 s.f.)   or $x = -0.449$ (3 s.f.)

Simultaneous equations involving one linear and one non-linear equation

Exercise 13.10 Solve the following quadratic equations using either the quadratic formula or by completing the square. For each question:
i give your answers in surd form
ii and give your answers to 2 d.p.

1 a $x^2 - x - 13 = 0$ b $x^2 + 4x - 11 = 0$
   c $x^2 + 5x - 7 = 0$ d $x^2 + 6x + 6 = 0$
   e $x^2 + 5x - 13 = 0$ f $x^2 - 9x + 19 = 0$
2 a $x^2 + 7x + 9 = 0$ b $x^2 - 35 = 0$
   c $x^2 + 3x - 3 = 0$ d $x^2 - 5x - 7 = 0$
   e $x^2 + x - 18 = 0$ f $x^2 - 8 = 0$
3 a $x^2 - 2x - 2 = 0$ b $x^2 - 4x - 11 = 0$
   c $x^2 - x - 5 = 0$ d $x^2 + 2x - 7 = 0$
   e $x^2 - 3x + 1 = 0$ f $x^2 - 8x + 3 = 0$
4 a $2x^2 - 3x - 4 = 0$ b $4x^2 + 2x - 5 = 0$
   c $5x^2 - 8x + 1 = 0$ d $-2x^2 - 5x - 2 = 0$
   e $3x^2 - 4x - 2 = 0$ f $-7x^2 - x + 15 = 0$

Simultaneous equations involving one linear and one non-linear equation

So far we have dealt with the solution of linear simultaneous equations and also the solution of quadratic equations. However, solving equations simultaneously need not only deal with two linear equations.

➔ Worked example

Solve the following linear and quadratic equations simultaneously.
$y = 2x + 3$ and $y = x^2 + x - 9$

Simultaneous equations involving one linear and one non-linear equation are solved using the method of substitution.

As both $2x + 3$ and $x^2 + x - 9$ are equal to y, then they must also be equal to each other.

$2x + 3 = x^2 + x - 9$

Rearranging the equation to collect all the terms on one side of the equation gives:

$x^2 - x - 12 = 0$

Note that equating a linear equation with a quadratic equation has produced a quadratic equation. This can therefore be solved in the normal way.

$x^2 - x - 12 = 0$
$(x - 4)(x + 3) = 0$
$x = 4$ and $x = -3$

Substituting these values of x into one of the original equations (the linear one is easier) will produce the corresponding y values.
When $x = 4$, $y = 2(4) + 3 = 11$
When $x = -3$, $y = 2(-3) + 3 = -3$

13 EQUATIONS AND INEQUALITIES

Exercise 13.11

Solve the following equations simultaneously:

1. $y = 4$ and $y = (x - 3)^2 + 4$
2. $y = -2x - 3$ and $y = x^2 + 5x + 3$
3. $y = 2x^2 + 4x + 1$ and $y = 3x + 11$
4. $y = 6x^2 - 5x + 2$ and $y = 5x + 6$
5. $y = -2x + 2$ and $y = 2(x - 3)^2 - 16$
6. $y = -5x^2 - 5x + 10$ and $y = 6x - 2$
7. $y = 11 - x$ and $y = \dfrac{30}{x}$
8. $y = 2x - 7 - \dfrac{3}{x}$ and $y = \dfrac{2}{x}$
9. Two rectangles are shown below.

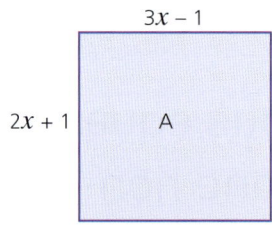
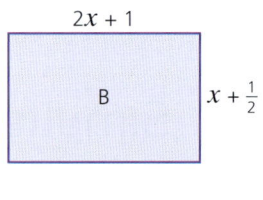

Calculate the value of x if the area of rectangle A and the perimeter of rectangle B have the same numerical value.

10. Two circles are shown below:

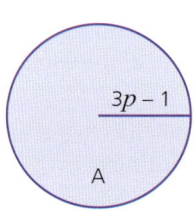
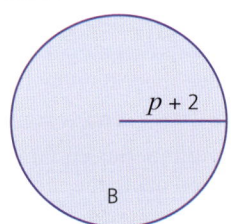

Calculate the value of p if the area of circle A has the same numerical value as the perimeter of circle B. Give your answer correct to 2 d.p.

Linear inequalities

The statement

    6 is less than 8

can be written as:

    $6 < 8$

This inequality can be manipulated in the following ways:

adding 2 to each side: $8 < 10$ this inequality is still true

subtracting 2 from each side: $4 < 6$ this inequality is still true

Linear inequalities

multiplying both sides by 2:	12 < 16	this inequality is still true
dividing both sides by 2:	3 < 4	this inequality is still true
multiplying both sides by −2:	−12 < −16	this inequality is not true
dividing both sides by −2:	−3 < −4	this inequality is not true

As can be seen, when both sides of an inequality are either multiplied or divided by a negative number, the inequality is no longer true. For it to be true, the inequality sign needs to be changed around.

i.e. $-12 > -16$ and $-3 > -4$

When solving **linear inequalities**, the procedure is very similar to that for solving linear equations.

→ Worked examples

Remember:

⟶ implies that the number is not included in the solution. It is associated with > and <.
⟶ implies that the number is included in the solution. It is associated with ⩾ and ⩽.

Solve the following inequalities and represent the solution on a number line:

a $15 + 3x < 6$
   $3x < -9$
   $x < -3$

b $17 \leqslant 7x + 3$
   $14 \leqslant 7x$
   $2 \leqslant x$ that is $x \geqslant 2$

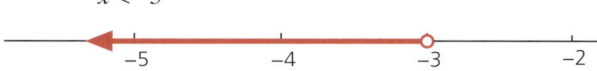

c $9 - 4x \geqslant 17$
   $-4x \geqslant 8$
   $x \leqslant -2$

Note the inequality sign has changed direction.

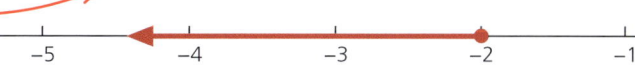

13 EQUATIONS AND INEQUALITIES

Exercise 13.12

Solve the following inequalities and illustrate your solution on a number line:

1. **a** $x + 3 < 7$ **b** $5 + x > 6$
   c $4 + 2x \leqslant 10$ **d** $8 \leqslant x + 1$
   e $5 > 3 + x$ **f** $7 < 3 + 2x$

2. **a** $x - 3 < 4$ **b** $x - 6 \geqslant -8$
   c $8 + 3x > -1$ **d** $5 \geqslant -x - 7$
   e $12 > -x - 12$ **f** $4 \leqslant 2x + 10$

3. **a** $\frac{x}{2} < 1$ **b** $4 \geqslant \frac{x}{3}$
   c $1 \leqslant \frac{x}{2}$ **d** $9x \geqslant -18$
   e $-4x + 1 < 3$ **f** $1 \geqslant -3x + 7$

→ Worked example

Find the range of values for which $7 < 3x + 1 \leqslant 13$ and illustrate the solutions on a number line.

This is in fact two inequalities, which can therefore be solved separately.

$7 < 3x + 1$ and $3x + 1 \leqslant 13$
$(-1) \to 6 < 3x$ $(-1) \to 3x \leqslant 12$
$(\div 3) \to 2 < x$ that is $x > 2$ $(\div 3) \to x \leqslant 4$

Exercise 13.13

Find the range of values for which the following inequalities are satisfied. Illustrate each solution on a number line:

1. **a** $4 < 2x \leqslant 8$ **b** $3 \leqslant 3x < 15$
   c $7 \leqslant 2x < 10$ **d** $10 \leqslant 5x < 21$

2. **a** $5 < 3x + 2 \leqslant 17$ **b** $3 \leqslant 2x + 5 < 7$
   c $12 < 8x - 4 < 20$ **d** $15 \leqslant 3(x - 2) < 9$

Student assessment 1

Solve the following equations:

1. **a** $y + 9 = 3$ **b** $3x - 5 = 13$
   c $12 - 5p = -8$ **d** $2.5y + 1.5 = 7.5$

2. **a** $5 - p = 4 + p$ **b** $8m - 9 = 5m + 3$
   c $11p - 4 = 9p + 15$ **b** $27 - 5r = r - 3$

3. **a** $\frac{p}{-2} = -3$ **b** $6 = \frac{2}{5}x$
   c $\frac{m - 7}{5} = 3$ **d** $\frac{4t - 3}{3} = 7$

4. **a** $\frac{2}{5}(t - 1) = 3$ **b** $5(3 - m) = 4(m - 6)$
   c $5 = \frac{2}{3}(x - 1)$ **d** $\frac{4}{5}(t - 2) = \frac{1}{4}(2t + 8)$
   e $\frac{4p}{9 - 3p} = \frac{20}{3}$ **f** $\frac{-5x}{3 - x} = \frac{5}{4}$

Linear inequalities

Solve the following simultaneous equations:

5 a $x + y = 11$ b $5p - 3q = -1$
   $x - y = 3$ $-2p - 3q = -8$
   c $3x + 5y = 26$ d $2m - 3n = -9$
   $x - y = 6$ $3m + 2n = 19$

6 A straight line $y = -3x + 5$ is plotted on the same axes as the quadratic $y = 4x^2 + 8x + 2$.
   Calculate the coordinates of the points of intersection.

Student assessment 2

1 The angles of a quadrilateral are x, $3x$, $(2x - 40)$ and $(3x - 50)$ degrees.
   a Construct an equation in terms of x.
   b Solve the equation.
   c Calculate the size of the four angles.

2 Three is subtracted from seven times a number. The result is multiplied by 5. If the answer is 55, calculate the value of the number by constructing an equation and solving it.

3 The interior angles of a pentagon are $9x$, $(5x + 10)$, $(6x + 5)$, $(8x - 25)$ and $(10x - 20)$ degrees. If the sum of the interior angles of a pentagon is 540°, find the size of each of the angles.

4 Solve the inequality below and illustrate your answer on a number line.
   $$6 < 2x \leqslant 10$$

5 Solve the following quadratic equation by factorisation:
   $$x^2 - x = 20$$

6 Solve the following quadratic equation by using the quadratic formula:
   $$2x^2 - 7 = 3x$$

7 Solve the following equations simultaneously:
   $$y = \frac{2}{x} + 1 \quad \text{and} \quad y = \frac{1}{2}x + 2$$

8 For what values of x is the following inequality true?
   $$\frac{7}{x - 5} > 0$$

13 EQUATIONS AND INEQUALITIES

Student assessment 3

1. The angles of a triangle are $x°$, $y°$ and $40°$. The difference between the two unknown angles is $30°$.
   a. Write down two equations from the information given above.
   b. What is the size of the two unknown angles?

2. The interior angles of a pentagon increase by $10°$ as you progress clockwise.
   a. Illustrate this information in a diagram.
   b. Write an expression for the sum of the interior angles.
   c. The sum of the interior angles of a pentagon is $540°$. Use this to calculate the largest **exterior** angle of the pentagon.
   d. Illustrate on your diagram the size of each of the five exterior angles.
   e. Show that the sum of the exterior angles is $360°$.

3. A flat sheet of card measures 12 cm by 10 cm. It is made into an open box by cutting a square of side x cm from each corner and then folding up the sides.
   a. Illustrate the box and its dimensions on a simple 3D sketch.
   b. Write an expression for the surface area of the outside of the box.
   c. If the surface area is 56 cm², form and solve a quadratic equation to find the value of x.

4. a. Show that $x - 2 = \dfrac{4}{x-3}$ can be written as $x^2 - 5x + 2 = 0$.
   b. Use the quadratic formula to solve $x - 2 = \dfrac{4}{x-3}$

5. A right-angled triangle ABC has side lengths as follows: $AB = x$ cm, AC is 2 cm shorter than AB, and BC is 2 cm shorter than AC.
   a. Illustrate this information on a diagram.
   b. Using this information, show that $x^2 - 12x + 20 = 0$.
   c. Solve the above quadratic and hence find the length of each of the three sides of the triangle.

14 Graphing inequalities and regions

Revision

An understanding of the following symbols is necessary:
- $>$ means 'is greater than'
- $\geqslant$ means 'is greater than or equal to'
- $<$ means 'is less than'
- $\leqslant$ means 'is less than or equal to'

 Exercise 14.1

1 Solve each of the following inequalities:
   a $15 + 3x < 21$
   b $18 \leqslant 7y + 4$
   c $19 - 4x \geqslant 27$
   d $2 \geqslant \frac{y}{3}$
   e $-4t + 1 < 1$
   f $1 \geqslant 3p + 10$

2 Solve each of the following inequalities:
   a $7 < 3y + 1 \leqslant 13$
   b $3 \leqslant 3p < 15$
   c $9 \leqslant 3(m - 2) < 15$
   d $20 < 8x - 4 < 28$

Graphing an inequality

The solution to an inequality can also be illustrated on a graph.

 Worked examples

a On a pair of axes, leave unshaded the region which satisfies the inequality $x \leqslant 3$.

   To do this, the line $x = 3$ is drawn.

   The region to the left of $x = 3$ represents the inequality $x \leqslant 3$ and therefore is unshaded as shown below.

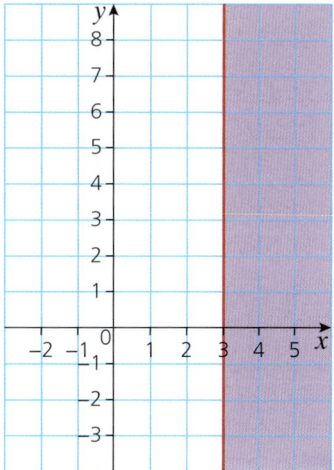

14 GRAPHING INEQUALITIES AND REGIONS

b On a pair of axes, leave unshaded the region which satisfies the inequality $y > 5$.

The line $y = 5$ is drawn first (in this case it is drawn as a broken line).

The region above the line $y = 5$ represents the inequality $y > 5$ and therefore is unshaded as shown (below left).

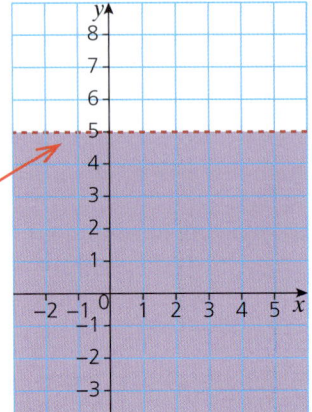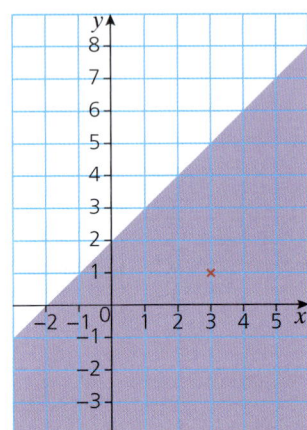

Note that a broken (dashed) line shows < or >, while a solid line shows ⩽ or ⩾.

c On a pair of axes, leave unshaded the region which satisfies the inequality $y \geqslant x + 2$.

The line $y = x + 2$ is drawn first (since it is included, this line is solid).

To know which region satisfies the inequality, and hence to know which side of the line to shade, the following steps are taken:

- Choose a point at random which does not lie on the line, e.g. (3, 1).
- Substitute those values of x and y into the inequality, i.e. $1 \geqslant 3 + 2$.
- If the inequality is false, then the region in which the point lies doesn't satisfy the inequality and can therefore be shaded as shown (above right).

Exercise 14.2

1 By drawing appropriate axes, leave unshaded the region which satisfies each of the following inequalities:

a $y \geqslant -x$
b $y \leqslant 2 - x$
c $x \geqslant y - 3$
d $x + y \geqslant 4$
e $2x - y \geqslant 3$
f $2y - x < 4$

Graphing more than one inequality

Several inequalities can be graphed on the same set of axes. If the regions which satisfy each inequality are left unshaded, then a solution can be found which satisfies all the inequalities, i.e. the region left unshaded by all the inequalities.

Graphing more than one inequality

 Worked example

On the same pair of axes, leave unshaded the regions which satisfy the following inequalities simultaneously:

$$x \leqslant 2 \quad y > -1 \quad y \leqslant 3 \quad y \leqslant x + 2$$

Hence find the region which satisfies all four inequalities.

If the four inequalities are graphed on separate axes, the solutions are as shown below:

$x \leqslant 2$

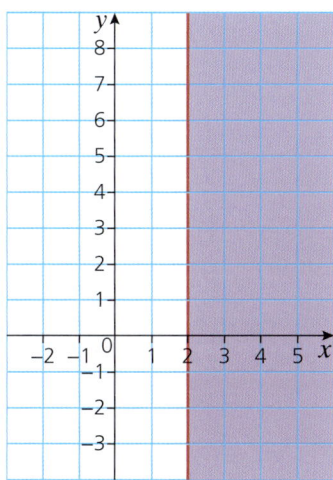

$y > -1$

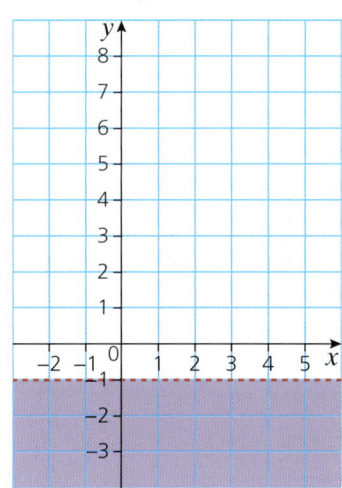

$y \leqslant 3$

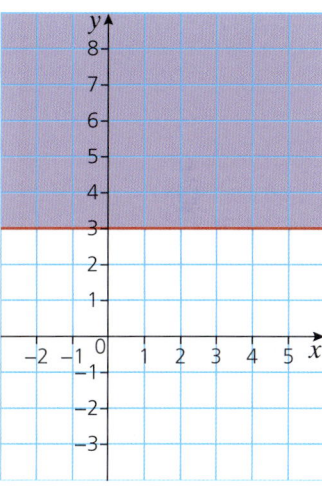

$y \leqslant x + 2$

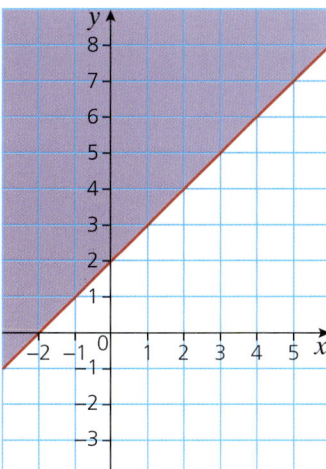

Combining all four on one pair of axes gives this diagram.

The unshaded region therefore gives a solution which satisfies all four inequalities.

14 GRAPHING INEQUALITIES AND REGIONS

Exercise 14.3

On the same pair of axes, plot the following inequalities and leave unshaded the region which satisfies all of them simultaneously.

1 $y \leq x$ $y > 1$ $x \leq 5$
2 $x + y \leq 6$ $y < x$ $y \geq 1$
3 $y \geq 3x$ $y \leq 5$ $x + y > 4$
4 $2y \geq x + 4$ $y \leq 2x + 2$ $y < 4$ $x \leq 3$

Student assessment 1

1 Solve the following inequalities:
   a $17 + 5x \leq 42$
   b $3 \geq \frac{y}{3} + 2$
2 Solve the following inequalities:
   a $5 + 6x \leq 47$
   b $4 \geq \frac{y+3}{3}$
3 Find the range of values for which:
   a $7 < 4y - 1 \leq 15$
   b $18 < 3(p + 2) \leq 30$
4 Find the range of values for which:
   a $3 \leq 3p < 12$
   b $24 < 8(x - 1) \leq 48$
5 Write the inequality which describes the unshaded region in the graph below.

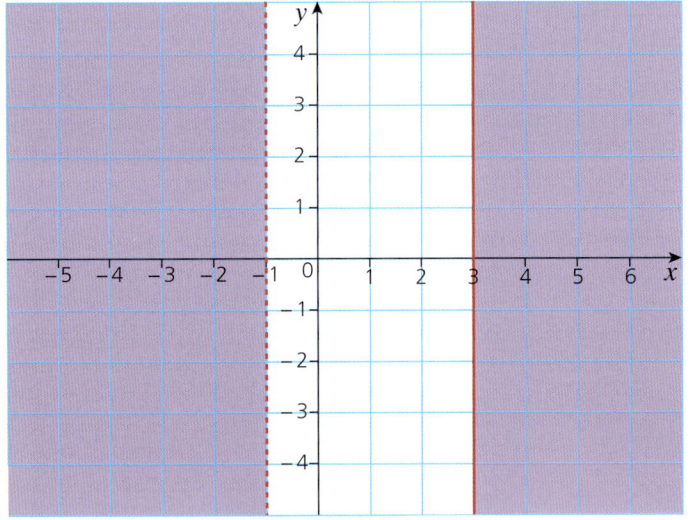

6 a On the same axes, graph the following inequalities, leaving unshaded the region which satisfies all three.
      $x \geq -1$ $y \geq -2$ $10y + 11x \leq 24$
   b Calculate the area of the unshaded region.
7 a On a grid draw the following lines:
      $y = \frac{3}{2}x + 3$ $x + y = 3$ $y = -3$
   b i On the grid, leave unshaded the region satisfying all of the following inequalities:
         $y \leq \frac{3}{2}x + 3$ $x + y \leq 3$ $y \geq -3$
      ii Calculate the area of the unshaded region.

15 Sequences

A **sequence** is a collection of terms arranged in a specific order, where each term is obtained according to a rule. Examples of some simple sequences are given below:

2, 4, 6, 8, 10   1, 4, 9, 16, 25   1, 2, 4, 8, 16
1, 1, 2, 3, 5, 8   1, 8, 27, 64, 125   10, 5, $\frac{5}{2}, \frac{5}{4}, \frac{5}{8}$

You could discuss with another student the rules involved in producing the sequences above.

The terms of a sequence can be expressed as $T_1, T_2, T_3, \ldots, T_n$ where:

T_1 is the first term
T_2 is the second term
T_n is the nth term

The use of subscripts is not on the Core syllabus and is Extended only, but it is a very useful way to describe terms in a sequence.

Therefore in the sequence 2, 4, 6, 8, 10, $T_1 = 2$, $T_2 = 4$, etc.

Linear sequences

In a **linear** sequence (also known as an arithmetic sequence) there is a **common difference** (d) between successive terms. Examples of some arithmetic sequences are given below:

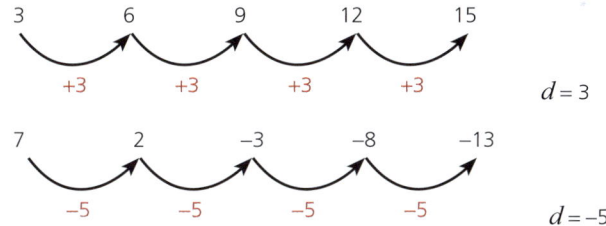

Formulas for the terms of a linear sequence

There are two main ways of describing a sequence.

1 A **term-to-term rule**.

   In the following sequence,

   the term-to-term rule is +5 and the first term is 7.

   This written using T_n notation is: $T_2 = T_1 + 5$, $T_3 = T_2 + 5$, etc. The general form is therefore written as $T_{n+1} = T_n + 5$, $T_1 = 7$, where T_n is the nth term and T_{n+1} the term after the nth term.

   Note: It is important to give one of the terms, e.g. T_1, so that the exact sequence can be generated.

15 SEQUENCES

2 A formula for the nth term of a sequence.

This type of rule links each term to its position in the sequence, for example:

Position	1	2	3	4	5	n
Term	7	12	17	22	27	

We can deduce from the figures above that each term can be calculated by multiplying its position number by 5 and adding 2. Algebraically this can be written as the formula for the nth term:

nth term $= 5n + 2$ or $T_n = 5n + 2$

This textbook focuses on the generation and use of the rule for the nth term.

With a linear sequence, the rule for the nth term can be deduced by looking at the common difference, for example:

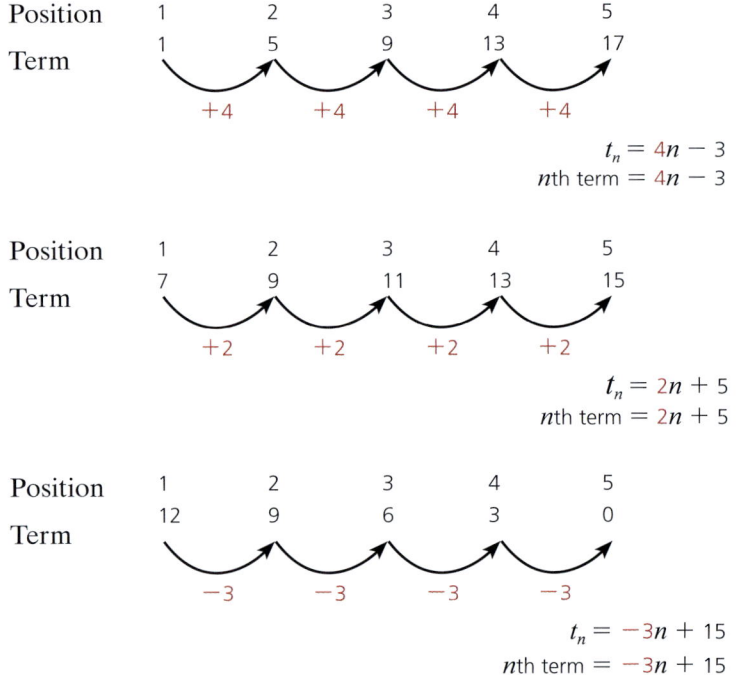

The common difference is the coefficient of n (i.e. the number by which n is multiplied). The constant is then worked out by calculating the number needed to make the term.

Linear sequences

 Exercise 15.1
1. For each of the sequences:
   i. Write down the next two terms.
   ii. Give an expression for the nth term.
   a. 4, 7, 10, 13, 16, ...
   b. 5, 9, 13, 17, 21, ...
   c. 4, 9, 14, 19, 24, ...
   d. 8, 10, 12, 14, 16, ...
   e. 29, 22, 15, 8, 1, ...
   f. 0, 4, 8, 12, 16, 20, ...
   g. 1, 10, 19, 28, 37, ...
   h. 15, 25, 35, 45, 55, ...
   i. 9, 20, 31, 42, 53, ...
   j. 1.5, 3.5, 5.5, 7.5, 9.5, 11.5, ...
   k. 0.25, 1.25, 2.25, 3.25, 4.25, ...
   l. 5, 4, 3, 2, 1, ...

2. The nth term for two different sequences are $2n + 8$ and $5n - 7$. Justifying your answers work out
   a. which sequence if any, the term 22 belongs to
   b. which sequence if any, the term 51 belongs to
   c. a term which belongs to both sequences.

 Worked examples

a. A sequence is given by the term-to-term rule $T_{n+1} = T_n + 3$, where $T_1 = 0$. Generate the terms T_2, T_3 and T_4.

$T_2 = T_1 + 3$, therefore $T_2 = 0 + 3 = 3$

$T_3 = T_2 + 3$, therefore $T_3 = 3 + 3 = 6$

$T_4 = T_3 + 3$, therefore $T_4 = 6 + 3 = 9$

b. Find the rule for the nth term of the sequence 12, 7, 2, –3, –8, ...

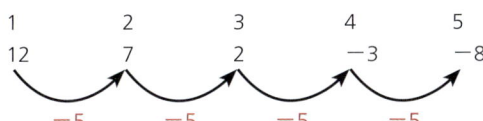

 Exercise 15.2
1. For each of the following sequences, the term-to-term rule and one of the terms are given.

   Calculate the terms required and state whether the sequence generated is linear or not.
   a. $T_{n+1} = T_n + 5$, where $T_1 = 2$. Calculate T_2, T_3 and T_4
   b. $T_{n+1} = T_n - 3$, where $T_1 = 4$. Calculate T_2, T_3 and T_4
   c. $T_{n+1} = T_n - 4$, where $T_3 = 2$. Calculate T_1, T_2 and T_4

15 SEQUENCES

Exercise 15.2 (cont)

d $T_{n+1} = 2T_n - 1$, where $T_2 = 0$. Calculate T_1, T_3 and T_4

e $T_{n+1} = \dfrac{T_n}{2} + 3$, where $T_1 = 2$. Calculate T_2, T_3 and T_4

f $T_{n+1} = \dfrac{T_n}{2} - 1$, where $T_3 = 10$. Calculate T_1, T_2 and T_4

2 For each of the following sequences:
  i describe the term-to-term rule in words
  ii write the term-to-term rule in its general form using the notation T_{n+1} and T_n
  iii deduce the rule for the nth term
  iv calculate the 10th term.
  a 5, 8, 11, 14, 17
  b 0, 4, 8, 12, 16
  c $\tfrac{1}{2}, 1\tfrac{1}{2}, 2\tfrac{1}{2}, 3\tfrac{1}{2}, 4\tfrac{1}{2}$
  d 6, 3, 0, −3, −6
  e −7, −4, −1, 2, 5
  f −9, −13, −17, −21, −25

3 Copy and complete each of the following tables of linear sequences:

a
Position	1	2	5		50	n
Term				45		$4n-3$

b
Position	1	2	5			n
Term				59	449	$6n-1$

c
Position	1				100	n
Term		0	−5	−47		$-n+3$

d
Position	1	2	3			n
Term	3	0	−3	−24	−294	

e
Position		5	7			n
Term	1	10	16	25	145	

f
Position	1	2	5		50	n
Term	−5.5	−7		−34		

4 For each of the following linear sequences:
  i deduce the common difference d
  ii write the term-to-term rule in its general form using the notation T_{n+1} and T_n
  iii give the formula for the nth term
  iv calculate the 50th term.

Sequences with quadratic and cubic rules

a 5, 9, 13, 17, 21
b 0, ..., 2, ..., 4
c −10, ..., ..., ..., 2
d $T_1 = 6, T_9 = 10$
e $T_3 = -50, T_{20} = 18$
f $T_5 = 60, T_{12} = 39$

5 The first four terms of a linear sequence are 7, 19, 31, 43.
   Decide whether each of the numbers below is in this linear sequence.
   Justify your answers.
   a −8
   b 67
   c 139
   d 245

Sequences with quadratic and cubic rules

So far all the sequences we have looked at have been linear i.e. the term-to-term rule has a common difference and the rule for the nth term is linear and takes the form $T_n = an + b$. The rule for the nth term can be found algebraically using the method of differences and this method is particularly useful for more complex sequences.

➡ Worked examples

a Deduce the rule for the nth term for the sequence 4, 7, 10, 13, 16, ...

Firstly, produce a table of the terms and their positions in the sequence:

Position	1	2	3	4	5
Term	4	7	10	13	16

Extend the table to look at the differences:

Position	1	2	3	4	5
Term	4	7	10	13	16
1st difference		3	3	3	3

As the row of 1st differences is constant, the rule for the nth term is linear and takes the form $T_n = an + b$.

By substituting the values of n into the rule, each term can be expressed in terms of a and b:

Position	1	2	3	4	5
Term	$a + b$	$2a + b$	$3a + b$	$4a + b$	$5a + b$
1st difference		a	a	a	a

Compare the two tables in order to deduce the values of a and b:

$a = 3$

$a + b = 4$ therefore $b = 1$

15 SEQUENCES

The rule for the nth term $T_n = an + b$ can be written as $T_n = 3n + 1$.

For a linear rule, this method is perhaps overcomplicated. However, it is very efficient for quadratic and cubic rules.

b Deduce the rule for the nth term for the sequence 0, 7, 18, 33, 52, …

Entering the sequence in a table gives:

Position	1	2	3	4	5
Term	0	7	18	33	52

Extending the table to look at the differences gives:

Position	1	2	3	4	5
Term	0	7	18	33	52
1st difference		7	11	15	19

The row of 1st differences is not constant, and so the rule for the nth term is not linear. Extend the table again to look at the row of 2nd differences:

Position	1	2	3	4	5
Term	0	7	18	33	52
1st difference		7	11	15	19
2nd difference			4	4	4

The row of 2nd differences is constant, and so the rule for the nth term is therefore a quadratic which takes the form $T_n = an^2 + bn + c$.

By substituting the values of n into the rule, each term can be expressed in terms of a, b and c as shown:

Position	1	2	3	4	5
Term	$a + b + c$	$4a + 2b + c$	$9a + 3b + c$	$16a + 4b + c$	$25a + 5b + c$
1st difference		$3a + b$	$5a + b$	$7a + b$	$9a + b$
2nd difference			$2a$	$2a$	$2a$

Comparing the two tables, the values of a, b and c can be deduced:

$2a = 4$ therefore $a = 2$
$3a + b = 7$ therefore $6 + b = 7$ giving $b = 1$
$a + b + c = 0$ therefore $2 + 1 + c = 0$ giving $c = -3$

The rule for the nth term $T_n = an^2 + bn + c$ can be written as $T_n = 2n^2 + n - 3$.

c Deduce the rule for the nth term for the sequence −6, −8, −6, 6, 34, …

Entering the sequence in a table gives:

Position	1	2	3	4	5
Term	−6	−8	−6	6	34

Sequences with quadratic and cubic rules

Extending the table to look at the differences:

Position	1	2	3	4	5
Term	−6	−8	−6	6	34
1st difference		−2	2	12	28

The row of 1st differences is not constant, and so the rule for the nth term is not linear. Extend the table again to look at the row of 2nd differences:

Position	1	2	3	4	5
Term	−6	−8	−6	6	34
1st difference		−2	2	12	28
2nd difference			4	10	16

The row of 2nd differences is not constant either, and so the rule for the nth term is not quadratic. Extend the table by a further row to look at the row of 3rd differences:

Position	1	2	3	4	5
Term	−6	−8	−6	6	34
1st difference		−2	2	12	28
2nd difference			4	10	16
3rd difference			6	6	

The row of 3rd differences is constant, and so the rule for the nth term is therefore cubic which takes the form $T_n = an^3 + bn^2 + cn + d$.

By substituting the values of n into the rule, each term can be expressed in terms of a, b, c and d as shown:

Position	1	2	3	4	5
Term	$a+b+c+d$	$8a+4b+2c+d$	$27a+9b+3c+d$	$64a+16b+4c+d$	$125a+25b+5c+d$
1st difference		$7a+3b+c$	$19a+5b+c$	$37a+7b+c$	$61a+9b+c$
2nd difference			$12a+2b$	$18a+2b$	$24a+2b$
3rd difference			$6a$	$6a$	

By comparing the two tables, equations can be formed and the values of a, b, c and d can be found:

$6a = 6$

therefore $\quad a = 1$

$12a + 2b = 4$

therefore $\quad 12 + 2b = 4 \quad$ giving $b = -4$

$7a + 3b + c = -2$

therefore $\quad 7 - 12 + c = -2 \quad$ giving $c = 3$

$a + b + c + d = -6$

therefore $\quad 1 - 4 + 3 + d = -6 \quad$ giving $d = -6$

Therefore, the equation for the nth term is $T_n = n^3 - 4n^2 + 3n - 6$.

15 SEQUENCES

 Exercise 15.3 By using a table if necessary, find the formula for the nth term of each of the following sequences:

1. 2, 5, 10, 17, 26
2. 0, 3, 8, 15, 24
3. 6, 9, 14, 21, 30
4. 9, 12, 17, 24, 33
5. −2, 1, 6, 13, 22
6. 4, 10, 20, 34, 52
7. 0, 6, 16, 30, 48
8. 5, 14, 29, 50, 77
9. 0, 12, 32, 60, 96
10. 1, 16, 41, 76, 121

 Exercise 15.4 Use a table to find the formula for the nth term of the following sequences:

1. 11, 18, 37, 74, 135
2. 0, 6, 24, 60, 120
3. −4, 3, 22, 59, 120
4. 2, 12, 36, 80, 150
5. 7, 22, 51, 100, 175
6. 7, 28, 67, 130, 223
7. 1, 10, 33, 76, 145
8. 13, 25, 49, 91, 157

> **Note**
>
> In Exercises 15.3 and 15.4, Core students can give their answers using nth terms whilst Extended students are expected to use T_n notation.

Exponential sequences

So far we have looked at sequences where there is a common difference between successive terms. There are, however, other types of sequences, e.g. 2, 4, 8, 16, 32. There is clearly a pattern to the way the numbers are generated as each term is double the previous term, but there is no common difference.

A sequence where there is a **common ratio** (r) between successive terms is known as an **exponential sequence** (or sometimes as a **geometric sequence**).

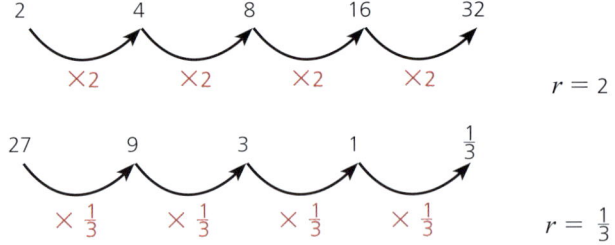

As with an arithmetic sequence, there are two main ways of describing an exponential sequence.

1. The term-to-term rule.
   For example, for the following sequence,

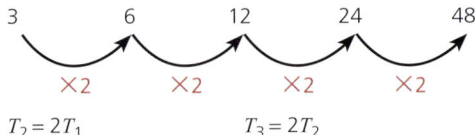

$T_2 = 2T_1$ $T_3 = 2T_2$

the general rule is $T_{n+1} = 2T_n$; $T_1 = 3$.

2   The formula for the *n*th term of an exponential sequence.
    As with an arithmetic sequence, this rule links each term to its
    position in the sequence,

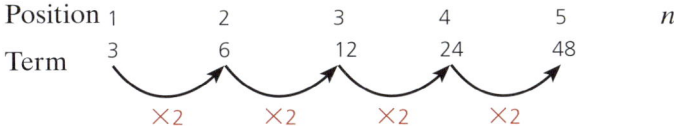

to reach the second term the calculation is 3×2 or 3×2^1
to reach the third term, the calculation is $3 \times 2 \times 2$ or 3×2^2
to reach the fourth term, the calculation is $3 \times 2 \times 2 \times 2$ or 3×2^3
to reach the *n*th term, the calculation is $3 \times 2^{n-1}$
In general therefore

$T_n = ar^{n-1}$

where *a* is the first term and *r* is the common ratio.

Applications of exponential sequences

In Chapter 8, simple and compound interest were shown as different ways that interest could be earned on money left in a bank account for a period of time. Here we look at compound interest as an example of an exponential sequence.

Compound interest

e.g. $100 is deposited in a bank account and left untouched. After 1 year the amount has increased to $110 as a result of interest payments. To work out the interest rate, calculate the multiplier from $100 → $110:

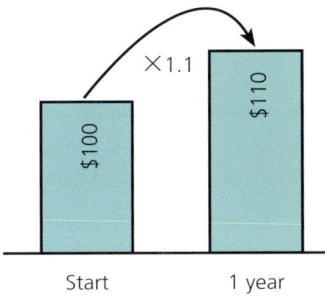

The multiplier is 1.1. This corresponds to a 10% increase. Therefore the simple interest rate is 10% in the first year.

Assume the money is left in the account and that the interest rate remains unchanged. Calculate the amount in the account after 5 years.

This is an example of an exponential sequence.

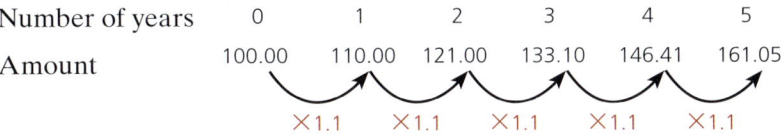

15 SEQUENCES

Alternatively, the amount after 5 years can be calculated using a variation of $T_n = ar^{n-1}$, i.e. $T_5 = 100 \times 1.1^5 = 161.05$. Note: As the number of years starts at 0, ×1.1 is applied 5 times to get to the fifth year.

This is an example of compound interest as the previous year's interest is added to the total and included in the following year's calculation.

→ Worked examples

a Jivan deposits $1500 in his savings account. The interest rate offered by the savings account is 6% each year for a 10-year period. Assuming Jivan leaves the money in the account, calculate how much interest he has gained after the 10 years.

An interest rate of 6% implies a common ratio of 1.06
Therefore $T_{10} = 1500 \times 1.06^{10} = 2686.27$
The amount of interest gained is $2686.27 - 1500 = \$1186.27$

b Adrienne deposits $2000 in her savings account. The interest rate offered by the bank for this account is 8% compound interest per year. Calculate the number of years Adrienne needs to leave the money in her account for it to double in value.

An interest rate of 8% implies a common ratio of 1.08
The amount each year can be found using the term-to-term rule
$T_{n+1} = 1.08 \times T_n$

$T_1 = 2000 \times 1.08 = 2160$
$T_2 = 2160 \times 1.08 = 2332.80$
$T_3 = 2332.80 \times 1.08 = 2519.42$
...
$T_9 = 3998.01$
$T_{10} = 4317.85$

Adrienne needs to leave the money in the account for 10 years in order for it to double in value.
Although answers have been rounded to 2 d.p. in some places, the full answer has been stored in the calculator and used in the next calculation. This avoids rounding errors affecting the final solution.

Exercise 15.5

1 Identify which of the following are exponential sequences and which are not.
   a 2, 6, 18, 54
   b 25, 5, 1, $\frac{1}{5}$
   c 1, 4, 9, 16
   d −3, 9, −27, 81
   e $\frac{1}{2}, \frac{2}{3}, \frac{3}{4}, \frac{4}{5}$
   f $\frac{1}{2}, \frac{2}{4}, \frac{4}{8}, \frac{8}{16}$

2 For the sequences in Question 1 that are exponential, calculate:
   i   the common ratio r
   ii the next two terms
   iii a formula for the nth term.

3 The nth term of an exponential sequence is given by the formula $T_n = -6 \times 2^{n-1}$.
  a Calculate T_1, T_2 and T_3.
  b What is the value of n if $T_n = -768$?

4 Part of an exponential sequence is given below:
  …, -1, …, …, 64, …   where $T_2 = -1$ and $T_5 = 64$.
  Calculate:
  a the common ratio r
  b the value of T_1
  c the value of T_{10}.

5 A homebuyer takes out a loan with a mortgage company for $200 000. The interest rate is 6% per year. If she is unable to repay any of the loan during the first 3 years, calculate the extra amount she will have to pay because of interest by the end of the third year.

6 A car is bought for $10 000. It loses value at a rate of 20% each year.
  a Explain why the car is not worthless after 5 years.
  b Calculate its value after 5 years.
  c Explain why a depreciation of 20% per year means, in theory, that the car will never be worthless.

Combinations of sequences

So far we have looked at linear and exponential sequences and explored different ways of working out the term-to-term rule and the rule for the nth term. However, sometimes sequences are just variations of well-known ones. Being aware of these can often save a lot of time and effort.

Consider the sequence 2, 5, 10, 17, 26. By looking at the differences it could be established that the 2nd differences are constant and, therefore, the formula for the nth term will involve an n^2 term, but this takes quite a lot of time.

On closer inspection, when compared with the sequence 1, 4, 9, 16, 25, we can see that each of the terms in the sequence 2, 5, 10, 17, 26 has had 1 added to it.

1, 4, 9, 16, 25 is the well-known sequence of square numbers. The formula for the nth term for the sequence of square numbers is $T_n = n^2$, therefore the formula for the nth term for the sequence 2, 5, 10, 17, 26 is $T_n = n^2 + 1$.

> **Note**
>
> It is good practice to be aware of key sequences and to always check whether a sequence you are looking at is a variation of one of those. The key sequences include:
>
> Square numbers 1, 4, 9, 16, 25….. where $T_n = n^2$
> Cube numbers 1, 8, 27, 64, 125….. where $T_n = n^3$
> Powers of two 2, 4, 8, 16, 32….. where $T_n = 2^n$
> Triangle numbers 1, 3, 6, 10, 15….. where $T_n = \frac{1}{2}n(n+1)$

15 SEQUENCES

Worked examples

a Consider the sequence below:

2, 8, 18, 32

  i By inspection, write the rule for the *n*th term of the sequence.

  The terms of the sequence are double those of the sequence of square numbers, therefore $T_n = 2n^2$.

  ii Write down the next two terms.

  50, 72

b For the sequence below, by inspection, write down the rule for the *n*th term and the next two terms.

1, 3, 7, 15, 31

The terms of the sequence are one less than the terms of the sequence of powers of two. Therefore $T_n = 2^n - 1$

The next two terms are 63, 127.

Exercise 15.6

In each of the questions below:
i write down the rule for the *n*th term of the sequence by inspection
ii write down the next two terms of the sequence.

1 2, 5, 10, 17
2 3, 10, 29, 66
3 $\frac{1}{2}$, 2, $4\frac{1}{2}$, 8
4 1, 2, 4, 8
5 2, 6, 12, 20
6 2, 12, 36, 80
7 0, 4, 18, 48
8 6, 12, 24, 48

Student assessment 1

1 For each of the sequences given below:
  i calculate the next two terms,
  ii explain the pattern in words.
  a 9, 18, 27, 36, …   b 54, 48, 42, 36, …
  c 18, 9, 4.5, …   d 12, 6, 0, −6, …
  e 216, 125, 64, …   f 1, 3, 9, 27, …

2 For each of the sequences shown below, give an expression for the *n*th term:
  a 6, 10, 14, 18, 22, …   b 13, 19, 25, 31, …
  c 3, 9, 15, 21, 27, …   d 4, 7, 12, 19, 28, …
  e 0, 10, 20, 30, 40, …   f 0, 7, 26, 63, 124, …

3 For each of the following linear sequences:
  i write down a formula for the *n*th term
  ii calculate the 10th term.
  a 1, 5, 9, 13, …   b 1, −2, −5, −8, …

Combinations of sequences

4 Copy and complete both of the following tables of linear sequences:

a
Position	1	2	3	10		n
Term	17	14			−55	

b
Position	2	6	10		n
Term	−4	−2		35	

Student assessment 2

1 George deposits $300 in a bank account. The bank offers 7% interest per year.
Assuming George does not take any money out of the account, calculate:
a the amount of money in the account after 8 years
b the minimum number of years the money must be left in the account, for the amount to be greater than $350.

2 A computer loses 35% of its value each year. If the computer cost $600 new, calculate:
a its value after 2 years
b its value after 10 years.

3 Part of an exponential sequence is given below:
…, …, 27, …, …, −1
where $T_3 = 27$ and $T_6 = -1$.
Calculate:
a the common ratio r
b the value T_1
c the value of n if $T_n = -\frac{1}{81}$

4 Using a table of differences if necessary, calculate the rule for the nth term of the sequence 8, 24, 58, 116, 204 …

5 Using a table of differences, calculate the rule for the nth term of the sequence 10, 23, 50, 97, 170 …

6 For both of the following, calculate T_5 and T_{100}:
a $T_n = 6n - 3$
b $T_n = -\frac{1}{2}n + 4$

16 Proportion

Direct proportion

Consider the tables below:

x	0	1	2	3	5	10
y	0	2	4	6	10	20

$y = 2x$

x	0	1	2	3	5	10
y	0	3	6	9	15	30

$y = 3x$

x	0	1	2	3	5	10
y	0	2.5	5	7.5	12.5	25

$y = 2.5x$

In each case y is **directly proportional** to x. This is written $y \propto x$. If any of these three tables is shown on a graph, the graph will be a straight line passing through the origin.

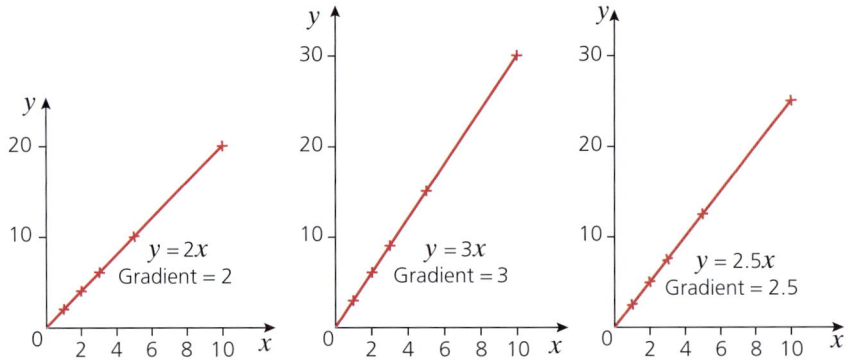

For any statement where $y \propto x$,

$y = kx$

where k is a constant equal to the gradient of the graph and is called the **constant of proportionality** or constant of variation.

Consider the tables below:

x	1	2	3	4	5
y	2	8	18	32	50

$y = 2x^2$

x	1	2	3	4	5
y	$\frac{1}{2}$	4	$13\frac{1}{2}$	32	$62\frac{1}{2}$

$y = \frac{1}{2}x^3$

Direct proportion

x	1	2	3	4	5	
y	1	$\sqrt{2}$	$\sqrt{3}$	2	$\sqrt{5}$	$y = \sqrt{x} = x^{\frac{1}{2}}$

In the cases above, y is directly proportional to x^n, where $n > 0$. This can be written as $y \propto x^n$.

The graphs of each of the three equations are shown below:

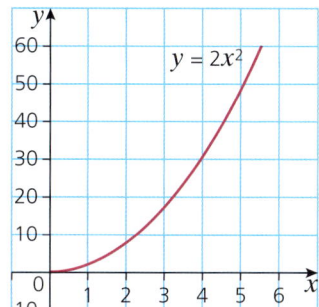 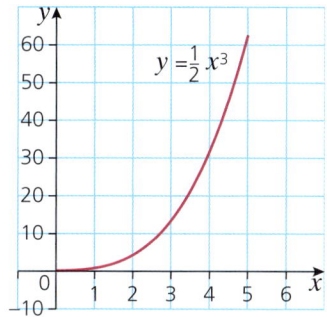 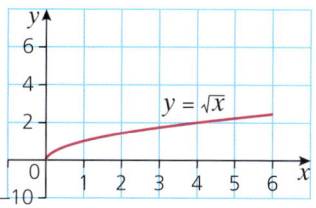

The graphs above, with (x, y) plotted, are not linear. However, if the graph of $y = 2x^2$ is plotted as (x^2, y), then the graph is linear and passes through the origin, demonstrating that $y \propto x^2$ as shown in the graph below.

x	1	2	3	4	5
x^2	1	4	9	16	25
y	2	8	18	32	50

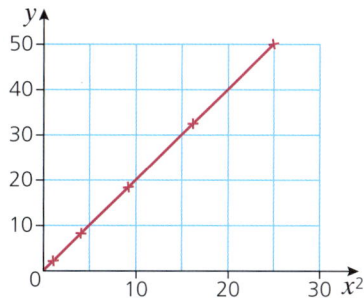

Similarly, the graph of $y = \frac{1}{2}x^3$ is curved when plotted as (x, y), but is linear and passes through the origin if it is plotted as (x^3, y) as shown:

x	1	2	3	4	5
x^3	1	8	27	64	125
y	$\frac{1}{2}$	4	$13\frac{1}{2}$	32	$62\frac{1}{2}$

16 PROPORTION

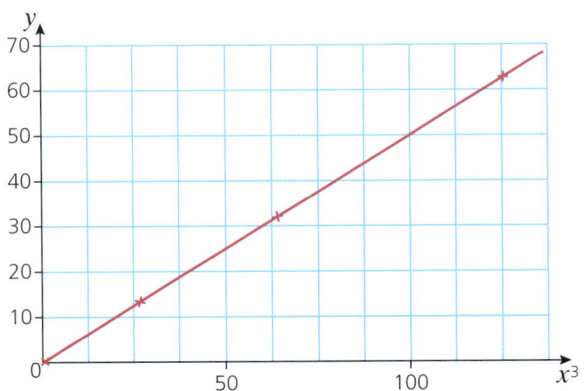

The graph of $y = \sqrt{x}$ is also linear if plotted as $(\sqrt{x}, y)$.

Inverse proportion

If y is **inversely proportional** to x, then $y \propto \frac{1}{x}$ and $y = \frac{k}{x}$.

If a graph of y against $\frac{1}{x}$ is plotted, this too will be a straight line passing through the origin.

→ Worked examples

a $y \propto x$. If $y = 7$ when $x = 2$, find y when $x = 5$.

$y = kx$

$7 = k \times 2$ so $k = 3.5$

When $x = 5$,

$y = 3.5 \times 5 = 17.5$

b $y \propto \frac{1}{x}$. If $y = 5$ when $x = 3$, find y when $x = 30$.

$y = \frac{k}{x}$

$5 = \frac{k}{3}$ so $k = 15$

When $x = 30$,

$y = \frac{15}{30} = 0.5$

Exercise 16.1

1   y is directly proportional to x. If $y = 6$ when $x = 2$, find:
   a   the constant of proportionality
   b   the value of y when $x = 7$
   c   the value of y when $x = 9$
   d   the value of x when $y = 9$
   e   the value of x when $y = 30$.

Inverse proportion

2 y is directly proportional to x^2. If $y = 18$ when $x = 6$, find:
  a the constant of proportionality
  b the value of y when $x = 4$
  c the value of y when $x = 7$
  d the value of x when $y = 32$
  e the value of x when $y = 128$.

3 y is inversely proportional to x^3. If $y = 3$ when $x = 2$, find:
  a the constant of proportionality
  b the value of y when $x = 4$
  c the value of y when $x = 6$
  d the value of x when $y = 24$.

4 y is inversely proportional to x^2. If $y = 1$ when $x = 0.5$, find:
  a the constant of proportionality
  b the value of y when $x = 0.1$
  c the value of y when $x = 0.25$
  d the value of x when $y = 64$.

Exercise 16.2

1 Write each of the following in the form:
  i $y \propto x$ ii $y = kx$.
  a y is directly proportional to x^3
  b y is inversely proportional to x^3
  c t is directly proportional to P
  d s is inversely proportional to t
  e A is directly proportional to r^2
  f T is inversely proportional to the square root of g.

2 If $y \propto x$ and $y = 6$ when $x = 2$, find y when $x = 3.5$.

3 If $y \propto \frac{1}{x}$ and $y = 4$ when $x = 2.5$ find:
  a y when $x = 20$
  b x when $y = 5$.

4 If $p \propto r^2$ and $p = 2$ when $r = 2$, find p when $r = 8$.

5 If $m \propto \frac{1}{r^3}$ and $m = 1$ when $r = 2$, find:
  a m when $r = 4$
  b r when $m = 125$.

6 If $y \propto x^2$ and $y = 12$ when $x = 2$, find y when $x = 5$.

Exercise 16.3

1 If a stone is dropped off the edge of a cliff, the height (h metres) of the cliff is proportional to the square of the time (t seconds) taken for the stone to reach the ground.
  A stone takes 5 seconds to reach the ground when dropped off a cliff 125 m high.
  a Write down a relationship between h and t, using k as the constant of proportionality.
  b Calculate the constant of proportionality.
  c Find the height of a cliff if a stone takes 3 seconds to reach the ground.
  d Find the time taken for a stone to fall from a cliff 180 m high.

16 PROPORTION

Exercise 16.3 (cont)

2. The velocity (v metres per second) of a body is known to be proportional to the square root of its kinetic energy (e joules). When the velocity of a body is 120 m/s, its kinetic energy is 1600 J.
   a. Write down a relationship between v and e, using k as the constant of proportionality.
   b. Calculate the value of k.
   c. If $v = 21$, calculate the kinetic energy of the body in joules.

3. The length (l cm) of an edge of a cube is proportional to the cube root of its mass (m grams). It is known that if $l = 15$, then $m = 125$. Let k be the constant of proportionality.
   a. Write down the relationship between l, m and k.
   b. Calculate the value of k.
   c. Calculate the value of l when $m = 8$.

4. The power (P) generated in an electrical circuit is proportional to the square of the current (I amps). When the power is 108 watts, the current is 6 amps.
   a. Write down a relationship between P, I and the constant of proportionality, k.
   b. Calculate the value of I when $P = 75$.

Student assessment 1

1. $y = kx$. When $y = 12$, $x = 8$.
   a. Calculate the value of k.
   b. Calculate y when $x = 10$.
   c. Calculate y when $x = 2$.
   d. Calculate x when $y = 18$.

2. $y = \dfrac{k}{x}$. When $y = 2$, $x = 5$.
   a. Calculate the value of k.
   b. Calculate y when $x = 4$.
   c. Calculate x when $y = 10$.
   d. Calculate x when $y = 0.5$.

3. $p = kq^3$. When $p = 9$, $q = 3$.
   a. Calculate the value of k.
   b. Calculate p when $q = 6$.
   c. Calculate p when $q = 1$.
   d. Calculate q when $p = 576$.

4. $m = \dfrac{k}{\sqrt{n}}$. When $m = 1$, $n = 25$.
   a. Calculate the value of k.
   b. Calculate m when $n = 16$.
   c. Calculate m when $n = 100$.
   d. Calculate n when $m = 5$.

5. $y = \dfrac{k}{x^2}$. When $y = 3$, $x = \dfrac{1}{3}$.
   a. Calculate the value of k.
   b. Calculate y when $x = 0.5$.
   c. Calculate both values of x when $y = \dfrac{1}{12}$.
   d. Calculate both values of x when $y = \dfrac{1}{3}$.

Student assessment 2

1. y is inversely proportional to x.
   a Copy and complete the table below:

x	1	2	4	8	16	32
y				4		

   b What is the value of x when $y = 20$?

2. Copy and complete the tables below:
   a $y \propto x$

x	1	2	4	5	10
y		10			

   b $y \propto \dfrac{1}{x}$

x	1	2	4	5	10
y	20				

   c $y \propto \sqrt{x}$

x	4	16	25	36	64
y	4				

3. The pressure (P) of a given mass of gas is inversely proportional to its volume (V) at a constant temperature. If $P = 4$ when $V = 6$, calculate:
   a P when $V = 30$
   b V when $P = 30$.

4. The gravitational force (F) between two masses is inversely proportional to the square of the distance (d) between them. If $F = 4$ when $d = 5$, calculate:
   a F when $d = 8$
   b d when $F = 25$.

17 Graphs in practical situations

Conversion graphs

A straight-line graph can be used to convert one set of units to another. Examples include converting from one currency to another, converting distance in miles to kilometres and converting temperature from degrees Celsius to degrees Fahrenheit.

→ Worked example

The graph below converts South African rand into euros based on an exchange rate of €1 = 8.80 rand.

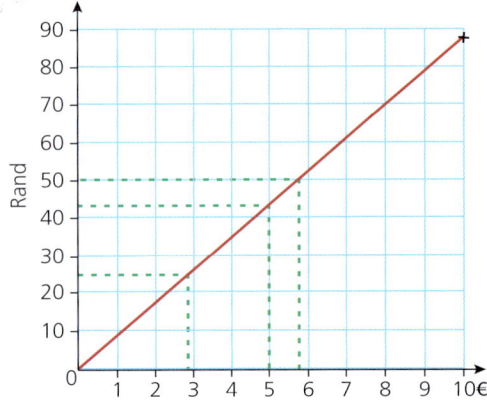

a Using the graph, estimate the number of rand equivalent to €5.

A line is drawn up from €5 until it reaches the plotted line, then across to the vertical axis.
From the graph it can be seen that €5 ≈ 44 rand.
(≈ is the symbol for 'is approximately equal to')

b Using the graph, find the cost in euros of a drink costing 25 rand.

A line is drawn across from 25 rand until it reaches the plotted line, then down to the horizontal axis.
From the graph it can be seen that the cost of the drink ≈ €2.80.

c If a meal costs 200 rand, use the graph to estimate its cost in euros.

The graph does not go up to 200 rand, therefore a factor of 200 needs to be used, e.g. 50 rand.
From the graph, 50 rand ≈ €5.70, therefore it can be deduced that 200 rand ≈ €22.80 (i.e. 4 × €5.70).

Speed, distance and time

Exercise 17.1

1. Given that 80 km = 50 miles, draw a **conversion graph** up to 100 km. Using your graph, estimate:
   a. how many miles is 50 km,
   b. how many kilometres is 80 miles,
   c. the speed in miles per hour (mph) equivalent to 100 km/h,
   d. the speed in km/h equivalent to 40 mph.

2. You can roughly convert temperature in degrees Celsius to degrees Fahrenheit by doubling the degrees Celsius and adding 30. Draw a conversion graph up to 50 °C. Use your graph to estimate the following:
   a. the temperature in °F equivalent to 25 °C,
   b. the temperature in °C equivalent to 100 °F,
   c. the temperature in °F equivalent to 0 °C,
   d. the temperature in °C equivalent to 120 °F.

3. Given that 0 °C = 32 °F and 50 °C = 122 °F, on the same graph as in Question 2, draw a true conversion graph.
   i. Use the true graph to calculate the conversions in Question 2.
   ii. Where would you say the rough conversion is most useful?

4. Long-distance calls from New York to Harare are priced at 85 cents/min off peak and $1.20/min at peak times.
   a. Draw, on the same axes, conversion graphs for the two different rates.
   b. From your graph, estimate the cost of an 8 minute call made off peak.
   c. Estimate the cost of the same call made at peak rate.
   d. A caller has $4 of credit on his phone. Estimate how much more time he can talk for if he rings at off peak instead of at peak times.

5. A maths exam is marked out of 120. Draw a conversion graph to change the following marks to percentages.
   a. 80   b. 110   c. 54   d. 72

Speed, distance and time

You may already be aware of the following formula:

$$\text{distance} = \text{speed} \times \text{time}$$

Rearranging the formula gives:

$$\text{speed} = \frac{\text{distance}}{\text{time}}$$

Where the speed is not constant:

$$\text{average speed} = \frac{\text{total distance}}{\text{total time}}$$

17 GRAPHS IN PRACTICAL SITUATIONS

Exercise 17.2

1. Find the average speed of an object moving:
   a 30 m in 5 s
   b 48 m in 12 s
   c 78 km in 2 h
   d 50 km in 2.5 h
   e 400 km in 2 h 30 min
   f 110 km in 2 h 12 min

2. How far will an object travel during:
   a 10 s at 40 m/s
   b 7 s at 26 m/s
   c 3 h at 70 km/h
   d 4 h 15 min at 60 km/h
   e 10 min at 60 km/h
   f 1 h 6 min at 20 m/s?

3. How long will it take to travel:
   a 50 m at 10 m/s
   b 1 km at 20 m/s
   c 2 km at 30 km/h
   d 5 km at 70 m/s
   e 200 cm at 0.4 m/s
   f 1 km at 15 km/h?

Travel graphs

The graph of an object travelling at a constant speed is a straight line as shown.

Gradient $= \dfrac{d}{t}$

The units of the gradient are m/s, hence the gradient of a distance–time graph represents the speed at which the object is travelling.

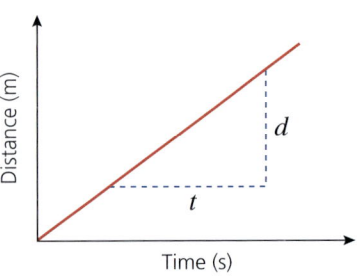

→ Worked example

The graph (right) represents an object travelling at constant speed.

a From the graph, calculate how long it took to cover a distance of 30 m.
   The time taken to travel 30 m is 3 seconds.

b Calculate the gradient of the graph.
   Taking two points on the line, gradient
   $= \dfrac{40}{4} = 10$.

c Calculate the speed at which the object was travelling.

   Gradient of a distance–time graph = speed.
   Therefore the speed is 10 m/s.

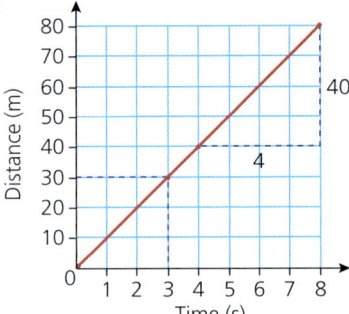

Travel graphs

Exercise 17.3

1. Draw a distance–time graph for the first 10 seconds of an object travelling at 6 m/s.

2. Draw a distance–time graph for the first 10 seconds of an object travelling at 5 m/s. Use your graph to estimate:
   a the time taken to travel 25 m,
   b how far the object travels in 3.5 seconds.

3. Two objects A and B set off from the same point and move in the same straight line. B sets off first, while A sets off 2 seconds later. Using the distance–time graph estimate:
   a the speed of each of the objects,
   b how far apart the objects would be 20 seconds after the start.

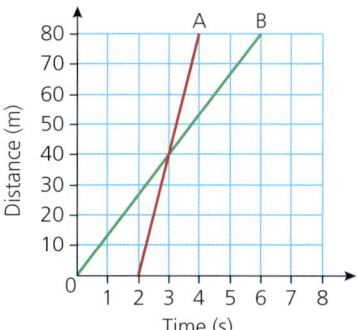

4. Three objects A, B and C move in the same straight line away from a point X. Both A and C change their speed during the journey, while B travels at the same constant speed throughout.

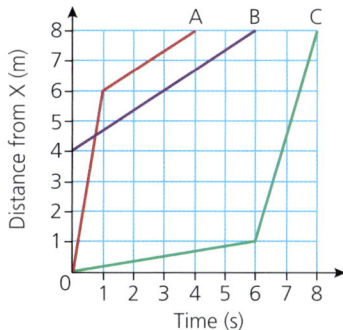

From the distance–time graph, estimate:
   a the speed of object B,
   b the two speeds of object A,
   c the average speed of object C,
   d how far object C is from X, 3 seconds from the start,
   e how far apart objects A and C are 4 seconds from the start.

17 GRAPHS IN PRACTICAL SITUATIONS

The graphs of two or more journeys can be shown on the same axes. The shape of the graph gives a clear picture of the movement of each of the objects.

→ Worked example

The journeys of two cars, X and Y, travelling between A and B are represented on the distance–time graph (right). Car X and Car Y both reach point B 100 km from A at 11 00.

a Calculate the speed of Car X between 07 00 and 08 00.

$$\text{speed} = \frac{\text{distance}}{\text{time}}$$

 $= \frac{60}{1} = \text{km/h} = 60 \text{ km/h}$

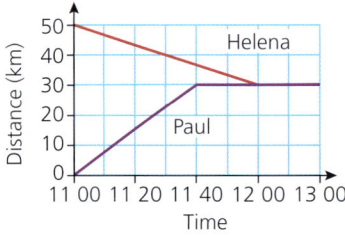

b Calculate the speed of Car Y between 09 00 and 11 00.

$\text{speed} = \frac{100}{2} \text{ km/h} = 50 \text{ km/h}$

c Explain what is happening to Car X between 08 00 and 09 00.

No distance has been travelled, therefore Car X is stationary.

Exercise 17.4

1 Two friends, Paul and Helena, arrange to meet for lunch at noon. They live 50 km apart and the restaurant is 30 km from Paul's home. The **travel graph** illustrates their journeys.

   a What is Paul's average speed between 11 00 and 11 40?
   b What is Helena's average speed between 11 00 and 12 00?
   c What does the horizontal part of Paul's line represent?

2 A car travels at a speed of 60 km/h for 1 hour. It then stops for 30 minutes and then continues at a constant speed of 80 km/h for a further 1.5 hours. Draw a distance–time graph for this journey.

3 Fadi cycles for 1.5 hours at 10 km/h. He then stops for an hour and then travels for a further 15 km in 1 hour. Draw a distance–time graph of Fadi's journey.

4 Two friends leave their houses at 16 00. The houses are 4 km apart and the friends travel towards each other on the same road. Fyodor walks at 7 km/h and Yin walks at 5 km/h.
   a On the same axes, draw a distance–time graph of their journeys.
   b From your graph, estimate the time at which they meet.
   c Estimate the distance from Fyodor's house to the point where they meet.

Speed–time graphs, acceleration and deceleration

5 A train leaves a station P at 18 00 and travels to station Q 150 km away. It travels at a steady speed of 75 km/h. At 18 10 another train leaves Q for P at a steady speed of 100 km/h.
   a On the same axes, draw a distance–time graph to show both journeys.
   b From the graph, estimate the time at which both trains pass each other.
   c At what distance from station Q do both trains pass each other?
   d Which train arrives at its destination first?

6 A train sets off from town P at 09 15 and heads towards town Q 250 km away. Its journey is split into the three stages, a, b and c. At 09 00 a second train leaves town Q heading for town P. Its journey is split into the two stages, d and e.

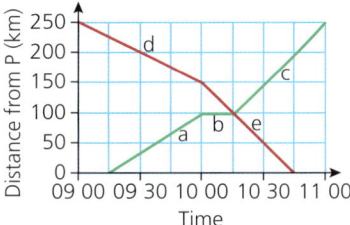

Using the graph, calculate the following:
   a the speed of the first train during stages a, b and c,
   b the speed of the second train during stages d and e.

Speed–time graphs, acceleration and deceleration

So far the graphs that have been dealt with have been like the one shown, i.e. distance–time graphs.

If the graph were of a girl walking, it would indicate that initially she was walking at a constant speed of 1.5 m/s for 10 seconds, then she stopped for 20 seconds and finally she walked at a constant speed of 0.5 m/s for 20 seconds.

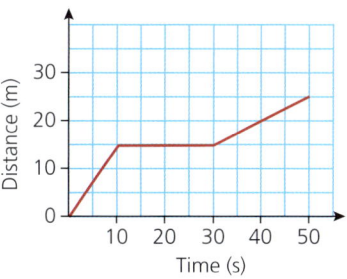

> **Note**
>
> For a distance–time graph, the following is true:
> - a straight line represents constant speed
> - a horizontal line indicates no movement (i.e. speed is zero)
> - the gradient of a line gives the speed.

This section also deals with the interpretation of travel graphs, but where the vertical axis represents the object's speed.

181

17 GRAPHS IN PRACTICAL SITUATIONS

> **Worked example**

The graph shows the speed of a car over a period of 16 seconds.

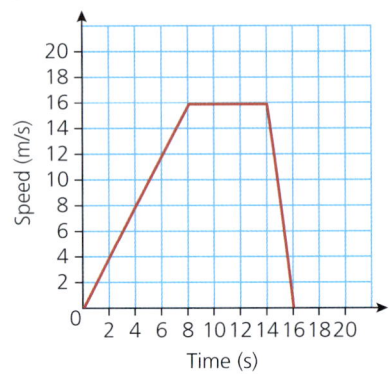

a   Explain the shape of the graph.
    For the first 8 seconds the speed of the car is increasing uniformly with time. This means it is **accelerating at a constant rate**. Between 8 and 14 seconds, the car is travelling at a constant speed of 16 m/s. Between 14 and 16 seconds, the speed of the car decreases uniformly. This means that it is **decelerating at a constant rate**.

b   Calculate the rate of acceleration during the first 8 seconds.
    From a speed–time graph, the acceleration is found by calculating the gradient of the line. Therefore:
    $$\text{acceleration} = \frac{16}{8} = 2\,\text{m/s}^2$$

c   Calculate the rate of deceleration between 14 and 16 seconds:
    $$\text{deceleration} = \frac{16}{2} = 8\,\text{m/s}^2$$

Exercise 17.5   Using the graphs below, calculate the acceleration/deceleration in each case.

1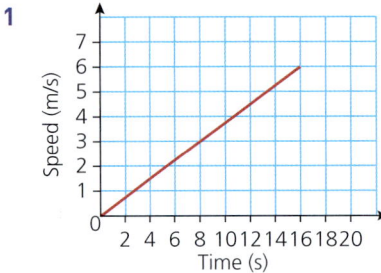

2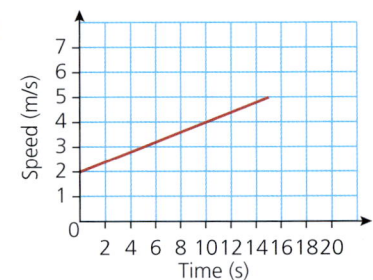

Speed–time graphs, acceleration and deceleration

3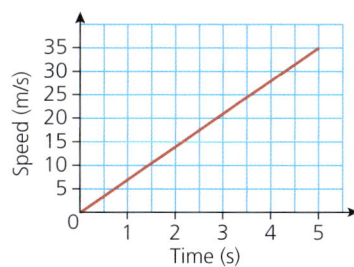

4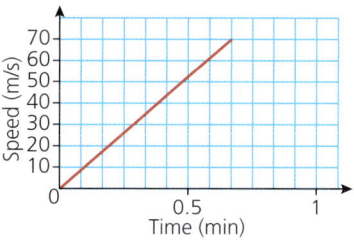

5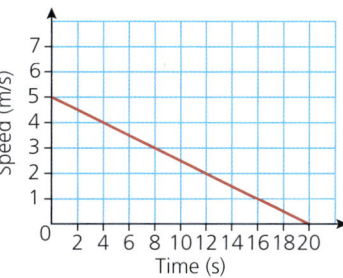

6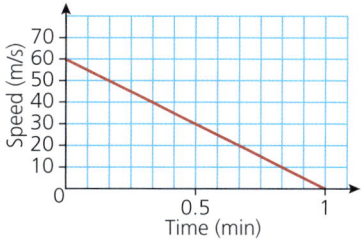

7 Sketch a graph to show an aeroplane accelerating from rest at a constant rate of 5 m/s² for 10 seconds.

8 A train travelling at 30 m/s starts to decelerate at a constant rate of 3 m/s². Sketch a speed–time graph showing the train's motion until it stops.

Exercise 17.6

1 The graph shows the speed–time graph of a boy running for 20 seconds.

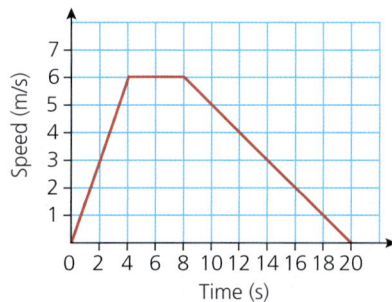

Calculate:
a the acceleration during the first four seconds,
b the acceleration during the second period of four seconds,
c the deceleration during the final twelve seconds.

17 GRAPHS IN PRACTICAL SITUATIONS

Exercise 17.6 (cont)

2 The speed–time graph represents a cheetah chasing a gazelle.

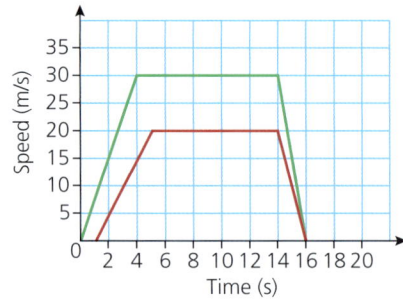

  a Does the top graph represent the cheetah or the gazelle?
  b Calculate the cheetah's acceleration in the initial stages of the chase.
  c Calculate the gazelle's acceleration in the initial stages of the chase.
  d Calculate the cheetah's deceleration at the end.

3 The speed–time graph represents a train travelling from one station to another.

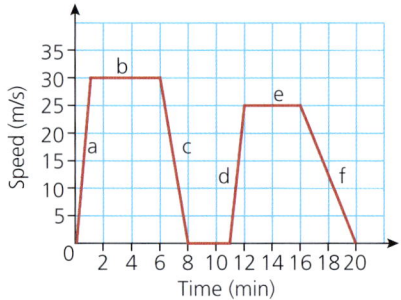

  a Calculate the acceleration during stage a.
  b Calculate the deceleration during stage c.
  c Calculate the deceleration during stage f.
  d Describe the train's motion during stage b.
  e Describe the train's motion 10 minutes from the start.

Area under a speed–time graph

The area under a speed–time graph gives the distance travelled.

> ### Worked example
>
> The table below shows the speed of a train over a 30-second period.
>
> | Time (s) | 0 | 5 | 10 | 15 | 20 | 25 | 30 |
> |---|---|---|---|---|---|---|---|
> | Speed (m/s) | 20 | 20 | 20 | 22.5 | 25 | 27.5 | 30 |

a Plot a speed–time graph for the first 30 seconds.

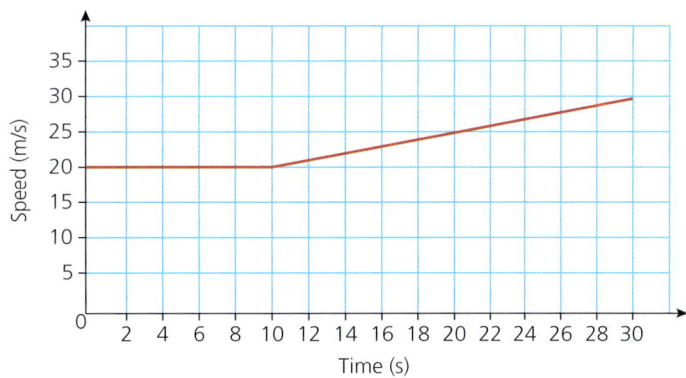

b Calculate the train's acceleration in the last 20 seconds.

Acceleration = $\frac{10}{20} = \frac{1}{2}$ m/s²

c Calculate the distance travelled during the 30 seconds.

This is calculated by working out the area under the graph. The graph can be split into two regions as shown below.

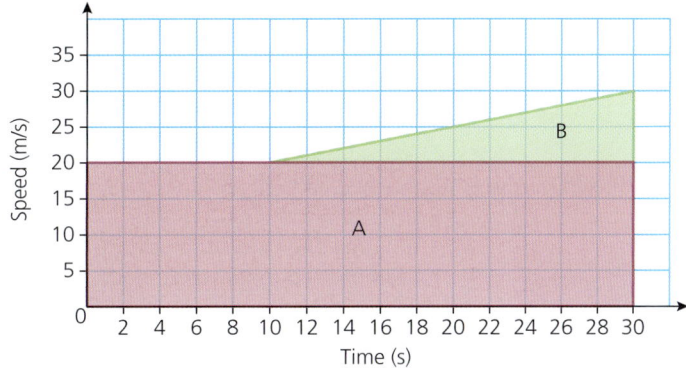

Distance represented by region A = (20×30) m
= 600 m

Distance represented by region B = $\left(\frac{1}{2} \times 20 \times 10\right)$ m
= 100 m

Total distance travelled = $(600 + 100)$ m
= 700 m

17 GRAPHS IN PRACTICAL SITUATIONS

Exercise 17.7

1 The table below gives the speed of a boat over a 10-second period.

Time (s)	0	2	4	6	8	10
Speed (m/s)	5	6	7	8	9	10

   a Plot a speed–time graph for the 10-second period.
   b Calculate the acceleration of the boat.
   c Calculate the total distance travelled during the 10 seconds.

2 A cyclist travelling at 6 m/s applies the brakes and decelerates at a constant rate of 2 m/s².
   a Copy and complete the table below.

Time (s)	0	0.5	1	1.5	2	2.5	3
Speed (m/s)	6						0

   b Plot a speed–time graph for the 3 seconds shown in the table above.
   c Calculate the distance travelled during the 3 seconds of deceleration.

3 A car accelerates as shown in the graph.

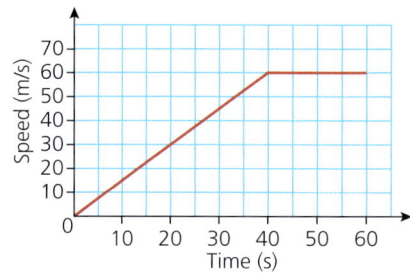

   a Calculate the rate of acceleration in the first 40 seconds.
   b Calculate the distance travelled over the 60 seconds shown.
   c After what time had the motorist travelled half the distance?

4 The graph represents the cheetah and gazelle chase from Question 2 in Exercise 17.6.

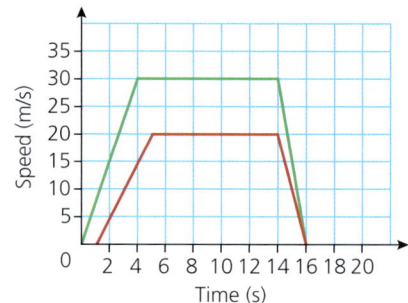

   a Calculate the distance run by the cheetah during the chase.
   b Calculate the distance run by the gazelle during the chase.

Non-linear travel graphs

5 The graph represents the train journey from Question 3 in Exercise 17.6. Calculate, in km, the distance travelled during the 20 minutes shown.

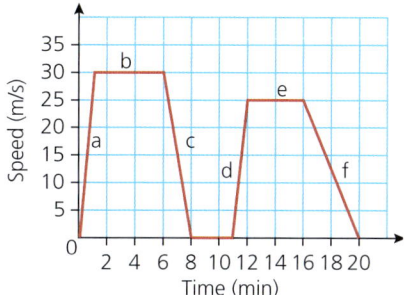

6 An aircraft accelerates uniformly from rest at a rate of $10\,\text{m/s}^2$ for 12 seconds before it takes off. Calculate the distance it travels along the runway.

7 The speed–time graph below depicts the motion of two motorbikes A and B over a 15-second period.

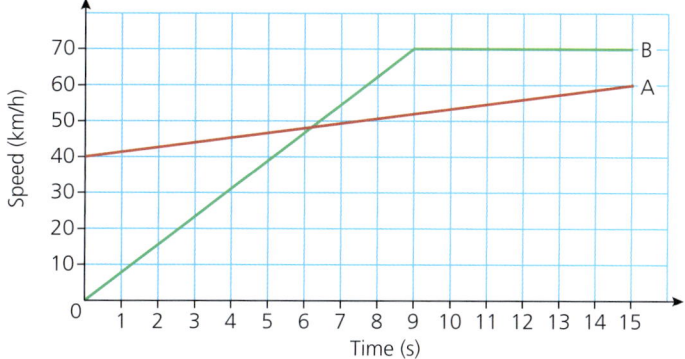

At the start of the graph, motorbike A overtakes a stationary motorbike B. Assume they then travel in the same direction.
  a Calculate motorbike A's acceleration over the 15 seconds in m/s^2.
  b Calculate motorbike B's acceleration over the first 9 seconds in m/s^2.
  c Calculate the distance travelled by A during the 15 seconds (give your answer to the nearest metre).
  d Calculate the distance travelled by B during the 15 seconds (give your answer to the nearest metre).
  e How far apart were the two motorbikes at the end of the 15-second period?

Non-linear travel graphs

So far, all the graphs investigated have involved straight lines. In real life, however, the motion of objects is unlikely to produce a straight line as changes in movement tend to be gradual rather than instantaneous.

17 GRAPHS IN PRACTICAL SITUATIONS

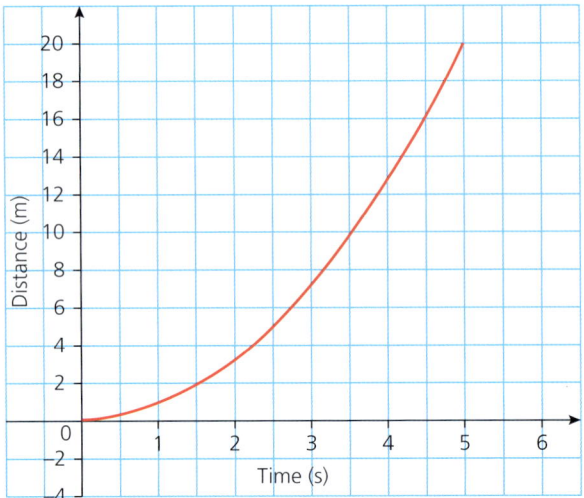

The gradient of a curved distance–time graph is not constant (i.e. it is always changing). To find the speed of an object, for example, 3 seconds after the start, it is necessary to calculate the gradient of the curve at that particular point. This is done by drawing a tangent to the curve at that point. The gradient of the tangent is the same as the gradient of the curve at the same point.

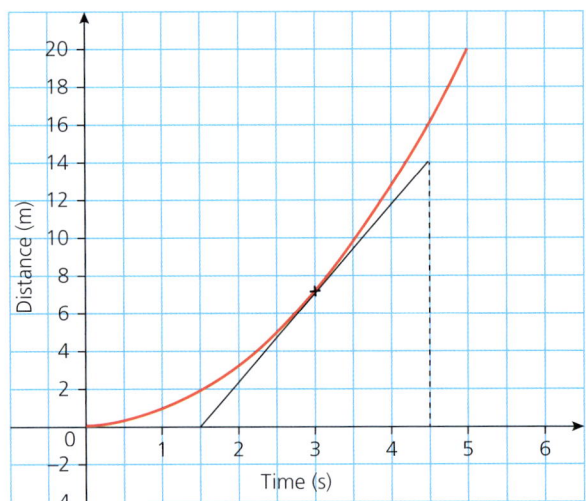

The gradient of the tangent is calculated as

$$\frac{\text{vertical height}}{\text{horizontal distance}} = \frac{14-0}{4.5-1.5} = \frac{14}{3} = 4\frac{2}{3}.$$

Therefore the speed of the object after 3 seconds is approximately $4\frac{2}{3}$ m/s.

Note: The calculation is only approximate as the tangent is drawn by eye and therefore the readings are not exact.

Non-linear travel graphs

Calculating the gradient of a curve is covered in more depth in Chapter 18: Graphs of functions.

The same method is used when calculating the acceleration of an object from a speed–time graph.

The graph above shows the motion of an object over a period of 6 seconds. To calculate the acceleration at a moment in time, for example, 4 seconds after the start, the gradient of the curve at that point needs to be calculated. This is done, once again, by drawing a tangent to the curve at that point and calculating its gradient.

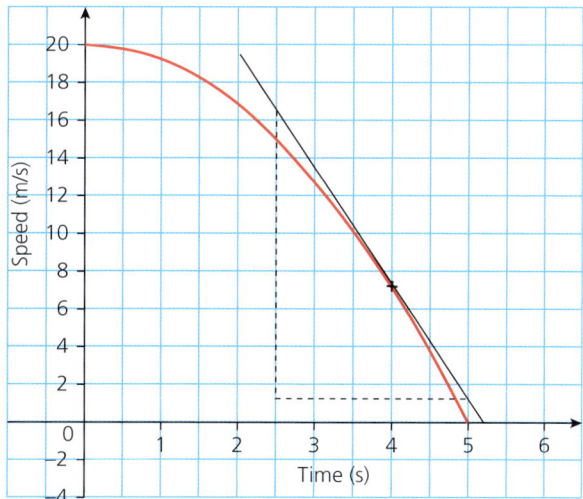

The gradient of the tangent $= \dfrac{17 - 1}{2.5 - 5} = \dfrac{16}{-2.5} = -6.4$.

Therefore, the acceleration after 4 seconds is $-6.4 \, \text{m/s}^2$ (i.e. the object is decelerating at $6.4 \, \text{m/s}^2$).

17 GRAPHS IN PRACTICAL SITUATIONS

Worked example

A cyclist starts cycling from rest. For the first 10 seconds, her distance from the start is given by the equation $D = 0.6t^2$, where D is the distance travelled in metres and t is the time in seconds.

a Complete a distance–time table of results for the first 10 seconds of motion.

Time (s)	0	1	2	3	4	5	6	7	8	9	10
Distance (m)	0	0.6	2.4	5.4	9.6	15	21.6	29.4	38.4	48.6	60

b Plot a graph of the cyclist's motion on a distance–time graph.

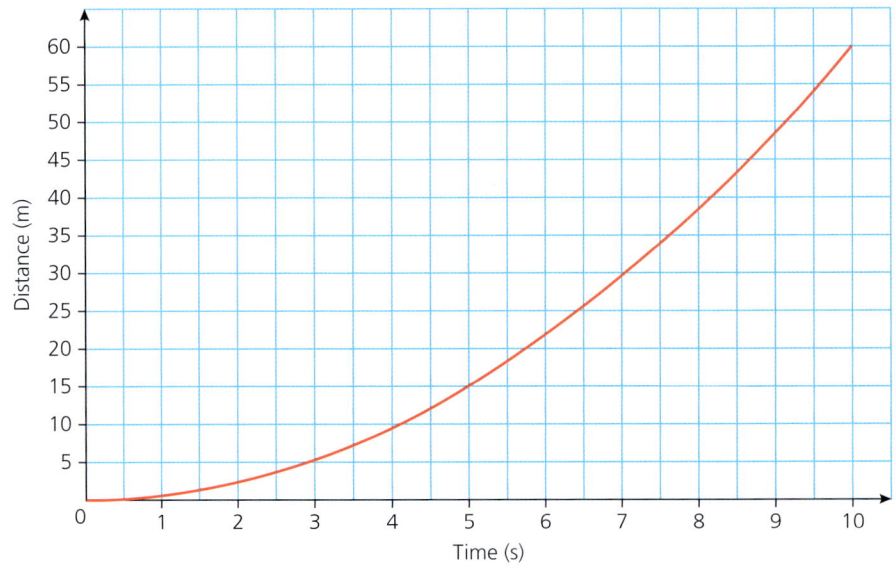

Non-linear travel graphs

c   Calculate the speed of the cyclist after 4s and after 8s.
    Tangents to the graph at 4s and 8s are drawn as shown:

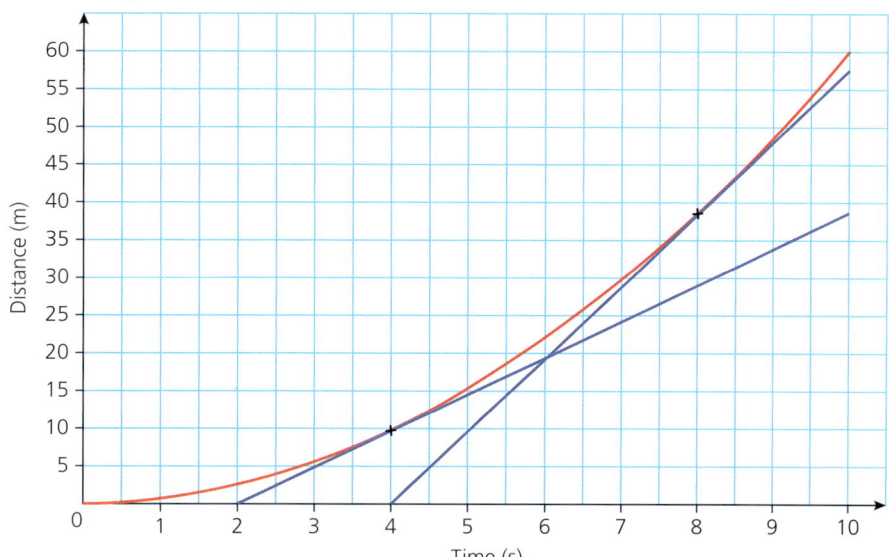

The gradient of the tangent to the curve at 4s = $\frac{36-0}{9.5-2} = \frac{36}{7.5} = 4.8$.

The speed after 4s is 4.8 m/s.

The gradient of the tangent to the curve at 8s = $\frac{48-0}{9-4} = \frac{48}{5} = 9.6$.

The speed after 8s is 9.6 m/s.

Exercise 17.8

1   A stone is dropped off a tall cliff. The distance it falls is given by the equation $d = 4.9t^2$, where d is the distance fallen in metres and t is the time in seconds.
    a   Complete the table of results below for the first 10 seconds.

Time (s)	0	2	4	6	8	10
Distance (m)	0		78.4			

    b   Plot the results in a graph.
    c   From your graph, calculate the speed the stone was travelling at after 5 seconds.

2   A car travelling at 20 m/s applies the brakes. The speed of the car is given by the equation $s = 20 - 0.75t^2$, where s is the speed in m/s and t is the time in seconds.
    a   Plot a graph of the car's motion from the moment the brakes are applied until the moment it comes to rest.
    b   Estimate from your graph the time it takes for the car to come to rest.
    c   Calculate the car's deceleration after 2 seconds.

17 GRAPHS IN PRACTICAL SITUATIONS

Exercise 17.8 (cont)

3 The distance of an object from its starting point is given by the equation $x = 2t^3 - t^2$, where x is the distance in metres from the start and t is the time in seconds.
   a Plot a distance–time graph for the first 6 seconds of motion.
   b Using your graph, estimate the speed of the object after 3.5 seconds.

Student assessment 1

1 Absolute zero (0 K) is equivalent to −273 °C and 0 °C is equivalent to 273 K. Draw a conversion graph which will convert K into °C. Use your graph to estimate:
   a the temperature in K equivalent to −40 °C,
   b the temperature in °C equivalent to 100 K.

2 A plumber has a call-out charge of $70 dollars and then charges a rate of $50 per hour.
   a Draw a conversion graph and estimate the cost of the following:
      i a job lasting $4\frac{1}{2}$ hours,
      ii a job lasting $6\frac{3}{4}$ hours.
   b If a job cost $245, estimate from your graph how long it took to complete.

3 A boy lives 3.5 km from his school. He walks home at a constant speed of 9 km/h for the first 10 minutes. He then stops and talks to his friends for 5 minutes. He finally runs the rest of his journey home at a constant speed of 12 km/h.
   a Illustrate this information on a distance–time graph.
   b Use your graph to estimate the total time it took the boy to get home that day.

4 Below are four distance–time graphs A, B, C and D. Two of them are not possible.
   a Which two graphs are impossible?
   b Explain why the two you have chosen are not possible.

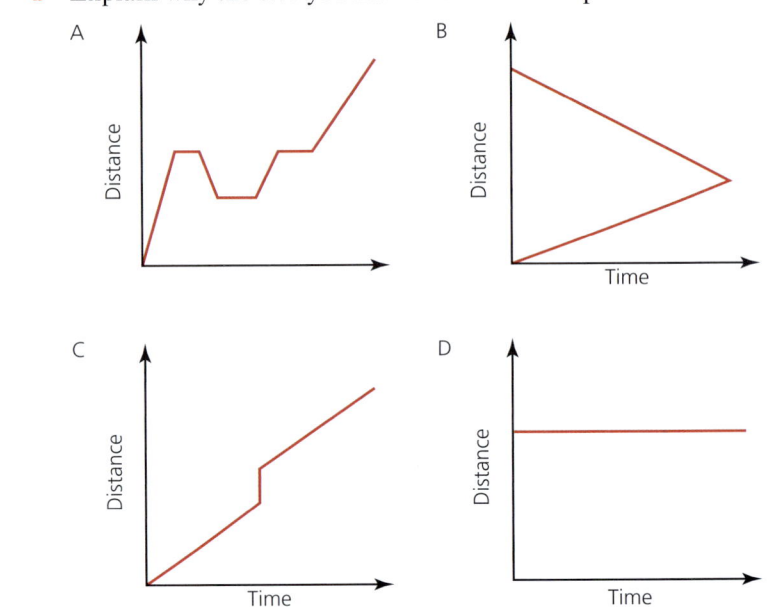

Non-linear travel graphs

Student assessment 2

1. The graph below is a speed–time graph for a car accelerating from rest.

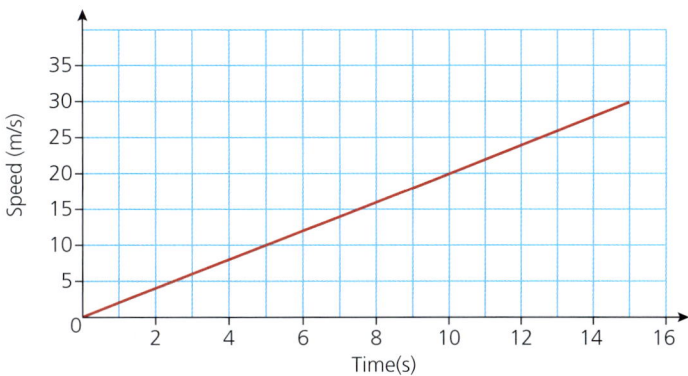

   a Calculate the car's acceleration in m/s².
   b Calculate, in metres, the distance the car travels in 15 seconds.
   c How long did it take the car to travel half the distance?

2. The speed–time graph represents a 100 m sprinter during a race.

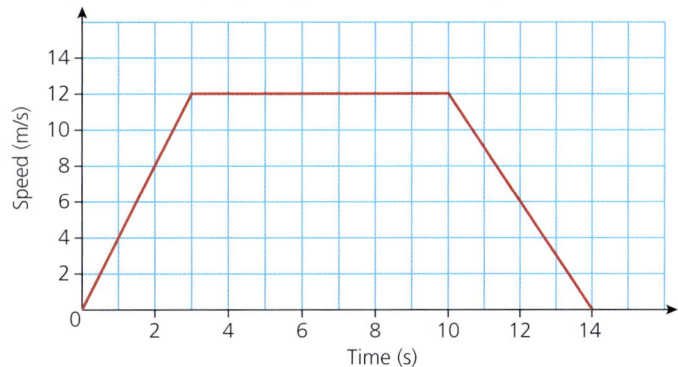

   a Calculate the sprinter's acceleration during the first two seconds of the race.
   b Calculate the sprinter's deceleration at the end of the race.
   c Calculate the distance the sprinter ran in the first 10 seconds.
   d Calculate the sprinter's time for the 100 m race. Give your answer to 3 s.f.

3. A motorcyclist accelerates uniformly from rest to 50 km/h in 8 seconds. She then accelerates to 110 km/h in a further 6 seconds.
   a Draw a speed–time graph for the first 14 seconds.
   b Use your graph to find the total distance the motorcyclist travels. Give your answer in metres.

17 GRAPHS IN PRACTICAL SITUATIONS

4 The graph shows the speed of a car over a period of 50 seconds.

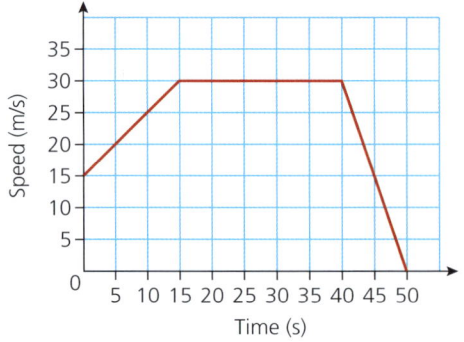

   a Calculate the car's acceleration in the first 15 seconds.
   b Calculate the distance travelled while the car moved at constant speed.
   c Calculate the total distance travelled.

5 The distance of an object from its starting position is given by the equation:
   $s = t^3 - 10t^2 + 21t + 25$, where s is the distance from the start in metres and t is the time in seconds.
   a Copy and complete the table of results:

Time (secs)	0	1	2	3	4	5	6
Distance (m)			35			5	

   b Plot a graph of distance from the start against time for the first 6 seconds of motion.
   c Estimate from your graph when the object is stationary.
   d Estimate the velocity of the object when $t = 5$ seconds.

Student assessment 3

1 The graph below is a speed–time graph for a car decelerating to rest.

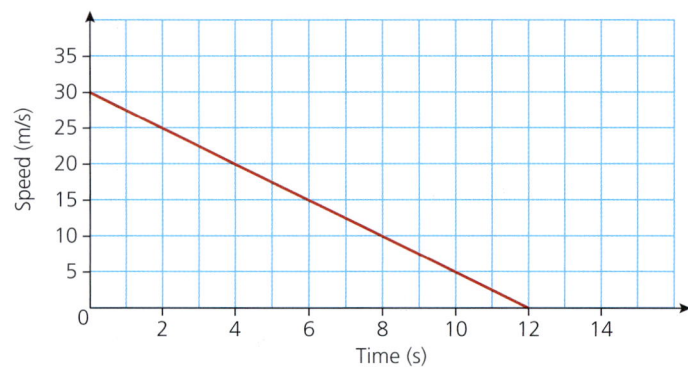

   a Calculate the car's deceleration in m/s^2.
   b Calculate, in metres, the distance the car travels in 12 seconds.
   c How long did it take the car to travel half the distance?

Non-linear travel graphs

2 The graph below shows the speeds of two cars A and B over a 15-second period.

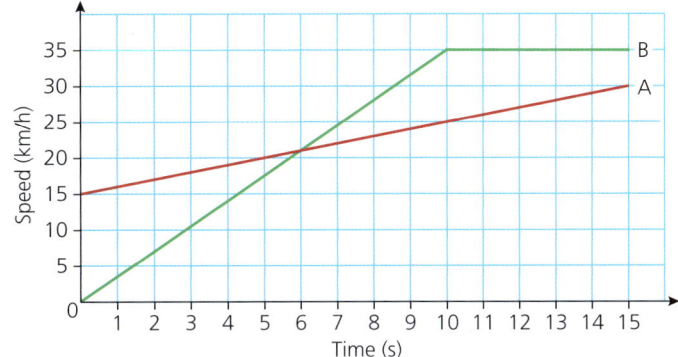

   a Calculate the acceleration of car A in m/s².
   b Calculate the distance travelled in metres during the 15 seconds by car A.
   c Calculate the distance travelled in metres during the 15 seconds by car B.

3 A motor cycle accelerates uniformly from rest to 30 km/h in 3 seconds. It then accelerates to 150 km/h in a further 6 seconds.
   a Draw a speed–time graph for the first 9 seconds.
   b Use your graph to find the total distance the motor cycle travels. Give your answer in metres.

4 Two cars X and Y are travelling in the same direction. The speed–time graph (below) shows their speeds over 12 seconds.

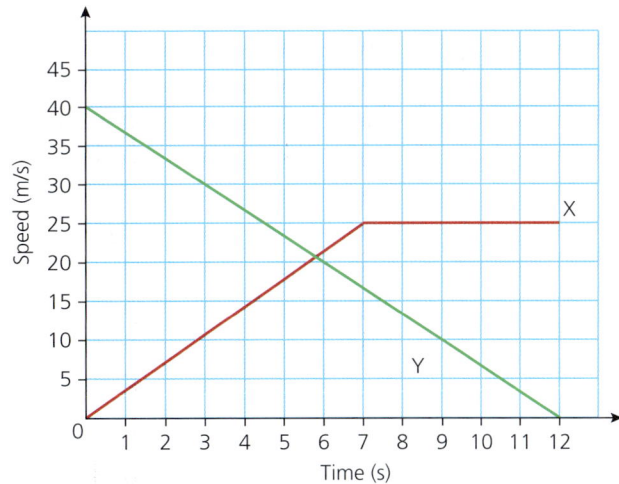

   a Calculate the deceleration of Y during the 12 seconds.
   b Calculate the distance travelled by Y in the 12 seconds.
   c Calculate the total distance travelled by X in the 12 seconds.

195

17 GRAPHS IN PRACTICAL SITUATIONS

5 The speed of an object is recorded over a period of 5 seconds. Its speed is given by the equation $v = t^3 - 4t^2 - t + 14$, where v is the speed of the object in m/s and t is the time in seconds.
  a Plot a speed–time graph for the object's motion in the first 5 seconds.
  b Explain the acceleration of the object at the graph's lowest point. Justify your answer.
  c Estimate the acceleration of the object when $t = 2$ seconds.

18 Graphs of functions

You should be familiar with the work covered in Chapter 21, Straight-line graphs, before working on this chapter.

Quadratic functions

The general expression for a **quadratic function** takes the form $ax^2 + bx + c$, where a, b and c are constants. Some examples of quadratic functions are:

$$y = 2x^2 + 3x + 12 \qquad y = x^2 - 5x + 6 \qquad y = 3x^2 + 2x - 3$$

A graph of a quadratic function produces a smooth curve called a **parabola**, for example:

$y = x^2$

x	−4	−3	−2	−1	0	1	2	3	4
y	16	9	4	1	0	1	4	9	16

$y = -x^2$

x	−4	−3	−2	−1	0	1	2	3	4
y	−16	−9	−4	−1	0	−1	−4	−9	−16

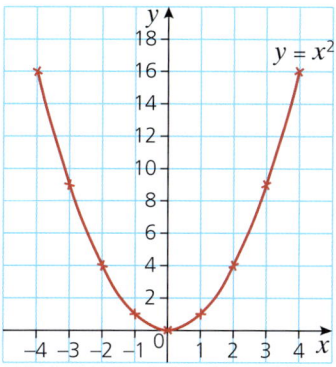
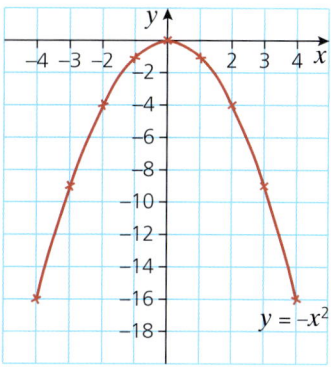

Notice how a quadratic function has a line of symmetry through its turning point.

18 GRAPHS OF FUNCTIONS

Worked examples

a Plot a graph of the function $y = x^2 - 5x + 6$ for $0 \leqslant x \leqslant 5$.

A table of values for x and y is given below:

x	0	1	2	3	4	5
y	6	2	0	0	2	6

These can then be plotted to give the graph:

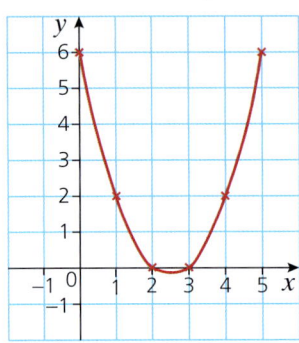

b Plot a graph of the function $y = -x^2 + x + 2$ for $-3 \leqslant x \leqslant 4$.

Drawing up a table of values gives:

x	−3	−2	−1	0	1	2	3	4
y	−10	−4	0	2	2	0	−4	−10

The graph of the function is given below:

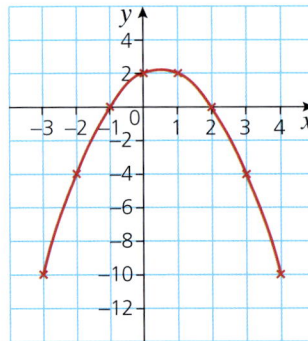

Graphical solution of a quadratic equation

 Exercise 18.1 For each of the following quadratic functions, construct a table of values and then draw the graph.

1. $y = x^2 + x - 2$, $-4 \leqslant x \leqslant 3$
2. $y = -x^2 + 2x + 3$, $-3 \leqslant x \leqslant 5$
3. $y = x^2 - 4x + 4$, $-1 \leqslant x \leqslant 5$
4. $y = -x^2 - 2x - 1$, $-4 \leqslant x \leqslant 2$
5. $y = x^2 - 2x - 15$, $-4 \leqslant x \leqslant 6$
6. $y = 2x^2 - 2x - 3$, $-2 \leqslant x \leqslant 3$
7. $y = -2x^2 + x + 6$, $-3 \leqslant x \leqslant 3$
8. $y = 3x^2 - 3x - 6$, $-2 \leqslant x \leqslant 3$
9. $y = 4x^2 - 7x - 4$, $-1 \leqslant x \leqslant 3$
10. $y = -4x^2 + 4x - 1$, $-2 \leqslant x \leqslant 3$

Graphical solution of a quadratic equation

 Worked example

a Draw a graph of $y = x^2 - 4x + 3$ for $-2 \leqslant x \leqslant 5$.

x	-2	-1	0	1	2	3	4	5
y	15	8	3	0	-1	0	3	8

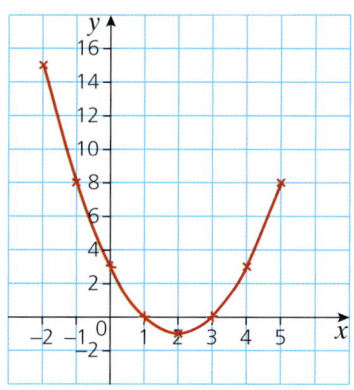

b Use the graph to solve the equation $x^2 - 4x + 3 = 0$.
   To solve the equation it is necessary to find the values of x when $y = 0$, i.e. where the graph crosses the x-axis.
   These points occur when $x = 1$ and $x = 3$ and are therefore the solutions.

199

18 GRAPHS OF FUNCTIONS

 Exercise 18.2 Solve each of the quadratic functions below by plotting a graph for the ranges of x stated.

1. $x^2 - x - 6 = 0$, $-4 \leqslant x \leqslant 4$
2. $-x^2 + 1 = 0$, $-4 \leqslant x \leqslant 4$
3. $x^2 - 6x + 9 = 0$, $0 \leqslant x \leqslant 6$
4. $-x^2 - x + 12 = 0$, $-5 \leqslant x \leqslant 4$
5. $x^2 - 4x + 4 = 0$, $-2 \leqslant x \leqslant 6$
6. $2x^2 - 7x + 3 = 0$, $-1 \leqslant x \leqslant 5$
7. $-2x^2 + 4x - 2 = 0$, $-2 \leqslant x \leqslant 4$
8. $3x^2 - 5x - 2 = 0$, $-1 \leqslant x \leqslant 3$

In the previous worked example, as $y = x^2 - 4x + 3$, a solution could be found to the equation $x^2 - 4x + 3 = 0$ by reading off where the graph crossed the x-axis. The graph can, however, also be used to solve other quadratic equations.

 Worked example

Use the graph of $y = x^2 - 4x + 3$ to solve the equation $x^2 - 4x + 1 = 0$.

$x^2 - 4x + 1 = 0$ can be rearranged to give:

2 has been added to both sides of the equation → $x^2 - 4x + 3 = 2$

Using the graph of $y = x^2 - 4x + 3$ and plotting the line $y = 2$ on the same graph, gives the graph shown below.

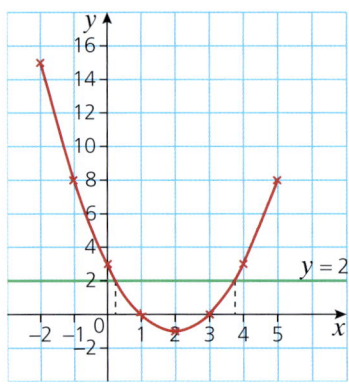

*The point at which a graph of a function changes from a negative gradient to a positive gradient, or vice versa (that is, the point where the gradient is zero) is called a **turning point**. See also stationary points on page 239.*

Where the curve and the line cross gives the solution to $x^2 - 4x + 3 = 2$ and hence also $x^2 - 4x + 1 = 0$.

Therefore the solutions to $x^2 - 4x + 1 = 0$ are

$x \approx 0.3$ and $x \approx 3.7$.

Exercise 18.3 Using the graphs that you drew in Exercise 18.2, solve the following quadratic equations. Show your method clearly.

1. $x^2 - x - 4 = 0$
2. $-x^2 - 1 = 0$
3. $x^2 - 6x + 8 = 0$
4. $-x^2 - x + 9 = 0$
5. $x^2 - 4x + 1 = 0$
6. $2x^2 - 7x = 0$
7. $-2x^2 + 4x = -1$
8. $3x^2 = 2 + 5x$

The completed square form and the graph of a quadratic equation

In Chapter 13, Exercise 13.10, we saw that a quadratic can be written in a form known as the completed square.

For example, $y = x^2 + 6x - 4$ can be written in completed square form as $y = (x + 3)^2 - 13$.

The general form of a quadratic equation written in completed square form is $y = a(x - b)^2 + c$, where a, b and c are constants.

To plot the graph of a quadratic written in completed square form, it is not necessary to expand the brackets first.

➔ Worked examples

a Plot the graph of the function $y = (x - 4)^2 + 2$ for $0 \leqslant x \leqslant 6$.

Drawing up a table of values gives:

x	0	1	2	3	4	5	6
y	18	11	6	3	2	3	6

The graph of the function can now be plotted:

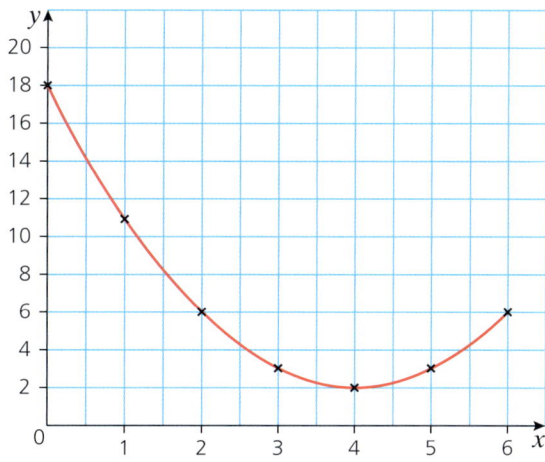

b Write the coordinates of the turning point of the graph.

The turning point in this case is the minimum point of the graph. It occurs at (4, 2).

In general, quadratic graphs take the parabola shape ∪ or ∩

For the ∪ shape the turning point is known as a minimum point, while for the ∩ the turning point is known as the maximum point.

18 GRAPHS OF FUNCTIONS

Exercise 18.4 For each of the quadratics in Questions 1–8:
- a Plot a graph of the function for the values of x given.
- b State whether the turning point is a maximum or a minimum point.
- c Write down the coordinates of the turning point.
- d In each case, write down the coordinates of the point on the curve which the curve's line of symmetry passes through.

1 $y = (x-1)^2 + 4;\quad -2 \leqslant x \leqslant 4$
2 $y = (x+3)^2 - 6;\quad -6 \leqslant x \leqslant 0$
3 $y = (x-4)^2 - 1;\quad 0 \leqslant x \leqslant 7$
4 $y = -(x-2)^2 + 3;\quad -1 \leqslant x \leqslant 5$
5 $y = -(x+1)^2 - 5;\quad -4 \leqslant x \leqslant 3$
6 $y = 2(x-1)^2 - 3;\quad -3 \leqslant x \leqslant 5$
7 $y = 3(x-3)^2 - 20;\quad 0 \leqslant x \leqslant 6$
8 $y = -2(x-5)^2 + 30;\quad 2 \leqslant x \leqslant 8$

9 a Describe any pattern you spot between the coordinates of the turning point and the equation of the quadratics in each of the graphs plotted in Questions 1–8 above.
   b If the general equation of a quadratic in completed square form is given as $y = a(x-b)^2 + c$, write down the coordinates of the turning point.

10 Without plotting the graph, deduce the coordinates of the turning point in each of the following:
   a $y = (x-6)^2 + 5$
   b $y = (x+8)^2 - 6$
   c $y = -(x+4)^2 + 12$
   d $y = -2(x-12)^2 + 1$

The reciprocal function

→ Worked example

Draw the graph of $y = \frac{2}{x}$ for $-4 \leqslant x \leqslant 4$.

x	−4	−3	−2	−1	0	1	2	3	4
y	−0.5	−0.7	−1	−2	—	2	1	0.7	0.5

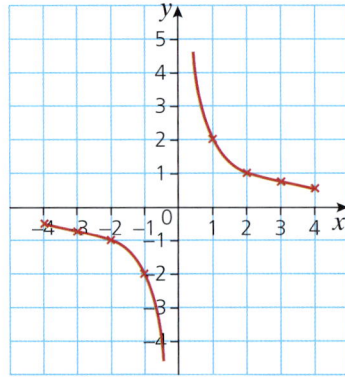

This is a **reciprocal function** giving a **hyperbola**.

Exercise 18.5

1. Plot the graph of the function $y = \frac{1}{x}$ for $-4 \leq x \leq 4$.

2. Plot the graph of the function $y = \frac{3}{x}$ for $-4 \leq x \leq 4$.

3. Plot the graph of the function $y = \frac{5}{2x}$ for $-4 \leq x \leq 4$.

Types of graph

Graphs of functions of the form ax^n take different forms depending on the values of a and n. The different types of line produced also have different names, as described below.

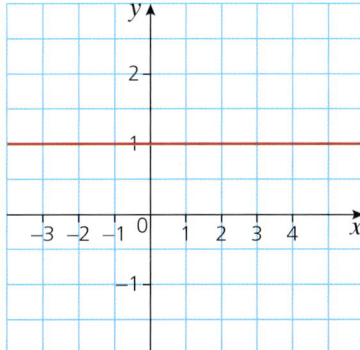

If $a = 1$ and $n = 0$, then $y = x^0$. This is a **linear** function giving a horizontal **straight line**.

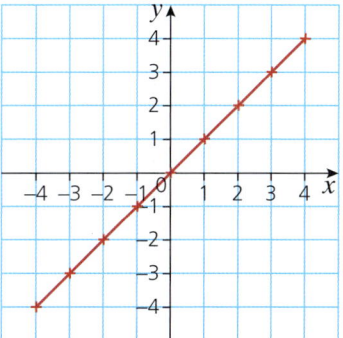

If $a = 1$ and $n = 1$, then $y = x^1$. This is a **linear** function giving a **straight line**.

Note

At Extended level, these four graphs can be referred to with function notation, $f(x) = x^0$, $f(x) = x^1$, $f(x) = x^2$, and $f(x) = \frac{1}{x}$.

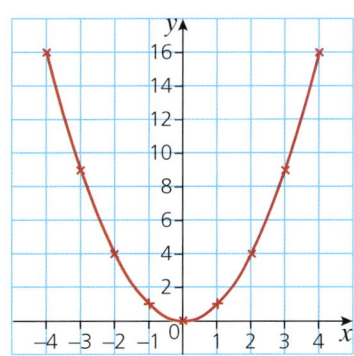

If $a = 1$ and $n = 2$, then $y = x^2$. This is a **quadratic** function giving a **parabola**.

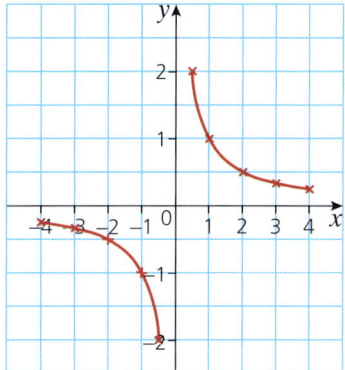

If $a = 1$ and $n = -1$, then $y = x^{-1}$ or $y = \frac{1}{x}$.

This is a **reciprocal** function giving a **hyperbola**.

18 GRAPHS OF FUNCTIONS

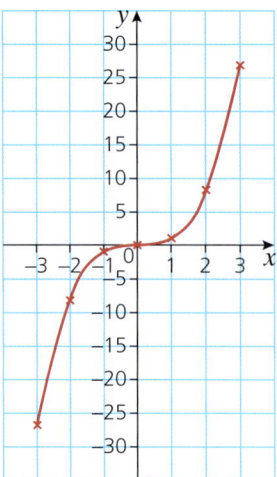

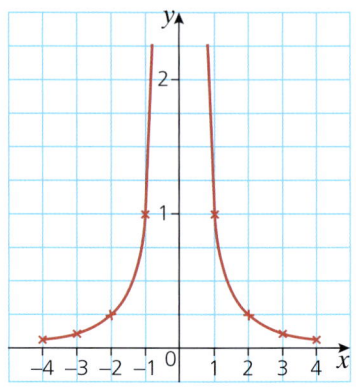

Note

In addition, there are three other types of graph at Extended level.

If $a = 1$ and $n = 3$, then $f(x) = x^3$. This is a **cubic** function giving a **cubic curve**.

If $a = 1$ and $n = -2$, then $f(x) = x^{-2}$ or $f(x) = \frac{1}{x^2}$.

This is a **reciprocal** function, shown on the graph above.

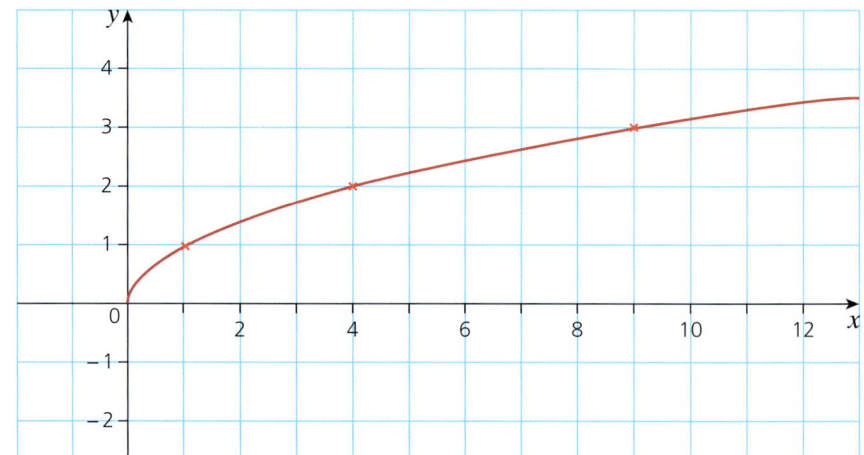

If $a = 1$ and $n = \frac{1}{2}$, then $f(x) = x^{\frac{1}{2}}$ or $f(x) = \sqrt{x}$.

This also produces a parabola.

Types of graph

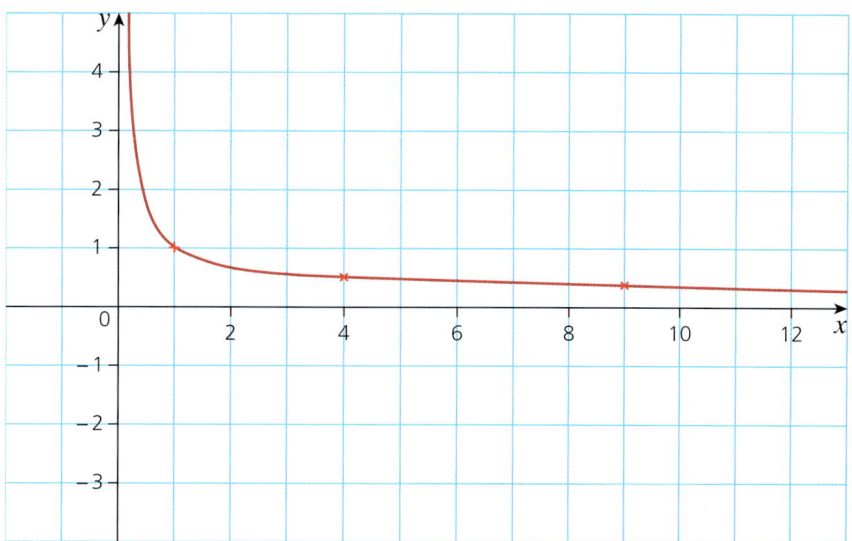

If $a = 1$ and $n = -\frac{1}{2}$, then $f(x) = x^{-\frac{1}{2}}$ or $f(x) = \frac{1}{\sqrt{x}}$.

This is also a reciprocal function.

➜ Worked example

Draw a graph of the function $y = 2x^2$ for $-3 \leqslant x \leqslant 3$.

x	−3	−2	−1	0	1	2	3
y	18	8	2	0	2	8	18

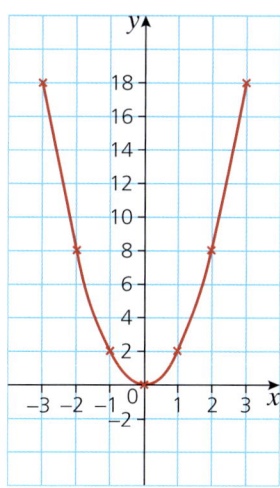

18 GRAPHS OF FUNCTIONS

Exercise 18.6

For each of the functions given below:
i draw up a table of values for x and $f(x)$,
ii plot the graph of the function.

1 $f(x) = \frac{1}{2}x + 4$, $-3 \leqslant x \leqslant 3$

2 $f(x) = -2x - 3$, $-4 \leqslant x \leqslant 2$

3 $f(x) = 2x^2 - 1$, $-3 \leqslant x \leqslant 3$

4 $f(x) = 0.5x^2 + x - 2$, $-5 \leqslant x \leqslant 3$

5 $f(x) = 3x^{-1}$, $-3 \leqslant x \leqslant 3$

6 $f(x) = \frac{1}{2}x^3 - 2x + 3$, $-3 \leqslant x \leqslant 3$

7 $f(x) = 2x^{-2}$, $-3 \leqslant x \leqslant 3$

8 $f(x) = \frac{1}{x^2} + 3x$, $-3 \leqslant x \leqslant 3$

9 $f(x) = \sqrt{x+2}$, $-2 \leqslant x \leqslant 14$ for $x = \{-2, 2, 7, 14\}$

10 $f(x) = \frac{1}{\sqrt{x}} - 4$, $-\frac{1}{16} \leqslant x \leqslant 9$ for $x = \{\frac{1}{16}, \frac{1}{4}, 1, 4, 9\}$

Exponential functions

Functions of the form $y = a^x$ are known as **exponential functions**. Plotting an exponential function is done in the same way as for other functions.

➔ Worked example

Plot the graph of the function $y = 2^x$ for $-3 \leqslant x \leqslant 3$.

x	-3	-2	-1	0	1	2	3
y	0.125	0.25	0.5	1	2	4	8

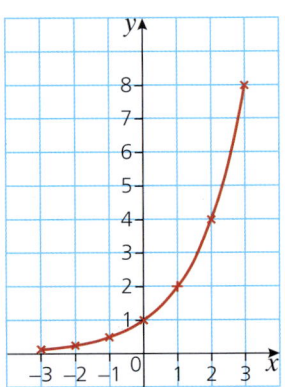

206

Gradients of curves

 Exercise 18.7 For each of the functions below:
  i draw up a table of values of x and $f(x)$,
  ii plot a graph of the function.

1 $f(x) = 3^x$, $-3 \leqslant x \leqslant 3$
2 $f(x) = 1$, $-3 \leqslant x \leqslant 3$
3 $f(x) = 2^x + 3$, $-3 \leqslant x \leqslant 3$
4 $f(x) = 3 \times 2^x + 2$, $-3 \leqslant x \leqslant 3$
5 $f(x) = 2^x - x$, $-3 \leqslant x \leqslant 3$
6 $f(x) = 3^x - x$, $-3 \leqslant x \leqslant 3$

Gradients of curves

The gradient of a straight line is constant and is calculated by considering the coordinates of two of the points on the line and then carrying out the calculation $\frac{y_2 - y_1}{x_2 - x_1}$ as shown below:

Gradient $= \frac{4 - 2}{4 - 0}$

$= \frac{1}{2}$

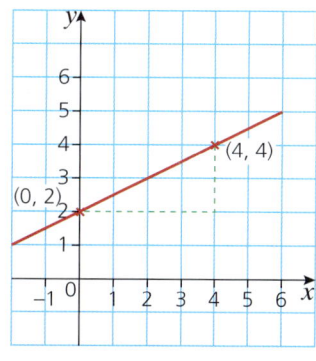

The gradient of a curve, however, is not constant: its slope changes. To calculate the gradient of a curve at a specific point, the following steps need to be taken:

» draw a tangent to the curve at that point,
» calculate the gradient of the tangent.

 Worked example

For the function $y = 2x^2$, calculate the gradient of the curve at the point where $x = 1$.

On a graph of the function $y = 2x^2$, identify the point on the curve where $x = 1$ and then draw a tangent to that point. This gives:

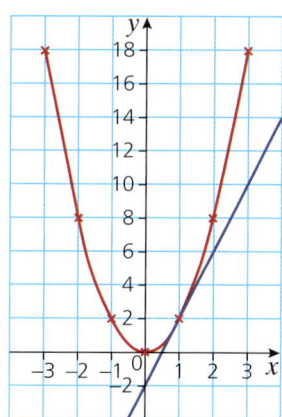

18 GRAPHS OF FUNCTIONS

Two points on the tangent are identified in order to calculate its gradient.

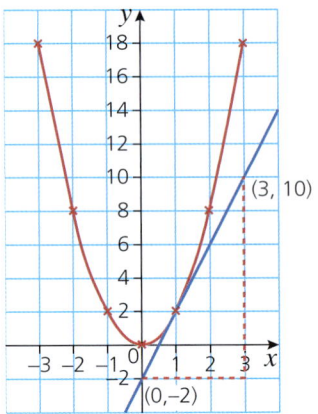

Gradient $= \frac{10-(-2)}{3-0}$

$= \frac{12}{3}$

$= 4$

Therefore the gradient of the function $y = 2x^2$ when $x = 1$ is 4.

Exercise 18.8

For each of the functions below:
i   plot a graph,
ii calculate the gradient of the function at the specified point.

1 $y = x^2$, $-4 \leq x \leq 4$,   gradient where $x = 1$

2 $y = \frac{1}{2}x^2$, $-4 \leq x \leq 4$,   gradient where $x = -2$

3 $y = x^3$, $-3 \leq x \leq 3$,   gradient where $x = 1$

4 $y = x^3 - 3x^2$, $-4 \leq x \leq 4$,   gradient where $x = -2$

5 $y = 4x^{-1}$, $-4 \leq x \leq 4$,   gradient where $x = -1$

6 $y = 2^x$, $-3 \leq x \leq 3$,   gradient where $x = 0$

Solving equations by graphical methods

As shown earlier in this chapter, if a graph of a function is plotted, then it can be used to solve equations.

Solving equations by graphical methods

→ Worked examples

a i Plot a graph of $y = 3x^2 - x - 2$ for $-3 \leqslant x \leqslant 3$.

x	–3	–2	–1	0	1	2	3
y	28	12	2	–2	0	8	22

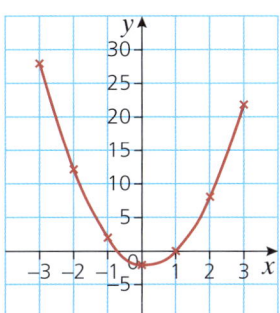

ii Use the graph to solve the equation $3x^2 - x - 2 = 0$.

To solve the equation, $y = 0$. Therefore where the curve intersects the x-axis gives the solution to the equation.

i.e. $3x^2 - x - 2 = 0$ when $x = -0.7$ and when $x = 1$.

iii Use the graph to solve the equation $3x^2 - 7 = 0$.
To be able to use the original graph, this equation needs to be manipulated in such a way that one side of the equation becomes:

$3x^2 - x - 2$.

Manipulating $3x^2 - 7 = 0$ gives:

$3x^2 - x - 2 = -x + 5$ (subtracting x from both sides, and adding 5 to both sides).

Hence finding where the curve $y = 3x^2 - x - 2$ intersects the line $y = -x + 5$ gives the solution to the equation $3x^2 - 7 = 0$.

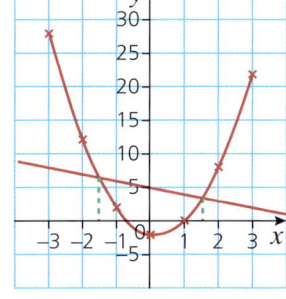

Therefore the solutions to $3x^2 - 7 = 0$ are $x \approx -1.5$ and $x \approx 1.5$.

b i Plot a graph of $y = \frac{1}{x} + x$ for $-4 \leqslant x \leqslant 4$.

x	–4	–3	–2	–1	0	1	2	3	4
y	–4.25	–3.3	–2.5	–2	—	2	2.5	3.3	4.25

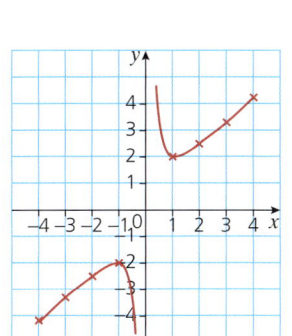

ii Use the graph to explain why $\frac{1}{x} + x = 0$ has no solution.

For $\frac{1}{x} + x = 0$, the graph will need to intersect the x-axis. From the plot opposite, it can be seen that the graph does not intersect the x-axis and hence the equation

$\frac{1}{x} + x = 0$ has no solution.

18 GRAPHS OF FUNCTIONS

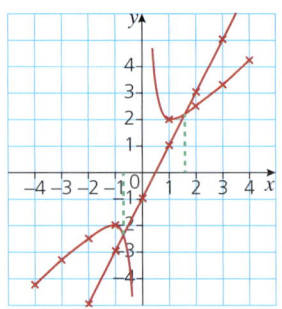

iii Use the graph to find the solution to $x^2 - x = 1$.

This equation needs to be manipulated in such a way that one side becomes $\frac{1}{x} + x$.

Manipulating $x^2 - x = 1$ gives:

$x - 1 = \frac{1}{x}$ (dividing both sides by x)

$2x - 1 = \frac{1}{x} + x$ (adding x to both sides)

Hence finding where the curve $y = \frac{1}{x} + x$ intersects the line $y = 2x - 1$ will give the solution to the equation $x^2 - x = 1$.

Therefore the solutions to the equation $x^2 - x = 1$ are $x \approx -0.6$ and $x \approx 1.6$.

Exercise 18.9

1 **a** Plot the function $y = \frac{1}{2}x^2 + 1$ for $-4 \leqslant x \leqslant 4$.
   b Showing your method clearly, use the graph to solve the equation $\frac{1}{2}x^2 = 4$.

2 **a** Plot the function $y = x^3 + x - 2$ for $-3 \leqslant x \leqslant 3$.
   b Showing your method clearly, use the graph to solve the equation $x^3 = 7 - x$.

3 **a** Plot the function $y = 2x^3 - x^2 + 3$ for $-2 \leqslant x \leqslant 2$.
   b Showing your method clearly, use the graph to solve the equation $2x^3 - 7 = 0$.

4 **a** Plot the function $y = \frac{2}{x^2} - x$ for $-4 \leqslant x \leqslant 4$.
   b Showing your method clearly, use the graph to solve the equation $4x^3 - 10x^2 + 2 = 0$.

5 An open box with a volume of $80\,\text{cm}^3$ is made by cutting $5\,\text{cm}$ squares from each corner of a square piece of metal and then folding up the sides as shown.

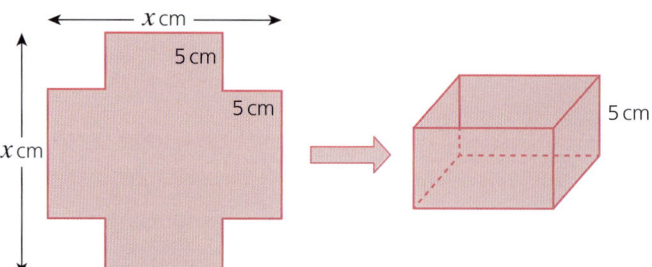

  a Write an expression for both the length and width of the box in terms of x.
  b Write an equation for the volume of the box in terms of x.
  c Calculate the possible dimensions of the original square metal sheet.

Solving equations by graphical methods

6 The cross-section of a bowl is shown on the axes below. y represents the bowl's depth and x its horizontal position.

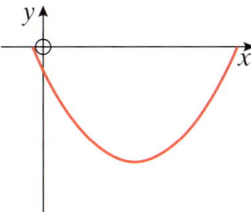

The equation of the surface of the bowl is given as $y = \frac{1}{2}x^2 - 4x - 2$. Calculate the depth of the bowl.

7 The cuboid below has dimensions as shown.

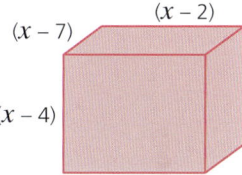

  a Write an equation for the volume V of the cuboid in terms of x.
  b Sketch a graph of the volume V of the cuboid for values of x in the range $2 \leqslant x \leqslant 7$.
  c Explain, using your graph, why the value of x must lie between 2 cm and 4 cm.
  d Using trial and improvement, deduce the value of x (to 1 d.p.) which produces the cuboid with the largest volume.

8 a Plot the function $y = 2^x - x$ for $-2 \leqslant x \leqslant 5$.
  b Showing your method clearly, use the graph to solve the equation $2^x = 2x + 2$.

9 A tap is dripping at a constant rate into a container. The level (l cm) of the water in the container is given by the equation $l = 2^t - 1$, where t is the time taken in hours.
  a Calculate the level of the water after 3 hours.
  b Calculate the level of the water in the container at the start.
  c Calculate the time taken for the level of the water to reach 31 cm.
  d Plot a graph showing the level of the water over the first 6 hours.
  e From your graph, estimate the time taken for the water to reach a level of 45 cm.

10 Draw a graph of $y = 4^x$ for values of x between -1 and 3. Use your graph to find approximate solutions to the following equations:
  a $4^x = 30$ **b** $4^x = \frac{1}{2}$

11 Draw a graph of $y = 2^x$ for values of x between -2 and 5. Use your graph to find approximate solutions to the following equations:
  a $2^x = 20$ **b** $2(x+2) = 40$

12 During an experiment, it is found that harmful bacteria grow at an exponential rate with respect to time. The approximate population of the bacteria, P, is modelled by the equation $P = 4^t + 100$, where t is the time in hours.
  a Calculate the approximate number of harmful bacteria at the start of the experiment.
  b Calculate the number of harmful bacteria after 5 hours. Give your answer to 3 significant figures.

18 GRAPHS OF FUNCTIONS

Exercise 18.9 (cont)

c Draw a graph of $P = 4^t + 100$, for values of t from 0 to 6.
d Estimate from your graph the time taken for the bacteria population to reach 600.

13 The population of a type of insect is falling at an exponential rate. The population P is known to be modelled by the equation $P = 1000 \times \left(\frac{1}{2}\right)^t$, where t is the time in weeks.
  a Copy and complete the following table of results, giving each value of P to the nearest whole number.

t	0	1	2	3	4	5	6	7	8	9	10
P			250								

  b Plot a graph for the table of results above.
  c Estimate from your graph the population of insects after $3\frac{1}{2}$ weeks.

Recognising and sketching functions

So far in this chapter, all graphs of functions have been plotted. In other words, values of x have been substituted into the equation, the corresponding y values have been calculated and the resulting (x, y) coordinates have been plotted.

However, plotting an accurate graph is time consuming and is not always necessary to answer a question. In many cases, a sketch of a graph is as useful as a plot and is considerably quicker.

When doing a sketch, certain key pieces of information need to be included. As a minimum, the points where the graph intersects both the x-axis and y-axis need to be given.

Sketching linear functions

Straight-line graphs can be sketched by working out where the line intersects both axes.

Worked example

Sketch the graph of $y = -3x + 5$

The graph intersects the y-axis when $x = 0$. This is therefore substituted into the equation

$y = -3(0) + 5$

so $y = 5$

The graph intersects the x-axis when $y = 0$. This is then substituted into the equation and solved.

$0 = -3x + 5$

$3x = 5$

Recognising and sketching functions

$x = \frac{5}{3}$ (or 1.6)

The sketch is therefore:

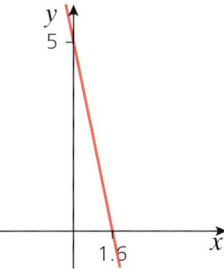

Note that the sketch below, although it looks very different to the one above, is also acceptable as it shows the same intersections with the axes.

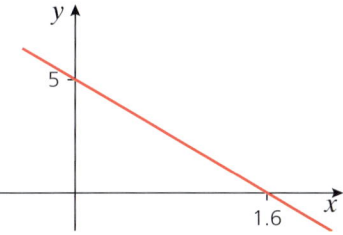

Sketching quadratic functions

With a quadratic function, the sketch should be a smooth parabola shape. Once again, the important points to include are where it intersects the y-axis and, if applicable, the x-axis. If it does intersect the x-axis, giving the coordinates of the turning point is often not necessary unless asked for. However, if the graph does not intersect the x-axis, the coordinates of the turning point should be included.

→ Worked examples

a Sketch the graph of $y = x^2 - 8x + 15$.

The graph intersects the y-axis when $x = 0$.
Substituting $x = 0$ gives $y = 15$.
The graph intersects the x-axis when $y = 0$.
Substituting $y = 0$ gives $x^2 - 8x + 15 = 0$, which needs to be solved.
As the quadratic factorises, this is the quickest method to use.
$x^2 - 8x + 15 = 0$
$(x - 5)(x - 3) = 0$

Therefore $x = 5$ or $x = 3$.

As the x^2 term is positive, the graph will be U-shaped. A possible sketch is shown below.

18 GRAPHS OF FUNCTIONS

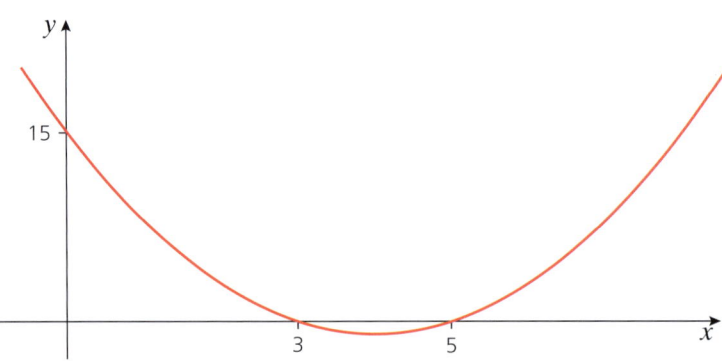

Knowledge of turning points is not specifically required for the Core syllabus.

If the coordinate of the turning point is needed, this can be calculated at this stage. A parabola is symmetrical, so the minimum point must occur midway between the intersections with the x-axis, i.e. when $x = 4$.

Substituting $x = 4$ into the equation of the quadratic gives
$y = (4)^2 - 8(4) + 15 = -1$.

Therefore the coordinate of the minimum point is $(4, -1)$.

b Sketch the graph of the quadratic $y = -x^2 - 4x - 9$.

The graph intersects the y-axis when $x = 0$.

Substituting $x = 0$ gives $y = -9$.

The graph intersects the x-axis when $y = 0$. Substituting $y = 0$ produces the equation $-x^2 - 4x - 9 = 0$, which needs to be solved. The equation does not factorise. If the quadratic formula was used no solutions would be found either, implying that the graph does not intersect the x-axis. If this is the case, the coordinates of the turning point must be found and the completed square form of the equation is the most useful form to use.

In completed square form, the equation $y = -x^2 - 4x - 9$ is written as $y = -(x+2)^2 - 5$. The coordinates of the turning point can be deduced from this as $(-2, -5)$

As the x^2 term is negative, the parabola is an inverted U-shape. A sketch of the graph is therefore:

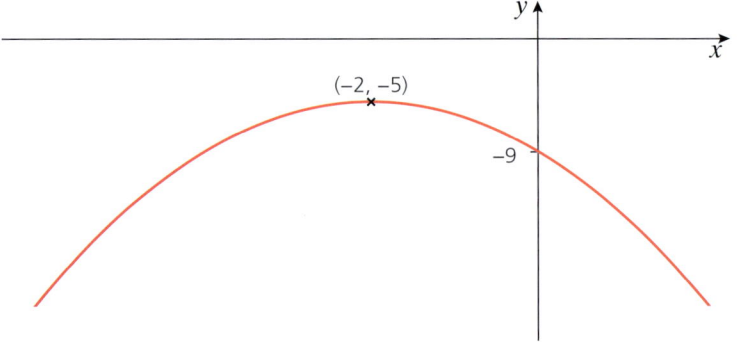

214

Recognising and sketching functions

Sketching cubic functions

Generally, a cubic function takes the form $y = ax^3 + bx^2 + cx + d$ (where $a \neq 0$).

The usual shape of a cubic equation is ∿ when the x^3 term is positive (i.e. $a > 0$), or ∽ when the x^3 term is negative (i.e. $a < 0$).

As a result of this, a cubic equation can intersect the x-axis up to three times.

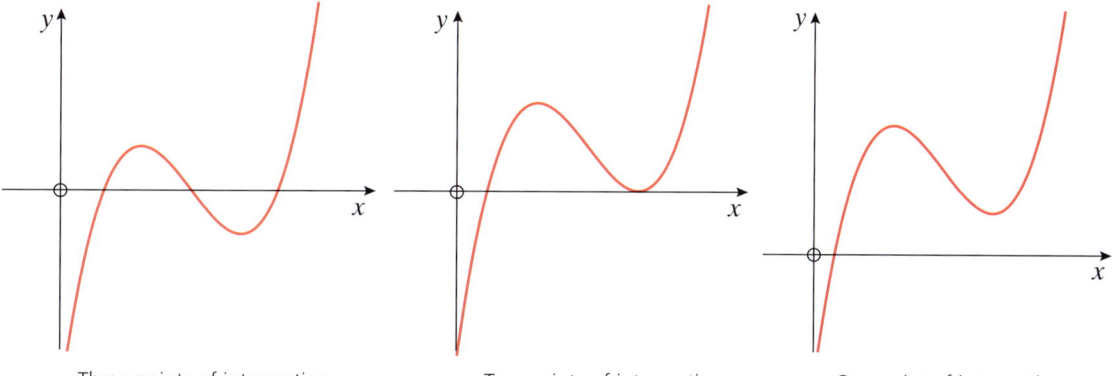

Three points of intersection   Two points of intersection   One point of intersection

To sketch a cubic function, the intersections with both the y-axis and x-axis must be given.

➡ Worked examples

a   Sketch the function $y = (x - 2)(x - 3)(x - 5)$.

Where the graph intersects the y-axis, $x = 0$. Substituting this into the equation gives:
$y = (0 - 2)(0 - 3)(0 - 5)$

$= (-2)(-3)(-5) = -30$.

Where the graph intersects the x-axis, $y = 0$. Substituting this into the equation gives:

$(x - 2)(x - 3)(x - 5) = 0$

Therefore $x = 2, 3$ or 5.

As the x^3 would be positive if the brackets were expanded, the shape of the graph must be ∿

Using this information, the cubic function can be sketched as shown:

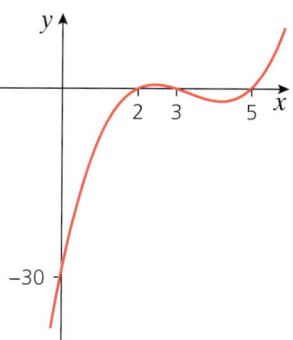

215

18 GRAPHS OF FUNCTIONS

b Sketch the graph of $y = (-x + 1)(x^2 - 6x + 9)$.

Where the graph intersects the y-axis, $x = 0$. Substituting this into the equation gives:

$y = (0 + 1)(0 - 0 + 9) = 9$.

To find where the graph intersects the x-axis substitute $y = 0$ into the equation.

$(-x + 1)(x^2 - 6x + 9) = 0$

The quadratic expression in the second bracket is more useful if written in factorised form.

i.e. $(x^2 - 6x + 9) = (x - 3)(x - 3)$ or $(x - 3)^2$.

The equation to be solved can now be written as $(-x + 1)(x - 3)^2 = 0$.

Therefore, $x = 1, 3$ or 3. It is more usual to just give the distinct values i.e. $x = 1$ or $x = 3$.

As the x^3 term is negative, the shape of the cubic must be ⌐

The repeated root of $x = 3$ implies that the graph just touches the x-axis at this point.

The sketch of the function is therefore:

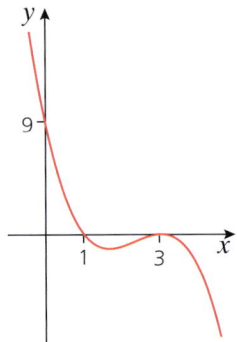

Exercise 18.10

1 Sketch the following linear functions showing clearly where the lines intersect both axes.

   a $y = 2x - 4$ **b** $y = \frac{1}{2}x + 6$ **c** $y = -2x - 3$

   d $y = -\frac{1}{3}x + 9$ **e** $2y + x - 2 = 0$ **f** $x = \frac{2y + 4}{3}$

2 Sketch the following quadratic functions, showing clearly where they intersect the y-axis and where/if they intersect the x-axis. Indicate also the coordinates of the turning point.

   a $y = (x - 4)(x - 6)$ **b** $y = (x + 2)(-x + 3)$ **c** $y = (x - 4)^2 + 1$

   d $y = -x^2 + 3x$ **e** $y = -x^2 + 4x - 4$ **f** $y = -x^2 - 12x - 37$

   g $y = x^2 + 6x - 5$ **h** $y = -3x^2 + 6x - 5$

3   Sketch the following cubic functions, showing clearly where they intersect the axes.
   a $y = (x + 1)(x - 2)(x - 4)$
   b $y = (x - 3)^2(x + 2)$
   c $y = (x)(x^2 - 10x + 25)$
   d $y = -x(-x + 4)(x - 6)$
   e $y = (2x - 1)(-2x^2 - 5x - 3)$

Sketching reciprocal functions

The reciprocal of x is $\frac{1}{x}$, similarly the reciprocal of x^2 is $\frac{1}{x^2}$. The reciprocal of an expression is 1 divided by that expression. In general therefore, reciprocal functions deal with functions of the form $y = \frac{1}{x^n}$. In this section we will look at how to sketch reciprocal functions.

Reciprocal functions of the form $y = \frac{1}{x}$ take one of two shapes.

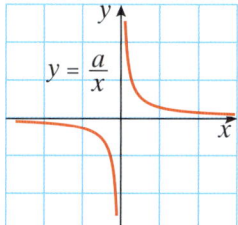
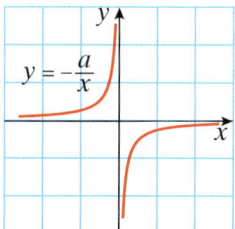

The shape of a positive function   The shape of a negative function

The graph has special properties which need to be highlighted when the function is sketched. As can be seen above, the graphs do not intersect either axis. As x increases, the graph gets closer and closer to the x-axis (because $\frac{1}{x}$ gets smaller as x increases). The x-axis is known as an **asymptote**. As x gets closer to 0, then $\frac{1}{x}$ gets bigger and as a result the graph gets closer to the y-axis. The y-axis is therefore also an asymptote of the graph.

The graph of $y = \frac{1}{x^2}$ has similar properties and takes one of two shapes.

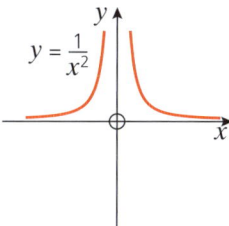
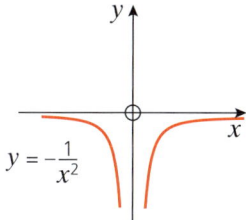

The shape of a positive function   The shape of a negative function

Here too, both the x- and y-axes are asymptotes.

18 GRAPHS OF FUNCTIONS

> **Worked examples**

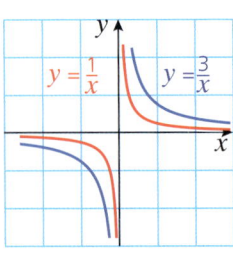

a Sketch on the same axes, the graphs of $y = \frac{1}{x}$ and $y = \frac{3}{x}$, labelling each graph clearly.

As the value of the numerator increases, the graph of $y = \frac{1}{x}$ is stretched in a direction parallel to the y-axis; this gives the appearance of the graph moving away from both axes.

b Sketch the graph of $y = \frac{1}{x^2} + 2$, stating clearly the equations of any asymptotes.

Compared with the graph of $y = \frac{1}{x^2}$, for a given x-value, the y-values for $y = \frac{1}{x^2} + 2$ have +2 added to them. This results in the whole graph of $y = \frac{1}{x^2}$ moving up two units in the y-direction (a translation of +2 in the y-direction).

The sketch is:

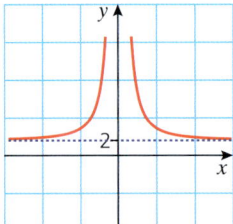

The asymptotes are therefore $y = 2$ and the y-axis.

Note that it is usual to indicate asymptotes other than the axes as a dotted line.

Exponential functions

Exponential functions take the general form $y = a^x$; examples therefore include $y = 2^x$ and $y = 3^x$.

These graphs also have a characteristic shape.

> **Worked example**

a On the same axes, plot the graphs of $y = 2^x$, $y = 3^x$ and $y = 4^x$.

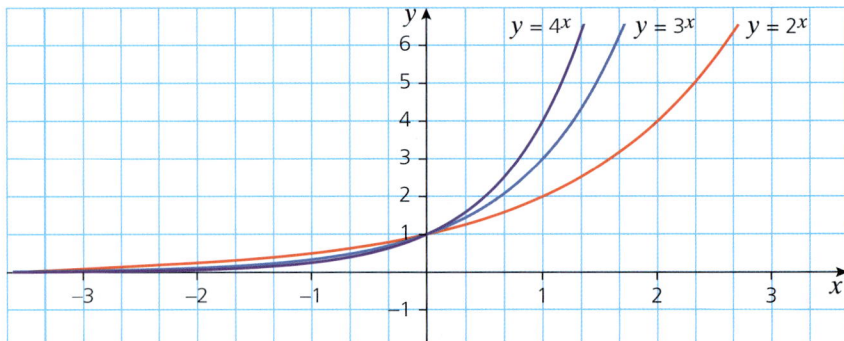

b Comment on any similarities between the three graphs.

The graphs all pass through the coordinate (0, 1) and the x-axis is an asymptote in all three cases.

All the graphs of the form $y = a^x$ pass through the point (0, 1). This is because when $x = 0$ (the intercept with the y-axis), the equation becomes $y = a^0$. From your knowledge of indices you will know that any number raised to the power of zero is one.

Similarly, the reason the x-axis is an asymptote can also be explained using indices. When x is negative, a^{-x} can be written as $\frac{1}{a^x}$. Therefore, as a^x increases in value $\frac{1}{a^x}$ gets closer and closer to zero, hence closer and closer to the x-axis.

The graphs of $y = a^x$ and $y = -a^x$ therefore take the general shapes as shown below:

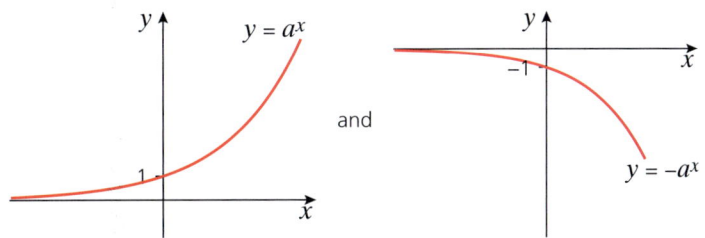

Exercise 18.11

1 Match each of the graphs below with a possible equation.

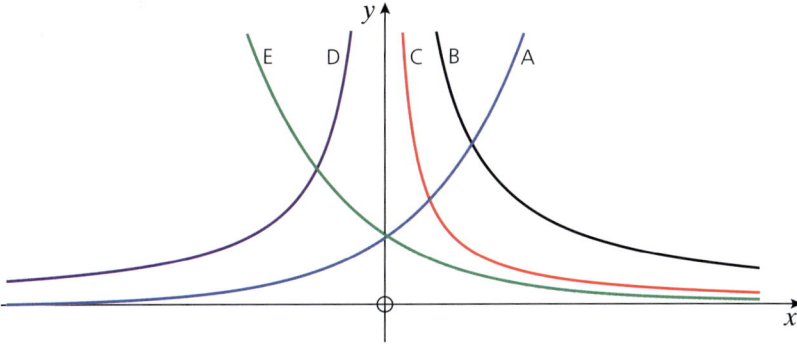

$y = \dfrac{3}{x}$ $y = 2^{-x}$ $y = \dfrac{1}{x}$ $y = -\dfrac{2}{x}$ $y = 2^x$

18 GRAPHS OF FUNCTIONS

Exercise 18.11 (cont)

2 Match each of the graphs below with a possible equation.

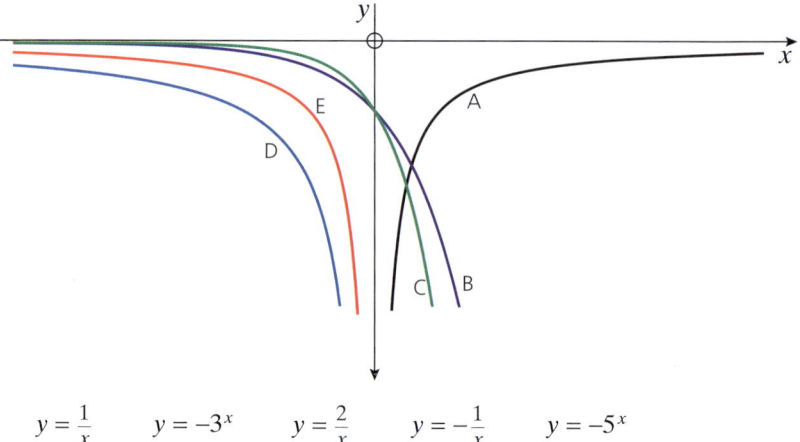

$y = \dfrac{1}{x}$ $y = -3^x$ $y = \dfrac{2}{x}$ $y = -\dfrac{1}{x}$ $y = -5^x$

Student assessment 1

1 Sketch the graph of the function $y = \dfrac{1}{x}$.

2 a Copy and complete the table below for the function $y = -x^2 - 7x - 12$.

x	−7	−6	−5	−4	−3	−2	−1	0	1	2
y		−6				−2				

   b Plot a graph of the function.

3 Plot a graph of each of the functions below between the given limits of x.
   a $y = x^2 - 3x - 10$, $-3 \leqslant x \leqslant 6$
   b $y = -x^2 - 4x - 4$, $-5 \leqslant x \leqslant 1$

4 a Plot the graph of the quadratic equation $y = -x^2 - x + 15$ for $-6 \leqslant x \leqslant 4$.
   b Showing your method clearly, use your graph to solve the following equations:
      i $10 = x^2 + x$
      ii $x^2 = x + 5$

5 a Plot the graph of $y = \dfrac{2}{x}$ for $-4 \leqslant x \leqslant 4$.
   b Showing your method clearly, use your graph to solve the equation $x^2 + x = 2$.

6 In each of the following equations:
   i State whether the turning point is maximum or a minimum, giving a reason for your answer.
   ii State the coordinate of the turning point.
   a $y = -(x + 3)^2 - 5$
   b $y = \dfrac{1}{2}(x - 6)^2 + 2$

7 Sketch the function $y = \frac{1}{2}x^2 - 2$, showing clearly where it intersects both axes and indicating the coordinates of the turning point.

8 a Sketch the function $y = (3x + 2)(-x^2 - x + 2)$, indicating clearly the intersections with both axes.
  b P is the point on the curve with coordinates $(-1, -2)$.
   A tangent to the curve is drawn at P and passes through $(0, 3)$.
   i Draw the tangent on your sketch above.
   ii Calculate the coordinate of the point where the tangent meets the x-axis.

Student assessment 2

1 a Name the types of graph shown below:
  i 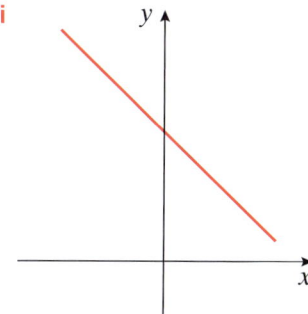   **ii**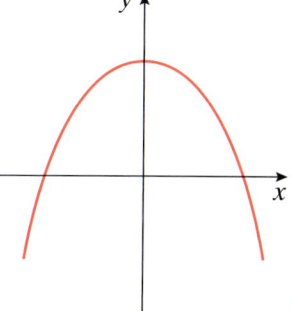

  b Give a **possible** equation for each of the graphs drawn.

2 For each of the functions below:
   i draw up a table of values,
   ii plot a graph of the function.
  a $f(x) = x^2 + 3x$, $\quad -5 \leqslant x \leqslant 2$
  b $f(x) = \frac{1}{x} + 3x$, $\quad -3 \leqslant x \leqslant 3$

3 a Plot the function $y = \frac{1}{2}x^3 + 2x^2$ for $-5 \leqslant x \leqslant 2$.
  b Calculate the gradient of the curve when:
   i $x = 1$ **ii** $x = -1$

4 a Plot a graph of the function $y = 2x^2 - 5x - 5$ for $-2 \leqslant x \leqslant 5$.
  b Use the graph to solve the equation $2x^2 - 5x - 5 = 0$.
  c Showing your method clearly, use the graph to solve the equation $2x^2 - 3x = 10$.

5 Sketch the function $\frac{y+2}{3} = x$.

6 Sketch the function $y = -2x^2 + x + 15$, stating clearly where it intersects the axes and indicating the coordinates of the turning point.

18 GRAPHS OF FUNCTIONS

7 A builder is constructing a fenced yard off the side of a house as shown below. The total length of the fence is 60 m.

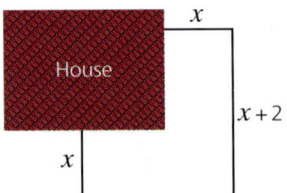

  a Write an expression for the length of the unmarked side in terms of x.
  b Write an equation for the area A of the yard.
  c Sketch the graph of the function for the area A of the yard.
  d From your graph deduce the value of x which gives the largest area for the yard.
  e Calculate the largest area possible for the yard.

Student assessment 3

1 a Name the types of graph shown below:

  i 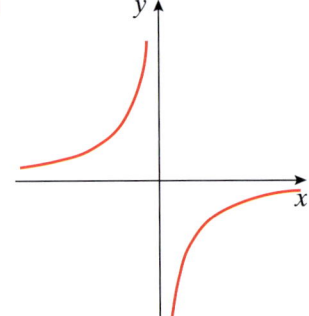   **ii**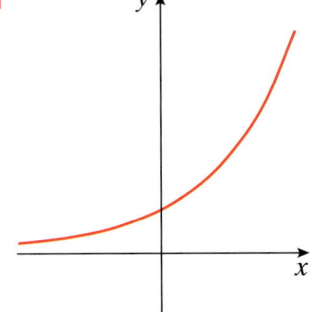

  b Give a **possible** equation for each of the graphs drawn.

2 For each of the functions below:
  i draw up a table of values,
  ii plot a graph of the function.
  a $f(x) = 2^x + x$, $-3 \leqslant x \leqslant 3$
  b $f(x) = 3^x - x^2$, $-3 \leqslant x \leqslant 3$

3 a Plot the function $y = -x^3 - 4x^2 + 5$ for $-5 \leqslant x \leqslant 2$.
  b Calculate the gradient of the curve when:
    i $x = 0$
    ii $x = -2$

Recognising and sketching functions

4 a Copy and complete the table below for the function $y = \dfrac{1}{x^2} - 5$.

x	−3	−2	−1	−0.5	−0.25	0	0.25	0.5	1	2	3
y				−1		—					

b Plot a graph of the function.
c Use the graph to solve the equation $\dfrac{1}{x^2} = 5$.
d Showing your method clearly, use your graph to solve the equation $\dfrac{1}{x^2} + x^2 = 7$.

5 a Sketch the graph of $y = -2^x$ indicating clearly any intersections with either axis.
b Give the equation of any asymptote(s).

6 a Sketch the graph of $y = 2^x - 4$ indicating clearly any intersections with either axis.
b Give the equation of any asymptote(s).

7 A cuboid with side lengths x cm, x cm and y cm is shown below. Its total surface area is 392 cm²

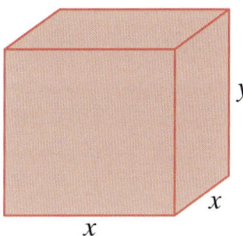

a Show that the length y can be written as $y = \dfrac{196 - x^2}{2x}$
b Write an equation for the volume V of the cuboid in terms of x.
c Sketch a graph for the volume V of the cuboid for values of x in the range $0 \leqslant x \leqslant 14$
d Using your graph as a reference, calculate the integer value of x which will produce the greatest volume.
e Using your answer to part **d**, calculate the maximum volume of the cuboid.

19 Differentiation and the gradient function

Calculus is the cornerstone of much of the mathematics studied at a higher level. Differential calculus deals with finding the gradient of a function. In this chapter, you will look at functions of the form $f(x) = ax^n + bx^{n-1} + ...$, where n is an integer.

The gradient of a straight line

You will already be familiar with calculating the gradient of a straight line.

The gradient of the line passing through points (x_1, y_1) and (x_2, y_2) is calculated by $\frac{y_2 - y_1}{x_2 - x_1}$.

Therefore, the gradient of the line passing through points P and Q is:

$$\frac{10 - 5}{11 - 1} = \frac{5}{10}$$
$$= \frac{1}{2}$$

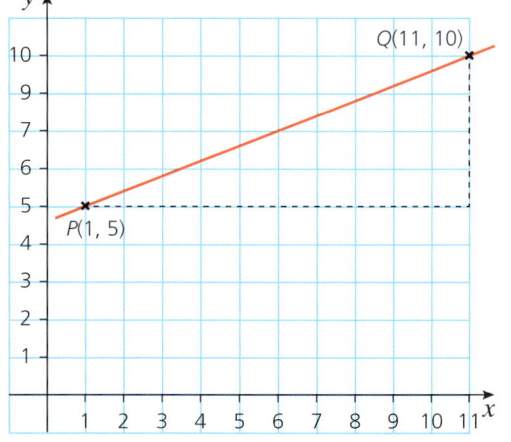

The gradient of a curve

The gradient of a straight line is constant, i.e. it is the same at any point on the line. However, not all functions are linear (straight lines). A function that produces a curved graph is more difficult to work with because the gradient of a curve is not constant.

The graph (right) shows the function $f(x) = x^2$.

Point P is on the curve at $(3, 9)$. If P moves along the curve to the right, the gradient of the curve becomes steeper.

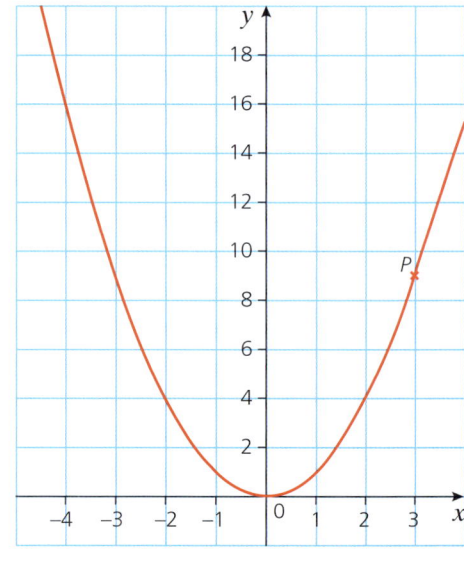

The gradient of a curve

If *P* moves along the curve towards the origin, the gradient of the curve becomes less steep.

The gradient of the function at the point *P*(1, 1) can be calculated as follows:

» Mark a point Q_1(3, 9) on the graph and draw the line **segment** PQ_1.

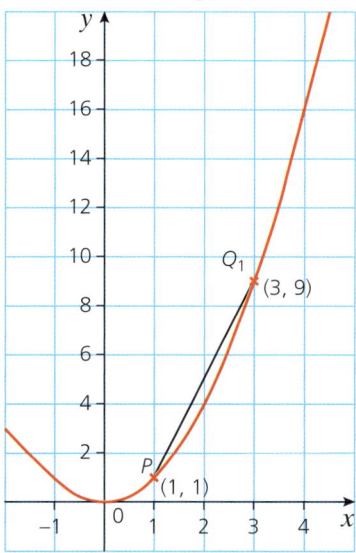

The gradient of the line segment PQ_1 is an approximation of the gradient of the curve at *P*.

Gradient of PQ_1 is $\frac{9-1}{3-1} = 4$

» Mark a point Q_2 closer to *P*, e.g. (2, 4), and draw the line segment PQ_2.

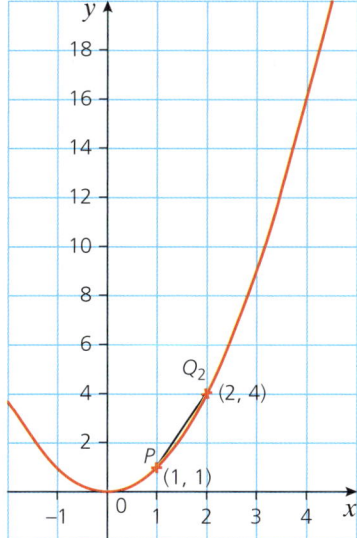

19 DIFFERENTIATION AND THE GRADIENT FUNCTION

The gradient of the line segment PQ_2 is still only an approximation of the gradient of the curve at P, but it is a better approximation than the gradient of PQ_1.

Gradient of PQ_2 is $\frac{4-1}{2-1} = 3$

» If a point $Q_3(1.5, 1.5^2)$ is chosen, the gradient PQ_3 will be an even better approximation.

Gradient of PQ_3 is $\frac{1.5^2 - 1}{1.5 - 1} = 2.5$

For the point $Q_4(1.25, 1.25^2)$, the gradient of PQ_4 is $\frac{1.25^2 - 1}{1.25 - 1} = 2.25$

For the point $Q_5(1.1, 1.1^2)$, the gradient of PQ_5 is $\frac{1.1^2 - 1}{1.1 - 1} = 2.1$

These results indicate that as point Q gets closer to P, the gradient of the line segment PQ gets closer to 2.

➡ Worked example

Prove that the gradient of the function $f(x) = x^2$ is 2 when $x = 1$.
Consider points P and Q on the function $f(x) = x^2$.
P is at $(1, 1)$ and Q, h units from P in the x-direction, has coordinates $(1+h, (1+h)^2)$.

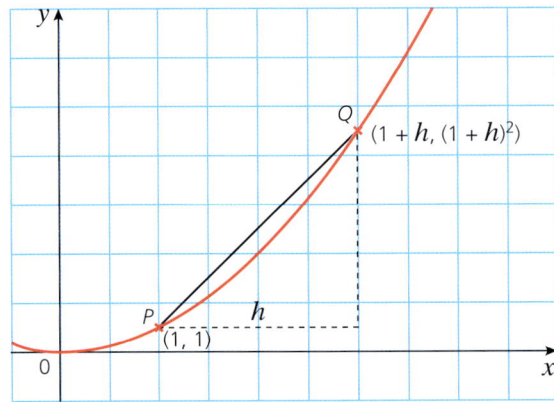

Gradient of line segment PQ is $\frac{(1+h)^2 - 1}{1 + h - 1} = \frac{1 + 2h + h^2 - 1}{h}$
$= \frac{h(2 + h)}{h}$
$= 2 + h$

As Q gets closer to P, h gets smaller and smaller (tends to 0), and the value of $2 + h$ becomes an even more accurate approximation of the gradient of the curve at point P.
As h tends to 0, the gradient $(2 + h)$ of the line segment PQ tends to 2.
This can be written as:

The gradient at $P(1, 1) = \lim_{h \to 0}(2 + h) = 2$

In other words, the limit of $2 + h$ as h tends to 0 is 2.

The gradient function

Exercise 19.1

Note

You will already be familiar with the fact that the gradient of a curve at the point P is the same as the gradient of the tangent to the curve at P.

1. **a** Using the proof above as a guide, find the gradient of the function $f(x) = x^2$ when:

   i $x = 2$ **ii** $x = 3$ **iii** $x = -1$.

   b Make a table of the values of x and the corresponding values of the gradient of the function $f(x)$.

   c Looking at the pattern in your results, complete the sentence below. For the function $f(x) = x^2$, the gradient is ...

2. **a** Find the gradient of the function $f(x) = 2x^2$ when:

   i $x = 1$ **ii** $x = 2$ **iii** $x = -2$.

   b Looking at the pattern in your results, complete the sentence below. For the function $f(x) = 2x^2$, the gradient is ...

3. **a** Find the gradient of the function $f(x) = \frac{1}{2}x^2$ when:

   i $x = 1$ **ii** $x = 2$ **iii** $x = 3$.

   b Looking at the pattern in your results, complete the sentence below. For the function $f(x) = \frac{1}{2}x^2$, the gradient is ...

The gradient function

You may have noticed a pattern in your answers to the previous exercise. In fact, there is a rule for calculating the gradient at any point on the particular curve. This rule is known as the **gradient function**,

The notation $f'(x)$ is shown here as it is used a lot in mathematics. However, knowledge of it is not part of the Extended syllabus.

$f'(x)$ or $\frac{dy}{dx}$.

The function $f(x) = x^2$ has a gradient function $f'(x) = 2x$

or $\frac{dy}{dx} = 2x$.

The above proof can be generalised for other functions $f(x)$.

Gradient of line segment PQ

$= \frac{f(x+h) - f(x)}{(x+h) - x}$

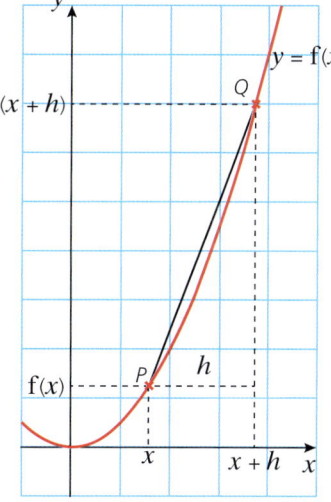

Gradient at $P = \lim_{h \to 0}$ (Gradient of line segment PQ)

$= \lim_{h \to 0} \frac{f(x+h) - f(x)}{h}$

This is known as finding the gradient function from **first principles**.

Note

Limit notation and proof of first principles are beyond the requirements of the syllabus. However, they are included here for interest.

19 DIFFERENTIATION AND THE GRADIENT FUNCTION

→ Worked example

Find, from first principles, the gradient function of $f(x) = x^2 + x$.

$$\frac{dy}{dx} = \lim_{h \to 0} \frac{f(x+h) - f(x)}{h}$$

$$= \lim_{h \to 0} \frac{((x+h)^2 + (x+h)) - (x^2 + x)}{h}$$

$$= \lim_{h \to 0} \frac{x^2 + 2xh + h^2 + x + h - x^2 - x}{h}$$

$$= \lim_{h \to 0} \frac{2xh + h^2 + h}{h}$$

$$= \lim_{h \to 0} (2x + h + 1)$$

$$= 2x + 1$$

So the gradient at any point $P(x, y)$ on the curve $y = x^2 + x$ is given by $2x + 1$.

Exercise 19.2

1 Find, from first principles, the gradient function of each function. Use the worked example above as a guide.
   a $f(x) = x^3$
   b $f(x) = 3x^2$
   c $f(x) = x^2 + 2x$
   d $f(x) = x^2 - 2$
   e $f(x) = 3x - 3$
   f $f(x) = 2x^2 - x + 1$

2 Copy and complete the table below using your gradient functions from the previous question and Exercise 19.1 Q1–3.

Function f(x)	Gradient function f′(x)
x^2	
$2x^2$	
$\frac{1}{2}x^2$	
$x^2 + x$	$2x + 1$
x^3	
$3x^2$	
$x^2 + 2x$	
$x^2 - 2$	
$3x - 3$	
$2x^2 - x + 1$	

3 Look at your completed table for Q2. Describe any patterns you notice between a function and its gradient function.

The functions used so far have all been **polynomials**. There is a relationship between a polynomial function and its gradient function. This is best summarised as follows:

If $f(x) = ax^n$, then $\frac{dy}{dx} = anx^{n-1}$.

So, to work out the gradient function of a polynomial, multiply the coefficient of x by the power of x and subtract 1 from the power.

> **Note**
> Another way of writing $\frac{dy}{dx}$ is $f'(x)$

Differentiation

> **Worked examples**
>
> a Calculate the gradient function of $f(x) = 2x^3$.
>
> $\frac{dy}{dx} = 3 \times 2x^{(3-1)} = 6x^2$
>
> b Calculate the gradient function of $y = 5x^4$.
>
> $\frac{dy}{dx} = 4 \times 5x^{(4-1)} = 20x^3$

Exercise 19.3

1 Calculate the gradient function of each of the following functions:
  a $f(x) = x^4$
  b $f(x) = x$
  c $f(x) = 3x^2$
  d $f(x) = 5x^3$
  e $f(x) = 6x^3$
  f $f(x) = 8x^7$

2 Calculate the gradient function of each of the following functions:
  a $f(x) = \frac{1}{3}x^3$
  b $f(x) = \frac{1}{4}x^4$
  c $f(x) = \frac{1}{4}x^2$
  d $f(x) = \frac{1}{2}x^4$
  e $f(x) = \frac{2}{5}x^3$
  f $f(x) = \frac{2}{9}x^3$

Differentiation

The process of finding the gradient function is known as **differentiation**. Differentiating a function produces the **derivative** or gradient function.

> **Worked examples**
>
> a Differentiate the function $f(x) = 3$ with respect to x.
>   The graph of $f(x) = 3$ is a horizontal line as shown:
>
>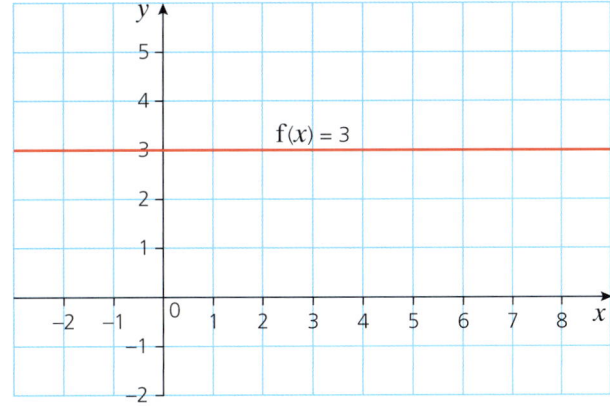
>
> A horizontal line has no gradient. Therefore
>
> $\Rightarrow$ for $f(x) = 3$, $\frac{dy}{dx} = 0$
>
> This can also be calculated using the rule for differentiation.
>
> $f(x) = 3$ can be written as $f(x) = 3x^0$.

19 DIFFERENTIATION AND THE GRADIENT FUNCTION

So $\frac{dy}{dx} = 0 \times 3x^{(0-1)}$

$= 0$

The derivative of a constant is zero.

If $f(x) = c \Rightarrow \frac{dy}{dx} = 0$

b Differentiate the function $f(x) = 2x$ with respect to x.

The graph of $f(x) = 2x$ is a straight line as shown:

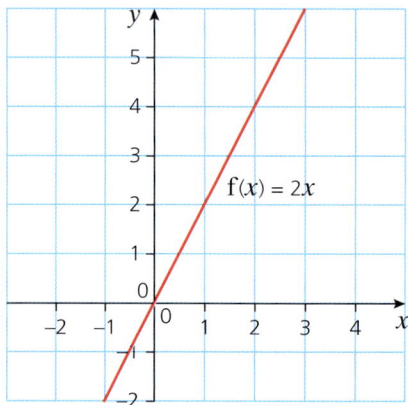

From earlier work on linear graphs, the gradient is known to be 2. Therefore

$\Rightarrow$ for $f(x) = 2x$, $\frac{dy}{dx} = 2$.

This too can be calculated using the rule for differentiation.

$f(x) = 2x$ can be written as $f(x) = 2x^1$.

So $\frac{dy}{dx} = 1 \times 2x^{(1-1)}$

$= 2x^0$

But $x^0 = 1$, therefore $\frac{dy}{dx} = 2$.

If $f(x) = ax \Rightarrow \frac{dy}{dx} = a$.

c Differentiate the function $f(x) = \frac{1}{3}x^3 - 2x + 4$ with respect to x.
The graphs of the function and its derivative are as follows:

Differentiation

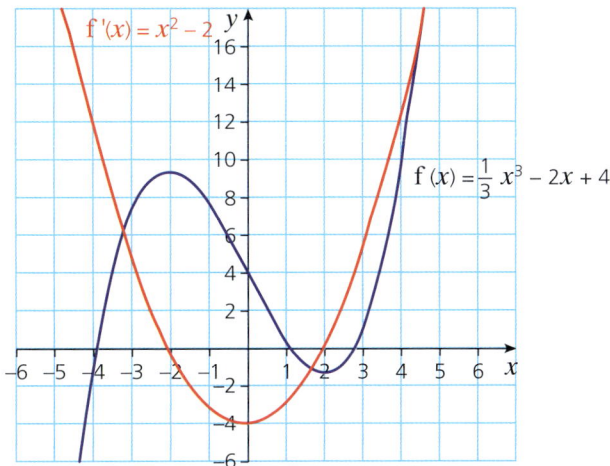

It can be seen that the derivative of the function $f(x)$ is a quadratic. The equation of this quadratic is $y = x^2 - 2$. The derivative of $f(x)$ is therefore $f'(x) = x^2 - 2$.

In general, the derivative of a polynomial function with several terms can be found by differentiating each of the terms individually.

d Differentiate the function $f(x) = \dfrac{2x^3 + x^2}{x}$ with respect to x.

A common error here is to differentiate each of the terms individually.

The derivative of $\dfrac{2x^3 + x^2}{x}$ is NOT $\dfrac{6x^2 + 2x}{1}$.

$\dfrac{2x^3 + x^2}{x}$ can be written as $\dfrac{2x^3}{x} + \dfrac{x^2}{x}$ and simplified to $2x^2 + x$.

Therefore $f(x) = \dfrac{2x^3 + x^2}{x}$

$= 2x^2 + x$

$\Rightarrow \dfrac{dy}{dx} = 4x + 1$

In general, rewrite functions as sums of terms in powers of x before differentiating.

Exercise 19.4

1 Differentiate each expression with respect to x.

a $5x^3$ b $7x^2$ c $4x^6$

d $\dfrac{1}{4}x^2$ e $\dfrac{2}{3}x^6$ f $\dfrac{3}{4}x^5$

g 5 h $6x$ i $\dfrac{1}{8}$

19 DIFFERENTIATION AND THE GRADIENT FUNCTION

Exercise 19.4 (cont)

2 Differentiate each expression with respect to x.
- **a** $3x^2 + 4x$
- **b** $5x^3 - 2x^2$
- **c** $10x^3 - \frac{1}{2}x^2$
- **d** $6x^3 - 3x^2 + x$
- **e** $12x^4 - 2x^2 + 5$
- **f** $\frac{1}{3}x^3 - \frac{1}{2}x^2 + x - 4$
- **g** $-3x^4 + 4x^2 - 1$
- **h** $-6x^5 + 3x^4 - x + 1$
- **i** $-\frac{3}{4}x^6 + \frac{2}{3}x^3 - 8$

3 Differentiate each expression with respect to x.
- **a** $\frac{x^3 + x^2}{x}$
- **b** $\frac{4x^3 - x^2}{x^2}$
- **c** $\frac{6x^3 + 2x^2}{2x}$
- **d** $\frac{x^3 + 2x^2}{4x}$
- **e** $3x(x+1)$
- **f** $2x^2(x-2)$
- **g** $(x+5)^2$
- **h** $(2x-1)(x+4)$
- **i** $(x^2+x)(x-3)$

So far we have only used the variables x and y when finding the gradient function. This does not always need to be the case. Sometimes, as demonstrated below, it is more convenient or appropriate to use other variables.

If a stone is thrown vertically upwards from the ground with a speed of 10 m/s, its distance s from its point of release is given by the formula $s = 10t - 4.9t^2$, where t is the time in seconds after the stone's release.

A graph plotted to show distance against time is shown below.

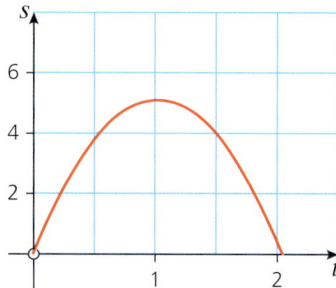

The velocity (v) of the stone at any point can be found by calculating the rate of change of distance with respect to time, i.e. $\frac{ds}{dt}$.

Therefore if $s = 10t - 4.9t^2$

$$v = \frac{ds}{dt}$$
$$= 10 - 9.8t$$

➡ Worked example

Calculate $\frac{ds}{dt}$ for the function $s = 6t^2 - 4t + 1$.

$\frac{ds}{dt} = 12t - 4$

Calculating the second derivative

Exercise 19.5

1 Differentiate each of the following with respect to t.
   a $y = 3t^2 + t$
   b $v = 2t^3 - t^2$
   c $m = 5t^3 - t^2$

2 Calculate the derivative of each of the following functions.
   a $y = x(x + 4)$
   b $r = t(1 - t)$
   c $v = t\left(\frac{1}{t} + t^2\right)$
   d $p = r^2\left(\frac{2}{r} - 3\right)$

3 Differentiate each of the following with respect to t.
   a $y = (t+1)(t-1)$
   b $r = (t-1)(2t+2)$
   c $v = \left(\frac{2t^2}{3} + 1\right)(t-1)$

Calculating the second derivative

In the previous section we considered the position of a stone thrown vertically upwards. Its velocity (v) at any point was found by differentiating the equation for the distance (s) with respect to t, i.e. $v = \frac{ds}{dt}$.

In this section, we extend this to consider acceleration (a) which is the rate of change of velocity with time, i.e. $a = \frac{dv}{dt}$.

Therefore as $\quad s = 10t - 4.9t^2$

$$v = \frac{ds}{dt}$$
$$= 10 - 9.8t$$

$$a = \frac{dv}{dt}$$
$$= -9.8 \quad \text{← acceleration due to gravity}$$

You will have noticed that the equation for the distance was differentiated twice to get the acceleration, i.e. the second derivative was obtained. Calculating the second derivative is a useful operation as will be seen later.

The notation used for the second derivative follows on from that used for the first derivative.

$\quad f(x) = ax^n \qquad\qquad\qquad\qquad y = ax^n$

$\Rightarrow f'(x) = anx^{n-1} \qquad$ or $\qquad \frac{dy}{dx} = anx^{n-1}$

$\Rightarrow f''(x) = an(n-1)x^{n-2} \qquad\qquad \frac{d^2y}{dx^2} = an(n-1)x^{n-2}$

The notation $f''(x)$ is shown here as it is used a lot in mathematics. However, knowledge of it is not part of the Extended syllabus.

Therefore either $f''(x)$ or $\frac{d^2y}{dx^2}$ are the most common forms of notation used for the second derivative, when differentiating with respect to x.

233

19 DIFFERENTIATION AND THE GRADIENT FUNCTION

Worked examples

a Find $\dfrac{d^2y}{dx^2}$ when $y = x^3 - 2x^2$.

$\dfrac{dy}{dx} = 3x^2 - 4x$

$\dfrac{d^2y}{dx^2} = 6x - 4$

b Work out $\dfrac{d^2s}{dt^2}$ for $s = 3t + \dfrac{1}{2}t^2$.

$\dfrac{ds}{dt} = 3 + t$

$\dfrac{d^2s}{dt^2} = 1$

Exercise 19.6

1 Find $\dfrac{d^2y}{dx^2}$ for each of the following.

a $y = 2x^3$
b $y = x^4 - \dfrac{1}{2}x^2$
c $y = \dfrac{1}{3}x^6$
d $y = 3x^2 - 2$
e $y = \dfrac{x^2}{4}$
f $y = 3x$

2 Find the second derivative of each function.

a $v = x^2(x - 3)$
b $P = \dfrac{1}{2}x^2(x^2 + x)$
c $t = x^{-1}(x + x^3)$
d $y = (x^2 + 1)(x^3 - x)$

Gradient of a curve at a point

You have seen that differentiating the equation of a curve gives the general equation for the gradient of any point on the curve. You can use this general equation to calculate the gradient at a specific point on the curve.

For the function $f(x) = \dfrac{1}{2}x^2 - 2x + 4$, the gradient function $f'(x) = x - 2$.

The gradient at any point on the curve can be calculated using this.

For example, when $x = 4$, $f'(4) = 4 - 2$

$\qquad\qquad\qquad\qquad\qquad = 2$

Therefore, the gradient of the curve $f(x) = \dfrac{1}{2}x^2 - 2x + 4$ is 2 when $x = 4$, as shown below.

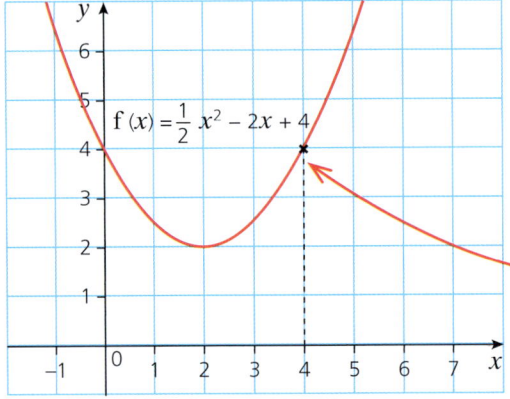

The gradient of the curve at $x = 4$ is 2.

Gradient of a curve at a point

> **Worked example**
>
> Calculate the gradient of the curve $f(x) = x^3 + x - 6$ when $x = -1$.
> The gradient function $f'(x) = 3x^2 + 1$.
> So when $x = -1$, $f'(-1) = 3(-1)^2 + 1$
> $= 4$
> Therefore the gradient is 4.

Exercise 19.7

1. Find the gradient of each function at the given value of x.
   a $f(x) = x^2$; $x = 3$
   b $f(x) = \frac{1}{2}x^2 - 2$; $x = -3$
   c $f(x) = 3x^3 - 4x^2 - 2$; $x = 0$
   d $f(x) = -x^2 + 2x - 1$; $x = 1$
   e $f(x) = -\frac{1}{2}x^3 + x - 3$; $x = -1, x = 2$
   f $f(x) = 6x$; $x = 5$

2. The number of newly infected people, N, on day t of a stomach bug outbreak is given by $N = 5t^2 - \frac{1}{2}t^3$.
   a Calculate the number of new infections N when:
      i $t = 1$
      ii $t = 3$
      iii $t = 6$
      iv $t = 10$.
   b Calculate the rate of new infections with respect to t, i.e. calculate $\frac{dN}{dt}$.
   c Calculate the rate of new infections at the following times:
      i $t = 1$
      ii $t = 3$
      iii $t = 6$
      iv $t = 10$.
   d Plot a graph of the equation $N = 5t^2 - \frac{1}{2}t^3$ for the values of t in the range $0 \leqslant t \leqslant 10$.
   e Explain your answers to part **a**, using your graph to support your explanation.
   f Explain your answers to part **c**, using your graph to support your explanation.

3. A weather balloon is released from the ground. Its height in metres, h, after time in hours t, is given by the formula:
   $h = 30t^2 - t^3$ when $t \leqslant 20$.
   a Calculate the height of the balloon when:
      i $t = 3$
      ii $t = 10$.
   b Calculate the rate at which the balloon is climbing with respect to time t.
   c Calculate the rate of ascent when:
      i $t = 2$
      ii $t = 5$
      iii $t = 20$.

19 DIFFERENTIATION AND THE GRADIENT FUNCTION

Exercise 19.7 (cont)

d The graph of h against t is shown below.

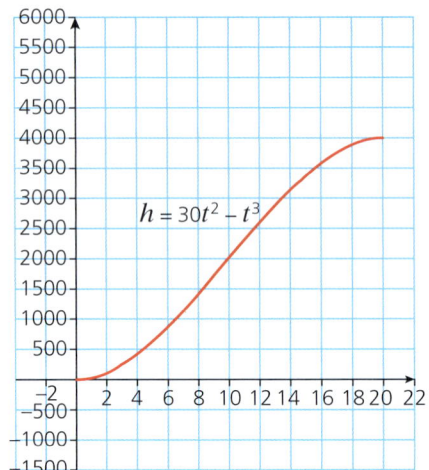

Referring to your graph, explain your answers to part **c**:

e Estimate from your graph the time when the balloon was climbing at its fastest rate. Explain your answer.

Calculating the value of x when the gradient is given

So far you have calculated the gradient of a curve for a given value of x. It is also possible to work backwards and calculate the value of x when the gradient of a point is given.

Consider the function $f(x) = x^2 - 2x + 1$.

It is known that the gradient at a particular point on the curve is 4, but the x-coordinate of that point is not known.

The gradient function of the curve is $f'(x) = 2x - 2$.

Since the gradient at this particular point is 4, you can form an equation:

$$f'(x) = 4$$

So $2x - 2 = 4$

$$\Rightarrow 2x = 6$$
$$\Rightarrow x = 3$$

Therefore $x = 3$ when the gradient of the curve is 4.

Calculating the value of x when the gradient is given

> ### Worked example
>
> The function $f(x) = x^3 - x^2 - 5$ has a gradient of 8 at a point P on the curve. Calculate the possible coordinates of point P.
>
> The gradient function $f'(x) = 3x^2 - 2x$
>
> At P, $3x^2 - 2x = 8$
>
> This can be rearranged into the quadratic $3x^2 - 2x - 8 = 0$ and solved algebraically.
>
> This requires the algebraic solution of the quadratic equation $3x^2 - 2x - 8 = 0$.
>
> Factorising gives $(3x + 4)(x - 2) = 0$
>
> Therefore $(3x + 4) = 0 \Rightarrow x = -\frac{4}{3}$ or $(x - 2) = 0 \Rightarrow x = 2$
>
> The values of $f(x)$ can be calculated by substituting the x-values in to the equation as shown:
>
> $f\left(-\frac{4}{3}\right) = \left(-\frac{4}{3}\right)^3 - \left(-\frac{4}{3}\right)^2 - 5 = -9\frac{4}{27}$
>
> $f(2) = 2^3 - 2^2 - 5 = -1$
>
> Therefore the possible coordinates of P are $\left(-1\frac{1}{3}, -9\frac{4}{27}\right)$ and $(2, -1)$

Exercise 19.8

1. Find the coordinate of the point P on each of the following curves, at the given gradient.
   a $f(x) = x^2 - 3$, gradient $= 6$
   b $f(x) = 3x^2 + 1$, gradient $= 15$
   c $f(x) = 2x^2 - x + 4$, gradient $= 7$
   d $f(x) = \frac{1}{2}x^2 - 3x - 1$, gradient $= -3$
   e $f(x) = \frac{1}{3}x^2 + 4x$, gradient $= 6$
   f $f(x) = -\frac{1}{5}x^2 + 2x + 1$, gradient $= 4$

2. Find the coordinate(s) of the point(s) on each of the following curves, at the given gradient.
   a $f(x) = \frac{1}{3}x^3 + \frac{1}{2}x^2 + 4x$, gradient $= 6$
   b $f(x) = \frac{1}{3}x^3 + 2x^2 + 6x$, gradient $= 3$
   c $f(x) = \frac{1}{3}x^3 - 2x^2$, gradient $= -4$
   d $f(x) = x^3 - x^2 + 4x$, gradient $= 5$

3. A stone is thrown vertically downwards off a tall cliff. The distance (s) it travels in metres is given by the formula $s = 4t + 5t^2$, where t is the time in seconds after the stone's release.
   a What is the rate of change of distance with time, $\frac{ds}{dt}$? (This represents the velocity.)
   b How many seconds after its release is the stone travelling at a velocity of 9 m/s?

19 DIFFERENTIATION AND THE GRADIENT FUNCTION

Exercise 19.8 (cont)

c The speed of the stone as it hits the ground is 34 m/s. How many seconds after its release did the stone hit the ground?
d Using your answer to part **c**, calculate the distance the stone falls and hence the height of the cliff.

4 The temperature inside a pressure cooker (T) in degrees Celsius is given by the formula
$$T = 20 + 12t^2 - t^3$$
where t is the time in minutes after the cooking started and $t \leq 8$.
a Calculate the initial (starting) temperature of the pressure cooker.
b What is the rate of temperature increase with time?
c What is the rate of temperature increase when:
   i $t = 1$ ii $t = 4$ iii $t = 8$?
d The pressure cooker was switched off when $\frac{dT}{dt} = 36$.

How long after the start could the pressure cooker have been switched off? Give both possible answers.

e What was the final temperature of the pressure cooker if it was switched off at the greater of the two times calculated in part **d**?

Equation of the tangent at a given point

You already know that the gradient of a tangent drawn at a point on a curve is equal to the gradient of the curve at that point.

> **Worked example**

Find the equation of the tangent of $f(x) = \frac{1}{2}x^2 + 3x + 1$ at a point P, where $x = 1$.

The function $f(x) = \frac{1}{2}x^2 + 3x + 1$ has a gradient function of $f'(x) = x + 3$

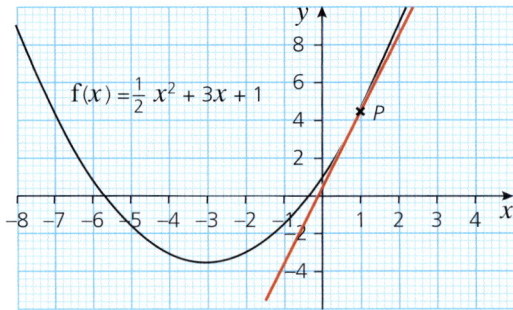

At point P, where $x = 1$, the gradient of the curve is 4.

The tangent drawn to the curve at P also has a gradient of 4.

The equation of the tangent can also be calculated. As it is a straight line, it must take the form $y = mx + c$. The gradient m is 4 as shown above.
Therefore $y = 4x + c$.

As the tangent passes through the point $P(1, 4\frac{1}{2})$, these values can be substituted for x and y so that c can be calculated.

$4\frac{1}{2} = 4 + c$

$\Rightarrow c = \frac{1}{2}$

The equation of the tangent is therefore $y = 4x + \frac{1}{2}$.

Exercise 19.9

1 For the function $f(x) = x^2 - 3x + 1$
   a Calculate the gradient function.
   b Calculate the gradient of the curve at the point $A(2, -1)$.
   A tangent is drawn to the curve at A.
   c What is the gradient of the tangent?
   d Calculate the equation of the tangent. Give your answer in the form $y = mx + c$.

2 For the function $f(x) = 2x^2 - 4x - 2$
   a Calculate the gradient of the curve at $x = 2$.
   A tangent is drawn to the curve at the point $(2, -2)$.
   b Calculate the equation of the tangent. Give your answer in the form $y = mx + c$.

3 A tangent is drawn to the curve $f(x) = \frac{1}{2}x^2 - 4x - 2$ at the point $P(0, -2)$.
   a Calculate the gradient of the tangent at P.
   b Calculate the equation of the tangent. Give your answer in the form $y = mx + c$.

4 A tangent, T_1, is drawn to the curve $f(x) = -x^2 + 4x + 1$ at point $A(4, 1)$.
   a Calculate the gradient of the tangent at A.
   b Calculate the equation of the tangent. Give your answer in the form $y = mx + c$.
   c Another tangent to the curve, T_2, is drawn at point $B(2, 5)$. Calculate the equation of T_2.

5 A tangent, T_1, is drawn to the curve $f(x) = -\frac{1}{4}x^2 - 3x + 1$ at point $P(-2, 6)$.
   a Calculate the equation of T_1.
   Another tangent to the curve, T_2, with equation $y = 10$, is drawn at point Q.
   b Calculate the coordinates of point Q.
   c T_1 and T_2 are extended so that they intersect. Calculate the coordinates of their point of intersection.

6 The equation of a tangent T, drawn to the curve $f(x) = -\frac{1}{2}x^2 - x - 4$ at P, has equation $y = -3x - 2$.
   a Calculate the gradient function of the curve.
   b What is the gradient of the tangent T?
   c What are the coordinates of point P?

Stationary points

There are times when the gradient of a point on a curve is zero, i.e. the tangent drawn at that point is horizontal. A point where the gradient of the curve is zero is known as a **stationary point**.

19 DIFFERENTIATION AND THE GRADIENT FUNCTION

> In the diagram, the stationary points A, B and C are also turning points (see page 200) because the gradient of the curve changes from positive to negative (or vice versa). Stationary point D is not a turning point because the gradient remains negative before and after the point.

There are different types of stationary point.

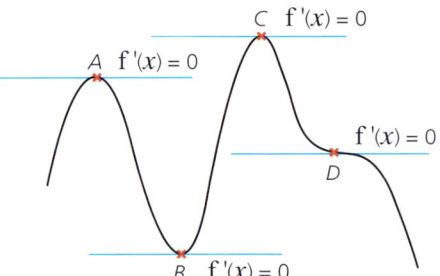

Points A and C are **local maxima**, point B is a **local minima** and point D is a **point of inflection**. This text covers local maximum and minimum points only.

As the worked example shows, it is not necessary to sketch a graph in order to find the position of any stationary points or to identify what type of stationary points they are.

→ Worked example

a A graph has equation $y = \frac{1}{3}x^3 - 4x + 5$. Find the coordinates of the stationary points on the graph.

If $y = \frac{1}{3}x^3 - 4x + 5$, $\frac{dy}{dx} = x^2 - 4$.

$x^2 - 4 = 0$

$x^2 = 4$

$x = \pm 2$

> At a stationary point $\frac{dy}{dx} = 0$, so solve $x^2 - 4 = 0$ to find the x-coordinate of any stationary point.

Substitute $x = 2$ and $x = -2$ into the equation of the curve to find the corresponding y-coordinates.

When $x = 2$, $\quad y = \frac{1}{3}(2)^3 - 4(2) + 5 = -\frac{1}{3}$.

When $x = -2$, $\quad y = \frac{1}{3}(-2)^3 - 4(-2) + 5 = 10\frac{1}{3}$.

The coordinates of the stationary points are $\left(2, -\frac{1}{3}\right)$ and $\left(-2, 10\frac{1}{3}\right)$.

b Determine the nature of each of the stationary points.
There are several methods that can be used to establish the type of stationary point.

Graphical deduction
As the curve is cubic and the coefficient of the x^3 term is positive, the shape of the curve is of the form

Stationary points

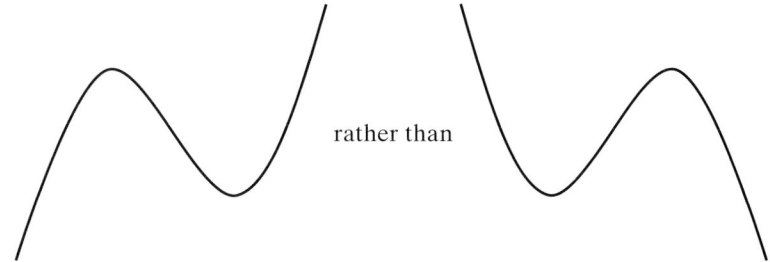

rather than

Therefore it can be deduced that the positions of the stationary points are:

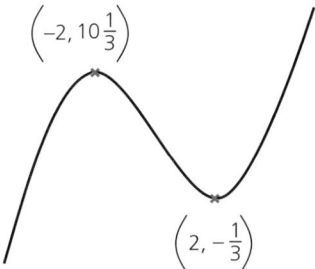

$\left(-2, 10\tfrac{1}{3}\right)$

$\left(2, -\tfrac{1}{3}\right)$

Hence $\left(-2, 10\tfrac{1}{3}\right)$ is a maximum point and $\left(2, -\tfrac{1}{3}\right)$ a minimum point.

Gradient inspection

The gradient of the curve either side of a stationary point can be calculated.
At the stationary point where $x = 2$, consider the gradient at $x = 1$ and at $x = 3$.

$\dfrac{dy}{dx} = x^2 - 4,$ so when $x = 1$, $\dfrac{dy}{dx} = -3$

and when $x = 3$, $\dfrac{dy}{dx} = 5$

As x increases, the gradient changes from negative to positive, therefore the stationary point must be a minimum.

At the stationary point where $x = -2$, consider the gradient at $x = -3$ and at $x = -1$.

$\dfrac{dy}{dx} = x^2 - 4,$ so when $x = -3$, $\dfrac{dy}{dx} = 5$

and when $x = -1$, $\dfrac{dy}{dx} = -3$

As x increases, the gradient changes from positive to negative, therefore the stationary point must be a maximum.

The second derivative

The second derivative, $\dfrac{d^2y}{dx^2}$, is usually the most efficient way of determining whether a stationary point is a maximum or minimum.

The proof is beyond the scope of this book. However, the general rule is that:

$\dfrac{d^2y}{dx^2} < 0 \Rightarrow$ a maximum point

$\dfrac{d^2y}{dx^2} > 0 \Rightarrow$ a minimum point.

19 DIFFERENTIATION AND THE GRADIENT FUNCTION

In this example, $\frac{dy}{dx} = x^2 - 4$ and $\frac{d^2y}{dx^2} = 2x$.

Substituting the *x*-values (−2 and 2) of the stationary points into $\frac{d^2y}{dx^2}$ gives:

$\frac{d^2y}{dx^2} = 2(-2) = -4$ (a maximum point)

and $\frac{d^2y}{dx^2} = 2(2) = 4$ (a minimum point).

Note: When $\frac{d^2y}{dx^2} = 0$, the stationary point could either be a maximum or a minimum point, so another method should be used.

Exercise 19.10

1 For each function, calculate:
  i the gradient function
  ii the coordinates of any stationary points.
  a $f(x) = x^2 - 6x + 13$
  b $f(x) = x^2 + 12x + 35$
  c $f(x) = -x^2 + 8x - 13$
  d $f(x) = -6x + 7$

2 For each function, calculate:
  i the gradient function
  ii the coordinates of any stationary points.
  a $f(x) = x^3 - 12x^2 - 58$
  b $f(x) = x^3 - 12x$
  c $f(x) = x^3 - 3x^2 - 45x + 8$
  d $f(x) = \frac{1}{3}x^3 + \frac{3}{2}x^2 - 4x - 5$

For Questions 3–6:
a Calculate the gradient function.
b Calculate the coordinates of any stationary points.
c Determine the nature of each stationary point.
d Calculate the value of the *y*-intercept.
e Sketch the graph of the function.

3 $f(x) = 1 - 4x - x^2$

4 $f(x) = \frac{1}{3}x^3 - 4x^2 + 12x - 3$

5 $f(x) = -\frac{2}{3}x^3 + 3x^2 - 4x$

6 $f(x) = x^3 - \frac{9}{2}x^2 - 30x + 4$

Stationary points

Student assessment 1

1 Find the gradient function of the following:
   a $y = x^3$
   b $y = 2x^2 - x$
   c $y = -\frac{1}{2}x^2 + 2x$
   d $y = \frac{2}{3}x^3 + 4x^2 - x$

2 Differentiate the following functions with respect to x.
   a $f(x) = x(x+2)$
   b $f(x) = (x+2)(x-3)$
   c $f(x) = \frac{x^3 - x}{x}$
   d $f(x) = \frac{x^3 + 2x^2}{2x}$

3 Find the values of p and q.
   $y = 3x^p - qx^2$ if $\frac{dy}{dx} = 12x^{q+1} - 4x$

4 Find the second derivative of the following functions:
   a $y = x^4 - 3x^2$
   b $s = 2t^5 - t^3$

5 Find the gradient of the following curves at the given values of x.
   a $f(x) = \frac{1}{2}x^2 + x;\ x = 1$
   b $f(x) = -x^3 + 2x^2 + x;\ x = 0$
   c $f(x) = (x-3)(x+8);\ x = \frac{1}{4}$

6 A stone is dropped from the top of a cliff. The distance (s) it falls is given by the equation $s = 5t^2$, where s is the distance in metres and t the time in seconds.
   a Calculate the velocity v, by differentiating the distance s with respect to time t.
   b Calculate the stone's velocity after 3 seconds.
   c The stone hits the ground travelling at 42 m s^{-1}
      i Calculate how long it took for the stone to hit the ground.
      ii Calculate the height of the cliff.

Student assessment 2

1 The function $f(x) = x^3 + x^2 - 1$ has a gradient of zero at points P and Q, where the x coordinate of P is less than that of Q.
   a Calculate the gradient function $f'(x)$.
   b Calculate the coordinates of P.
   c Calculate the coordinates of Q.
   d Determine which of the points P or Q is a maximum. Explain your method clearly.

19 DIFFERENTIATION AND THE GRADIENT FUNCTION

2 a Explain why the point $A(1, 1)$ lies on the curve $y = x^3 - x^2 + x$
  b Calculate the gradient of the curve at A.
  c Calculate the equation of the tangent to the curve at A.
  d Calculate the equation of the normal to the curve at A.

3 For the function $f(x) = (x - 2)^2 + 3$
  a Calculate $f'(x)$.
  b Determine the coordinates of the stationary point.

4 For the function $f(x) = x^4 - 2x^2$
  a Calculate $f'(x)$.
  b Determine the coordinates of any stationary points.
  c Determine the nature of any stationary point.
  d Find where the graph intersects or touches:
   i the y-axis
   ii the x-axis
  e Sketch the graph of f(x).

20 Functions

Functions as a mapping

Consider the equation $y = 2x + 3$. It describes the relationship between two variables x and y. In this case, 3 is added to twice the value of x to produce y.

A function is a particular type of relationship between two variables. It has certain characteristics.

Consider the equation $y = 2x + 3$ for values of x within $-1 \leqslant x \leqslant 3$.

A table of results can be constructed and a mapping drawn.

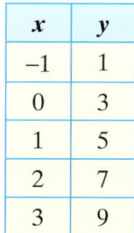 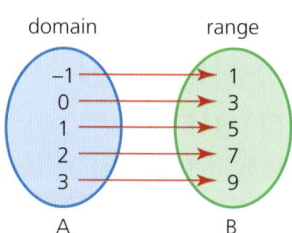

With a function, each value in set B (the **range**) is produced from one value in set A (the **domain**). The relationship can be written as a function:

$$f(x) = 2x + 3; -1 \leqslant x \leqslant 3$$

It is also usual to include the domain after the function, as a different domain will produce a different range.

The mapping from A to B can be a **one-to-one** mapping or a **many-to-one** mapping.

The function above, $f(x) = 2x + 3; -1 \leqslant x \leqslant 3$, is a one-to-one function as one value in the domain maps onto one value in the range. However, the function $f(x) = x^2$; where x is an integer, is a many-to-one function, as a value in the range can be generated by more than one value in the domain, as shown.

It is important to understand that one value in the domain (set A) maps to only one value in the range (set B). Therefore the mapping shown is the function $f(x) = x^2$; where x is an integer.

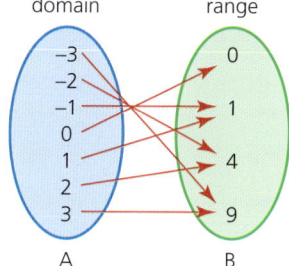

Some mappings will not represent functions; for example, consider the relationship $y = \pm\sqrt{x}$.

20 FUNCTIONS

The following table and mapping diagram can be produced:

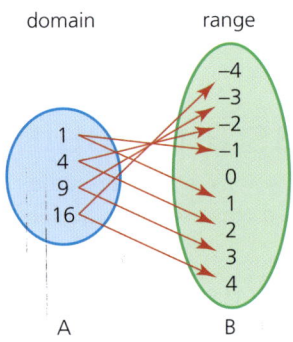

x	y
1	± 1
4	± 2
9	± 3
16	± 4

This relationship is not a function, as a value in the domain produces more than one value in the range.

Note: If a domain is not specified then it is assumed to be all real values.

Calculating the range from the domain

The domain is the set of input values and the range is the set of output values for a function. (Note that the range is not the difference between the greatest and least values as in statistics.) The range is therefore not only dependent on the function itself, but also on the domain.

➡ Worked example

Calculate the range for the following functions:

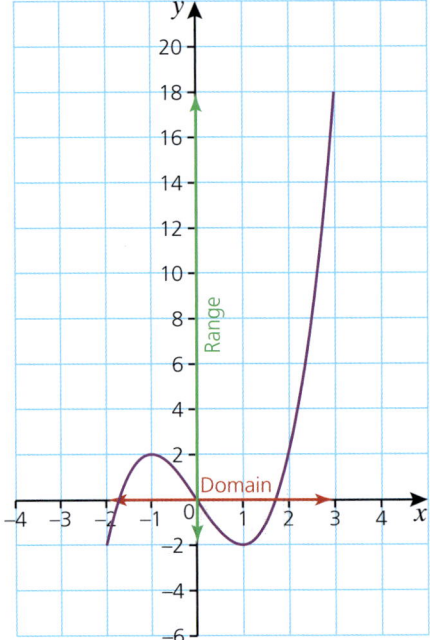

Calculating the range from the domain

a f(x) = $x^3 - 3x$; $-2 \leq x \leq 3$

The graph of the function is shown above. As the domain is restricted to $-2 \leq x \leq 3$, the range is limited from -2 to 18.

This is written as: Range $-2 \leq f(x) \leq 18$.

b f(x) = $x^3 - 3x$; x is a real number

The graph will be similar to the one above except that the domain is not restricted. As the domain is for all real values of x, this implies that any real number can be an input value. As a result, the range will also be all real values.

This is written as: Range f(x) is all real numbers.

Exercise 20.1

1 Which of the following mappings show a function?

a domain   range

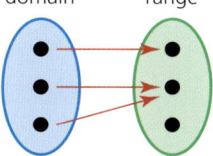

b domain   range

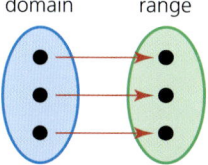

c domain   range

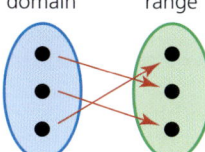

d domain   range

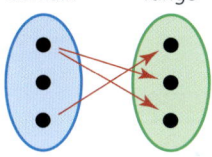

Give the domain and range of each of the functions in Questions 2–8.

2 f(x) = $2x - 1$; $-1 \leq x \leq 3$
3 f(x) = $3x + 2$; $-4 \leq x \leq 0$
4 f(x) = $-x + 4$; $-4 \leq x \leq 4$
5 f(x) = $x^2 + 2$; $-3 \leq x \leq 3$
6 f(x) = $x^2 + 2$; x is a real number
7 f(x) = $-x^2 + 2$; $0 \leq x \leq 4$
8 f(x) = $x^3 - 2$; $-3 \leq x \leq 1$

Just as with formulas, values can be substituted into functions.

➡ Worked examples

a For the function f(x) = $3x - 5$, evaluate:

i f(2)
f(2) = $3 \times 2 - 5$
    = $6 - 5$
    = 1

ii f(0)
f(0) = $3 \times 0 - 5$
    = $0 - 5$
    = -5

iii f(−2)
f(−2) = $3 \times (-2) - 5$
     = $-6 - 5$
     = -11

20 FUNCTIONS

b For the function f: $x \mapsto \frac{2x+6}{3}$, evaluate:

    i   f(3)
$$f(3) = \frac{2 \times 3 + 6}{3}$$
$$= \frac{6+6}{3}$$
$$= 4$$

    ii   f(1.5)
$$f(1.5) = \frac{2 \times 1.5 + 6}{3}$$
$$= \frac{3+6}{3}$$
$$= 3$$

    iii   f(–1)
$$f(-1) = \frac{2 \times (-1) + 6}{3}$$
$$= \frac{-2+6}{3}$$
$$= \frac{4}{3}$$

c For the function $f(x) = x^2 + 4$, evaluate:

    i   f(2)
$$f(2) = 2^2 + 4$$
$$= 4 + 4$$
$$= 8$$

    ii   f(6)
$$f(6) = 6^2 + 4$$
$$= 36 + 4$$
$$= 40$$

    iii   f(–1)
$$f(-1) = (-1)^2 + 4$$
$$= 1 + 4$$
$$= 5$$

Exercise 20.2

1 If $f(x) = 2x + 2$, calculate:
   a f(2)   **b** f(4)   **c** f(0.5)   **d** f(1.5)
   e f(0)   **f** f(–2)   **g** f(–6)   **h** f(–0.5)

2 If $f(x) = 4x - 6$, calculate:
   a f(4)   **b** f(7)   **c** f(3.5)   **d** f(0.5)
   e f(0.25)   **f** f(–3)   **g** f(–4.25)   **h** f(0)

3 If $g(x) = -5x + 2$, calculate:
   a g(0)   **b** g(6)   **c** g(4.5)   **d** g(3.2)
   e g(0.1)   **f** g(–2)   **g** g(–6.5)   **h** g(–2.3)

4 If $h(x) = -3x - 7$, calculate:
   a h(4)   **b** h(6.5)   **c** h(0)   **d** h(0.4)
   e h(–9)   **f** h(–5)   **g** h(–2)   **h** h(–3.5)

Exercise 20.3

1 If $f(x) = \frac{3x+2}{4}$, calculate:
   a f(2)   **b** f(8)   **c** f(2.5)   **d** f(0)
   e f(–0.5)   **f** f(–6)   **g** f(–4)   **h** f(–1.6)

2 If $g(x) = \frac{5x-3}{3}$, calculate:
   a g(3)   **b** g(6)   **c** g(0)   **d** g(–3)
   e g(–1.5)   **f** g(–9)   **g** g(–0.2)   **h** g(–0.1)

Calculating the range from the domain

3 If h: $x \mapsto \frac{-6x+8}{4}$, calculate:
   - a h(1)
   - b h(0)
   - c h(4)
   - d h(1.5)
   - e h(–2)
   - f h(–0.5)
   - g h(–22)
   - h h(–1.5)

4 If $f(x) = \frac{-5x-7}{-8}$, calculate:
   - a f(5)
   - b f(1)
   - c f(3)
   - d f(–1)
   - e f(–7)
   - f $f\left(-\frac{3}{5}\right)$
   - g f(–0.8)
   - h f(0)

Exercise 20.4

1 If $f(x) = x^2 + 3$, calculate:
   - a f(4)
   - b f(7)
   - c f(1)
   - d f(0)
   - e f(–1)
   - f f(0.5)
   - g f(–3)
   - h $f(\sqrt{2})$

2 If $f(x) = 3x^2 - 5$, calculate:
   - a f(5)
   - b f(8)
   - c f(1)
   - d f(0)
   - e f(–2)
   - f $f(\sqrt{3})$
   - g $f\left(-\frac{1}{2}\right)$
   - h $f\left(-\frac{1}{3}\right)$

3 If $g(x) = -2x^2 + 4$, calculate:
   - a g(3)
   - b $g\left(\frac{1}{2}\right)$
   - c g(0)
   - d g(1.5)
   - e g(–4)
   - f g(–1)
   - g $g(\sqrt{5})$
   - h g(–6)

4 If $h(x) = \frac{-5x^2 + 15}{-2}$, calculate:
   - a h(1)
   - b h(4)
   - c $h(\sqrt{3})$
   - d h(0.5)
   - e h(0)
   - f h(–3)
   - g $h\left(\frac{1}{\sqrt{2}}\right)$
   - h h(–2.5)

5 If $f(x) = -6x(x - 4)$, calculate:
   - a f(0)
   - b f(2)
   - c f(4)
   - d f(0.5)
   - e $f\left(-\frac{1}{2}\right)$
   - f $f\left(-\frac{1}{6}\right)$
   - g f(–2.5)
   - h $f(\sqrt{2})$

6 If $g(x) = \frac{(x+2)(x-4)}{-x}$, calculate:
   - a g(1)
   - b g(4)
   - c g(8)
   - d g(0)
   - e g(–2)
   - f g(–10)
   - g $g\left(-\frac{3}{2}\right)$
   - h g(–8)

Exercise 20.5

1 If $f(x) = 2x + 1$, write the following in their simplest form:
   - a f(x + 1)
   - b f(2x – 3)
   - c $f(x^2)$
   - d $f\left(\frac{x}{2}\right)$
   - e $f\left(\frac{x}{4}+1\right)$
   - f f(x) – x

2 If $g(x) = 3x^2 - 4$, write the following in their simplest form:
   - a g(2x)
   - b $g\left(\frac{x}{4}\right)$
   - c $g(\sqrt{2x})$
   - d g(3x) + 4
   - e g(x – 1)
   - f g(2x + 2)

3 If $f(x) = 4x^2 + 3x - 2$, write the following in their simplest form:
   - a f(x) + 4
   - b f(2x) + 2
   - c f(x + 2) – 20
   - d f(x – 1) + 1
   - e $f\left(\frac{x}{2}\right)$
   - f f(3x + 2)

20 FUNCTIONS

Inverse functions

The **inverse** of a function is its reverse, i.e. it 'undoes' the function's effects. The inverse of the function f(x) is written as f^{-1}(x). To find the inverse of a function:

» rewrite the function, replacing f(x) with y
» interchange x and y
» rearrange the equation to make y the subject.

Worked examples

a Find the inverse of each of the following functions:

  i f(x) = x + 2
  $y = x + 2$
  $x = y + 2$
  $y = x - 2$

  So f^{-1}(x) = $x - 2$

  ii g(x) = 2x – 3
  $y = 2x - 3$
  $x = 2y - 3$
  $y = \frac{x+3}{2}$

  So g^{-1}(x) = $\frac{x+3}{2}$

b If f(x) = $\frac{x-3}{3}$ calculate:

  i f^{-1}(2)
  ii f^{-1}(–3)

  First calculate the **inverse function** f^{-1}(x):

  $y = \frac{x-3}{3}$

  $x = \frac{y-3}{3}$

  $y = 3x + 3$

  So f^{-1}(x) = 3x + 3

  i f^{-1}(2) = 3(2) + 3 = 9
  ii f^{-1}(–3) = 3(–3) + 3 = –6

Exercise 20.6

Find the inverse of each of the following functions:

1 a f(x) = x + 3
   b f(x) = x + 6
   c f(x) = x – 5
   d g(x) = x
   e h(x) = 2x
   f p(x) = $\frac{x}{3}$

2 a f(x) = 4x
   b f(x) = 2x + 5
   c f(x) = 3x – 6
   d f(x) = $\frac{x+4}{2}$
   e g(x) = $\frac{3x-2}{4}$
   f g(x) = $\frac{8x+7}{5}$

3 a f(x) = $\frac{1}{2}x$ + 3
   b g(x) = $\frac{1}{4}x$ – 2
   c h(x) = 4(3x – 6)
   d p(x) = 6(x + 3)
   e q(x) = –2(–3x + 2)
   f f(x) = $\frac{2}{3}$(4x – 5)

Composite functions

 Exercise 20.7

1 If $f(x) = x - 4$, evaluate:
   a $f^{-1}(2)$ b $f^{-1}(0)$ c $f^{-1}(-5)$

2 If $f(x) = 2x + 1$, evaluate:
   a $f^{-1}(5)$ b $f^{-1}(0)$ c $f^{-1}(-11)$

3 If $g(x) = 6(x - 1)$, evaluate:
   a $g^{-1}(12)$ b $g^{-1}(3)$ c $g^{-1}(6)$

4 If $g(x) = \frac{2x+4}{3}$, evaluate:
   a $g^{-1}(4)$ b $g^{-1}(0)$ c $g^{-1}(-6)$

5 If $h(x) = \frac{1}{3}x - 2$, evaluate:
   a $h^{-1}\left(-\frac{1}{2}\right)$ b $h^{-1}(0)$ c $h^{-1}(-2)$

6 If $f(x) = \frac{4x-2}{5}$, evaluate:
   a $f^{-1}(6)$ b $f^{-1}(-2)$ c $f^{-1}(0)$

Composite functions

 Worked examples

a If $f(x) = x + 2$ and $g(x) = x + 3$, find $fg(x)$.
   $fg(x) = f(x + 3)$
   $\quad\quad = (x + 3) + 2$
   $\quad\quad = x + 5$

b If $f(x) = 2x - 1$ and $g(x) = x - 2$, find $fg(x)$.
   $fg(x) = f(x - 2)$
   $\quad\quad = 2(x - 2) - 1$
   $\quad\quad = 2x - 4 - 1$
   $\quad\quad = 2x - 5$

c If $f(x) = 2x + 3$ and $g(x) = 2x$, evaluate $fg(3)$.
   $fg(x) = f(2x)$
   $\quad\quad = 2(2x) + 3$
   $\quad\quad = 4x + 3$
   $fg(3) = 4 \times 3 + 3$
   $\quad\quad = 15$

fg(x) implies substituting the function g(x) into the function f(x)

Exercise 20.8

1 Write a formula for $fg(x)$ in each of the following:
   a $f(x) = x - 3$ $g(x) = x + 5$
   b $f(x) = x + 4$ $g(x) = x - 1$
   c $f(x) = x$ $g(x) = 2x$
   d $f(x) = \frac{x}{2}$ $g(x) = 2x$

Exercise 20.8 (cont)

2 Write a formula for pq(x) in each of the following:
   a p(x) = 2x   q(x) = x + 4
   b p(x) = 3x + 1   q(x) = 2x
   c p(x) = 4x + 6   q(x) = (2x − 1)²
   d p(x) = −x + 4   q(x) = (x + 2)²

3 Write a formula for jk(x) in each of the following:
   a $j(x) = \frac{x-2}{4}$ k(x) = 4x
   b j(x) = 3x + 2 $k(x) = \frac{x-3}{2}$
   c $j(x) = \frac{2x+5}{3}$ $k(x) = \frac{1}{2}x + 1$
   d $j(x) = \frac{1}{4}(x-3)$ $k(x) = \frac{8x+2}{5}$

4 Evaluate fg(2) in each of the following:
   a f(x) = x − 4   g(x) = x + 3
   b f(x) = 2x   g(x) = −x + 6
   c f(x) = 3x   g(x) = 6x + 1
   d $f(x) = \frac{x}{2}$ g(x) = −2x

5 Evaluate gh(−4) in each of the following:
   a g(x) = 3x + 2   h(x) = −4x
   b $g(x) = \frac{1}{2}(3x - 1)$ $h(x) = \frac{2x}{5}$
   c g(x) = 4(−x + 2) $h(x) = \frac{2x+6}{4}$
   d $g(x) = \frac{4x+4}{5}$ $h(x) = -\frac{1}{3}(-x + 5)$

Student assessment 1

1 For the function f(x) = 5x − 1, evaluate:
   a f(2)   b f(0)   c f(−3)

2 For the function $g(x) = \frac{3x-2}{2}$, evaluate:
   a g(4)   b g(0)   c g(−3)

3 For the function $f(x) = \frac{(x+3)(x-4)}{2}$, evaluate:
   a f(0)   b f(−3)   c f(−6)

4 Find the inverse of each of the following functions:
   a f(x) = −x + 4   b $g(x) = \frac{3(x-6)}{2}$

5 If $h(x) = \frac{3}{2}(-x + 3)$, evaluate:
   a $h^{-1}(-3)$ b $h^{-1}\left(\frac{3}{2}\right)$

6 If f(x) = 4x + 2 and g(x) = −x + 3, find fg(x).

7 A function is given as f(x) = (x + 4)(x − 2)²; −5 ⩽ x ⩽ 3
   a Sketch f(x) for the given domain.
   b Deduce the range of f(x).

Student assessment 2

1. For the function $f(x) = 3x + 1$, evaluate:
   a. $f(4)$
   b. $f(-1)$
   c. $f(0)$

2. For the function $g(x) = \frac{-x-2}{3}$, evaluate:
   a. $g(4)$
   b. $g(-5)$
   c. $g(1)$

3. For the function $f(x) = x^2 - 3x$, evaluate:
   a. $f(1)$
   b. $f(3)$
   c. $f(-3)$

4. Find the inverse of the following functions:
   a. $f(x) = -3x + 9$
   b. $g(x) = \frac{(x-2)}{4}$

5. If $h(x) = -5(-2x + 4)$, evaluate:
   a. $h^{-1}(-10)$
   b. $h^{-1}(0)$

6. If $f(x) = 8x + 2$ and $g(x) = 4x - 1$, find $fg(x)$.

7. A function is given as $g(x) = -(x + 3)^2 + 5; -6 \leqslant x \leqslant 1$
   a. Sketch $g(x)$ for the given domain.
   b. Deduce the range of $g(x)$.

Mathematical investigations and ICT 2

House of cards

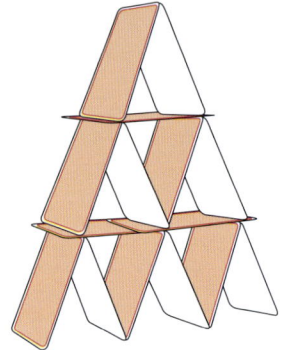

The drawing shows a house of cards three layers high. Fifteen cards are needed to construct it.
1. How many cards are needed to construct a house ten layers high?
2. The world record is for a house 75 layers high. How many cards are needed to construct this house of cards?
3. Show that the general formula for a house n layers high requiring c cards is:

$c = \frac{1}{2}n(3n + 1)$

Chequered boards

A chessboard is an 8×8 square grid consisting of alternating black and white squares as shown:

There are 64 unit squares of which 32 are black and 32 are white.

Consider boards of different sizes. The examples below show rectangular boards, each consisting of alternating black and white unit squares.

Total number of unit squares is 30.

Number of black squares is 15.

Number of white squares is 15.

Total number of unit squares is 21.

Number of black squares is 10.

Number of white squares is 11.

1. Investigate the number of black and white unit squares on different rectangular boards.
   Note: For consistency you may find it helpful to always keep the bottom right-hand square the same colour.
2. What are the numbers of black and white squares on a board $m \times n$ units?

Modelling: Stretching a spring

A spring is attached to a clamp stand as shown below.

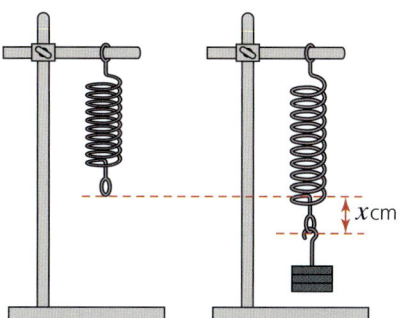

Different weights are attached to the end of the spring. The mass (m) in grams is noted as is the amount by which the spring stretches (x) in centimetres.

The data collected is shown in the table below:

Mass (g)	50	100	150	200	250	300	350	400	450	500
Extension (cm)	3.1	6.3	9.5	12.8	15.4	18.9	21.7	25.0	28.2	31.2

1. Plot a graph of mass against extension.
2. Describe the approximate relationship between the mass and the extension.
3. Draw a line of best fit through the data.
4. Calculate the equation of the line of best fit.
5. Use your equation to predict what the length of the spring would be for a mass of 275 g.
6. Explain why it is unlikely that the equation would be useful to find the extension if a mass of 5 kg was added to the spring.

ICT activity

You have seen that it is possible to solve some exponential equations by applying the laws of indices.

Use a graphics calculator and appropriate graphs to solve the following exponential equations:

1. $4^x = 40$
2. $3^x = 17$
3. $5^{x-1} = 6$
4. $3^{-x} = 0.5$

TOPIC 3

Coordinate geometry

Contents
Chapter 21 Straight-line graphs (E3.1, E3.2, E3.3, E3.4, E3.5, E3.6, E3.7)

Learning objectives

E3.1 Coordinates
Use and interpret Cartesian coordinates in two dimensions.

E3.2 Drawing linear graphs
Draw straight-line graphs for linear equations.

E3.3 Gradient of linear graphs
1   Find the gradient of a straight line.
2   Calculate the gradient of a straight line from the coordinates of two points on it.

E3.4 Length and midpoint
1   Calculate the length of a line segment.
2   Find the coordinates of the midpoint of a line segment.

E3.5 Equations of linear graphs
Interpret and obtain the equation of a straight-line graph.

E3.6 Parallel lines
Find the gradient and equation of a straight line parallel to a given line.

E3.7 Perpendicular lines
Find the gradient and equation of a straight line perpendicular to a given line.

The French

In the middle of the seventeenth century there were three great French mathematicians, René Descartes, Blaise Pascal and Pierre de Fermat.

René Descartes (1596–1650) was a philosopher and a mathematician. His book *The Meditations* asks 'How and what do I know?' His work in mathematics made a link between algebra and geometry. He thought that all nature could be explained in terms of mathematics. Although he was not considered as talented a mathematician as Pascal and Fermat, he has had greater influence on modern thought. The (x, y) coordinates we use are called Cartesian coordinates after Descartes.

Blaise Pascal (1623–1662) was a genius who studied geometry as a child. When he was 16 he stated and proved Pascal's Theorem, which relates any six points on any conic section. The Theorem is sometimes called the 'Cat's Cradle'. He founded probability theory and made contributions to the invention of calculus. He is best known for Pascal's Triangle.

Pierre de Fermat (1601–1665) was a brilliant mathematician and, along with Descartes, one of the most influential. Fermat invented number theory and worked on calculus. He discovered probability theory with his friend Pascal. It can be argued that Fermat was at least Newton's equal as a mathematician.

Fermat's most famous discovery in number theory includes 'Fermat's Last Theorem'. This theorem is derived from Pythagoras' theorem, which states that for a right-angled triangle, $x^2 = y^2 + z^2$ where x is the length of the hypotenuse. Fermat said that if the index (power) was greater than two and x, y, z were all whole numbers, then the equation was never true. (This theorem was only proved in 1995 by the English mathematician Andrew Wiles.)

René Descartes (1596–1650)

21 Straight-line graphs

Coordinates

To fix a point in two dimensions (2D), its position is given in relation to a point called the **origin**. Through the origin, axes are drawn perpendicular to each other. The horizontal axis is known as the *x*-**axis**, and the vertical axis is known as the *y*-**axis**.

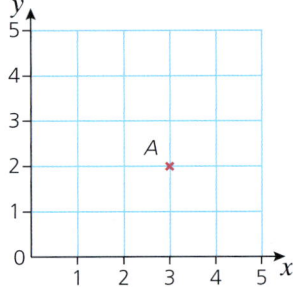

The *x*-axis is numbered from left to right. The *y*-axis is numbered from bottom to top.

The position of point *A* is given by two coordinates: the *x*-coordinate first, followed by the *y*-coordinate. So the coordinates of point *A* are (3, 2).

A number line can extend in both directions by extending the *x*- and *y*-axes below zero, as shown in the grid below:

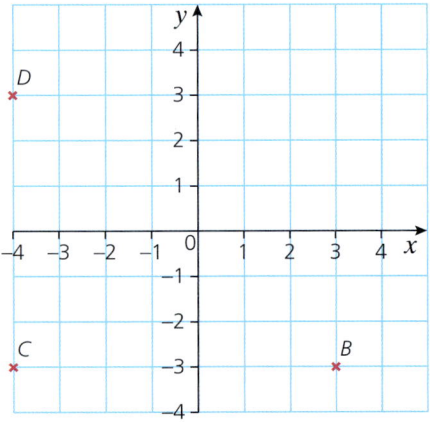

Points *B*, *C* and *D* can be described by their coordinates:

Point *B* is at (3, −3),
Point *C* is at (−4, −3),
Point *D* is at (−4, 3).

Exercise 21.1

1 Draw a pair of axes with both x and y from -8 to $+8$.
  Mark each of the following points on your grid:
  a $A = (5, 2)$ b $B = (7, 3)$ c $C = (2, 4)$
  d $D = (-8, 5)$ e $E = (-5, -8)$ f $F = (3, -7)$
  g $G = (7, -3)$ h $H = (6, -6)$

2 $A = (3, 2)$ $B = (3, -4)$ $C = (-2, -4)$ $D = (-2, 2)$

Draw a separate grid for each of Questions 2–4 with x- and y-axes from -6 to $+6$. Plot and join the points in order to name each shape drawn.

3 $E = (1, 3)$ $F = (4, -5)$ $G = (-2, -5)$

4 $H = (-6, 4)$ $I = (0, -4)$ $J = (4, -2)$ $K = (-2, 6)$

Exercise 21.2

Draw a pair of axes with both x and y from -10 to $+10$.

1 Plot the points $P = (-6, 4)$, $Q = (6, 4)$ and $R = (8, -2)$.
  Plot point S such that $PQRS$ when drawn is a parallelogram.
  a Draw diagonals PR and QS. What are the coordinates of their point of intersection?
  b What is the area of $PQRS$?

2 On the same axes, plot point M at $(-8, 4)$ and point N at $(4, 4)$.
  a Join points $MNRS$. What shape is formed?
  b What is the area of $MNRS$?
  c Explain your answer to Question **2b**.

3 On the same axes, plot point J where point J has y-coordinate $+10$ and JRS, when joined, forms an isosceles triangle. What is the x-coordinate of all points on the line of symmetry of triangle JRS?

Exercise 21.3

1 a On a grid with axes numbered from -10 to $+10$ draw a hexagon $ABCDEF$ with centre $(0, 0)$, points $A(0, 8)$ and $B(7, 4)$ and two lines of symmetry.
  b Write down the coordinates of points C, D, E and F.

2 a On a similar grid to Question 1, draw an octagon $PQRSTUVW$ which has point $P(2, -8)$, point $Q(-6, -8)$ and point $R(-7, -5)$. $PQ = RS = TU = VW$ and $QR = ST = UV = WP$.
  b List the coordinates of points S, T, U, V and W.
  c What are the coordinates of the centre of rotational symmetry of the octagon?

Reading scales

Exercise 21.4

1 The points A, B, C and D are not at whole number points on the number line. Point A is at 0.7.

  What are the positions of points B, C and D?

2 On this number line, point E is at 0.4.

  What are the positions of points F, G and H?

21 STRAIGHT-LINE GRAPHS

Exercise 21.4 (cont)

3 What are the positions of points I, J, K, L and M?

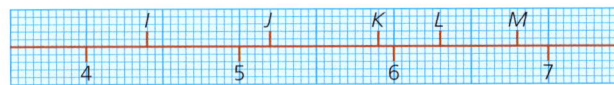

4 Point P is at position 0.4 and point W is at position 9.8. What are the positions of points Q, R, S, T, U and V?

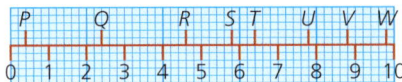

Exercise 21.5

Give the coordinates of points A, B, C, D, E, F, G and H.

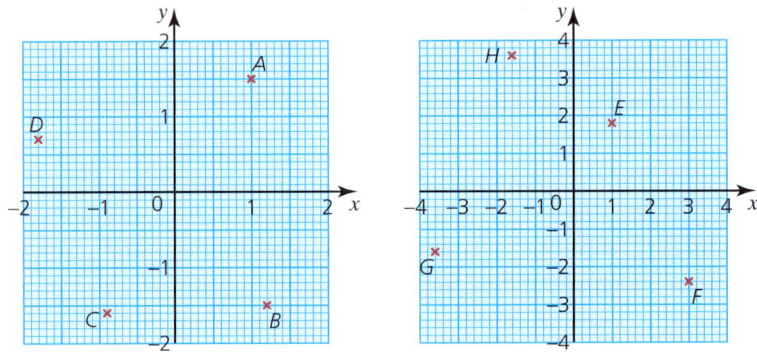

The gradient of a straight line

Lines are made of an infinite number of points. This chapter looks at those whose points form a straight line.

The graph below shows three straight lines.

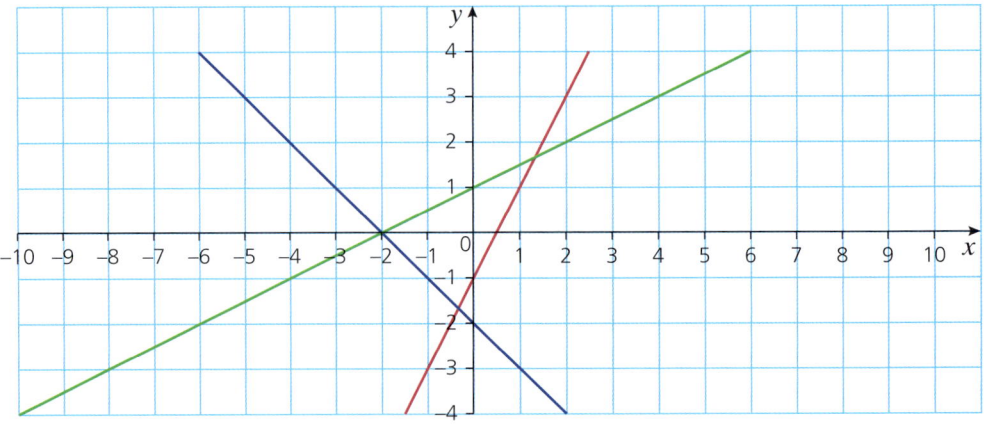

The gradient of a straight line

The lines have some properties in common (e.g. they are straight), but also have differences. One of their differences is that they have different slopes. The slope of a line is called its **gradient**.

The gradient of a straight line is constant, i.e. it does not change. The gradient of a straight line can be calculated by considering the coordinates of any two points on the line.

On the line below, two points A and B have been chosen.

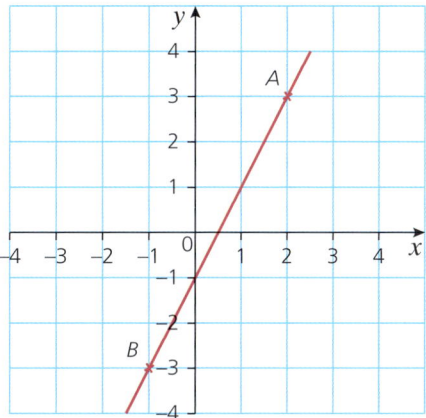

The coordinates of the points are $A(2, 3)$ and $B(-1, -3)$. The gradient is calculated using the following formula:

$$\text{Gradient} = \frac{\text{vertical distance between two points}}{\text{horizontal distance between two points}}$$

Graphically this can be represented as follows:

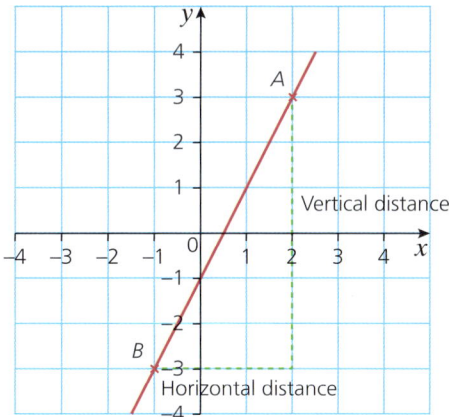

Therefore, gradient $= \frac{3-(-3)}{2-(-1)} = \frac{6}{3} = 2$

In general, therefore, if the two points chosen have coordinates (x_1, y_1) and (x_2, y_2), the gradient is calculated as:

$$\text{Gradient} = \frac{y_2 - y_1}{x_2 - x_1}$$

21 STRAIGHT-LINE GRAPHS

Worked example

Calculate the gradient of the line shown below.

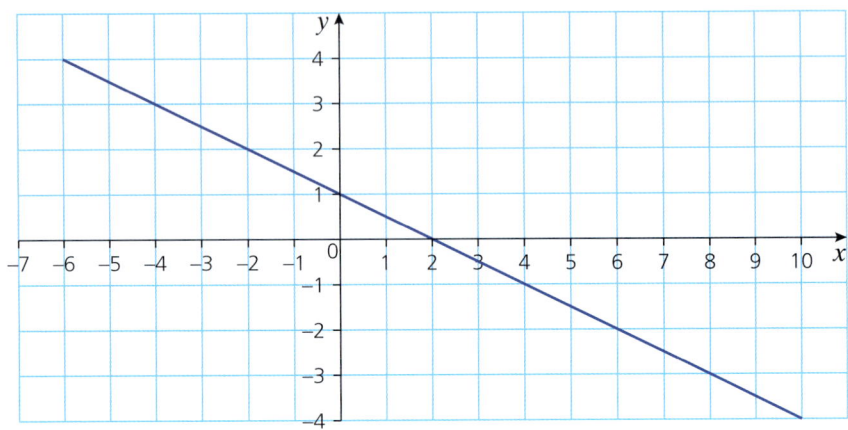

Choose two points on the line, e.g. (−4, 3) and (8, −3).

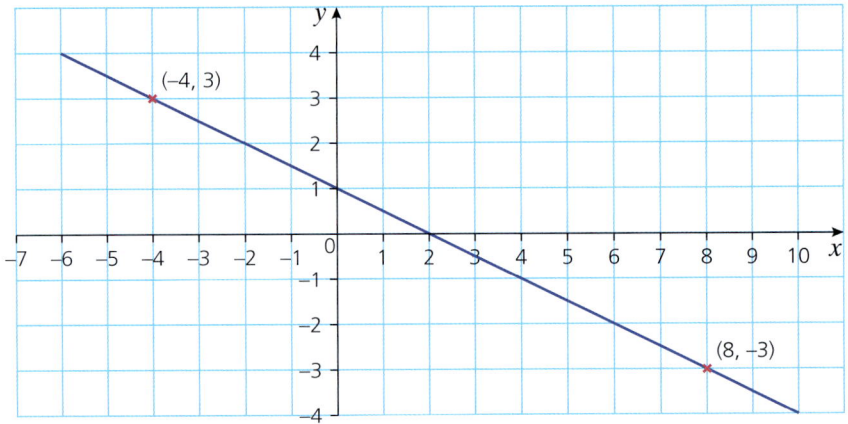

Let point 1 be (−4, 3) and point 2 be (8, −3).

Gradient $= \dfrac{y_2 - y_1}{x_2 - x_1} = \dfrac{-3 - 3}{8 - (-4)}$

$= \dfrac{-6}{12} = -\dfrac{1}{2}$

Note: The gradient is not affected by which point is chosen as point 1 and which is chosen as point 2. In the example above, if point 1 was (8, −3) and point 2 (−4, 3), the gradient would be calculated as:

Gradient $= \dfrac{y_2 - y_1}{x_2 - x_1} = \dfrac{3 - (-3)}{-4 - 8}$

$= \dfrac{6}{-12} = -\dfrac{1}{2}$

The gradient of a straight line

To check if the sign of the gradient is correct, the following guideline is useful.

A line sloping this way will have a positive gradient

A line sloping this way will have a negative gradient

A large value for the gradient implies that the line is steep. The line on the right below will have a greater value for the gradient than the line on the left as it is steeper.

Exercise 21.6

1 For each of the following lines, select two points on the line and then calculate its gradient.

a
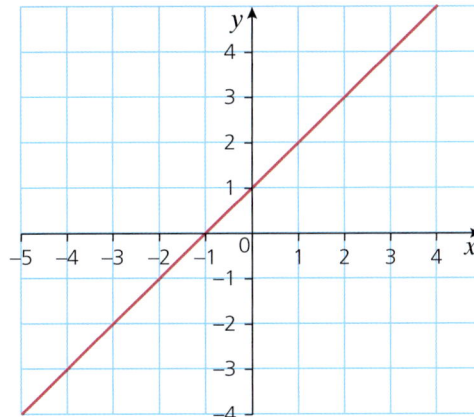

b
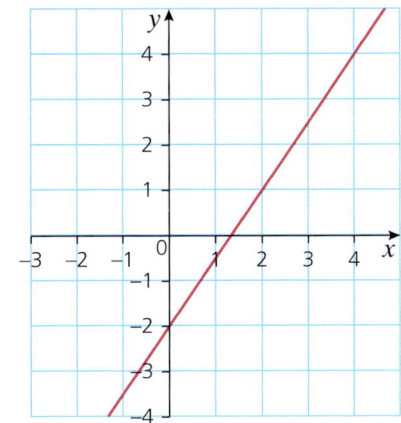

Exercise 21.6 (cont)

c
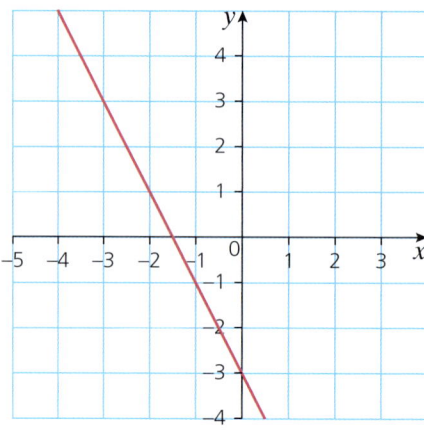

d
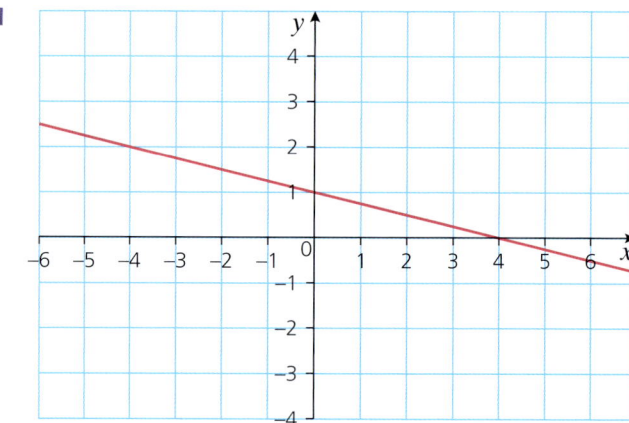

e
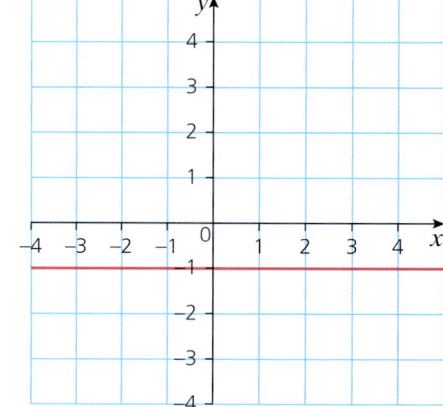

The equation of a straight line

f

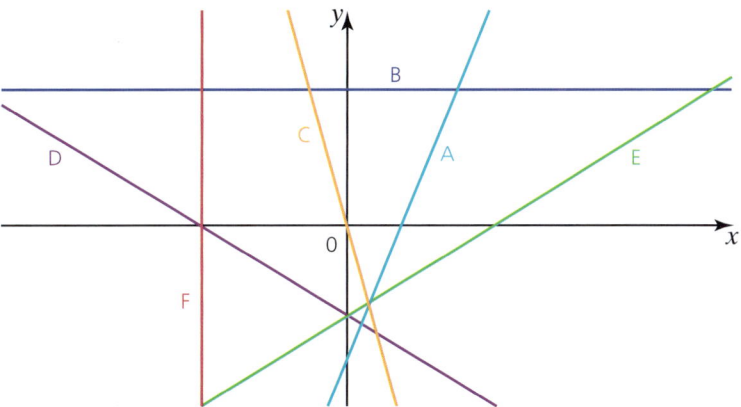

2. From your answers to Question 1e, what conclusion can you make about the gradient of any horizontal line?

3. From your answers to Question 1f, what conclusion can you make about the gradient of any vertical line?

4. The graph below shows six straight lines labelled A–F.

Six gradients are given below. Deduce which line has which gradient.

Gradient = $\frac{1}{2}$ Gradient is undefined   Gradient = 2

Gradient = -3 Gradient = 0   Gradient = $-\frac{1}{2}$

The equation of a straight line

The coordinates of every point on a straight line all have a common relationship. This relationship when expressed algebraically as an equation in terms of x and/or y is known as the **equation of the straight line**.

21 STRAIGHT-LINE GRAPHS

> **Worked examples**

a By looking at the coordinates of some of the points on the line below, establish the equation of the straight line.

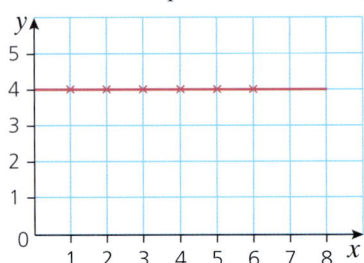

x	y
1	4
2	4
3	4
4	4
5	4
6	4

Some of the points on the line have been identified and their coordinates entered in a table above. By looking at the table it can be seen that the only rule all the points have in common is that $y = 4$.
Hence the equation of the straight line is $y = 4$.

b By looking at the coordinates of some of the points on the line, establish the equation of the straight line.

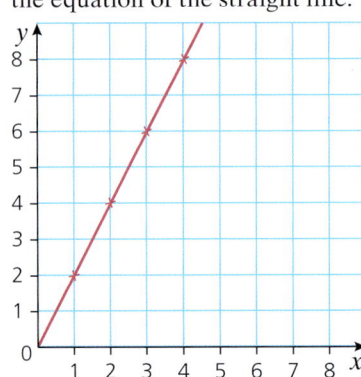

x	y
1	2
2	4
3	6
4	8

Once again, by looking at the table it can be seen that the relationship between the x- and y-coordinates is that each y-coordinate is twice the corresponding x-coordinate.
Hence the equation of the straight line is $y = 2x$.

 Exercise 21.7

1 In each of the following, identify the coordinates of some of the points on the line and use these to find the equation of the straight line.

a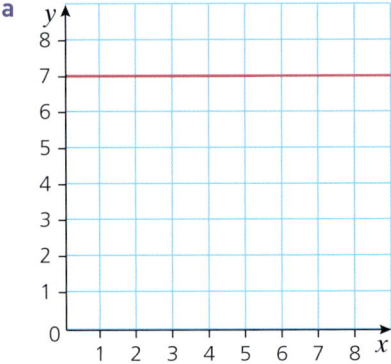

b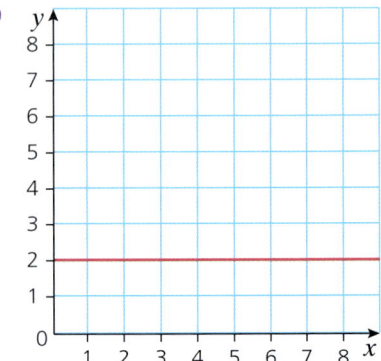

The equation of a straight line

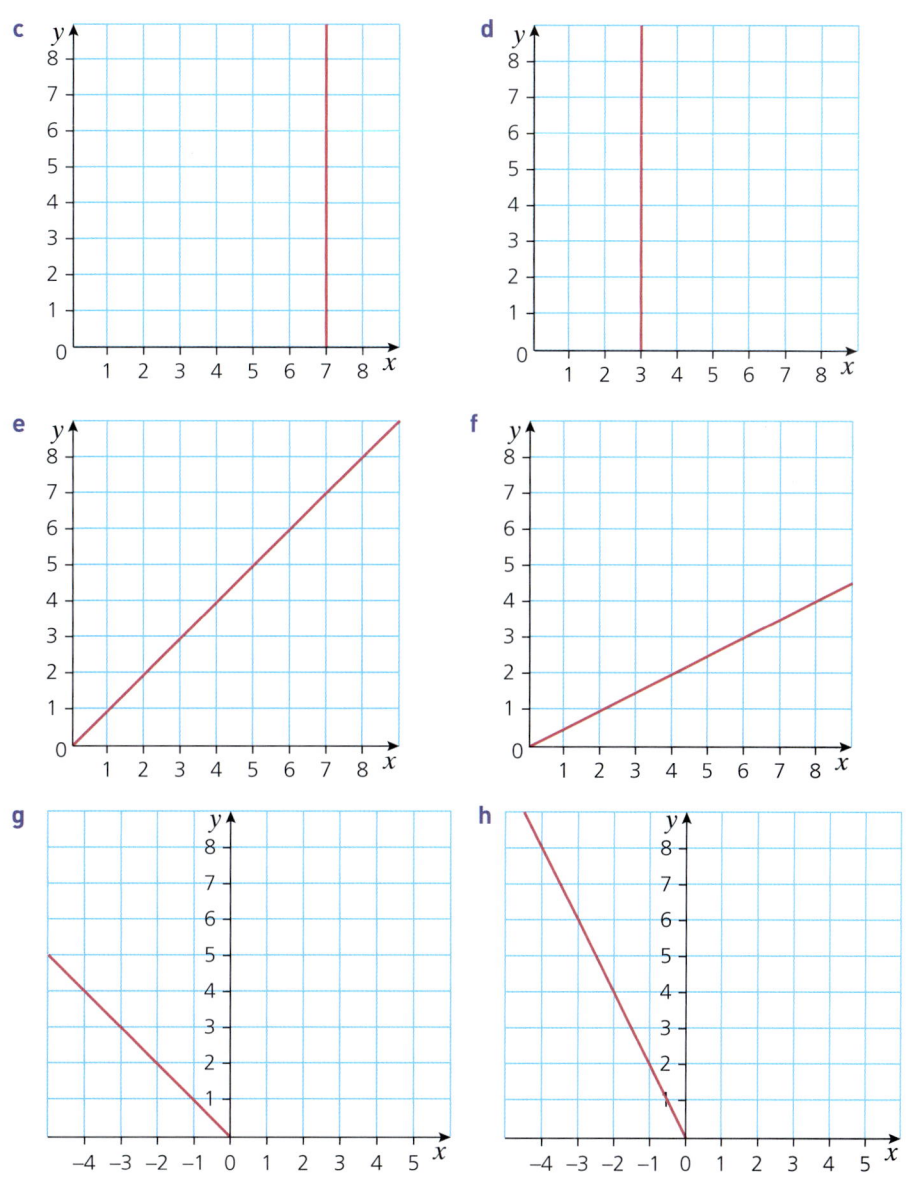

21 STRAIGHT-LINE GRAPHS

 Exercise 21.8

1 In each of the following, identify the coordinates of some of the points on the line and use these to find the equation of the straight line.

a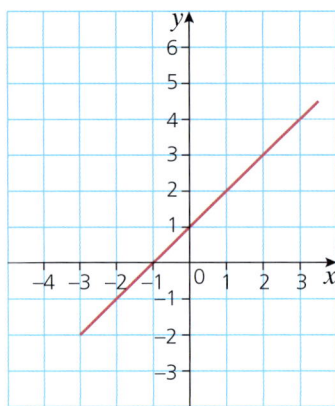

b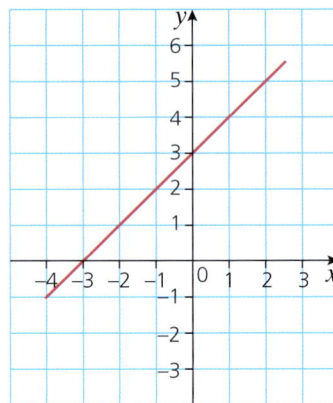

c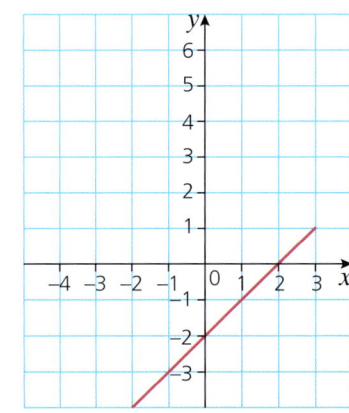

d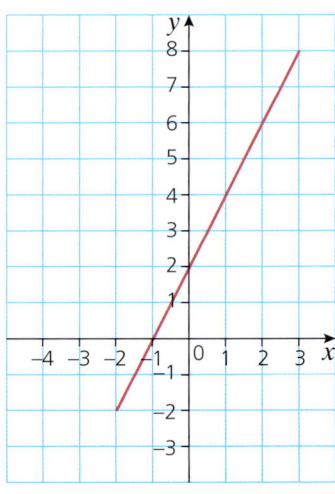

e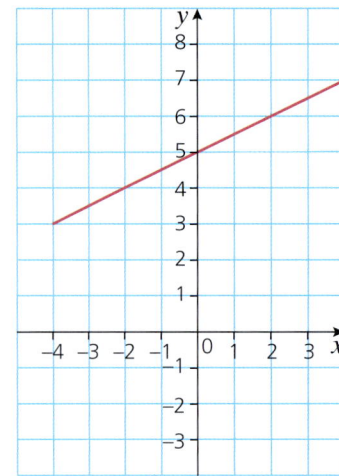

f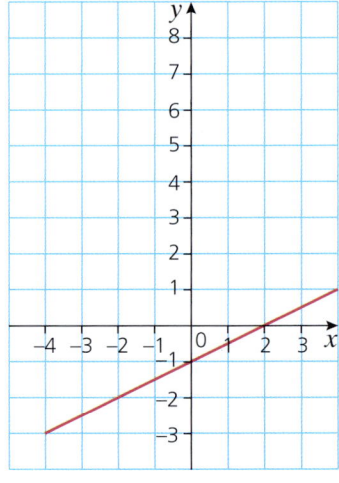

The equation of a straight line

2 In each of the following, identify the coordinates of some of the points on the line and use these to find the equation of the straight line.

a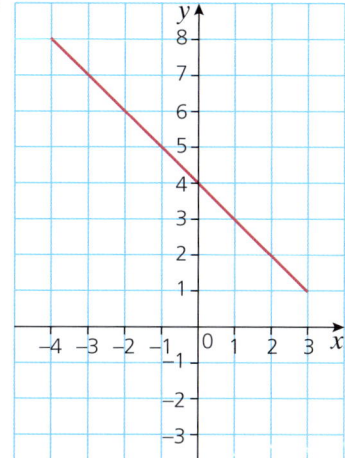

b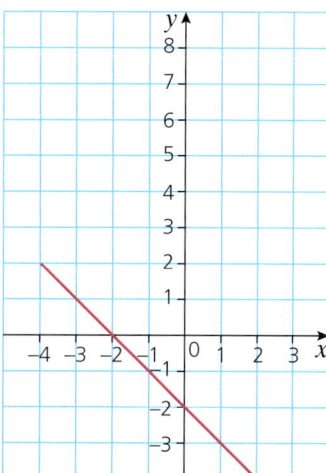

c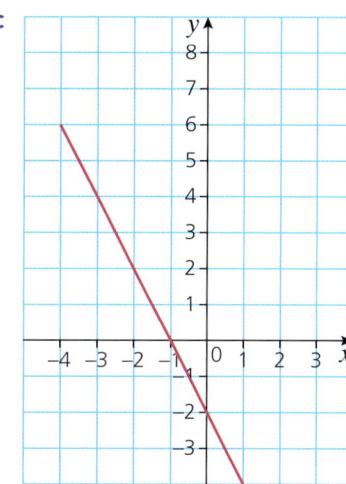

d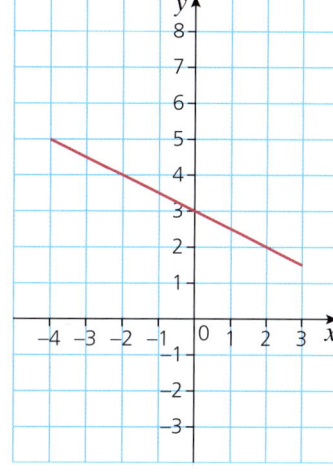

e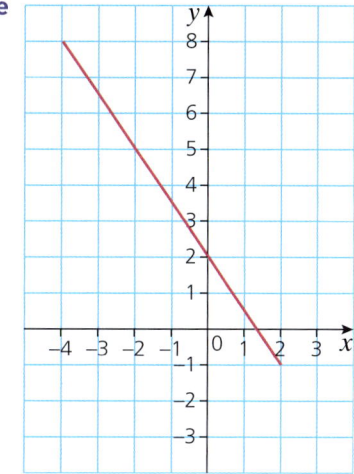

f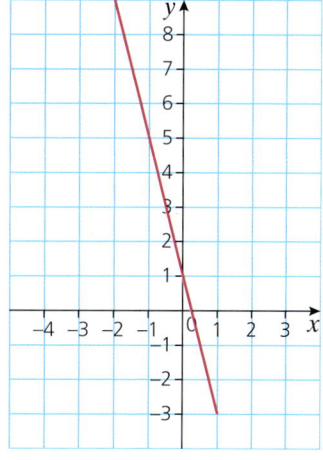

21 STRAIGHT-LINE GRAPHS

Exercise 21.8 (cont)

3 a For each of the graphs in Questions 1 and 2, calculate the gradient of the straight line.
   b What do you notice about the gradient of each line and its equation?
   c What do you notice about the equation of the straight line and where the line intersects the *y*-axis?

4 Copy the diagrams in Question 1. Draw two lines on each diagram parallel to the given line.
   a Write the equation of these new lines in the form $y = mx + c$.
   b What do you notice about the equations of these new parallel lines?

5 In Question 2, you have an equation for these lines in the form $y = mx + c$. Change the value of the intercept c and then draw the new line. What do you notice about this new line and the first line?

The general equation of a straight line

In general the equation of any straight line can be written in the form:

$y = mx + c$

where m represents the gradient of the straight line and c the intercept with the *y*-axis. This is shown in the diagram below.

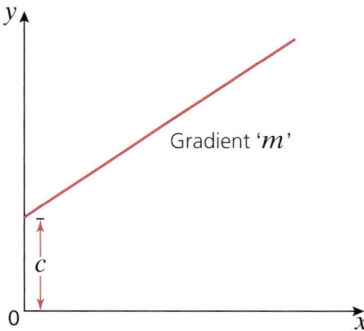

By looking at the equation of a straight line written in the form $y = mx + c$, it is therefore possible to deduce the line's gradient and intercept with the *y*-axis without having to draw it.

→ Worked examples

a Calculate the gradient and *y*-intercept of the following straight lines:
   i $y = 3x - 2$ gradient = 3
                              y-intercept = −2
   ii $y = -2x + 6$ gradient = −2
                              y-intercept = 6

b Calculate the gradient and *y*-intercept of the following straight lines:
   i $2y = 4x + 2$
       This needs to be rearranged into **gradient-intercept** form (i.e. $y = mx + c$).
       $y = 2x + 1$ gradient = 2
                              y-intercept = 1

ii $y - 2x = -4$
Rearranging into gradient-intercept form, we have:
$y = 2x - 4$ gradient = 2
              y-intercept = -4

iii $-4y + 2x = 4$
Rearranging into gradient-intercept form, we have:
$y = \frac{1}{2}x - 1$ gradient = $\frac{1}{2}$
              y-intercept = -1

iv $\frac{y+3}{4} = -x + 2$

Rearranging into gradient-intercept form, we have:
$y + 3 = -4x + 8$
$y = -4x + 5$ gradient = -4
              y-intercept = 5

Equation of a straight line in the form $ax + by + c = 0$

The general equation of a straight line takes the form $y = mx + c$; however, this is not the only form the equation of a straight line can take.

A different form is to rearrange the equation so that all the terms are on one side of the equation and equal to zero and to write it so that there are no fractions.

Worked examples

a The equation of a straight line is given as $y = \frac{1}{2}x + 3$. Write this in the form $ax + by + c = 0$, where a, b and c are integers.

Rearranging the equation so that all the terms are on one side produces $\frac{1}{2}x - y + 3 = 0$.

However, there is still a fraction in the equation and the question stated that a, b and c are integers.

To eliminate the fraction, both sides of the equation are multiplied by two. This gives: $x - 2y + 6 = 0$

b The equation of a straight line is given as $y = \frac{2}{5}x - \frac{1}{3}$. Write this in the form $ax + by + c = 0$, where a, b and c are integers.

Rearranging the equation so that all the terms are on the same side gives:
$\frac{2}{5}x - y - \frac{1}{3} = 0$

To eliminate the fractions, multiply both sides by 15 to give $6x - 15y - 5 = 0$.

21 STRAIGHT-LINE GRAPHS

Exercise 21.9

For the following linear equations, calculate both the gradient and y-intercept in each case.

1. a $y = 2x + 1$
   b $y = 3x + 5$
   c $y = x - 2$
   d $y = \frac{1}{2}x + 4$
   e $y = -3x + 6$
   f $y = -\frac{2}{3}x + 1$
   g $y = -x$
   h $y = -x - 2$
   i $y = -(2x - 2)$

2. a $y - 3x = 1$
   b $y + \frac{1}{2}x - 2 = 0$
   c $y + 3 = -2x$
   d $y + 2x + 4 = 0$
   e $y - \frac{1}{4}x - 6 = 0$
   f $-3x + y = 2$
   g $2 + y = x$
   h $8x - 6 + y = 0$
   i $-(3x + 1) + y = 0$

3. a $2y = 4x - 6$
   b $2y = x + 8$
   c $\frac{1}{2}y = x - 2$
   d $\frac{1}{4}y = -2x + 3$
   e $3y - 6x = 0$
   f $\frac{1}{3}y + x = 1$
   g $6y - 6 = 12x$
   h $4y - 8 + 2x = 0$
   i $2y - (4x - 1) = 0$

4. a $2x - y = 4$
   b $x - y + 6 = 0$
   c $-2y = 6x + 2$
   d $12 - 3y = 3x$
   e $5x - \frac{1}{2}y = 1$
   f $-\frac{2}{3}y + 1 = 2x$
   g $9x - 2 = -y$
   h $-3x + 7 = -\frac{1}{2}y$
   i $-(4x - 3) = -2y$

5. a $\frac{y+2}{4} = \frac{1}{2}x$
   b $\frac{y-3}{x} = 2$
   c $\frac{y-x}{8} = 0$
   d $\frac{2y-3x}{2} = 6$
   e $\frac{3y-2}{x} = -3$
   f $\frac{\frac{1}{2}y - 1}{x} = -2$
   g $\frac{3x-y}{2} = 6$
   h $\frac{6-2y}{3} = 2$
   i $\frac{-(x+2y)}{5x} = 1$

6. a $\frac{3x-y}{y} = 2$
   b $\frac{-x+2y}{4} = y + 1$
   c $\frac{y-x}{x+y} = 2$
   d $\frac{1}{y} = \frac{1}{x}$
   e $\frac{-(6x+y)}{2} = y + 1$
   f $\frac{2x-3y+4}{4} = 4$

7. a $\frac{y+1}{x} + \frac{3y-2}{2x} = -1$
   b $\frac{x}{y+1} + \frac{1}{2y+2} = 3$
   c $\frac{-(-y+3x)}{-(6x-2y)} = 1$
   d $\frac{-(x-2y)-(-x-2y)}{4+x-y} = -2$

8. Write each of the following equations in the form $ax + by + c = 0$ where a, b and c are integers.
   a $y = \frac{1}{3}x + 1$
   b $y = \frac{2}{5}x - 2$
   c $3y = \frac{3}{2}x - \frac{1}{2}$
   d $-\frac{1}{2}y = x + \frac{1}{3}$
   e $\frac{y+1}{2} = \frac{3}{4}x - 1$
   f $-\frac{3}{5}y - 4 = x$

9. Write each of the following equations in the form $ax + by + c = 0$ where a, b and c are integers.
   a $\frac{y-1}{x} - \frac{y-2}{3x} = 1$
   b $\frac{2y+x}{x-y} = \frac{1}{2}$
   c $\frac{3}{y} = -\frac{2}{5x}$
   d $\frac{-(2x+3y)}{3x} = \frac{2}{3}$
   e $\frac{3(x-2y)}{x} - 4 = 0$
   f $\frac{2x}{5y} + \frac{x}{-y} = -1$

Parallel lines and their equations

Lines that are parallel, by their very definition, must have the same gradient. Similarly, lines with the same gradient must be parallel. So a straight line with equation $y = -3x + 4$ must be parallel to a line with equation $y = -3x - 2$ as both have a gradient of -3.

➡ Worked examples

a A straight line has equation $y = 2x + 4$. Another straight line has equation $y = -2x + 4$. Explain, giving reasons, whether the two lines are parallel to each other or not.

They are not parallel as one has a gradient of 2, the other has a gradient of -2.

b A straight line has equation $4x - 2y + 1 = 0$.
Another straight line has equation $\frac{2x - 4}{y} = 1$.
Explain, giving reasons, whether the two lines are parallel to each other or not.
Rearranging the equations into gradient-intercept form gives:

$4x - 2y + 1 = 0$ $\qquad\qquad$ $\frac{2x - 4}{y} = 1$

$2y = 4x + 1$ $\qquad\qquad\qquad$ $y = 2x - 4$

$y = 2x + \frac{1}{2}$

With both equations written in gradient-intercept form, it is possible to see that both lines have a gradient of 2 and are therefore parallel.

c A straight line A has equation $y = -3x + 6$. A second line B is parallel to line A and passes through the point with coordinates $(-4, 10)$. Calculate the equation of line B.

As line B is a straight line it must take the form $y = mx + c$.

As it is parallel to line A, its gradient must be -3.
Because line B passes through the point $(-4, 10)$, these values can be substituted into the general equation of the straight line to give:

$10 = -3 \times (-4) + c$

Rearranging to find c gives: $c = -2$
The equation of line B is therefore $y = -3x - 2$.

Exercise 21.10

1 A straight line has equation $y = x + \frac{4}{3}$. Write down the equation of another straight line parallel to it.
2 A straight line has equation $y = -x + 6$. Which of the following lines is/are parallel to it?
   a $y = -x - 2$ $\qquad\qquad\qquad$ b $y = 8 - x$
   c $y = x - 6$ $\qquad\qquad\qquad\;\,$ d $y = -x$
3 A straight line has equation $3y - 3x = 4$. Write down the equation of another straight line parallel to it.
4 A straight line has equation $y = -x + 6$. Which of the following lines is/are parallel to it?
   a $2(y + x) = -5$ $\qquad\qquad$ b $-3x - 3y + 7 = 0$
   c $2y = -x + 12$ $\qquad\qquad\;\,$ d $y + x = \frac{1}{10}$

21 STRAIGHT-LINE GRAPHS

Exercise 21.10 (cont)

5 Find the equation of the line parallel to $y = 4x - 1$ that passes through $(0, 0)$.

6 Find the equations of lines parallel to $y = -3x + 1$ that pass through each of the following points:
   a $(0, 4)$
   b $(-2, 4)$
   c $\left(-\frac{5}{2}, 4\right)$

7 Find the equations of lines parallel to $x - 2y = 6$ that pass through each of the following points:
   a $(-4, 1)$
   b $\left(\frac{1}{2}, 0\right)$

Drawing straight-line graphs

To draw a straight-line graph only two points need to be known. Once these have been plotted, the line can be drawn between them and extended if necessary at both ends.

Worked examples

a Plot the line $y = x + 3$.

To identify two points, simply choose two values of x. Substitute these into the equation and calculate their corresponding y values.

When $x = 0$, $y = 3$

When $x = 4$, $y = 7$

Therefore two of the points on the line are $(0, 3)$ and $(4, 7)$.

The straight line $y = x + 3$ is plotted below.

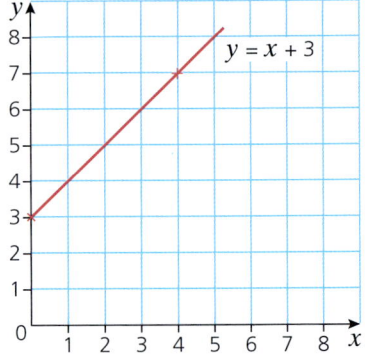

b Plot the line $y = -2x + 4$.

When $x = 2$, $y = 0$

When $x = -1$, $y = 6$

The coordinates of two points on the line are $(2, 0)$ and $(-1, 6)$.

Note that, in questions of this sort, it is often easier to rearrange the equation into gradient-intercept form first.

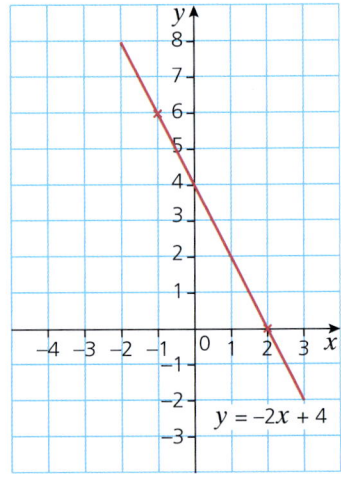

Graphical solution of simultaneous equations

 **Exercise 21.11**

1 Plot the following straight lines:
   a $y = 2x + 3$
   b $y = x - 4$
   c $y = 3x - 2$
   d $y = -2x$
   e $y = -x - 1$
   f $-y = x + 1$
   g $-y = 3x - 3$
   h $2y = 4x - 2$
   i $y - 4 = 3x$

2 Plot the following straight lines:
   a $-2x + y = 4$
   b $-4x + 2y = 12$
   c $3y = 6x - 3$
   d $2x = x + 1$
   e $3y - 6x = 9$
   f $2y + x = 8$
   g $x + y + 2 = 0$
   h $3x + 2y - 4 = 0$
   i $4 = 4y - 2x$

3 Plot the following straight lines:
   a $\frac{x+y}{2} = 1$
   b $x + \frac{y}{2} = 1$
   c $\frac{x}{3} + \frac{y}{2} = 1$
   d $y + \frac{x}{2} = 3$
   e $\frac{y}{5} + \frac{x}{3} = 0$
   f $\frac{-(2x+y)}{4} = 1$
   g $\frac{y - (x - y)}{3x} = -1$
   h $\frac{y}{2x+3} - \frac{1}{2} = 0$
   i $-2(x + y) + 4 = -y$

Graphical solution of simultaneous equations

When solving two equations simultaneously, the aim is to find a solution which works for both equations. In Chapter 13 it was shown how to arrive at the solution algebraically. It is, however, possible to arrive at the same solution graphically.

➡ Worked example

a By plotting both of the following equations on the same axes, find a common solution.

$x + y = 4$ (1)
$x - y = 2$ (2)

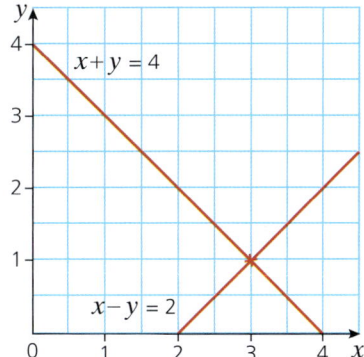

When both lines are plotted, the point at which they cross gives the common solution as it is the only point which lies on both lines.

Therefore the common solution is the point (3, 1).

275

21 STRAIGHT-LINE GRAPHS

b Check the result obtained above by solving the equations algebraically.
$x + y = 4$ (1)
$x - y = 2$ (2)

Adding equations (1) + (2) → $2x = 6$
$x = 3$

Substituting $x = 3$ into equation (1) we have:
$3 + y = 4$
$y = 1$

Therefore the common solution occurs at (3, 1) so $x = 3, y = 1$.

 Exercise 21.12

Solve the simultaneous equations below:
i by graphical means,
ii by algebraic means.

1 a $x + y = 5$
$x - y = 1$
b $x + y = 7$
$x - y = 3$
c $2x + y = 5$
$x - y = 1$
d $2x + 2y = 6$
$2x - y = 3$
e $x + 3y = -1$
$x - 2y = -6$
f $x - y = 6$
$x + y = 2$

2 a $3x - 2y = 13$
$2x + y = 4$
b $4x - 5y = 1$
$2x + y = -3$
c $x + 5 = y$
$2x + 3y - 5 = 0$
d $x = y$
$x + y + 6 = 0$
e $2x + y = 4$
$4x + 2y = 8$
f $y - 3x = 1$
$y = 3x - 3$

Calculating the length of a line segment

A line segment is formed when two points are joined by a straight line. To calculate the **distance between two points**, and therefore the length of the line segment, their coordinates need to be given. Once these are known, **Pythagoras' theorem** can be used to calculate the distance.

→ Worked example

The coordinates of two points are (1, 3) and (5, 6). Draw a pair of axes, plot the given points and calculate the distance between them.

By dropping a vertical line from the point (5, 6) and drawing a horizontal line from (1, 3), a right-angled triangle is formed. The length of the hypotenuse of the triangle is the length we wish to find.

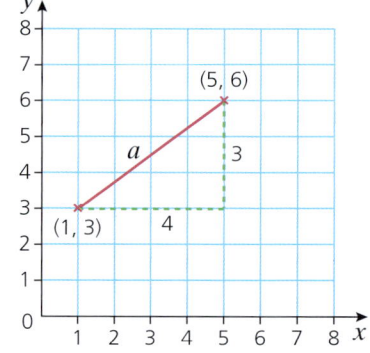

Using Pythagoras' theorem, we have:
$a^2 = 3^2 + 4^2 = 25$
$a = \sqrt{25} = 5$

The length of the line segment is 5 units.

The midpoint of a line segment

To find the distance between two points directly from their coordinates, use the following formula:

$$d = \sqrt{(x_1 - x_2)^2 + (y_1 - y_2)^2}$$

➡ Worked example

Without plotting the points, calculate the distance between the points (1, 3) and (5, 6).

$$d = \sqrt{(1 - 5)^2 + (3 - 6)^2}$$
$$= \sqrt{(-4)^2 + (-3)^2}$$
$$= \sqrt{25} = 5$$

The distance between the two points is 5 units.

The midpoint of a line segment

To find the **midpoint** of a line segment, use the coordinates of its end points. To find the *x*-coordinate of the midpoint, find the mean of the *x*-coordinates of the end points. Similarly, to find the *y*-coordinate of the midpoint, find the mean of the *y*-coordinates of the end points.

➡ Worked examples

a Find the coordinates of the midpoint of the line segment AB where A is (1, 3) and B is (5, 6).

The *x*-coordinate of the midpoint will be $\frac{1+5}{2} = 3$

The *y*-coordinate of the midpoint will be $\frac{3+6}{2} = 4.5$

So the coordinates of the midpoint are (3, 4.5).

b Find the coordinates of the midpoint of a line segment PQ where P is (−2, −5) and Q is (4, 7).

The *x*-coordinate of the midpoint will be $\frac{-2+4}{2} = 1$

The *y*-coordinate of the midpoint will be $\frac{-5+7}{2} = 1$

So the coordinates of the midpoint are (1, 1).

Exercise 21.13

1 **i** Plot each of the following pairs of points.
   ii Calculate the distance between each pair of points.
   iii Find the coordinates of the midpoint of the line segment joining the two points.

   a (5, 6) (1, 2)   b (6, 4) (3, 1)   c (1, 4) (5, 8)
   d (0, 0) (4, 8)   e (2, 1) (4, 7)   f (0, 7) (−3, 1)
   g (−3, −3) (−1, 5) h (4, 2) (−4, −2)   i (−3, 5) (4, 5)
   j (2, 0) (2, 6)   k (−4, 3) (4, 5)   l (3, 6) (−3, −3)

21 STRAIGHT-LINE GRAPHS

Exercise 21.13 (cont)

2 Without plotting the points:
  i calculate the distance between each of the following pairs of points
  ii find the coordinates of the midpoint of the line segment joining the two points.
  a (1, 4) (4, 1)
  b (3, 6) (7, 2)
  c (2, 6) (6, −2)
  d (1, 2) (9, −2)
  e (0, 3) (−3, 6)
  f (−3, −5) (−5, −1)
  g (−2, 6) (2, 0)
  h (2, −3) (8, 1)
  i (6, 1) (−6, 4)
  j (−2, 2) (4, −4)
  k (−5, −3) (6, −3)
  l (3, 6) (5, −2)

The equation of a line through two points

The equation of a straight line can be deduced once the coordinates of two points on the line are known.

Worked example

Calculate the equation of the straight line passing through the points (−3, 3) and (5, 5).
The equation of any straight line can be written in the general form $y = mx + c$.
Here we have:

$$\text{gradient} = \frac{5 - 3}{5 - (-3)} = \frac{2}{8}$$

$$\text{gradient} = \frac{1}{4}$$

The equation of the line now takes the form $y = \frac{1}{4}x + c$.

Since the line passes through the two given points, their coordinates must satisfy the equation. So to calculate the value of 'c' the x and y coordinates of one of the points are substituted into the equation. Substituting (5, 5) into the equation gives:

$$5 = \frac{1}{4} \times 5 + c$$

$$5 = \frac{5}{4} + c$$

Therefore $c = 5 - 1\frac{1}{4} = 3\frac{3}{4}$

The equation of the straight line passing through (−3, 3) and (5, 5) is:

$$y = \frac{1}{4}x + 3\frac{3}{4}$$

Exercise 21.14

Find the equation of the straight line which passes through each of the following pairs of points:
1 a (1, 1) (4, 7)
   b (1, 4) (3, 10)
   c (1, 5) (2, 7)
   d (0, −4) (3, −1)
   e (1, 6) (2, 10)
   f (0, 4) (1, 3)
   g (3, −4) (10, −18)
   h (0, −1) (1, −4)
   i (0, 0) (10, 5)

2   a (−5, 3) (2, 4)   b (−3, −2) (4, 4)   c (−7, −3) (−1, 6)
    d (2, 5) (1, −4)   e (−3, 4) (5, 0)   f (6, 4) (−7, 7)
    g (−5, 2) (6, 2)   h (1, −3) (−2, 6)   i (6, −4) (6, 6)

Perpendicular lines

The two lines shown below are perpendicular to each other.

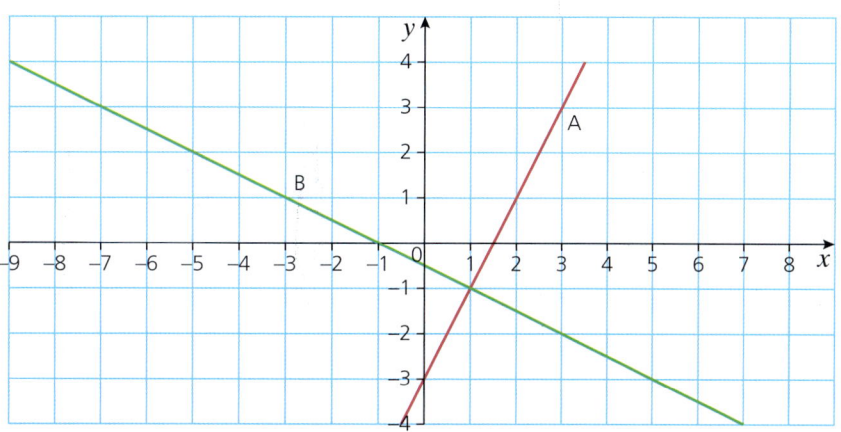

Line A has a gradient of 2.

Line B has a gradient of $-\frac{1}{2}$.

The diagram below also shows two lines perpendicular to each other.

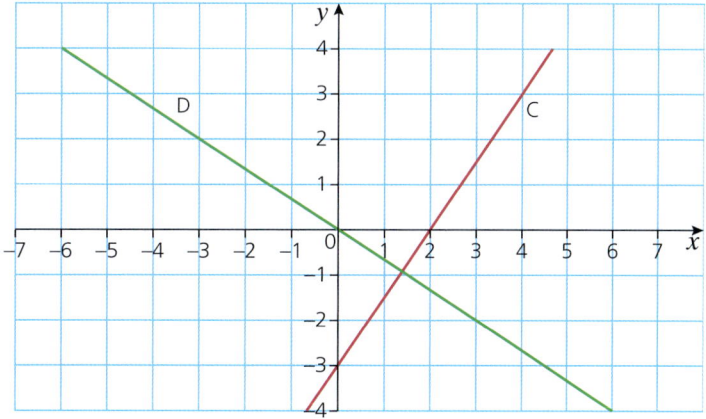

Line C has a gradient of $\frac{3}{2}$.

Line D has a gradient of $-\frac{2}{3}$.

Notice that in both cases, the product of the two gradients is equal to −1.

In the first example $2 \times (-\frac{1}{2}) = -1$.

In the second example $\frac{3}{2} \times (-\frac{2}{3}) = -1$.

21 STRAIGHT-LINE GRAPHS

This is in fact the case for the gradients of any two perpendicular lines.

If two lines L_1 and L_2 are perpendicular to each other, the product of their gradients m_1 and m_2 is -1.

i.e. $m_1 m_2 = -1$

Therefore the gradient of one line is the negative reciprocal of the other line.

i.e. $m_1 = -\dfrac{1}{m_2}$

→ Worked examples

a **i** Calculate the gradient of the line joining the two points $(3, 6)$ and $(1, -6)$.

Gradient $= \dfrac{6-(-6)}{3-1} = \dfrac{12}{2} = 6$

ii Calculate the gradient of a line perpendicular to the one in part **i** above.

$m_1 = -\dfrac{1}{m_2}$, therefore the gradient of the perpendicular line is $-\dfrac{1}{6}$.

iii The perpendicular line also passes through the point $(-1, 6)$. Calculate the equation of the perpendicular line.

The equation of the perpendicular line will take the form $y = mx + c$.

As its gradient is $-\dfrac{1}{6}$ and it passes through the point $(-1, 6)$, this can be substituted into the equation to give:

$6 = -\dfrac{1}{6} \times (-1) + c$

Therefore $c = \dfrac{35}{6}$.

The equation of the perpendicular line is $y = -\dfrac{1}{6}x + \dfrac{35}{6}$.

b **i** Show that the point $(-4, -1)$ lies on the line $y = -\dfrac{1}{4}x - 2$.

If the point $(-4, -1)$ lies on the line, its values of x and y will satisfy the equation. Substituting the values of x and y into the equation gives:

$-1 = -\dfrac{1}{4} \times (-4) - 2$

$-1 = -1$

Therefore the point lies on the line.

ii Deduce the gradient of a line perpendicular to the one given in part **i** above.

$m_1 = -\dfrac{1}{m_2}$ therefore $m_1 = -\dfrac{1}{-\frac{1}{4}} = 4$

Therefore the gradient of the perpendicular line is 4.

iii The perpendicular line also passes through the point $(-4, -1)$. Calculate its equation.

The equation of the perpendicular line takes the general form $y = mx + c$.

Substituting in the values of x, y and m gives:

$-1 = 4 \times (-4) + c$

Therefore $c = 15$.

The equation of the perpendicular line is $y = 4x + 15$.

Perpendicular lines

Exercise 21.15

1 Calculate:
   i   the gradient of the line joining the following pairs of points
   ii the gradient of a line perpendicular to this line
   iii the equation of the perpendicular line if it passes through the second point each time.

   a (1, 4) (4, 1)   b (3, 6) (7, 2)   c (2, 6) (6, −2)
   d (1, 2) (9, −2)   e (0, 3) (−3, 6)   f (−3, −5) (−5, −1)
   g (−2, 6) (2, 0)   h (2, −3) (8, 1)   i (6, 1) (−6, 4)
   j (−2, 2) (4, −4)   k (−5, −3) (6, −3) l (3, 6) (5, −2)

2 Calculate the gradient of lines perpendicular to each of the following straight lines.
   a $2y = 3x + 2$ b $4x - 5y = 10$ c $\frac{1}{2}y + 4x - 3 = 0$ d $\frac{2}{3}y - \frac{1}{2}x = 1$

3 The diagram below shows a square $ABCD$. The coordinates of A and B are given.

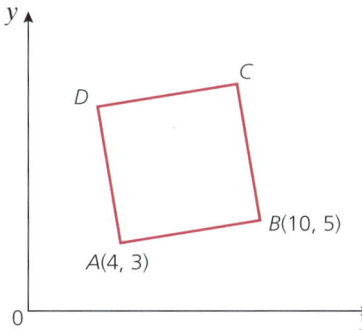

Calculate:
a the gradient of the line AB
b the equation of the line passing through A and B
c the gradient of the line AD
d the equation of the line passing through A and D
e the equation of the line passing through B and C
f the coordinates of C
g the coordinates of D
h the equation of the line passing through C and D
i the length of the sides of the square to 1 d.p.
j the coordinates of the midpoint of the line segment AC.

4 The diagram below shows a right-angled isosceles triangle ABC, where $AB = AC$.
The coordinates of A and B are given.

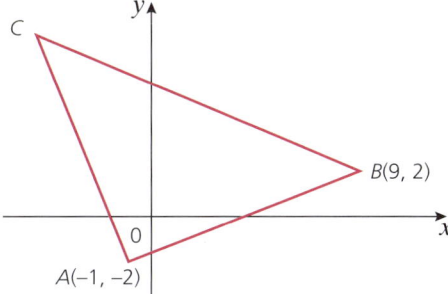

21 STRAIGHT-LINE GRAPHS

Exercise 21.15 (cont)

a Calculate:
  i the equation of the line passing through the points A and B
  ii the equation of the line passing through A and C
  iii the length of the line segment BC to 1 d.p.
  iv the coordinates of the midpoints of all three sides of the triangle.
b A perpendicular bisector of the line AB is a line which is at right angles to AB and passes through its midpoint.
  Calculate the equation of the perpendicular bisector of AB.

Student assessment 1

1 For each of the following lines, select two points on the line and then calculate its gradient.

a
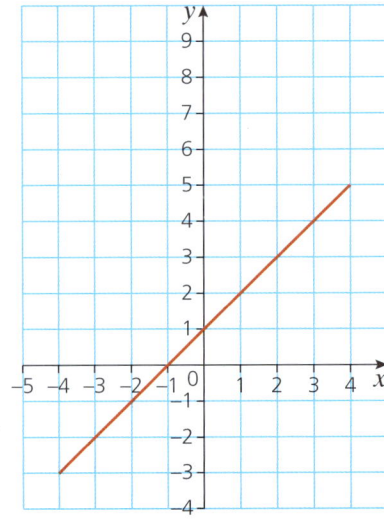

b
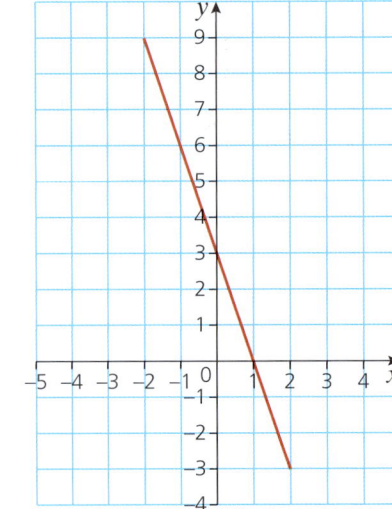

Perpendicular lines

2 Find the equation of the straight line for each of the following:
   a
   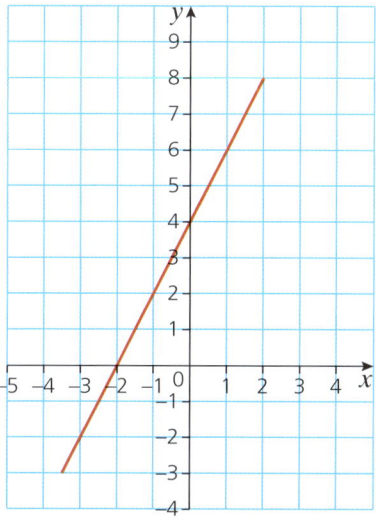

   b
   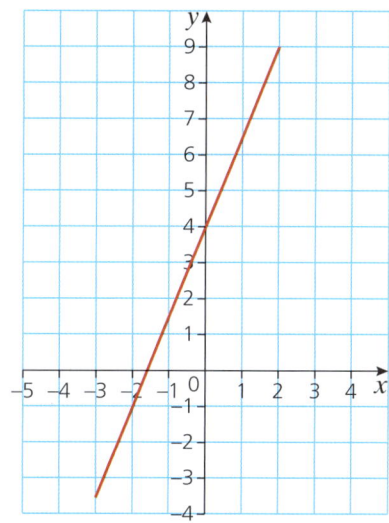

3 Write down the equation of the line parallel to the line $y = -\frac{2}{3}x + 4$ which passes through the point (6, 2).

4 Plot the following graphs on the same pair of axes, labelling each clearly.
   a $x = -2$
   b $y = 3$
   c $y = 2x$
   d $y = -\frac{x}{2}$

5 Calculate the gradient and y-intercept for each of the following linear equations:
   a $y = -3x + 4$
   b $\frac{1}{3}y - x = 2$
   c $2x + 4y - 6 = 0$

21 STRAIGHT-LINE GRAPHS

6 Solve the following pairs of simultaneous equations graphically:
   a $x + y = 4$
       $x - y = 0$
   b $3x + y = 2$
       $x - y = 2$
   c $y + 4x + 4 = 0$
       $x + y = 2$
   d $x - y = -2$
       $3x + 2y + 6 = 0$

7 The coordinates of the end points of two line segments are given below.
For each line segment calculate:
   i the length
   ii the midpoint.
   a $(-6, -1)$ $(6, 4)$
   b $(1, 2)$ $(7, 10)$

8 Find the equation of the straight line which passes through each of the following pairs of points:
   i in the form $y = mx + c$
   ii in the form $ax + by + c = 0$
   a $(1, -1)$ $(4, 8)$
   b $(0, 7)$ $(3, 1)$

9 A line L_1 passes through the points $(-2, 5)$ and $(5, 3)$.
   a Write down the equation of the line L_1.

Another line L_2 is perpendicular to L_1 and also passes through the point $(-2, 5)$.
   b Write down the equation of the line L_2.

10 The diagram below shows a **rhombus** $ABCD$. The coordinates of A, B and D are given.

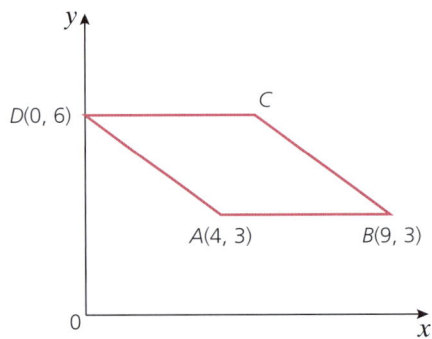

   a Calculate:
      i the coordinate of the point C
      ii the equation of the line passing through A and C
      iii the equation of the line passing through B and D.
   b Are the diagonals of the rhombus perpendicular to each other? Justify your answer.

3 Mathematical investigations and ICT 3

Plane trails

In an aircraft show, planes are made to fly with a coloured smoke trail. Depending on the formation of the planes, the trails can intersect in different ways.

In the diagram below, the three smoke trails do not cross, as they are parallel.

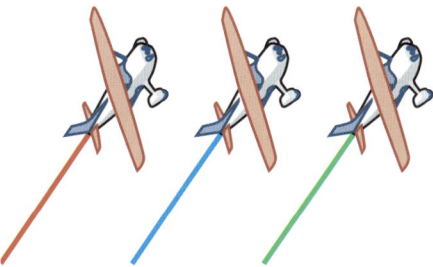

In the following diagram, there are two crossing points.

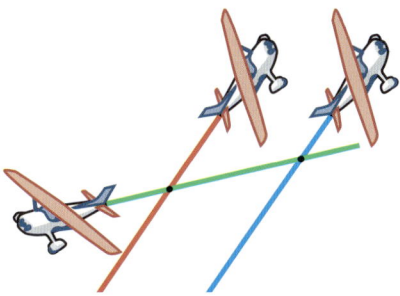

By flying differently, the three planes can produce trails that cross at three points.

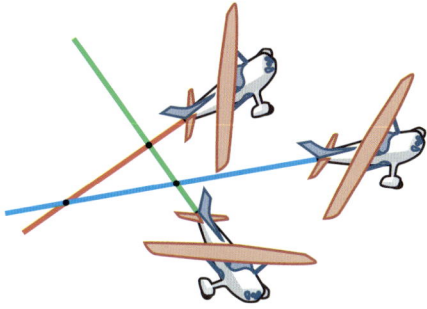

1 Investigate the connection between the maximum number of crossing points and the number of planes.

MATHEMATICAL INVESTIGATIONS AND ICT 3

2 Record the results of your investigation in an ordered table.

3 Write an algebraic rule linking the number of planes (p) and the maximum number of crossing points (n).

Hidden treasure

A television show sets up a puzzle for its contestants to try and solve. Some buried treasure is hidden on a 'treasure island'. The treasure is hidden in one of the 12 treasure chests shown (right). Each contestant stands by one of the treasure chests.

The treasure is hidden according to the following rule:
  It is not hidden in chest 1.
  Chest 2 is left empty for the time being.
  It is not hidden in chest 3.
  Chest 4 is left empty for the time being.
  It is not hidden in chest 5.

The pattern of crossing out the first chest and then alternate chests is continued until only one chest is left. This will involve going around the circle several times, continuing the pattern. The treasure is hidden in the last chest left.

The diagrams below show how the last chest is chosen:

After the first round, chests 1, 3, 5, 7, 9 and 11 have been discounted.

After the second round, chests 2, 6 and 10 have also been discounted.

After the third round, chests 4 and 12 have also been discounted. This leaves only chest 8.

The treasure is therefore hidden in chest 8.

Unfortunately for participants, the number of contestants changes each time.

1 Investigate which treasure chest you would choose if there are:
  a 4 contestants
  b 5 contestants
  c 8 contestants
  d 9 contestants
  e 15 contestants.

2 Investigate the winning treasure chest for other numbers of contestants and enter your results in an ordered table.

3 State any patterns you notice in your table of results.

4 Use your patterns to predict the winning chest for 31, 32 and 33 contestants.

5 Write a rule linking the winning chest x and the number of contestants n.

ICT activity

For each question, use a graphing package to plot the inequalities on the same pair of axes. Leave unshaded the region which satisfies all of them simultaneously.

1 $y \leq x$ $y > 0$ $x \leq 3$
2 $x + y > 3$ $y \leq 4$ $y - x > 2$
3 $2y + x \leq 5$ $y - 3x - 6 < 0$ $2y - x > 3$

TOPIC 4
Geometry

Contents
Chapter 22 Geometrical vocabulary and construction (E4.1, E4.2, E4.3)
Chapter 23 Similarity and congruence (E4.4)
Chapter 24 Symmetry (E4.5, E4.8)
Chapter 25 Angle properties (E4.6, E4.7)

Learning objectives

E4.1 Geometrical terms
1. Use and interpret the following geometrical terms:
   - point
   - vertex
   - line
   - plane
   - parallel
   - perpendicular
   - perpendicular bisector
   - bearing
   - right angle
   - acute, obtuse and reflex angles
   - interior and exterior angles
   - similar
   - congruent
   - scale factor.
2. Use and interpret the vocabulary of:
   - triangles
   - special quadrilaterals
   - polygons
   - nets
   - solids.
3. Use and interpret the vocabulary of a circle.

E4.2 Geometrical constructions
1. Measure and draw lines and angles.
2. Construct a triangle, given the three sides, using a ruler and pair of compasses only.
3. Draw, use and interpret nets.

E4.3 Scale drawings
1. Draw and interpret scale drawings.
2. Use and interpret three-figure bearings.

E4.4 Similarity
1. Calculate lengths of similar shapes.
2. Use the relationships between lengths and areas of similar shapes and lengths, surface areas and volumes of similar solids.
3. Solve problems and give simple explanations involving similarity.

E4.5 Symmetry
1. Recognise line symmetry and order of rotational symmetry in two dimensions.
2. Recognise symmetry properties of prisms, cylinders, pyramids and cones.

E4.6 Angles
1. Calculate unknown angles and give simple explanations using the following geometrical properties:
   - sum of angles at a point = 360°
   - sum of angles at a point on a straight line = 180°
   - vertically opposite angles are equal
   - angle sum of a triangle = 180° and angle sum of a quadrilateral = 360°.
2. Calculate unknown angles and give geometric explanations for angles formed within parallel lines:
   - corresponding angles are equal
   - alternate angles are equal
   - co-interior angles sum to 180° (supplementary).
3. Know and use angle properties of regular and irregular polygons.

E4.7 Circle theorems I
Calculate unknown angles and give explanations using the following geometrical properties of circles:
- angle in a semicircle = 90°
- angle between tangent and radius = 90°
- angle at the centre is twice the angle at the circumference
- angles in the same segment are equal
- opposite angles of a cyclic quadrilateral sum to 180° (supplementary)
- alternate segment theorem.

E4.8 Circle theorems II
Use the following symmetry properties of circles:
- equal chords are equidistant from the centre
- the perpendicular bisector of a chord passes through the centre
- tangents from an external point are equal in length.

The Greeks

Many of the great Greek mathematicians came from the Greek Islands, from cities such as Ephesus or Miletus (which are in present day Turkey) or from Alexandria in Egypt. This section briefly mentions some of the Greek mathematicians of 'The Golden Age'. You may wish to find out more about them.

Thales of Alexandria invented the 365-day calendar and predicted the dates of eclipses of the Sun and the Moon.

Pythagoras of Samos founded a school of mathematicians and worked with geometry. His successor as leader was Theano, the first woman to hold a major role in mathematics.

Eudoxus of Asia Minor (Turkey) worked with irrational numbers like pi and discovered the formula for the volume of a cone.

Euclid of Alexandria formed what would now be called a university department. His book became the set text in schools and universities for 2000 years.

Apollonius of Perga (Turkey) worked on, and gave names to, the parabola, the hyperbola and the ellipse.

Archimedes is accepted today as the greatest mathematician of all time. However, he was so far ahead of his time that his influence on his contemporaries was limited by their lack of understanding.

Archimedes (287–212BCE)

22 Geometrical vocabulary and construction

Angles and lines

Different types of angle have different names:

acute angles lie between 0° and 90°

right angles are exactly 90°

obtuse angles lie between 90° and 180°

reflex angles lie between 180° and 360°

To find the shortest distance between two points, you measure the length of the **straight line** which joins them.

Two lines which meet at right angles are **perpendicular** to each other.

So in the diagram below, CD is **perpendicular** to AB and AB is perpendicular to CD.

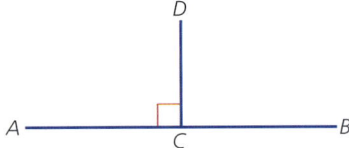

If the lines AD and BD are drawn to form a triangle, the line CD can be called the **height** or **altitude** of the triangle ABD.

Parallel lines are straight lines which can be continued to infinity in either direction without meeting.

Railway lines are an example of parallel lines. Parallel lines are marked with arrows as shown:

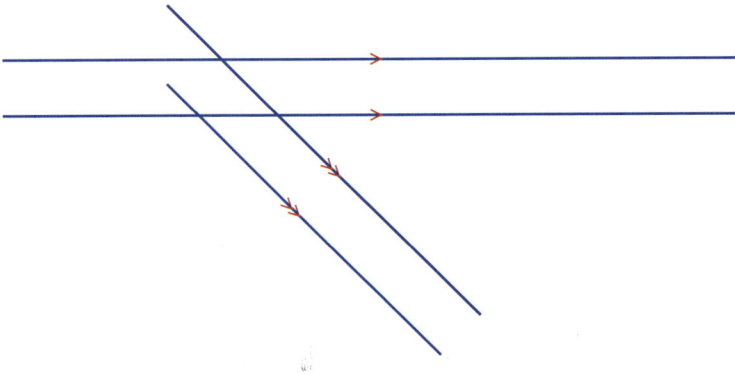

Triangles

Triangles can be described in terms of their sides or their angles, or both.

An **acute-angled** triangle has all its angles less than 90°.

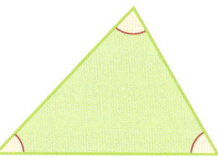

A **right-angled** triangle has an angle of 90°.

An **obtuse-angled** triangle has one angle greater than 90°.

An **isosceles** triangle has two sides of equal length, and the angles opposite the equal sides are equal.

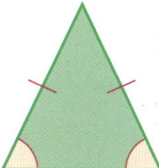

An **equilateral** triangle has three sides of equal length and three equal angles.

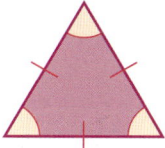

A **scalene** triangle has three sides of different lengths and all three angles are different.

22 GEOMETRICAL VOCABULARY AND CONSTRUCTION

Congruent triangles

Congruent triangles are **identical**. They have corresponding sides of the same length and **corresponding angles** which are equal.

> **Note**
> All diagrams are not drawn to scale.

Similar triangles

If the angles of two triangles are the same, then their corresponding sides will also be in proportion to each other. When this is the case, the triangles are said to be **similar**.

In the diagram below, triangle ABC is similar to triangle XYZ. Similar shapes are covered in more detail in Chapter 23.

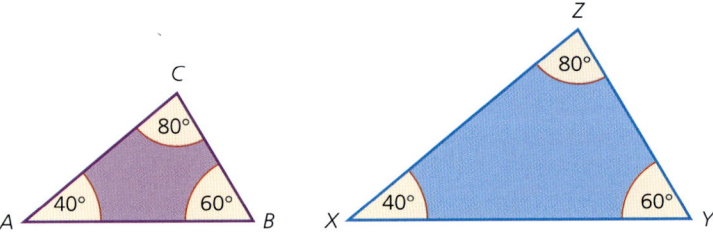

Exercise 22.1

1 In the diagrams below, identify pairs of congruent triangles.

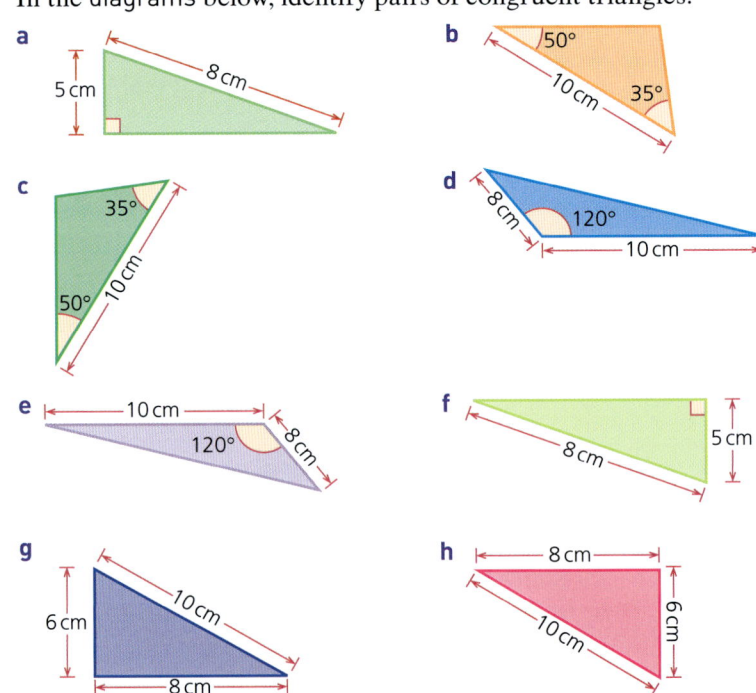

Circles

> **Note**
> Half a circle is known as a semicircle.

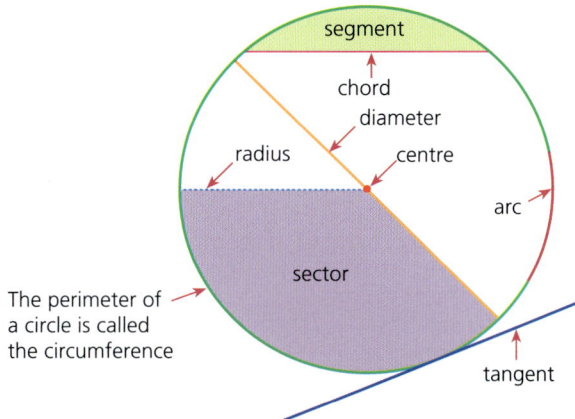

The perimeter of a circle is called the circumference

The diagram shows two arcs. The major arc is green and the minor arc is pink.

Quadrilaterals

A **quadrilateral** is a plane (two-dimensional) shape consisting of four angles and four sides. There are several types of quadrilateral. The main ones and their properties are described below.

Two pairs of parallel sides.

All sides are equal.

All angles are equal.

Diagonals intersect at right angles.

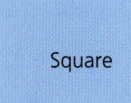

Two pairs of parallel sides.

Opposite sides are equal.

All angles are equal.

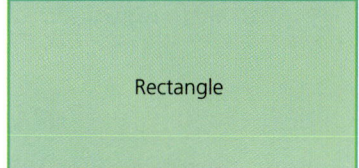

Two pairs of parallel sides.

All sides are equal.

Opposite angles are equal.

Diagonals intersect at right angles.

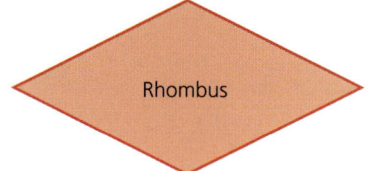

22 GEOMETRICAL VOCABULARY AND CONSTRUCTION

Two pairs of parallel sides.

Opposite sides are equal.

Opposite angles are equal.

One pair of parallel sides.

An **isosceles trapezium** has one pair of parallel sides and the other pair of sides are equal in length.

Two pairs of equal sides.

One pair of equal angles.

Diagonals intersect at right angles.

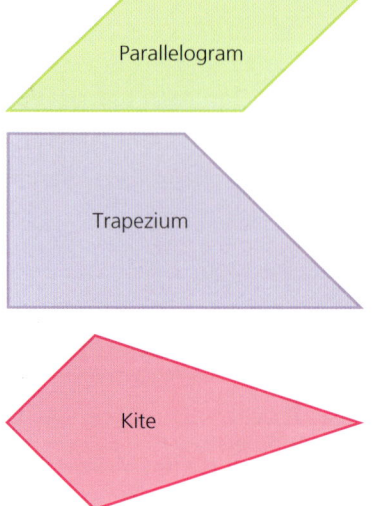

Exercise 22.2

1 Copy and complete the following table. The first line has been started for you.

	Rectangle	Square	Parallelogram	Kite	Rhombus	Equilateral triangle
Opposite sides equal in length	Yes	Yes				
All sides equal in length						
All angles right angles						
Both pairs of opposite sides parallel						
Diagonals equal in length						
Diagonals intersect at right angles						
All angles equal						

Polygons

Any two-dimensional closed figure made up of straight lines is called a **polygon**.

If the sides are the same length and the interior angles are equal, the figure is called a **regular polygon**. If the sides and angles are not all of equal length and size, it is known as an **irregular polygon**.

> **Note**
>
> Heptagon, nonagon and dodecagon are not part of the syllabus, but are included here for interest.

The names of the common polygons are:

3 sides	**tri**angle
4 sides	**quad**rilateral
5 sides	**penta**gon
6 sides	**hexa**gon
7 sides	**hepta**gon
8 sides	**octa**gon
9 sides	**nona**gon
10 sides	**deca**gon
12 sides	**dodeca**gon

Two polygons are said to be **similar** if
a their angles are the same
b corresponding sides are in proportion.

Nets

The diagram below is the **net** of a cube. It shows the faces of the cube opened out into a two-dimensional plan. The net of a three-dimensional shape can be folded up to make that shape.

Exercise 22.3

Draw the following on squared paper:
1. Two other possible nets of a cube.
2. The net of a cuboid (rectangular prism).
3. The net of a triangular prism.
4. The net of a cylinder.
5. The net of a square-based **pyramid**.
6. The net of a tetrahedron.

22 GEOMETRICAL VOCABULARY AND CONSTRUCTION

Constructing triangles

Triangles can be drawn accurately by using a ruler and a pair of compasses. This is called **constructing** a triangle.

 Worked example

The sketch shows the triangle ABC.

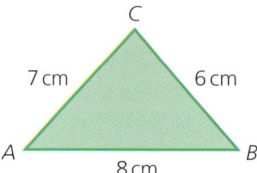

Construct the triangle ABC given that:
$AB = 8$ cm, $BC = 6$ cm and $AC = 7$ cm

- Draw the line AB using a ruler:

- Open up a pair of compasses to 6 cm. Place the compass point on B and draw an arc:

- Note that every point on the arc is 6 cm away from B.

- Open up the pair of compasses to 7 cm. Place the compass point on A and draw another arc, with centre A and radius 7 cm, ensuring that it intersects with the first arc. Every point on the second arc is 7 cm from A. Where the two arcs intersect is point C, as it is both 6 cm from B and 7 cm from A.

Scale drawings

- Join C to A and C to B:

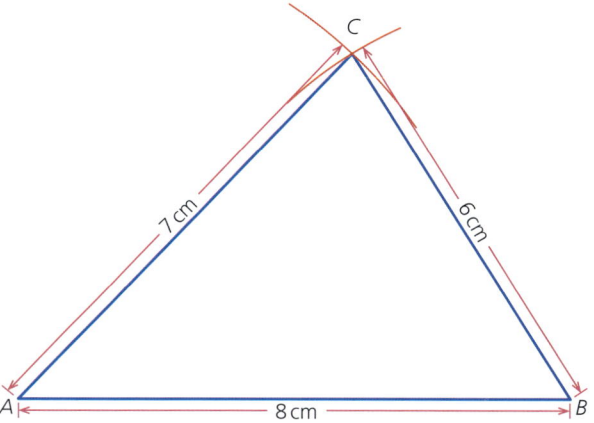

Exercise 22.4

Using only a ruler and a pair of compasses, construct the following triangles:

1. Triangle ABC where $AB = 10\,\text{cm}$, $AC = 7\,\text{cm}$ and $BC = 9\,\text{cm}$
2. Triangle LMN where $LM = 4\,\text{cm}$, $LN = 8\,\text{cm}$ and $MN = 5\,\text{cm}$
3. Triangle PQR, an equilateral triangle of side length $7\,\text{cm}$
4. **a** Triangle ABC where $AB = 8\,\text{cm}$, $AC = 4\,\text{cm}$ and $BC = 3\,\text{cm}$
   b Is this triangle possible? Explain your answer.

Scale drawings

Scale drawings are used when an accurate diagram, drawn in proportion, is needed. Common uses of scale drawings include maps and plans. The use of scale drawings involves understanding how to scale measurements.

➔ Worked examples

a A map is drawn to a scale of $1:10000$. If two objects are $1\,\text{cm}$ apart on the map, how far apart are they in real life? Give your answer in metres.
A scale of $1:10000$ means that $1\,\text{cm}$ on the map represents $10000\,\text{cm}$ in real life.
Therefore the distance $= 10000\,\text{cm}$
$\phantom{\text{Therefore the distance }} = 100\,\text{m}$

b A model boat is built to a scale of $1:50$. If the length of the real boat is $12\,\text{m}$, calculate the length of the model boat in cm.
A scale of $1:50$ means that $50\,\text{cm}$ on the real boat is $1\,\text{cm}$ on the model boat.
$\quad 12\,\text{m} = 1200\,\text{cm}$
Therefore the length of the model boat $= 1200 \div 50\,\text{cm}$
$\phantom{\text{Therefore the length of the model boat }} = 24\,\text{cm}$

297

22 GEOMETRICAL VOCABULARY AND CONSTRUCTION

c i Construct, to a scale of 1:1, a triangle ABC such that $AB = 6\,\text{cm}$, $AC = 5\,\text{cm}$ and $BC = 4\,\text{cm}$.

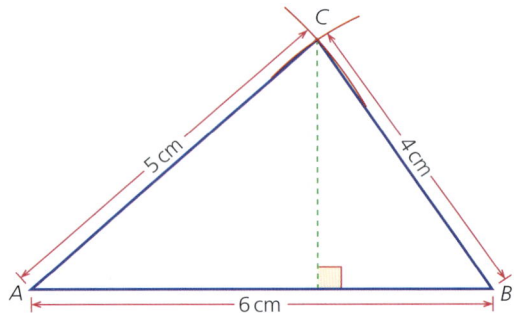

ii Measure the perpendicular length of C from AB.
Perpendicular length is 3.3 cm.

iii Calculate the area of the triangle.

$$\text{Area} = \frac{\text{base length} \times \text{perpendicular height}}{2}$$

$$\text{Area} = \frac{6 \times 3.3}{2}\,\text{cm} = 9.9\,\text{cm}^2$$

Exercise 22.5

1 In the following questions, both the scale to which a map is drawn and the distance between two objects on the map are given.
Find the real distance between the two objects, giving your answer in metres.

a 1:10 000   3 cm
b 1:10 000   2.5 cm
c 1:20 000   1.5 cm
d 1:8000   5.2 cm

2 In the following questions, both the scale to which a map is drawn and the true distance between two objects are given.
Find the distance between the two objects on the map, giving your answer in cm.

a 1:15 000   1.5 km
b 1:50 000   4 km
c 1:10 000   600 m
d 1:25 000   1.7 km

3 A rectangular pool measures 20 m by 36 m as shown below:

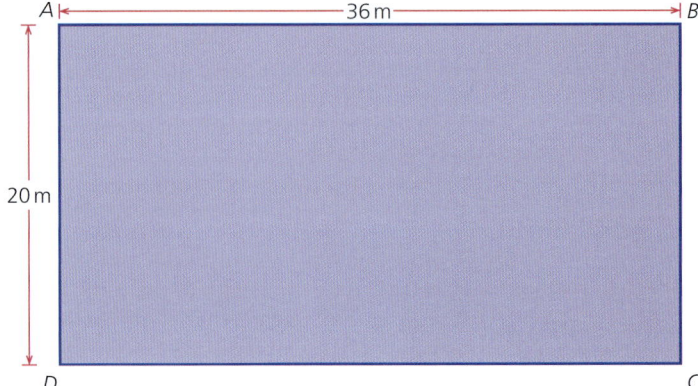

a Construct a scale drawing of the pool, using 1 cm for every 4 m.
b A boy swims across the pool from *D* in a straight line so that he arrives at a point which is 40 m from *D* and 30 m from *C*. Work out the distance the boy swam.

4 A triangular enclosure is shown in the diagram below:

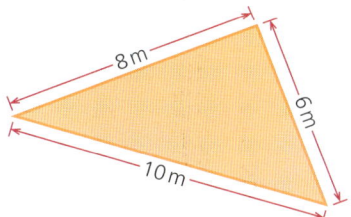

a Using a scale of 1 cm for each metre, construct a scale drawing of the enclosure.
b Calculate the true area of the enclosure.

Student assessment 1

1 Are the angles below acute, obtuse, reflex or right angles?

a
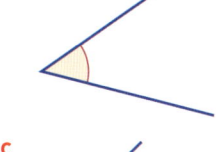

b
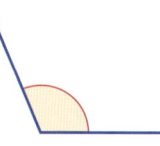

c
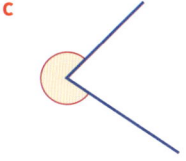

d

2 Draw and label two pairs of intersecting parallel lines.

3 Identify the types of triangles below in two ways (for example, obtuse-angled scalene triangle):

a

b
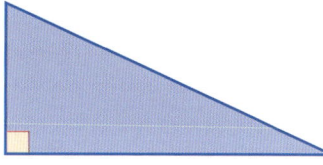

4 Draw a circle of radius 3 cm. Mark on it:
  a a diameter   **b** a chord   **c** a sector.

5 Draw a rhombus and write down three of its properties.

6 On squared paper, draw the net of a triangular prism.

22 GEOMETRICAL VOCABULARY AND CONSTRUCTION

7 Construct triangle ABC such that $AB = 8\,\text{cm}$, $AC = 6\,\text{cm}$ and $BC = 12\,\text{cm}$.

8 A plan of a living room is shown below:

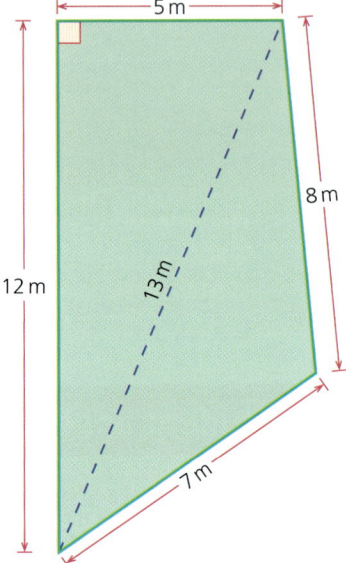

  a Using a pair of compasses, construct a scale drawing of the room using 1 cm for every metre.
  b Using a set square if necessary, calculate the total area of the actual living room.

9 In the following questions, both the scale to which a map is drawn and the true distance between two objects are given. Find the distance between the two objects on the map, giving your answer in cm.
  a 1 : 20 000   4.4 km   **b** 1 : 50 000   12.2 km

23 Similarity and congruence

Similar shapes

Note

All diagrams are not drawn to scale.

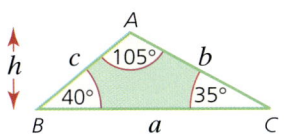
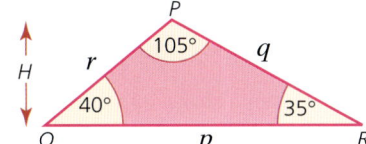

Two polygons are said to be **similar** if a) they are equi-angular and b) corresponding sides are in proportion.

For triangles, being equi-angular implies that corresponding sides are in proportion. The converse is also true.

In the diagrams triangle ABC and triangle PQR are similar.

For similar figures the ratios of the lengths of the sides are the same and represent the **scale factor**, i.e.

$\frac{p}{a} = \frac{q}{b} = \frac{r}{c} = k$ (where k is the scale factor of enlargement)

The heights of similar triangles are proportional also:

$\frac{H}{h} = \frac{p}{a} = \frac{q}{b} = \frac{r}{c} = k$

The ratio of the areas of similar triangles (the **area factor**) is equal to the square of the scale factor.

$\frac{\text{Area of } PQR}{\text{Area of } ABC} = \frac{\frac{1}{2} H \times p}{\frac{1}{2} h \times a} = \frac{H}{h} \times \frac{p}{a} = k \times k = k^2$

Exercise 23.1

1. **a** Explain why the two triangles (below) are similar.
   b Calculate the scale factor which reduces the larger triangle to the smaller one.
   c Calculate the value of x and the value of y.

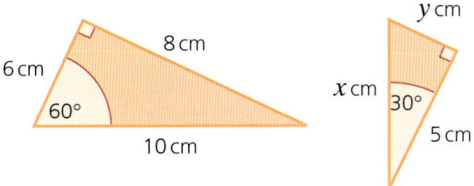

23 SIMILARITY AND CONGRUENCE

Exercise 23.1 (cont)

2 Which of the triangles below are similar?

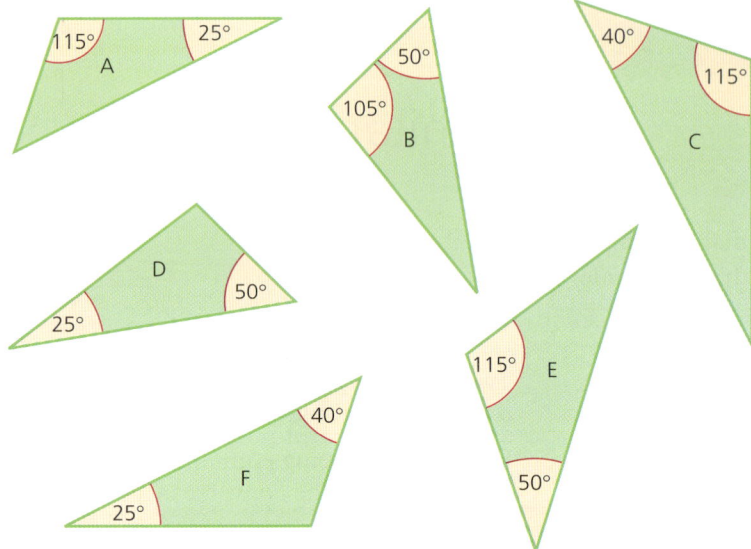

3 The triangles below are similar.

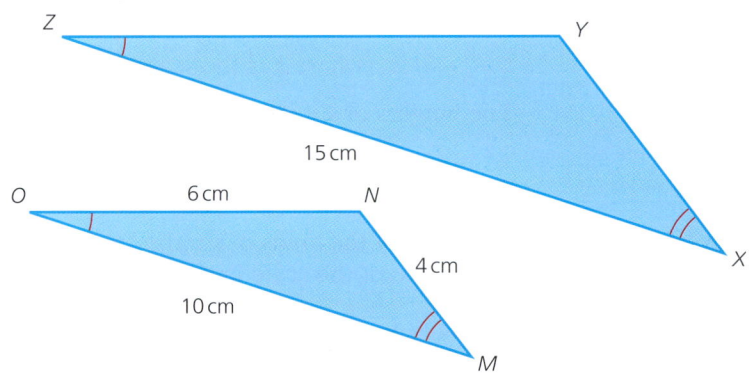

   a Calculate the length XY.
   b Calculate the length YZ.

4 In the triangle calculate the lengths of sides p, q and r.

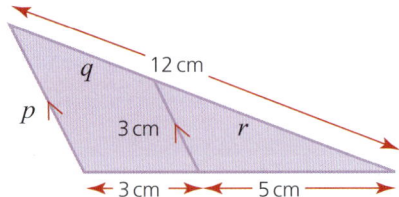

Similar shapes

5 In the trapezium calculate the lengths of the sides *e* and *f*.

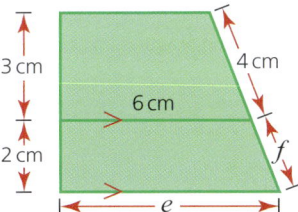

6 The triangles PQR and LMN are similar.

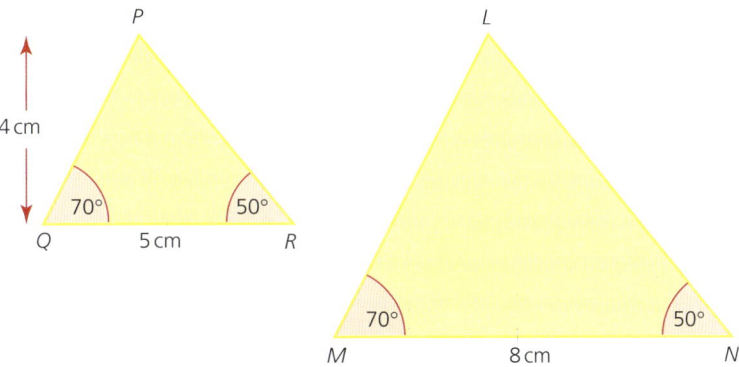

Calculate:
- **a** the area of triangle PQR
- **b** the scale factor of enlargement
- **c** the area of triangle LMN.

7 The triangles ABC and XYZ below are similar.

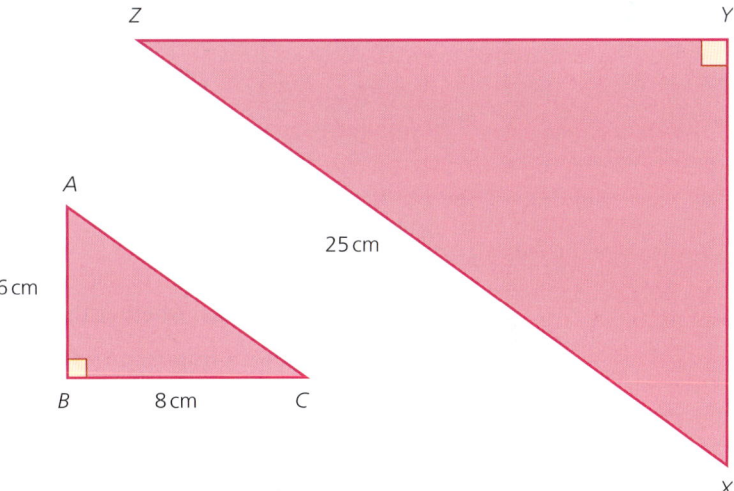

- **a** Using Pythagoras' theorem calculate the length of AC.
- **b** Calculate the scale factor of enlargement.
- **c** Calculate the area of triangle XYZ.

23 SIMILARITY AND CONGRUENCE

Exercise 23.1 (cont)

8 The triangle ADE shown has an area of $12\,\text{cm}^2$.
   a Calculate the area of triangle ABC.
   b Calculate the length BC.

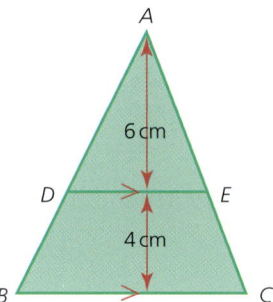

9 The parallelograms below are similar.

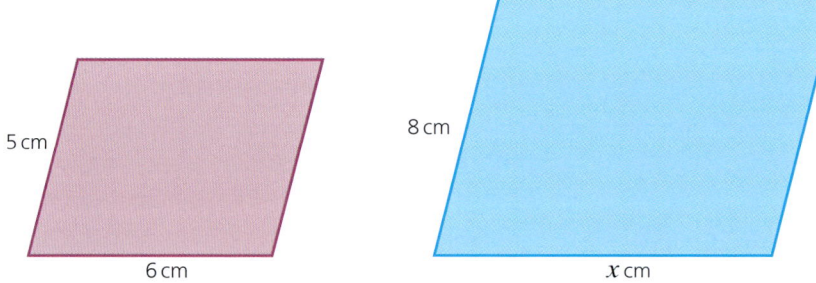

Calculate the length of the side marked x.

10 The diagram below shows two rhombuses.

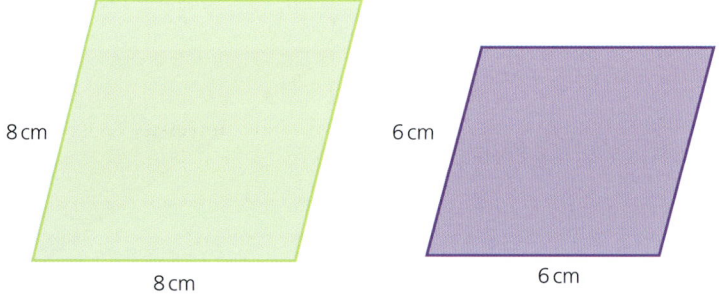

Explain, giving reasons, whether the two rhombuses are similar.

11 The diagram shows a trapezium within a trapezium. Explain, giving reasons, whether the two trapezia are similar.

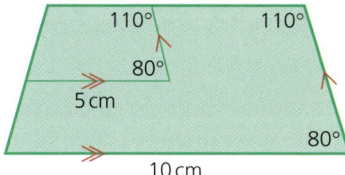

Area and volume of similar shapes

Exercise 23.2

1 In the hexagons below, hexagon B is an enlargement of hexagon A by a scale factor of 2.5.

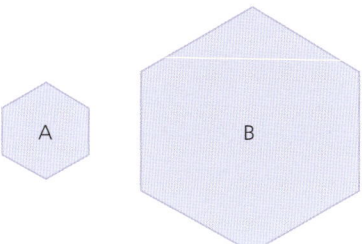

If the area of A is 8 cm², calculate the area of B.

2 P and Q are two regular pentagons. Q is an enlargement of P by a scale factor of 3. If the area of pentagon Q is 90 cm², calculate the area of P.

3 The diagram below shows four triangles A, B, C and D. Each is an enlargement of the previous one by a scale factor of 1.5.

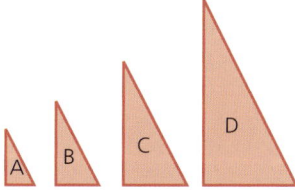

   a If the area of C is 202.5 cm², calculate the area of:
      i triangle D   ii triangle B   iii triangle A.
   b If the triangles were to continue in this sequence, which letter triangle would be the first to have an area greater than 15 000 cm²?

4 A square is enlarged by increasing the length of its sides by 10%. If the length of its sides was originally 6 cm, calculate the area of the enlarged square.

5 A square of side length 4 cm is enlarged by increasing the length of its sides by 25% and then increasing the lengths again by a further 50%. Calculate the area of the final square.

6 An equilateral triangle has an area of 25 cm². If the length of its sides is reduced by 15%, calculate the area of the reduced triangle.

Area and volume of similar shapes

Earlier in the chapter we found the following relationship between the scale factor and the area factor of enlargement:

Area factor = (scale factor)²

A similar relationship can be stated for volumes of similar shapes:

i.e. **Volume factor** = (scale factor)³

23 SIMILARITY AND CONGRUENCE

Exercise 23.3

1 The diagram shows a scale model of a garage. Its width is 5 cm, its length 10 cm and the height of its walls 6 cm.

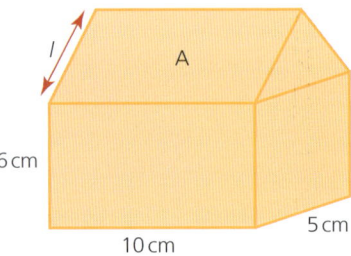

  a If the width of the real garage is 4 m, calculate:
    i the length of the real garage
    ii the real height of the garage wall.
  b If the apex of the roof of the real garage is 2 m above the top of the walls, use Pythagoras' theorem to find the real slant length l.
  c What is the area of the roof section A on the model?

2 A cuboid has dimensions as shown in the diagram.

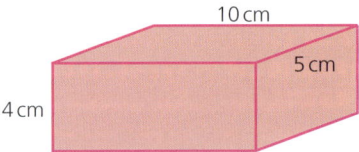

  If the cuboid is enlarged by a scale factor of 2.5, calculate:
  a the total surface area of the original cuboid
  b the total surface area of the enlarged cuboid
  c the volume of the original cuboid
  d the volume of the enlarged cuboid.

3 A cube has side length 3 cm.
  a Calculate its total surface area.
  b If the cube is enlarged and has a total surface area of 486 cm², calculate the scale factor of enlargement.
  c Calculate the volume of the enlarged cube.

4 Two cubes P and Q are of different sizes. If n is the ratio of their corresponding sides, express in terms of n:
  a the ratio of their surface areas
  b the ratio of their volumes.

5 The cuboids A and B shown below are similar.

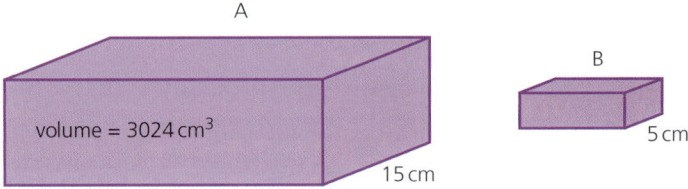

Calculate the volume of cuboid B.

Area and volume of similar shapes

6 Two similar troughs X and Y are shown below.

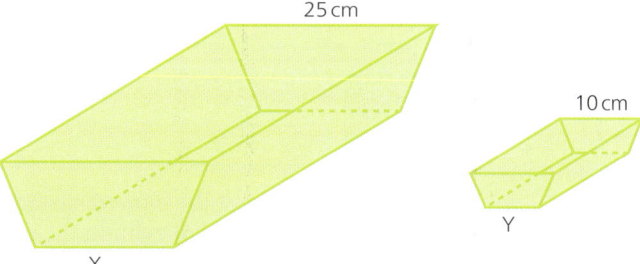

If the capacity of X is 10 litres, calculate the capacity of Y.

Exercise 23.4

1 The two cylinders L and M shown below are similar.

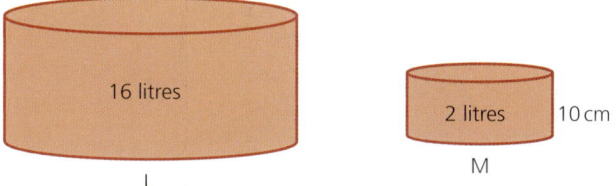

If the height of cylinder M is 10 cm, calculate the height of cylinder L.

2 A square-based pyramid (below) is cut into two shapes by a cut running parallel to the base and made halfway up.

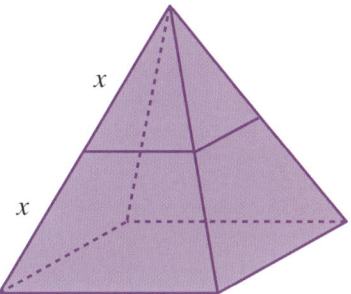

a Calculate the ratio of the volume of the smaller pyramid to that of the original one.
b Calculate the ratio of the volume of the small pyramid to that of the truncated base.

3 The two cones A and B are similar. Cone B is an enlargement of A by a scale factor of 4.
If the volume of cone B is 1024 cm³, calculate the volume of cone A.

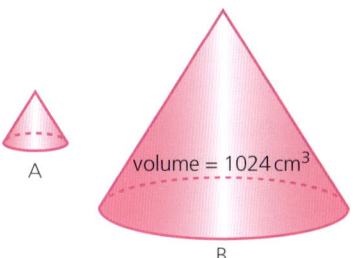

23 SIMILARITY AND CONGRUENCE

Exercise 23.4 (cont)

4 a Stating your reasons clearly, decide whether the two cylinders shown are similar or not.

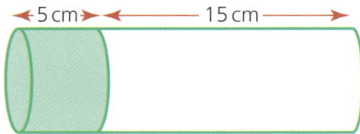

b What is the ratio of the curved surface area of the shaded cylinder to that of the unshaded cylinder?

5 The diagram shows a triangle.
   a Calculate the area of triangle RSV.
   b Calculate the area of triangle QSU.
   c Calculate the area of triangle PST.

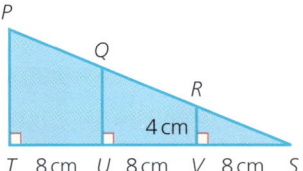

6 The area of an island on a map is $30\,\text{cm}^2$. The scale used on the map is $1:100000$.
   a Calculate the area in square kilometres of the real island.
   b An airport on the island is on a rectangular piece of land measuring $3\,\text{km}$ by $2\,\text{km}$. Calculate the area of the airport on the map in cm^2.

7 The two packs of cheese X and Y are similar.
   The total surface area of pack Y is four times that of pack X.
   Calculate:
   a the dimensions of pack Y
   b the mass of pack X if pack Y has a mass of $800\,\text{g}$.

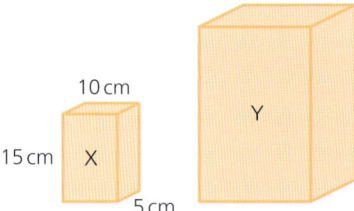

Congruent shapes

Two shapes are **congruent** if their corresponding sides are the same length and their corresponding angles are the same size, i.e. the shapes are exactly the same size and shape.

Shapes X and Y are congruent:

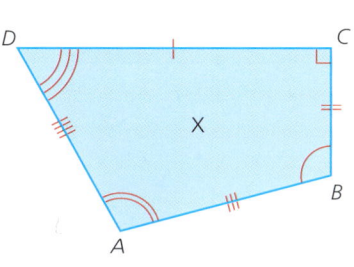
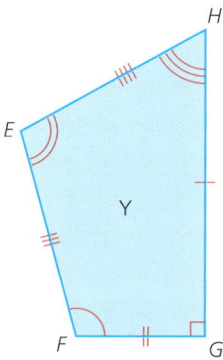

They are congruent as $AB = EF$, $BC = FG$, $CD = GH$ and $DA = HE$. Also angle DAB = angle HEF, angle ABC = angle EFG, angle BCD = angle FGH and angle CDA = angle GHE.

Congruent shapes can therefore be reflections and rotations of each other.

Note: Congruent shapes are, by definition, also similar, but similar shapes are not necessarily congruent.

→ Worked example

Triangles ABC and DEF are congruent:

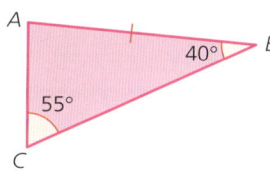
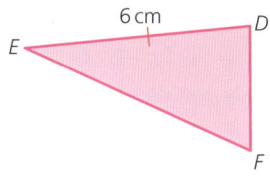

In the triangle DEF, the angle at D is defined as angle FDE.

a Calculate the size of angle FDE.
   As the two triangles are congruent angle FDE = angle CAB
   angle $CAB = 180° - 40° - 55° = 85°$
   Therefore angle $FDE = 85°$

b Deduce the length of AB.
   As $AB = DE$, $AB = 6$ cm

23 SIMILARITY AND CONGRUENCE

Exercise 23.5

1. Look at the shapes on the grid below. Which shapes are congruent to shape A?

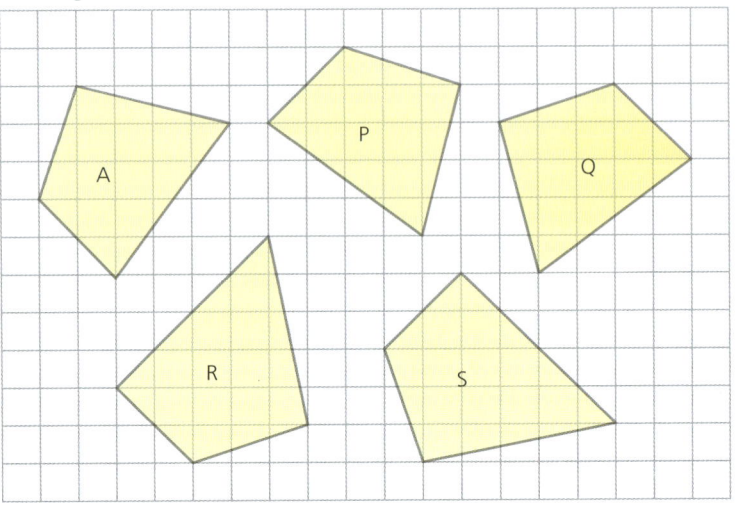

2. The two shapes below are congruent:

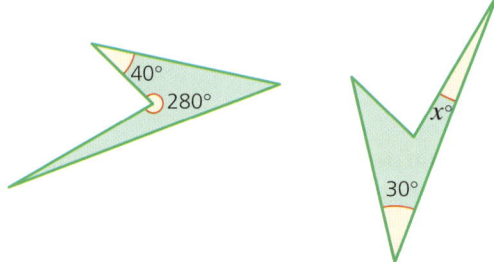

   Calculate the size of x.

3. A quadrilateral is plotted on a pair of axes. The coordinates of its four vertices are (0, 1), (0, 5), (3, 4) and (3, 3).
   Another quadrilateral, congruent to the first, is also plotted on the same axes. Three of its vertices have coordinates (6, 5), (5, 2) and (4, 2). Calculate the coordinates of the fourth vertex.

Congruent shapes

4 Triangle P is drawn on a graph. One side of another triangle, Q, is also shown.

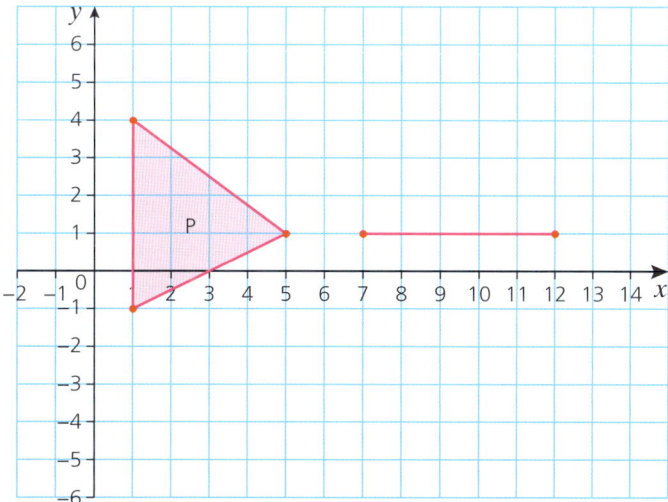

If triangles P and Q are congruent, give all the possible coordinates for the position of the missing vertex.

5 Regular hexagon *ABCDEF* is shown below:

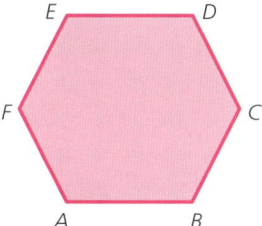

a Identify two triangles congruent to triangle *ACD*.
b Identify two triangles congruent to triangle *DEF*.

6 Parallelogram *PQRS* is shown below:

Explain, justifying your answer, whether triangle *PQS* is congruent to triangle *QRS*.

23 SIMILARITY AND CONGRUENCE

Student assessment 1

1 The two triangles (below) are similar.

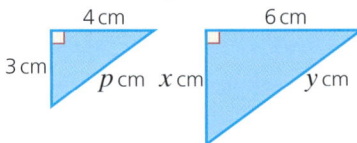

   a Using Pythagoras' theorem, calculate the value of p.
   b Calculate the values of x and y.

2 Cones M and N are similar.
   a Express the ratio of their surface areas in the form, area of M : area of N.
   b Express the ratio of their volumes in the form, volume of M : volume of N.

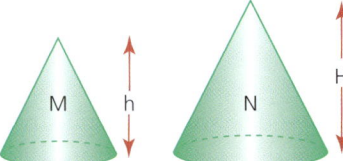

3 Calculate the values of x, y and z in the triangle below.

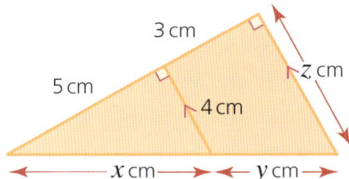

4 The tins A and B are similar. The capacity of tin B is three times that of tin A. If the label on tin A has an area of 75 cm², calculate the area of the label on tin B.

5 A cube of side 4 cm is enlarged by a scale factor of 2.5.
   a Calculate the volume of the enlarged cube.
   b Calculate the surface area of the enlarged cube.

6 The two troughs X and Y are similar. The scale factor of enlargement from Y to X is 4. If the capacity of trough X is 1200 cm³, calculate the capacity of trough Y.

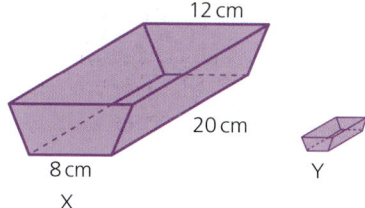

7 The rectangular floor plan of a house measures 8 cm by 6 cm. If the scale of the plan is 1 : 50, calculate:
   a the dimensions of the actual floor,
   b the area of the actual floor in m².

8 The volume of the cylinder below is 400 cm³.
   Calculate the volume of a similar cylinder formed by enlarging the one shown by a scale factor of 2.

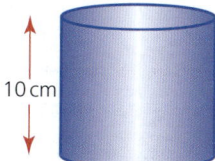

9 The diagram below shows an equilateral triangle ABC. The midpoints L, M and N of each side are also joined.

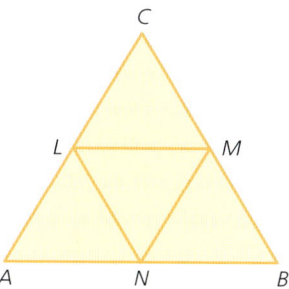

   a Identify a trapezium congruent to trapezium $BCLN$.
   b Identify a triangle similar to triangle LMN.

10 Decide whether each of the following statements is true or false.
   a All circles are similar.
   b All squares are similar.
   c All rectangles are similar.
   d All equilateral triangles are congruent.

24 Symmetry

Symmetry in two- and three-dimensional shapes

> **Note**
> All diagrams are not drawn to scale.

A **line of symmetry** divides a two-dimensional (flat) shape into two congruent (identical) shapes.

e.g.

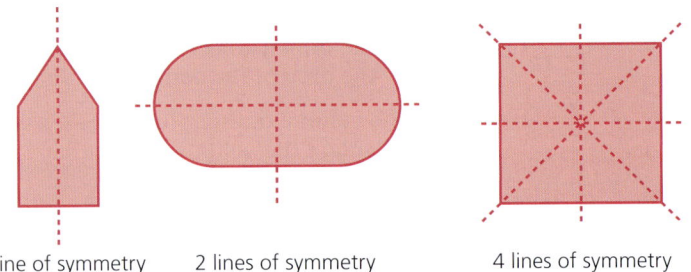

1 line of symmetry   2 lines of symmetry   4 lines of symmetry

A **plane of symmetry** divides a three-dimensional (solid) shape into two congruent solid shapes.

e.g.

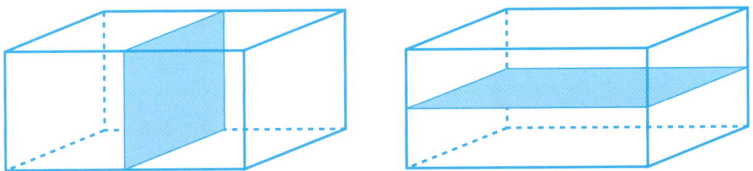

A cuboid has at least three planes of symmetry, two of which are shown above.

A shape has **reflective symmetry** if it has one or more lines or planes of symmetry.

A two-dimensional shape has **rotational symmetry** if, when rotated about a central point, it is identical to the original shape and orientation. The number of times it does this during a complete revolution is called the **order of rotational symmetry**.

e.g.

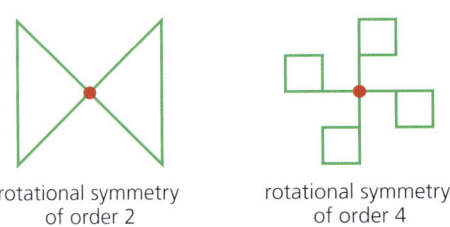

rotational symmetry   rotational symmetry
of order 2   of order 4

Symmetry in two- and three-dimensional shapes

A three-dimensional shape has **rotational symmetry** if, when rotated about a central axis, it looks the same at certain intervals.
e.g.

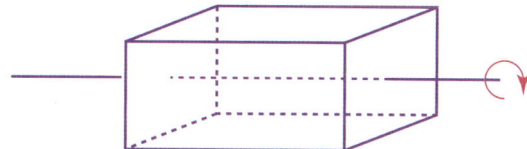

This cuboid has rotational symmetry of order 2 about the axis shown.

Exercise 24.1

Note

Heptagon and nonagon are not part of the Extended syllabus, but are included here for interest.

1 Copy and complete the following table of the symmetry properties for the given regular polygons. It has been started for you.

Regular polygon	Number of lines of symmetry	Order of rotational symmetry
Equilateral triangle	3	
Square		
Pentagon		
Hexagon		6
Heptagon		
Octagon		
Nonagon		
Decagon		

2 Draw each of the solid shapes below twice, then:
   i on each drawing of the shape, draw a different plane of symmetry,
   ii state how many planes of symmetry the shape has in total.

a
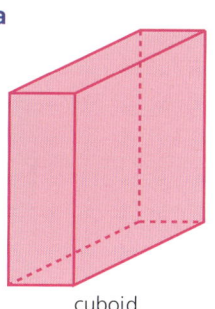
cuboid

b
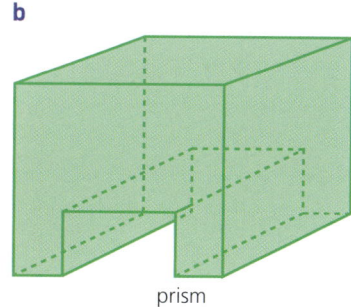
prism

24 SYMMETRY

Exercise 24.1 (cont)

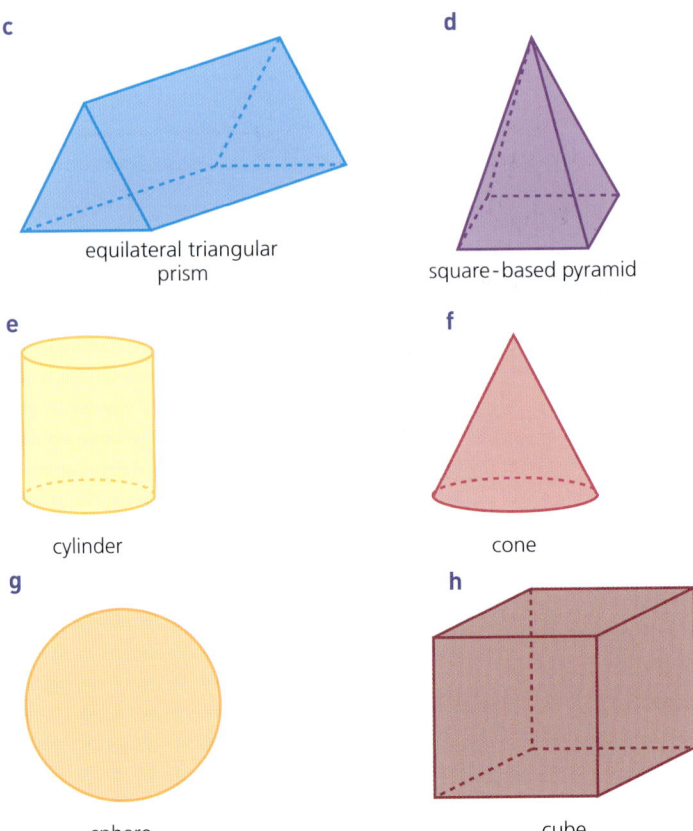

c equilateral triangular prism
d square-based pyramid
e cylinder
f cone
g sphere
h cube

3 For each of the solid shapes shown below, determine the order of rotational symmetry about the axis shown.

a
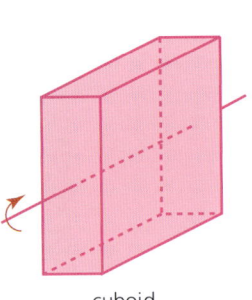
cuboid

b
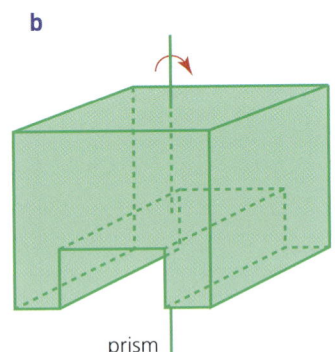
prism

Symmetry in two- and three-dimensional shapes

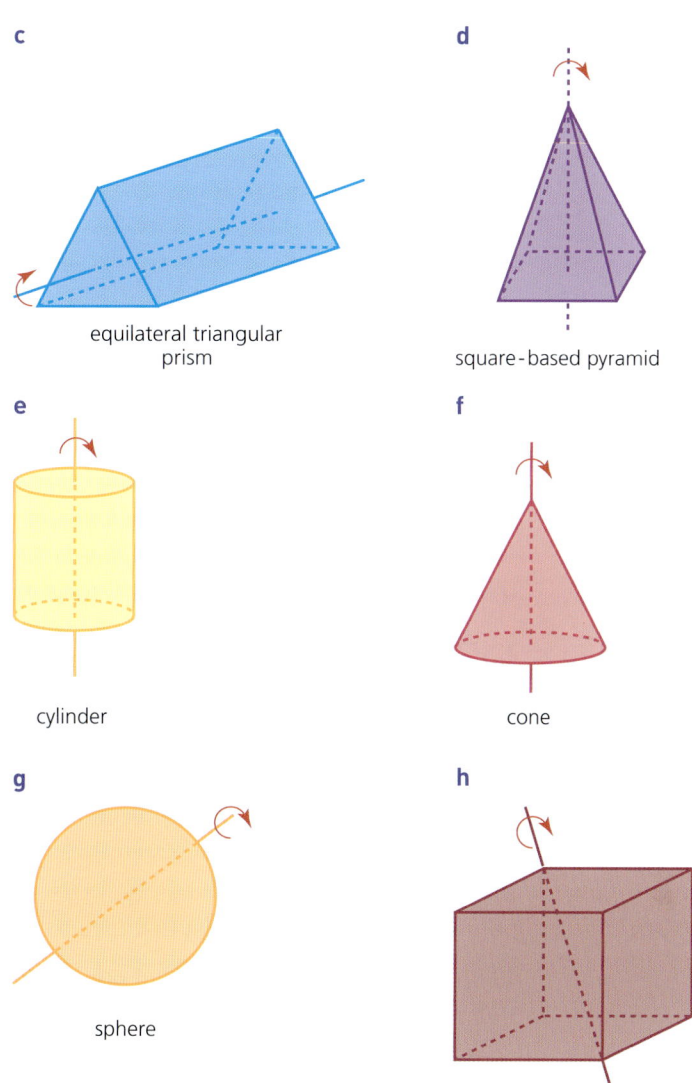

c equilateral triangular prism

d square-based pyramid

e cylinder

f cone

g sphere

h cube

24 SYMMETRY

Circle properties

Equal chords and perpendicular bisectors

If chords AB and XY are of equal length, then, since OA, OB, OX and OY are radii, the triangles OAB and OXY are congruent isosceles triangles.

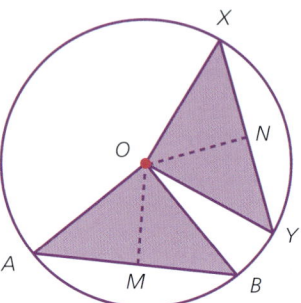

It follows that:

» the section of a line of symmetry OM through triangle OAB is the same length as the section of a line of symmetry ON through triangle OXY,

» OM and ON are **perpendicular bisectors** of AB and XY respectively.

This is one of the circle theorems, namely, that the perpendicular bisector of a chord passes through the centre of the circle.

This also demonstrates the fact that the perpendicular distance from a point to a line is the shortest distance to the line.

Therefore it can be deduced that equal chords are equidistant from the centre.

Exercise 24.2

1 In the diagram, O is the centre of the circle, PQ and RS are chords of equal length and M and N are their respective midpoints.

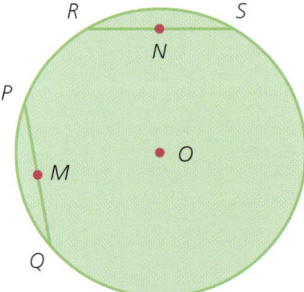

a   What kind of triangle is triangle POQ?
b   Describe the line ON in relation to RS.
c   If angle POQ is 80°, calculate angle OQP.
d   Calculate angle ORS.

Circle properties

Exercise 24.2 (cont)

e If *PQ* is 6 cm, calculate the length *OM*.
f Calculate the diameter of the circle.

2 In the diagram, *O* is the centre of the circle. *AB* and *CD* are equal chords and the points *R* and *S* are their midpoints respectively.

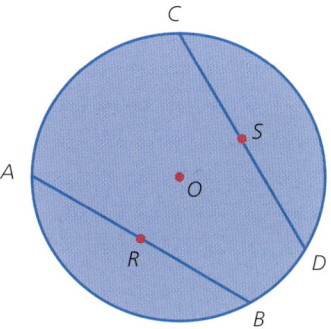

State whether the statements below are true or false, giving reasons for your answers.
a angle *COD* = 2 × angle *AOR*.
b *OR* = *OS*.
c If angle *ROB* is 60° then triangle *AOB* is equilateral.
d *OR* and *OS* are perpendicular bisectors of *AB* and *CD* respectively.

3 Using the diagram, state whether the following statements are true or false, giving reasons for your answer.

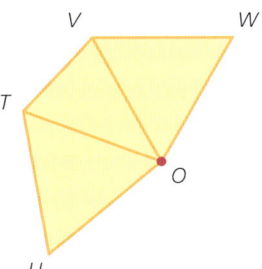

a If triangle *VOW* and triangle *TOU* are isosceles triangles, then *T*, *U*, *V* and *W* would all lie on the circumference of a circle with its centre at *O*.
b If triangle *VOW* and triangle *TOU* are congruent isosceles triangles, then *T*, *U*, *V* and *W* would all lie on the circumference of a circle with its centre at *O*.

Tangents from an external point

Triangles *OAC* and *OBC* are congruent since angle *OAC* and angle *OBC* are right angles, *OA* = *OB* because they are both radii, and *OC* is common to both triangles. Hence *AC* = *BC*.

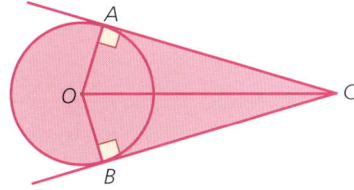

319

24 SYMMETRY

In general, therefore, tangents being drawn to the same circle from an external point are equal in length.

Exercise 24.3

1 Copy each of the diagrams below and calculate the size of the angle marked $x°$ in each case. Assume that the lines drawn from points on the circumference are tangents.

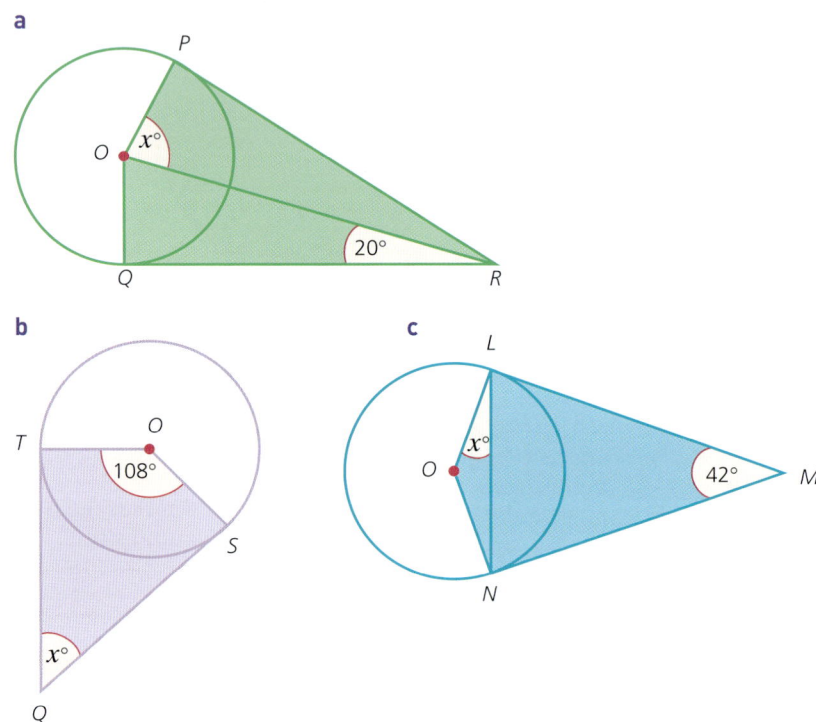

2 Copy each of the diagrams below and calculate the length of the side marked y cm in each case. Assume that the lines drawn from points on the circumference are tangents.

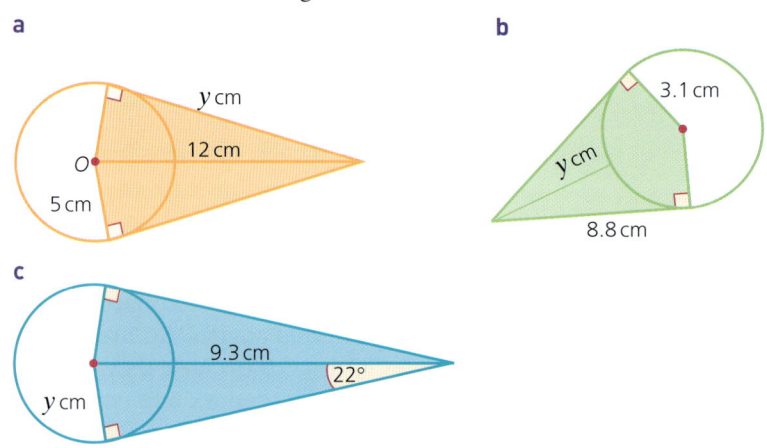

Circle properties

Student assessment 1

1 Draw a two-dimensional shape with exactly:
  a rotational symmetry of order 2,
  b rotational symmetry of order 4,
  c rotational symmetry of order 6.

2 Draw and name a three-dimensional shape with the following orders of rotational symmetry around one axis. Mark the position of this axis of symmetry clearly.
  a Order 2
  b Order 3
  c Order 8

3 In the diagram, OM and ON are perpendicular bisectors of AB and XY respectively. $OM = ON$.
Prove that AB and XY are chords of equal length.

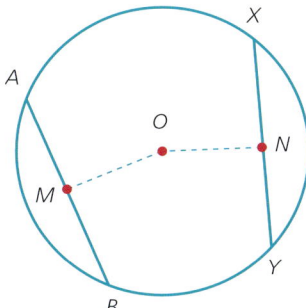

4 In the diagram, XY and YZ are both tangents to the circle with centre O.
  a Calculate angle OZX.
  b Calculate the length XZ.

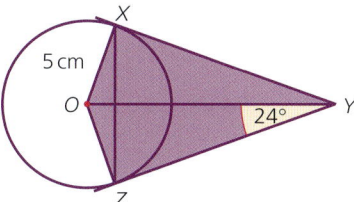

5 In the diagram, LN and MN are both tangents to the circle centre O. If angle LNO is 35°, calculate the circumference of the circle.

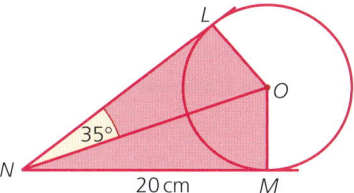

25 Angle properties

> **Note**
> All diagrams are not drawn to scale.

Angles at a point and on a line

One complete revolution is equivalent to a rotation of 360° about a point. Similarly, half a complete revolution is equivalent to a rotation of 180° about a point. These facts can be seen clearly by looking at either a circular angle measurer or a semicircular protractor.

Worked examples

a Calculate the size of the angle x in the diagram below:

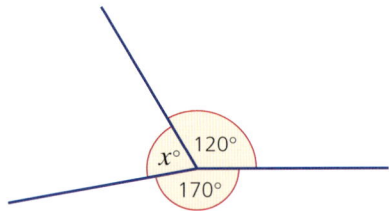

The sum of all the angles around a point is 360°. Therefore:

$120 + 170 + x = 360$

$x = 360 - 120 - 170$

$x = 70$

Therefore angle x is 70°.

Note that the size of the angle x is **calculated** and **not measured**.

b Calculate the size of the angle a in the diagram below:

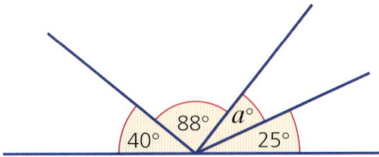

The sum of all the **angles at a point** on a straight line is 180°. Therefore:

$40 + 88 + a + 25 = 180$

$a = 180 - 40 - 88 - 25$

$a = 27$

Therefore angle a is 27°.

Angles formed within parallel lines

> **Note**
>
> Pairs of angles which add up to 180° are also called **supplementary** angles.

When two straight lines cross, it is found that the angles opposite each other are the same size. They are known as **vertically opposite angles**. By using the fact that angles at a point on a straight line add up to 180°, it can be shown why vertically opposite angles must always be equal in size.

$a + b = 180°$
$c + b = 180°$

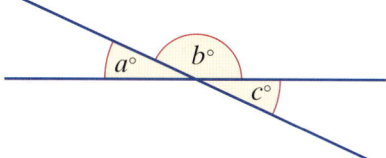

Therefore, a is equal to c.

When a line intersects two parallel lines, as in the diagram below, it is found that certain angles are the same size.

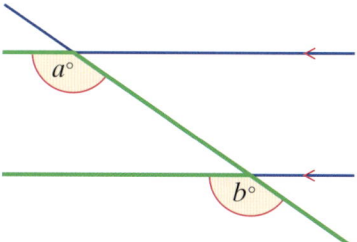

The angles a and b are equal and are known as **corresponding angles**. Corresponding angles can be found by looking for an 'F' formation in a diagram.

A line intersecting two parallel lines also produces another pair of equal angles, known as **alternate angles**. These can be shown to be equal by using the fact that both vertically opposite and corresponding angles are equal.

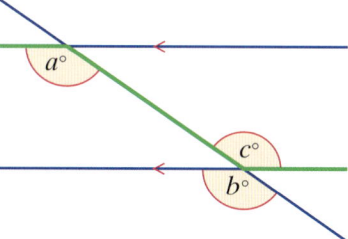

In the diagram above, $a = b$ (corresponding angles). But $b = c$ (vertically opposite). It can therefore be deduced that $a = c$.

Angles a and c are alternate angles. These can be found by looking for a 'Z' formation in a diagram.

25 ANGLE PROPERTIES

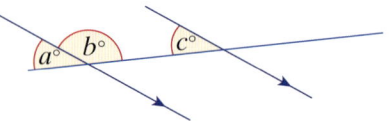

As $a° = c°$ (corresponding angles) and $a° + b° = 180°$ (angles on a straight line add up to 180°) then $b° + c° = 180°$.

b and c are **co-interior** angles as they face each other between parallel lines. Co-interior angles therefore add up to 180°.

Exercise 25.1 In each of the following questions, some of the angles are given. Deduce, giving your reasons, the size of the other labelled angles.

1

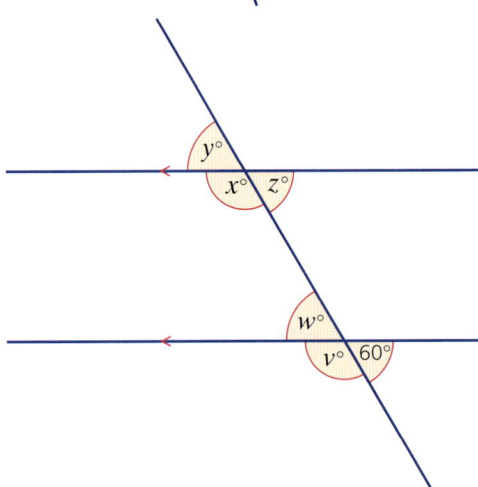

2

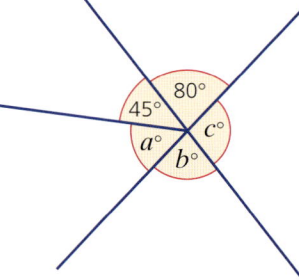

3

4

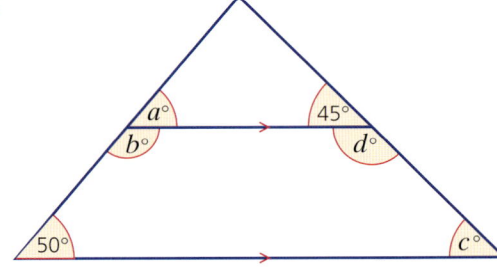

5

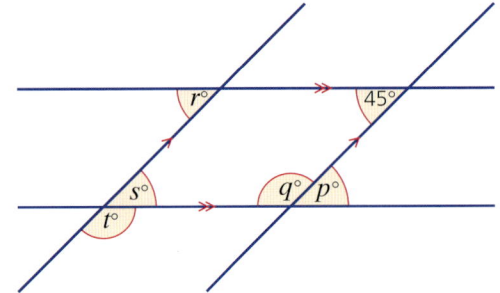

6

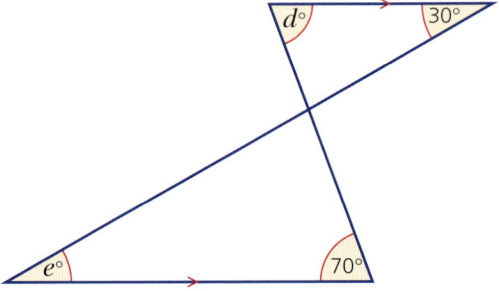

324

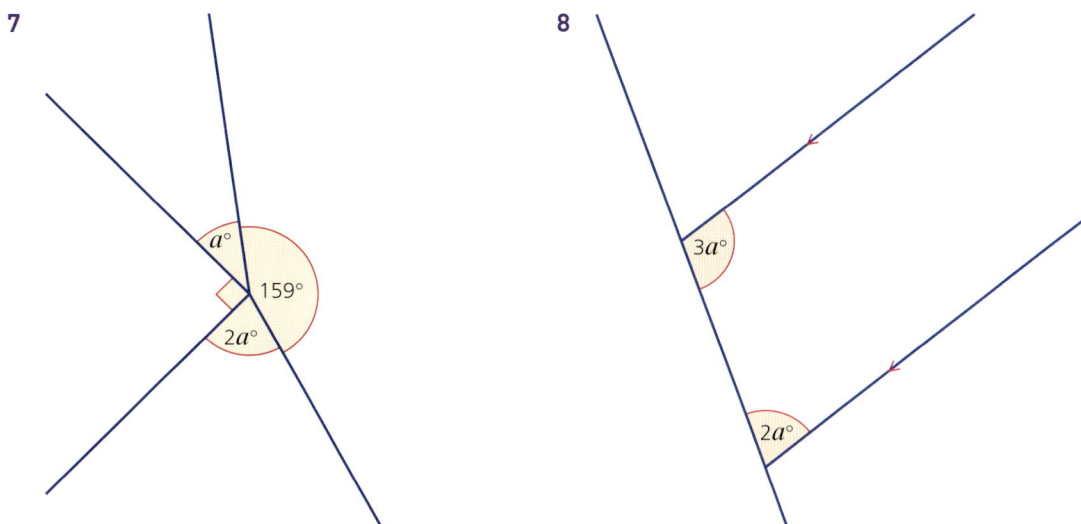

Angles in a triangle

The sum of the angles in a triangle is 180°.

→ Worked example

Calculate the size of the angle x in the triangle below:

$37 + 64 + x = 180$
$x = 180 - 37 - 64$

Therefore angle x is 79°.

25 ANGLE PROPERTIES

Exercise 25.2

1 For each of the triangles below, use the information given to calculate the size of angle x.

a
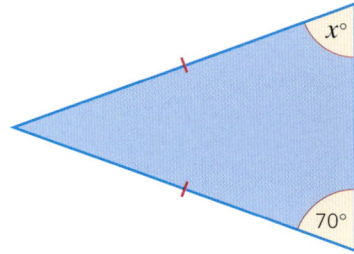

b
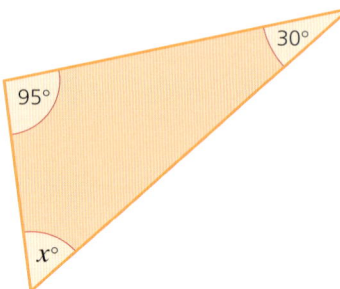

c
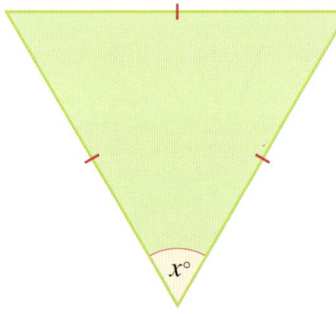

d
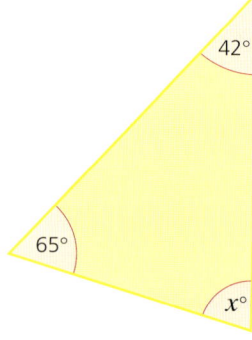

e
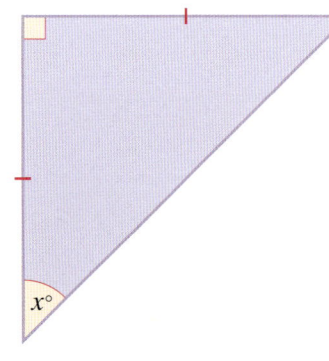

f

2 In each of the diagrams below, calculate the size of the labelled angles.

a
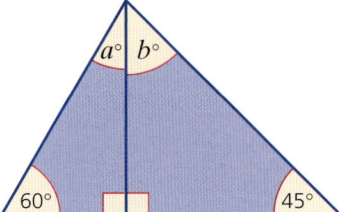

b
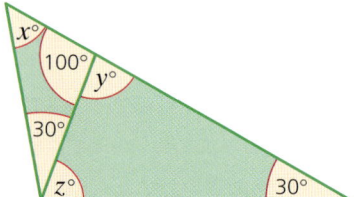

c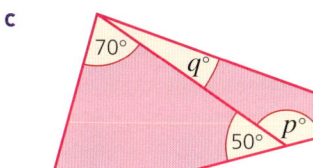

d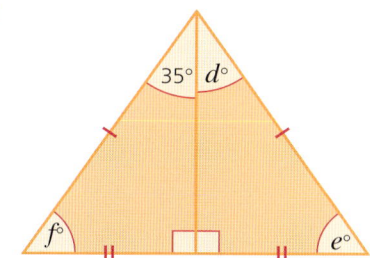

e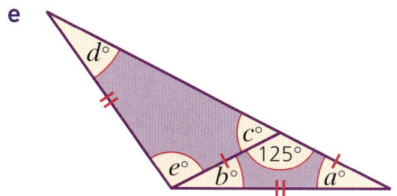

f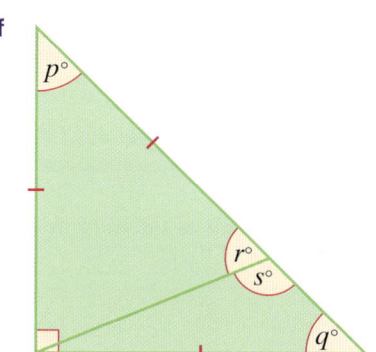

Angles in a quadrilateral

In the quadrilaterals below, a straight line is drawn from one of the corners (vertices) to the opposite corner. The result is to split the quadrilaterals into two triangles.

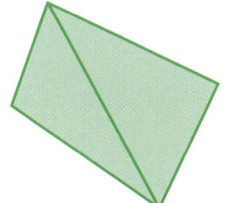 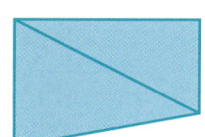  

The sum of the angles of a triangle is 180°. Therefore, as a quadrilateral can be drawn as two triangles, the sum of the four angles of any quadrilateral must be 360°.

25 ANGLE PROPERTIES

> **Worked example**
>
> Calculate the size of angle *p* in the quadrilateral below:
>
>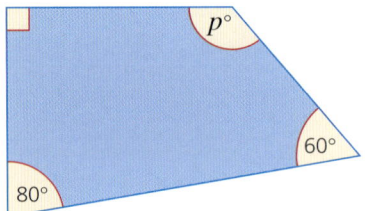
>
> $$90 + 80 + 60 + p = 360$$
> $$p = 360 - 90 - 80 - 60$$
>
> Therefore angle *p* is 130°.

Exercise 25.3 In each of the diagrams below, calculate the size of the labelled angles.

1 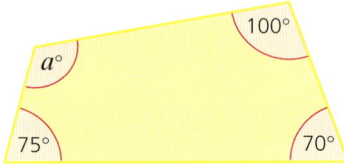   2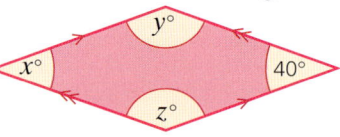

3 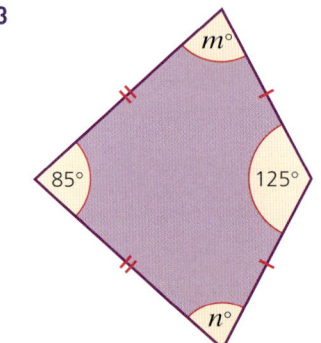   4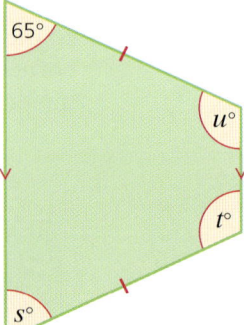

5    6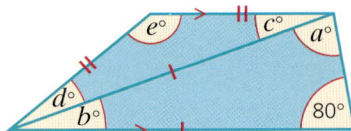

7 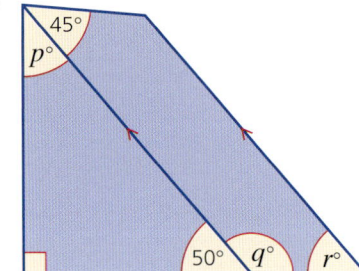   8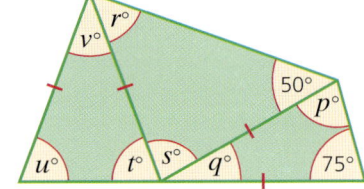

Polygons

A **regular polygon** is distinctive in that all its sides are of equal length and all its angles are of equal size. Below are some examples of regular polygons.

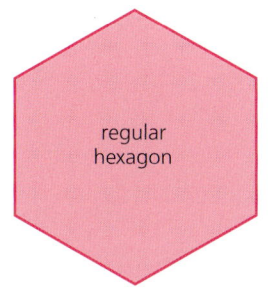

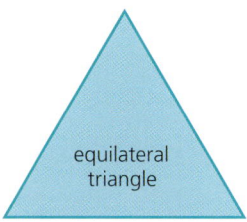

The sum of the interior angles of a polygon

In the polygons below, a straight line is drawn from each vertex to vertex A.

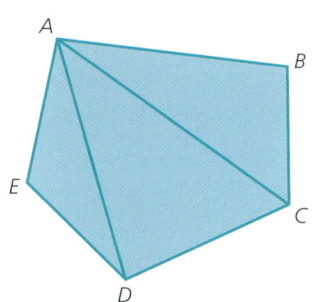
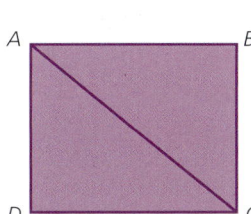
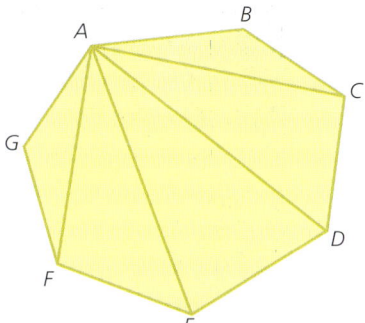

As can be seen, the number of triangles is always two less than the number of sides the polygon has, i.e. if there are n sides, there will be $(n-2)$ triangles.

Since the angles of a triangle add up to 180°, the sum of the interior angles of a polygon is therefore $180(n-2)$ degrees.

➡ Worked example

Find the sum of the interior angles of a regular pentagon and hence the size of each interior angle.
For a pentagon, $n = 5$.
Therefore the sum of the interior angles $= 180(5-2)°$
$\qquad\qquad\qquad\qquad\qquad\qquad\quad = 180 \times 3°$
$\qquad\qquad\qquad\qquad\qquad\qquad\quad = 540°$
For a regular pentagon the interior angles are of equal size.
Therefore each angle $= \frac{540°}{5} = 108°$.

25 ANGLE PROPERTIES

The sum of the exterior angles of a polygon

The angles marked a, b, c, d, e and f in the diagram below represent the exterior angles of a regular hexagon.

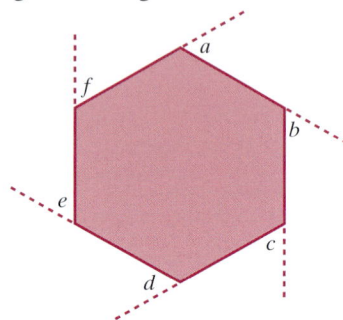

For any convex polygon the sum of the exterior angles is 360°.

If the polygon is regular and has n sides, then each exterior angle $= \frac{360°}{n}$.

→ Worked examples

a Find the size of an exterior angle of a regular nine-sided polygon.

$\frac{360}{9} = 40°$

b Calculate the number of sides a regular polygon has if each exterior angle is 15°.

$n = \frac{360}{15}$
$= 24$

The polygon has 24 sides.

Exercise 25.4

Note

Heptagon, nonagon and dodecagon are not part of the syllabus. They have 7, 9 and 12 sides respectively.

1 Find the sum of the interior angles of the following polygons:
   a a hexagon   b a nonagon   c a heptagon

2 Find the value of each interior angle of the following regular polygons:
   a an octagon   b a square
   c a decagon   d a dodecagon

3 Find the size of each exterior angle of the following regular polygons:
   a a pentagon   b a dodecagon   c a heptagon

4 The exterior angles of regular polygons are given below. In each case calculate the number of sides the polygon has.
   a 20°   b 36°   c 10°
   d 45°   e 18°   f 3°

5 The interior angles of regular polygons are given below. In each case calculate the number of sides the polygon has.
   a 108°   b 150°   c 162°
   d 156°   e 171°   f 179°

6 Calculate the number of sides a regular polygon has if an interior angle is five times the size of an exterior angle.

330

The angle in a semicircle

In the diagram below, if AB represents the diameter of the circle, then the angle at C is 90°.

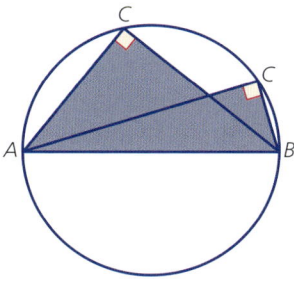

Exercise 25.5 In each of the following diagrams, O marks the centre of the circle. Calculate the value of x in each case.

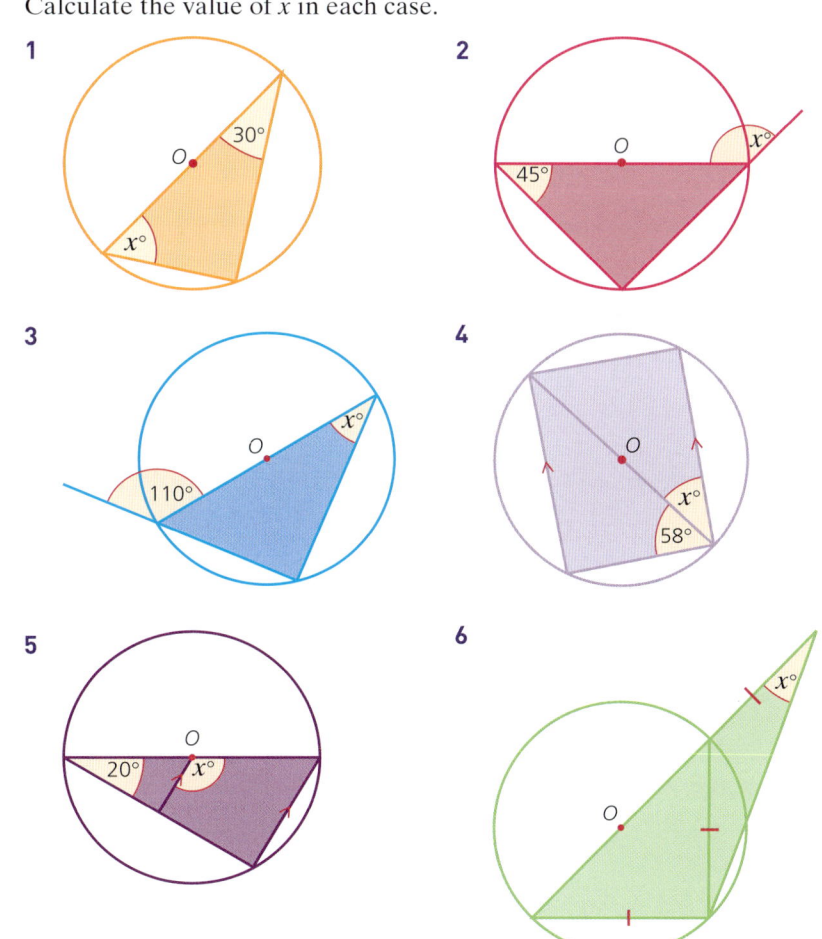

25 ANGLE PROPERTIES

The angle between a tangent and a radius of a circle

The angle between a tangent at a point and the radius to the same point on the circle is a right angle.

In the diagram below, triangles OAC and OBC are congruent as angle OAC and angle OBC are right angles, $OA = OB$ because they are both radii, and OC is common to both triangles.

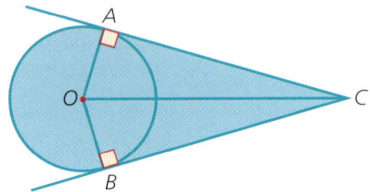

Exercise 25.6 In each of the following diagrams, O marks the centre of the circle. Calculate the value of x in each case.

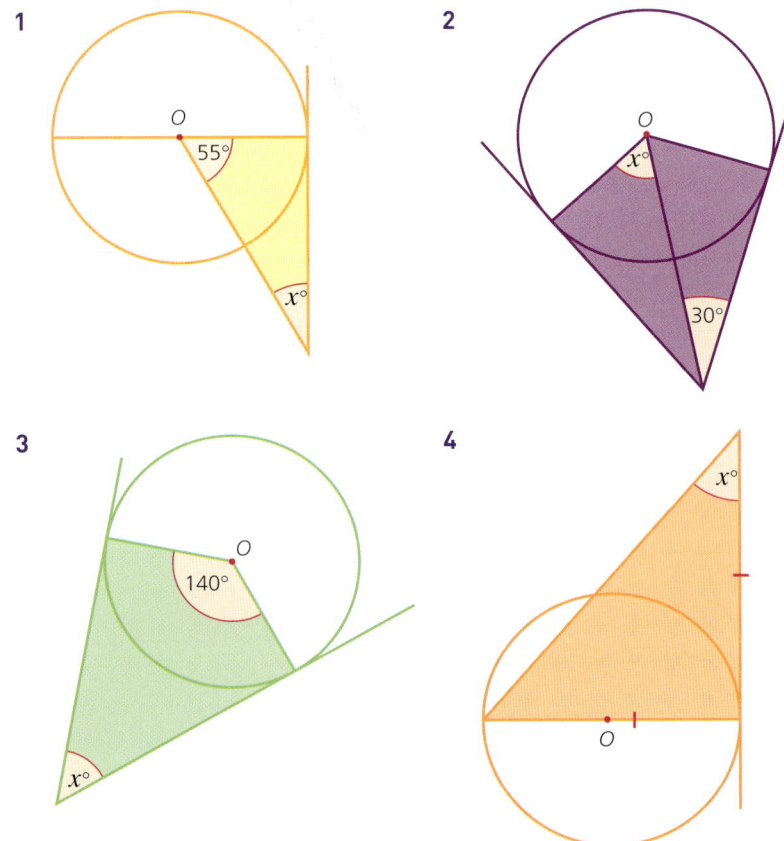

Angle properties of irregular polygons

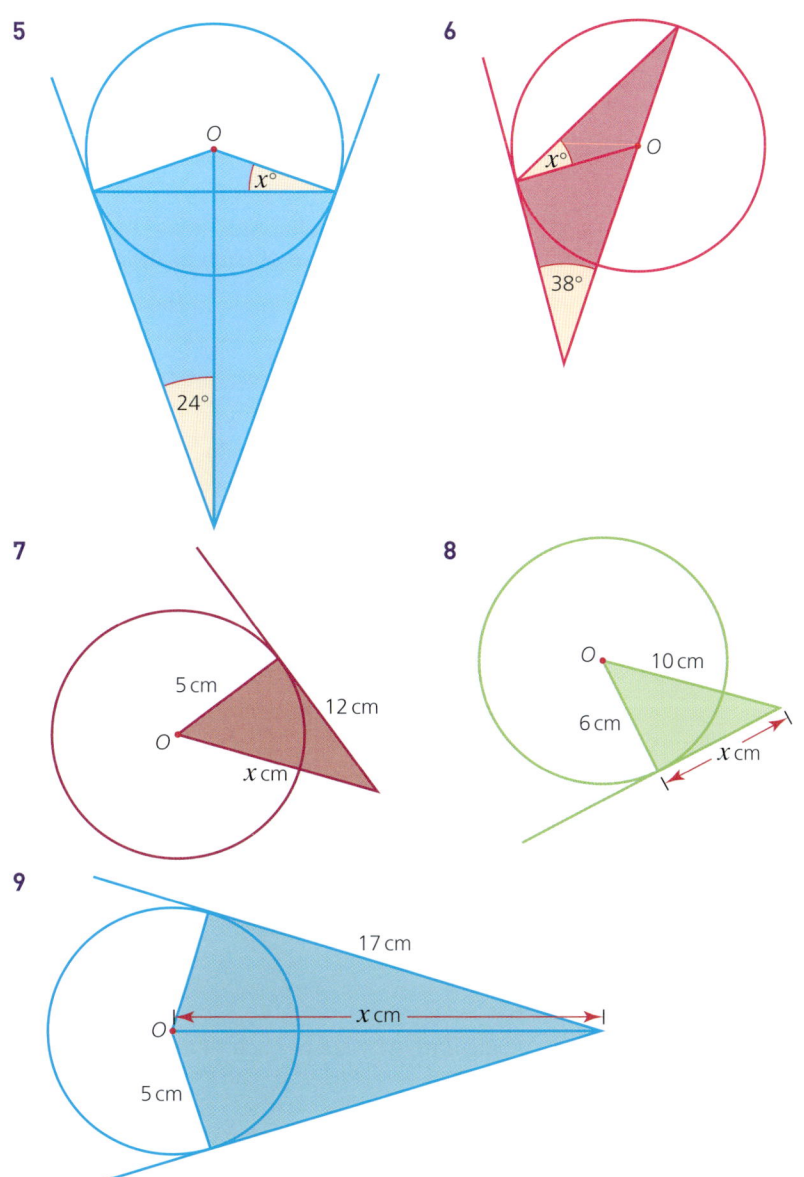

Angle properties of irregular polygons

As explained earlier in this chapter, the sum of the interior angles of a polygon is given by $180(n - 2)°$, where n represents the number of sides of the polygon. The sum of the exterior angles of any polygon is 360°.

Both of these rules also apply to irregular polygons, i.e. those where the lengths of the sides and the sizes of the interior angles are not all equal.

25 ANGLE PROPERTIES

Exercise 25.7

1 For the pentagon:

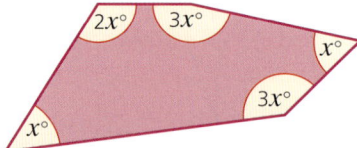

  a calculate the value of *x*,
  b calculate the size of each of the angles.

2 Find the size of each angle in the octagon (below).

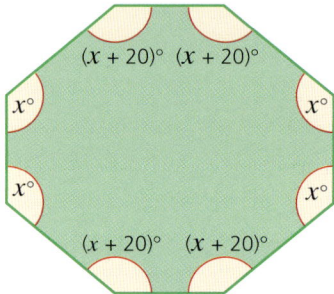

3 Calculate the value of *x* for the pentagon shown.

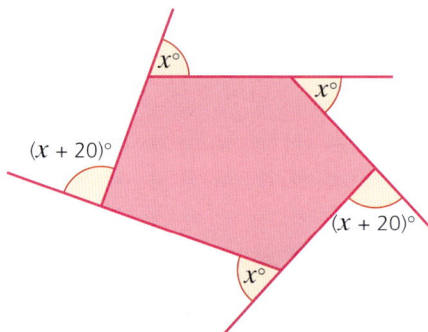

4 Calculate the size of each of the angles *a*, *b*, *c*, *d* and *e* in the hexagon.

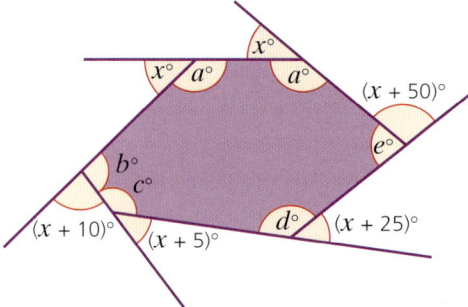

Angle at the centre of a circle

The angle subtended at the centre of a circle by an arc is twice the size of the angle on the circumference subtended by the same arc.

These diagrams illustrate this theorem:

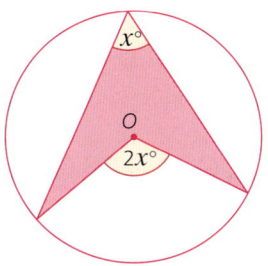 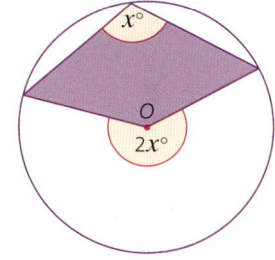

Exercise 25.8 In each of the following diagrams, O marks the centre of the circle. Calculate the size of the marked angles:

1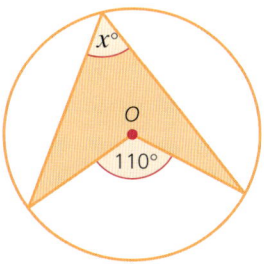

2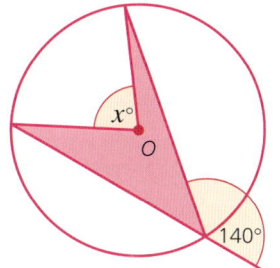

3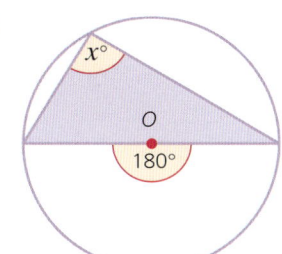

4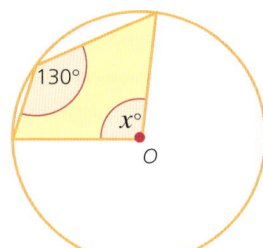

5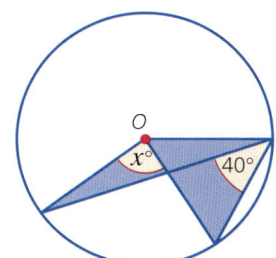

6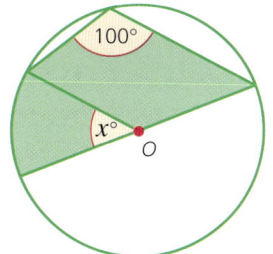

25 ANGLE PROPERTIES

Exercise 25.8 (cont)

7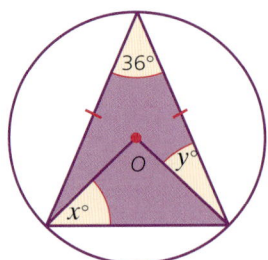

8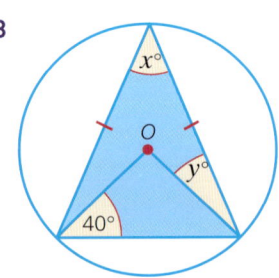

Angles in the same segment

Angles in the same segment of a circle are equal.

This can be explained simply by using the theorem that the angle subtended at the centre is twice the angle on the circumference. Looking at the diagram, if the angle at the centre is $2x°$, then each of the angles at the circumference must be equal to $x°$.

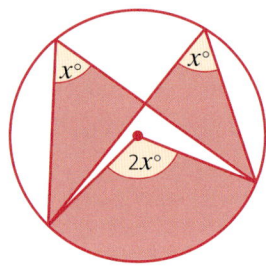

Exercise 25.9 Calculate the marked angles in the following diagrams:

1 2
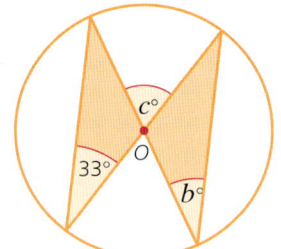

3 4
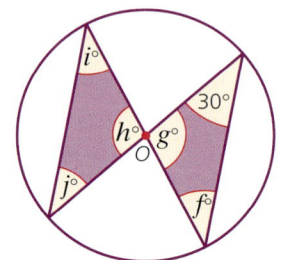

5 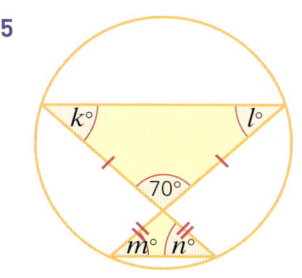   6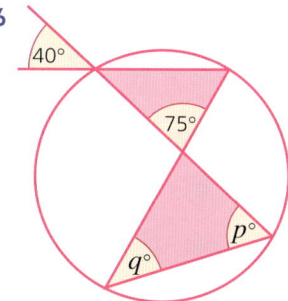

Angles in opposite segments

Points P, Q, R and S all lie on the circumference of the circle (below). They are called concyclic points. Joining the points P, Q, R and S produces a **cyclic quadrilateral**.

The opposite angles are **supplementary**, i.e. they add up to 180°. Since $p° + r° = 180°$ (supplementary angles) and $r° + t° = 180°$ (**angles on a straight line**) it follows that $p° = t°$.

Therefore the exterior angle of a cyclic quadrilateral is equal to the interior opposite angle.

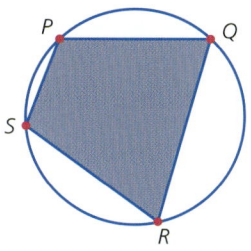 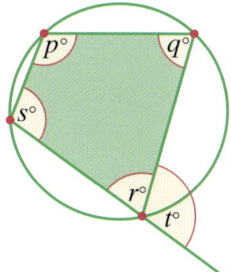

Alternate segment theorem

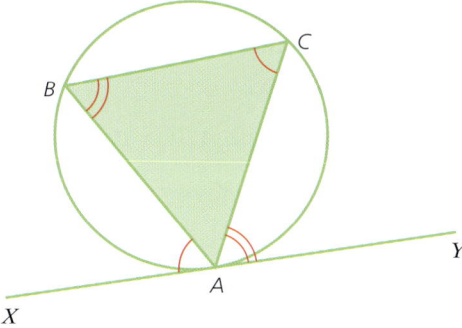

XY is a tangent to the circle at point A. AB is a chord of the circle.

25 ANGLE PROPERTIES

The alternate segment theorem states that the angle between a tangent and a chord at their point of contact is equal to the angle in the alternate segment.

i.e. angle BAX = angle ACB, similarly, angle CAY = angle ABC

This can be proved as follows:

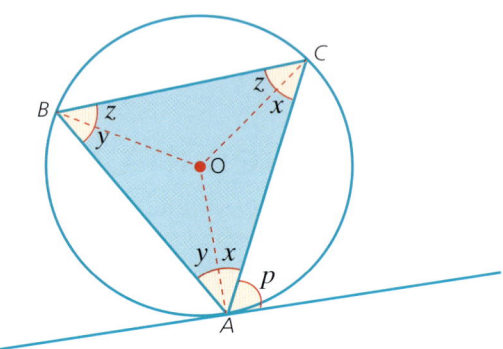

Point O is the centre of the circle.

Drawing radii to points A, B and C creates isosceles triangles AOC, BOC and AOB.

Angle p is the angle between the chord AC and the tangent to the circle at A.

$p = 90 - x$ because the angle between a radius and a tangent to a circle at a point is a right angle.

In triangle ABC, $(x + y) + (y + z) + (x + z) = 180°$

Therefore $$2x + 2y + 2z = 180°$$
$$2(x + y + z) = 180°$$
$$x + y + z = 90°$$

If $x + y + z = 90°$ then $y + z = 90 - x$

As angle $ABC = y + z$, then angle $ABC = 90 - x$

As $p = 90 - x$

Then angle $ABC = p$

Alternate segment theorem

Exercise 25.10 For each Question 1–8, calculate the size of the marked angles.

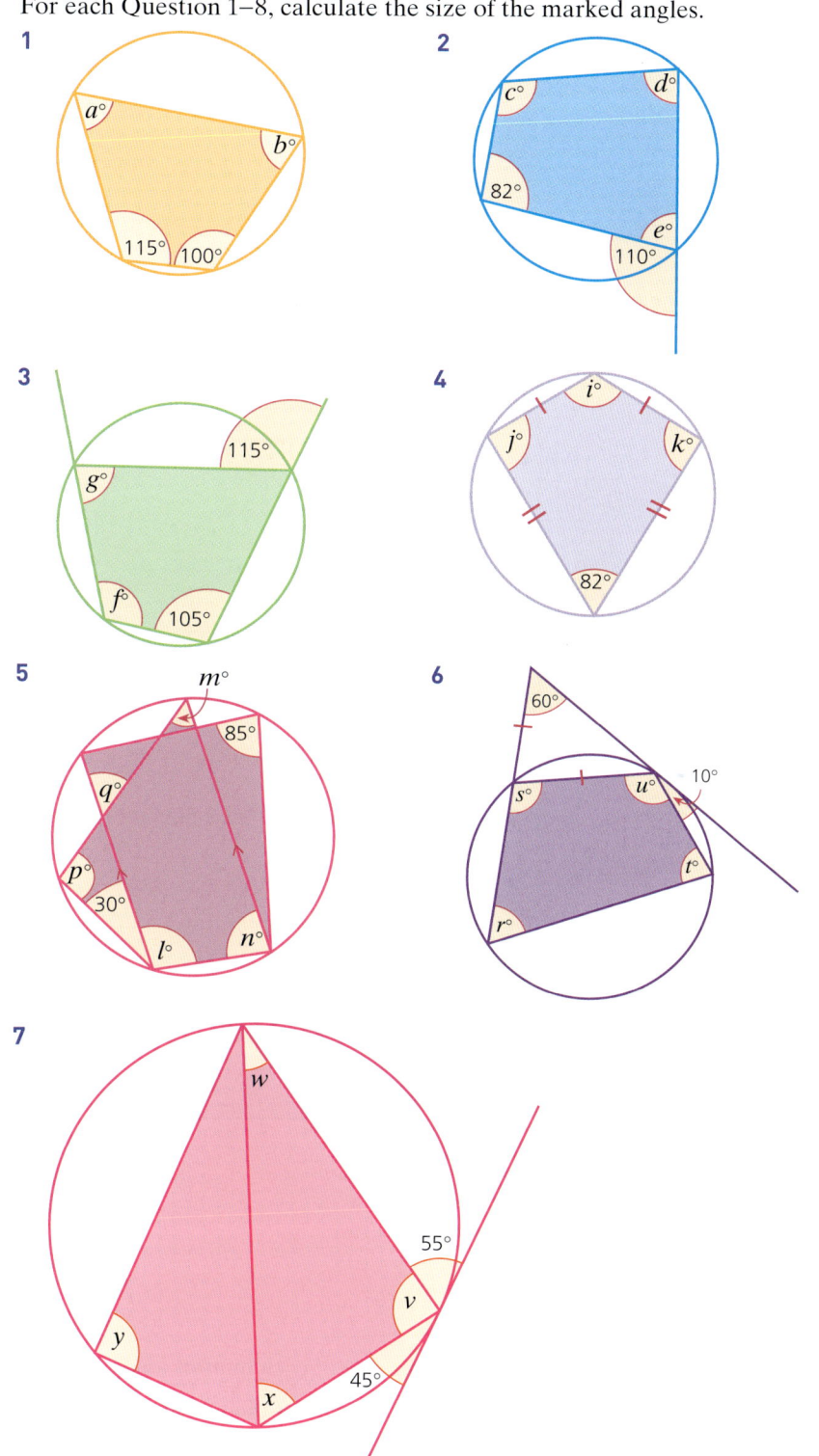

25 ANGLE PROPERTIES

Exercise 25.10 (cont)

8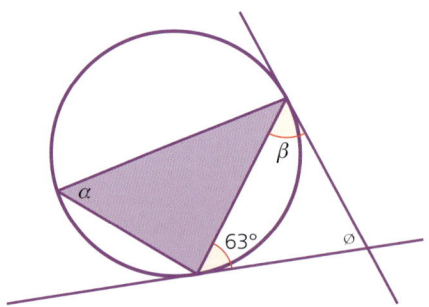

Student assessment 1

1 For the diagrams below, calculate the size of the labelled angles.

a   b

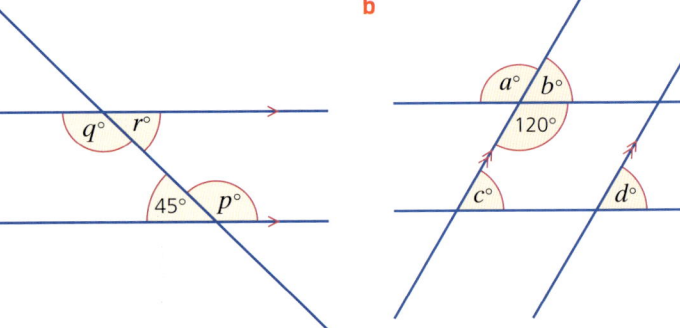

2 For the diagrams below, calculate the size of the labelled angles.

a   b

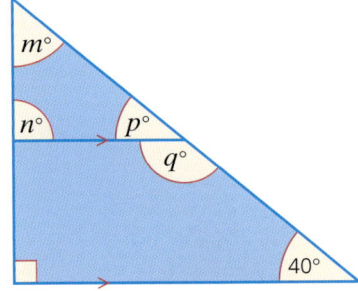
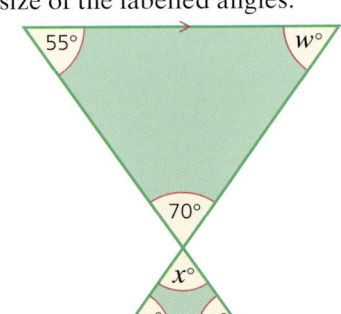

c

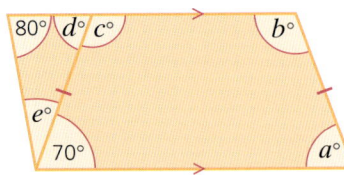

3 Find the size of each interior angle of a 20-sided regular polygon.

4 What is the sum of the interior angles of a nine-sided polygon?

5 What is the sum of the exterior angles of a regular polygon?

6 What is the size of each exterior angle of a regular pentagon?

7 A regular polygon has interior angles of size 156°. Calculate the number of sides it has.

8 If AB is the diameter of the circle and $AC = 5\,\text{cm}$ and $BC = 12\,\text{cm}$, calculate:
   a the size of angle ACB,
   b the length of the radius of the circle.

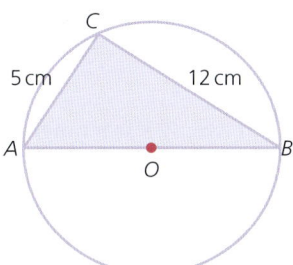

In Questions 9–12, O marks the centre of the circle. Calculate the size of the angle marked x in each case.

9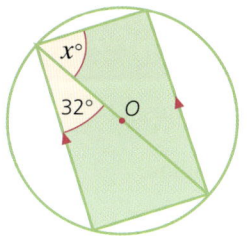

10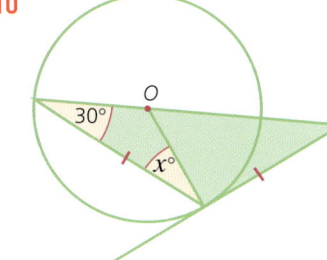

11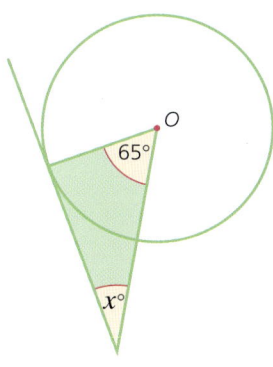

12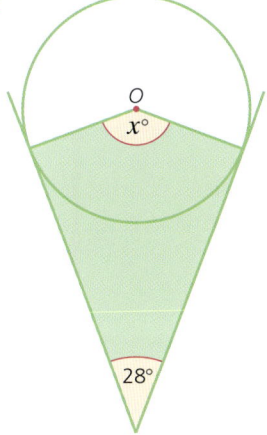

25 ANGLE PROPERTIES

13

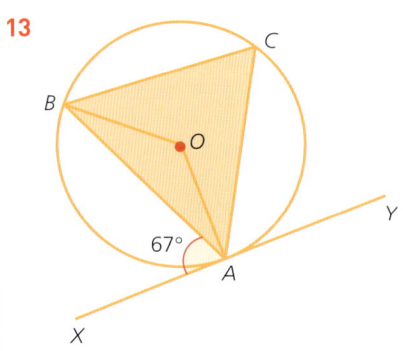

In the diagram above, XY is a tangent to the circle at A. O is the centre of the circle. Calculate each of the following angles, justifying each of your answers.
- **a** angle ACB.
- **b** reflex angle AOB.
- **c** angle ABO.

Student assessment 2

1 In the following diagrams, O marks the centre of the circle. Identify which angles are:
- **i** supplementary angles,
- **ii** right angles,
- **iii** equal.

a

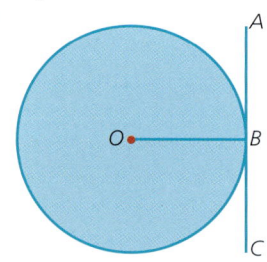

b

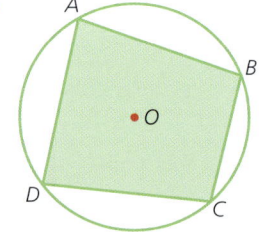

c

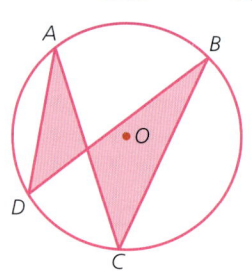

d

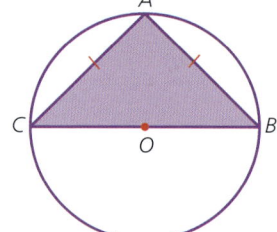

Alternate segment theorem

2 If angle $POQ = 84°$ and O marks the centre of the circle in the diagram, calculate the following:
  a angle PRQ **b** angle OQR

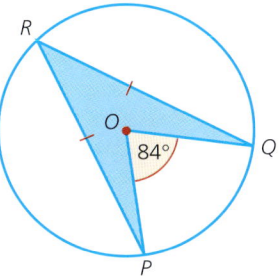

3 Calculate angle DAB and angle ABC in the diagram below.

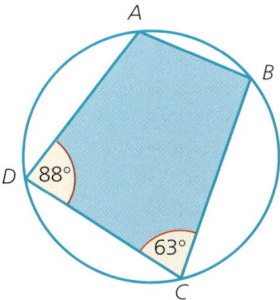

4 If DC is a diameter and O marks the centre of the circle, calculate angles BDC and DAB.

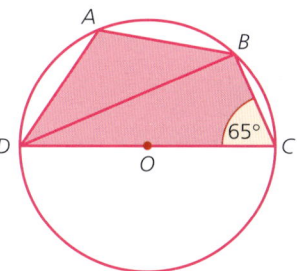

5 Calculate as many angles as possible in the diagram below. O marks the centre of the circle.

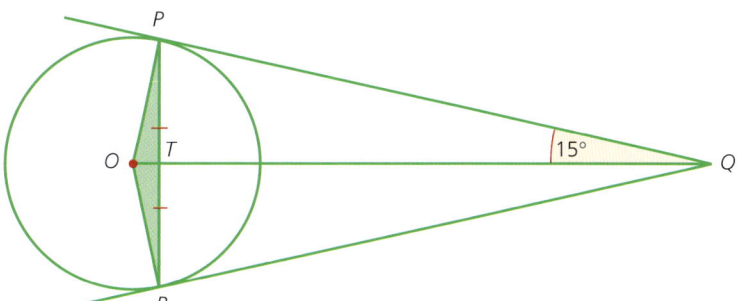

343

25 ANGLE PROPERTIES

6 Calculate the values of c, d and e.

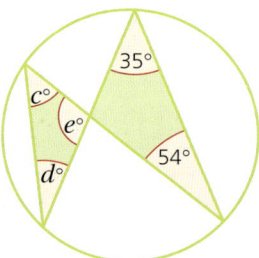

7 Calculate the values of f, g, h and i.

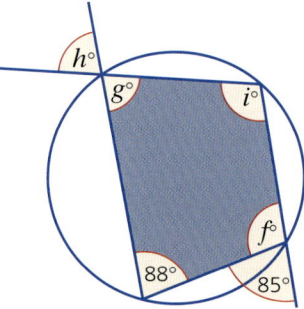

8 In the diagram below, XY is a tangent to the circle at A. All four vertices of the quadrilateral $ABCD$ lie on the circumference of the circle.
If angle $XAC = 124°$ and angle $BCA = 73°$, calculate, giving reasons, the size of the angle BAC.

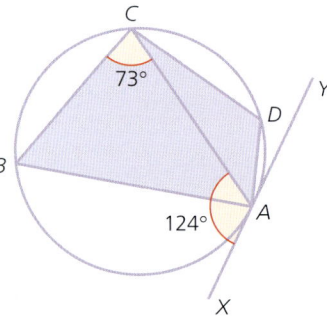

4 Mathematical investigations and ICT 4

Fountain borders

The Alhambra Palace in Granada, Spain, has many fountains which pour water into pools. Many of the pools are surrounded by beautiful ceramic tiles. This investigation looks at the number of square tiles needed to surround a particular shape of pool.

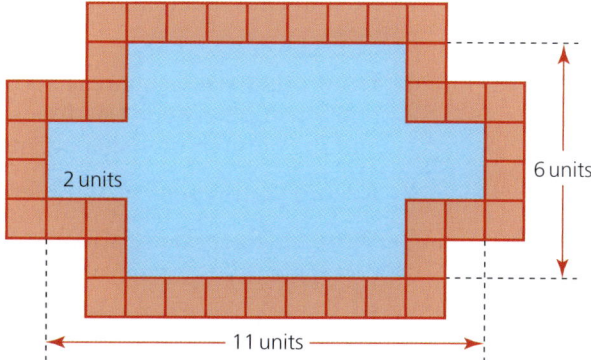

The diagram above shows a rectangular pool 11×6 units, in which a square of dimension 2×2 units is taken from each corner.

The total number of unit square tiles needed to surround the pool is 38.

The shape of the pools can be generalised as shown below:

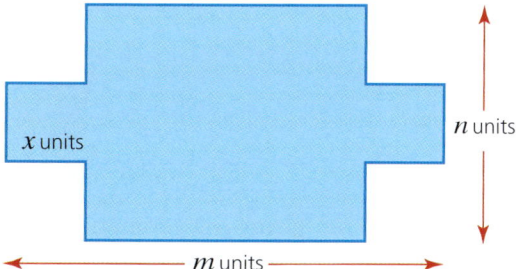

1. Investigate the number of unit square tiles needed for different-sized pools. Record your results in an ordered table.
2. From your results write an algebraic rule in terms of m, n and x for the number of tiles T needed to surround a pool.
3. Justify, in words and using diagrams, why your rule works.

MATHEMATICAL INVESTIGATIONS AND ICT 4

Tiled walls

Many cultures have used tiles to decorate buildings. Putting tiles on a wall takes skill. These days, to make sure that each tile is in the correct position, 'spacers' are used between the tiles.

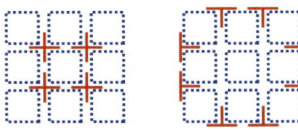

You can see from the diagrams that there are + shaped spacers (used where four tiles meet) and T shaped spacers (used at the edges of a pattern).

1 Draw other-sized squares and rectangles and investigate the relationship between the dimensions of the shape (length and width) and the number of + shaped and T shaped spacers.

2 Record your results in an ordered table.

3 Write an algebraic rule for the number of + shaped spacers c in a rectangle l tiles long by w tiles wide.

4 Write an algebraic rule for the number of T shaped spacers t in a rectangle l tiles long by w tiles wide.

ICT activity 1

In this activity, you will be using a dynamic geometry package such as Cabri or GeoGebra to demonstrate that for the triangle below:

$$\frac{AB}{ED} = \frac{AC}{EC} = \frac{BC}{DC}$$

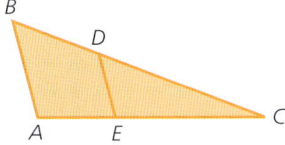

1 a Using the geometry package, construct the triangle ABC.
  b Construct the line segment ED such that it is parallel to AB. (You will need to construct a line parallel to AB first and then attach the line segment ED to it.)
  c Using a 'measurement' tool, measure each of the lengths AB, AC, BC, ED, EC and DC.
  d Using a 'calculator' tool, calculate the ratios $\frac{AB}{ED}$, $\frac{AC}{EC}$, $\frac{BC}{DC}$.

2 Comment on your answers to Question 1d.

3 a Grab vertex *B* and move it to a new position. What happens to the ratios you calculated in Question 1d?
  b Grab the vertices *A* and *C* in turn and move them to new positions. What happens to the ratios? Explain why this happens.

4 Grab point *D* and move it to a new position along the side *BC*. Explain, giving reasons, what happens to the ratios.

ICT activity 2

Using a geometry package, such as Cabri or GeoGebra, demonstrate the following angle properties of a circle:

1 The angle subtended at the centre of a circle by an arc is twice the size of the angle on the circumference subtended by the same arc. The diagram below demonstrates the construction that needs to be formed:

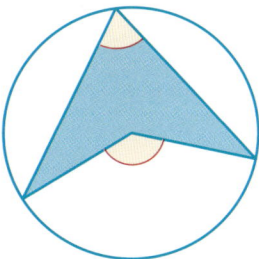

2 The angles in the same segment of a circle are equal.

3 The exterior angle of a cyclic quadrilateral is equal to the interior opposite angle.

TOPIC 5

Mensuration

Contents

Chapter 26 Measures (E5.1)
Chapter 27 Perimeter, area and volume (E5.2, E5.3, E5.4, E5.5)

> **Learning objectives**
>
> **E5.1 Units of measure**
> Use metric units of mass, length, area, volume and capacity in practical situations and convert quantities into larger or smaller units.
>
> **E5.2 Area and perimeter**
> Carry out calculations involving the perimeter and area of a rectangle, triangle, parallelogram and trapezium.
>
> **E5.3 Circles, arcs and sectors**
> 1 Carry out calculations involving the circumference and area of a circle.
> 2 Carry out calculations involving arc length and sector area as fractions of the circumference and area of a circle.
>
> **E5.4 Surface area and volume**
> Carry out calculations and solve problems involving the surface area and volume of a:
> - cuboid
> - prism
> - cylinder
> - sphere
> - pyramid
> - cone.
>
> **E5.5 Compound shapes and parts of shapes**
> 1 Carry out calculations and solve problems involving perimeters and areas of:
>   - compound shapes
>   - parts of shapes.
> 2 Carry out calculations and solve problems involving surface areas and volumes of:
>   - compound solids
>   - parts of solids.

The Egyptians

The Egyptians must have been very talented architects and surveyors to have planned the many large buildings and monuments built in that country thousands of years ago.

Evidence of the use of mathematics in Egypt in the Old Kingdom (about 2500BCE) is found on a wall near Meidum (south of Cairo); it gives guidelines for the slope of the stepped pyramid built there. The lines in the diagram are spaced at a distance of one cubit. A cubit is the distance from the tip of the finger to the elbow (about 50 cm). These lines show the use of that unit of measurement.

The earliest true mathematical documents date from about 1800BCE. The Moscow Mathematical Papyrus, the Egyptian Mathematical Leather Roll, the Kahun Papyri and the Berlin Papyrus all date to this period.

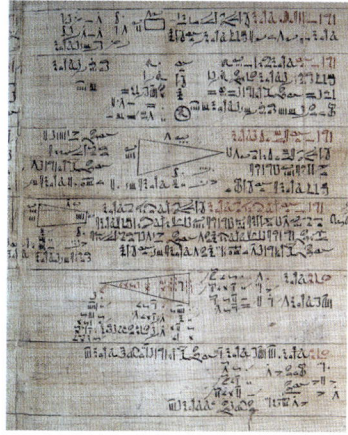

The Rhind Mathematical Papyrus

The Rhind Mathematical Papyrus, which was written in about 1650BCE, is said to be based on an older mathematical text. The Moscow Mathematical Papyrus and Rhind Mathematical Papyrus are so-called mathematical problem texts. They consist of a collection of mainly mensuration problems with solutions. These could have been written by a teacher for students to solve similar problems to the ones you will work on in this topic.

During the New Kingdom (about 1500–100BCE) papyri record land measurements. In the worker's village of Deir el-Medina, several records have been found that record volumes of dirt removed while digging the underground tombs.

26 Measures

Metric units

The metric system uses a variety of units for length, mass and capacity.
- The common **units of length** are: kilometre (km), metre (m), centimetre (cm) and millimetre (mm).
- The common **units of mass** are: tonne (t), kilogram (kg), gram (g) and milligram (mg).
- The common **units of capacity** are: litre (L or l) and millilitre (ml).

Note: 'centi' comes from the Latin *centum* meaning hundred (a centimetre is one hundredth of a metre);

'milli' comes from the Latin *mille* meaning thousand (a millimetre is one thousandth of a metre);

'kilo' comes from the Greek *khilloi* meaning thousand (a kilometre is one thousand metres).

It may be useful to have some practical experience of estimating lengths, volumes and capacities before starting the following exercises.

Exercise 26.1 Copy and complete the sentences below:

1. a There are ... centimetres in one metre.
   b There are ... millimetres in one metre.
   c One metre is one ... of a kilometre.
   d One milligram is one ... of a gram.
   e One thousandth of a litre is one ... .

2. Which of the units below would be used to measure the following?
   mm, cm, m, km, mg, g, kg, t, ml, litres
   a your height
   b the length of your finger
   c the mass of a shoe
   d the amount of liquid in a cup
   e the height of a van
   f the mass of a ship
   g the capacity of a swimming pool
   h the length of a highway
   i the mass of an elephant
   j the capacity of the petrol tank of a car

Converting from one unit to another

Length

1 km = 1000 m

Therefore 1 m = $\frac{1}{1000}$ km

1 m = 1000 mm

Therefore 1 mm = $\frac{1}{1000}$ m

1 m = 100 cm

Therefore 1 cm = $\frac{1}{100}$ m

1 cm = 10 mm

Therefore 1 mm = $\frac{1}{10}$ cm

Worked examples

a Change 5.8 km into m.
Since 1 km = 1000 m,
5.8 km is 5.8 × 1000 m
5.8 km = 5800 m

b Change 4700 mm to m.
Since 1 m is 1000 mm,
4700 mm is 4700 ÷ 1000 m
4700 mm = 4.7 m

c Convert 2.3 km into cm.
2.3 km is 2.3 × 1000 m = 2300 m
2300 m is 2300 × 100 cm
2.3 km = 230 000 cm

 **Exercise 26.2**

1 Put in the missing unit to make the following statements correct:
 a 300 ... = 30 cm
 b 6000 mm = 6 ...
 c 3.2 m = 3200 ...
 d 4.2 ... = 4200 mm
 e 2.5 km = 2500 ...

2 Convert the following to millimetres:
 a 8.5 cm
 b 23 cm
 c 0.83 m
 d 0.05 m
 e 0.004 m

3 Convert the following to metres:
 a 560 cm
 b 6.4 km
 c 96 cm
 d 0.004 km
 e 12 mm

4 Convert the following to kilometres:
 a 1150 m
 b 250 000 m
 c 500 m
 d 70 m
 e 8 m

26 MEASURES

Mass

1 tonne is 1000 kg

Therefore 1 kg = $\frac{1}{1000}$ tonne

1 kilogram is 1000 g

Therefore 1 g = $\frac{1}{1000}$ kg

1 g is 1000 mg

Therefore 1 mg = $\frac{1}{1000}$ g

→ Worked examples

a Convert 8300 kg to tonnes.
   Since 1000 kg = 1 tonne, 8300 kg is 8300 ÷ 1000 tonnes
   8300 kg = 8.3 tonnes

b Convert 2.5 g to mg.
   Since 1 g is 1000 mg, 2.5 g is 2.5 × 1000 mg
   2.5 g = 2500 mg

Exercise 26.3

1 Convert the following:
   a 3.8 g to mg
   b 28 500 kg to tonnes
   c 4.28 tonnes to kg
   d 320 mg to g
   e 0.5 tonnes to kg

Capacity

1 litre is 1000 millilitres

Therefore 1 ml = $\frac{1}{1000}$ litre

$1 \text{ m}^3 = 1000$ litres

$1 \text{ cm}^3 = 1$ ml

Exercise 26.4

1 Calculate the following and give the totals in ml:
   a 3 litres + 1500 ml
   b 0.88 litre + 650 ml
   c 0.75 litre + 6300 ml
   d 450 ml + 0.55 litre

2 Calculate the following and give the total in litres:
   a 0.75 litre + 450 ml
   b 850 ml + 490 ml
   c 0.6 litre + 0.8 litre
   d 80 ml + 620 ml + 0.7 litre

Area and volume conversions

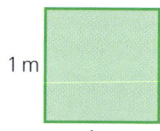

Converting between units for area and volume is not as straightforward as converting between units for length.

The diagram (left) shows a square of side length 1 m.

Area of the square = $1\,m^2$

However, if the lengths of the sides are written in cm, each of the sides are 100 cm.

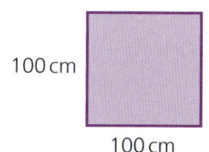

Area of the square = $100 \times 100 = 10\,000\,cm^2$

Therefore an area of $1\,m^2 = 10\,000\,cm^2$.

Similarly, a square of side length 1 cm is the same as a square of side length 10 mm. Therefore an area of $1\,cm^2$ is equivalent to an area of $100\,mm^2$.

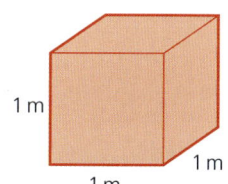

The diagram (left) shows a cube of side length 1 m.

Volume of the cube = $1\,m^3$

Once again, if the lengths of the sides are written in cm, each of the sides are 100 cm.

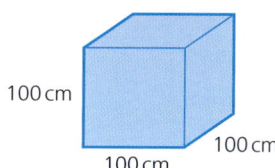

Volume of the cube = $100 \times 100 \times 100 = 1\,000\,000\,cm^3$

Therefore a volume of $1\,m^3 = 1\,000\,000\,cm^3$.

Similarly, a cube of side length 1 cm is the same as a cube of side length 10 mm.

Therefore a volume of $1\,cm^3$ is equivalent to a volume of $1000\,mm^3$.

 Exercise 26.5

1 Convert the following areas:
   a $10\,m^2$ to cm^2 b $2\,m^2$ to mm^2 c $5\,km^2$ to m^2
   d $3.2\,km^2$ to m^2 e $8.3\,cm^2$ to mm^2

2 Convert the following areas:
   a $500\,cm^2$ to m^2 b $15\,000\,mm^2$ to cm^2 c $1000\,m^2$ to km^2
   d $40\,000\,mm^2$ to m^2 e $2\,500\,000\,cm^2$ to km^2

3 Convert the following volumes:
   a $2.5\,m^3$ to cm^3 b $3.4\,cm^3$ to mm^3 c $2\,km^3$ to m^3
   d $0.2\,m^3$ to cm^3 e $0.03\,m^3$ to mm^3

4 Convert the following volumes:
   a $150\,000\,cm^3$ to m^3 b $24\,000\,mm^3$ to cm^3 c $850\,000\,m^3$ to km^3
   d $300\,mm^3$ to cm^3 e $15\,cm^3$ to m^3

5 Convert the following volumes and capacities:
   a 1.2 litres to cm^3 b $0.5\,m^3$ to litres
   c 4250 ml to cm^3 d 220 litres to m^3

6 A water tank in the shape of a cuboid has dimensions $120 \times 80 \times 50\,cm$. It is 75% full. How many litres of water are in the tank?

353

26 MEASURES

Student assessment 1

1. Convert the following lengths into the units indicated:
   a 2.6 cm to mm
   b 0.88 m to cm
   c 6800 m to km
   d 0.875 km to m

2. Convert the following masses into the units indicated:
   a 4.2 g to mg
   b 3940 g to kg
   c 4.1 kg to g
   d 0.72 tonnes to kg

3. Convert the following liquid measures into the units indicated:
   a 1800 ml to litres
   b 0.083 litre to ml
   c 3.2 litres to ml
   d 250 000 ml to litres

4. Convert the following areas:
   a 56 cm² to mm²
   b 2.05 m² to cm²

5. Convert the following volumes:
   a 8670 cm³ to m³
   b 444 000 cm³ to m³

Student assessment 2

1. Convert the following lengths into the units indicated:
   a 3100 mm to cm
   b 6.4 km to m
   c 0.4 cm to mm
   d 460 mm to cm

2. Convert the following masses into the units indicated:
   a 3.6 mg to g
   b 550 mg to g
   c 6500 g to kg
   d 1510 kg to tonnes

3. Convert the following measures of capacity to the units indicated:
   a 3400 ml to litres
   b 6.7 litres to ml
   c 0.73 litre to ml
   d 300 000 ml to litres

4. Convert the following areas:
   a 0.03 m² to mm²
   b 0.005 km² to m²

5. Convert the following volumes:
   a 100 400 cm³ to m³
   b 5005 m³ to km³

27 Perimeter, area and volume

The perimeter and area of a rectangle

The **perimeter** of a shape is the distance around the outside of the shape. Perimeter can be measured in mm, cm, m, km, etc.

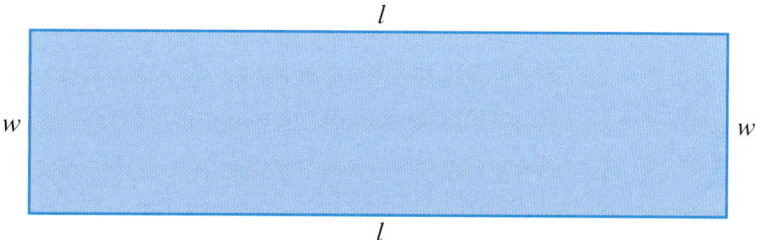

The perimeter of the rectangle above of length l and width w is therefore:

Perimeter $= l + w + l + w$

This can be rearranged to give:

Perimeter $= 2l + 2w$

This in turn can be factorised to give:

Perimeter $= 2(l + w)$

The **area** of a shape is the amount of surface that it covers.

Area is measured in mm^2, cm^2, m^2, km^2, etc.

The area A of the rectangle above is given by the formula:

$A = lw$

➜ Worked example

Calculate the width of a rectangle of area $200\,cm^2$ and length $25\,cm$.

$A = lw$
$200 = 25w$
$w = 8$

So the width is $8\,cm$.

27 PERIMETER, AREA AND VOLUME

The area of a triangle

Rectangle *ABCD* has a triangle *CDE* drawn inside it.

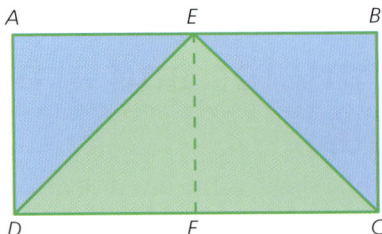

Point *E* is said to be a **vertex** of the triangle.

EF is the **height** or **altitude** of the triangle.

CD is the **length** of the rectangle, but is called the **base** of the triangle.

It can be seen from the diagram that triangle *DEF* is half the area of the rectangle *AEFD*.

Also triangle *CFE* is half the area of rectangle *EBCF*.

It follows that triangle *CDE* is half the area of rectangle *ABCD*.

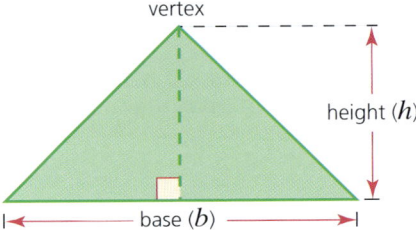

The **area of a triangle** $A = \frac{1}{2}bh$, where *b* is the base and *h* is the height.

Note: It does not matter which side is called the base, but the height must be measured at right angles from the base to the opposite vertex.

 **Exercise 27.1**

1 Calculate the areas of the triangles below:

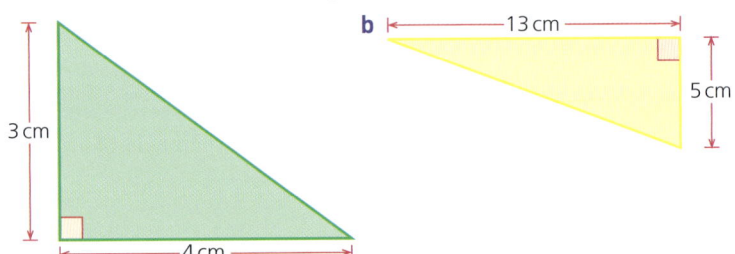

The area of a triangle

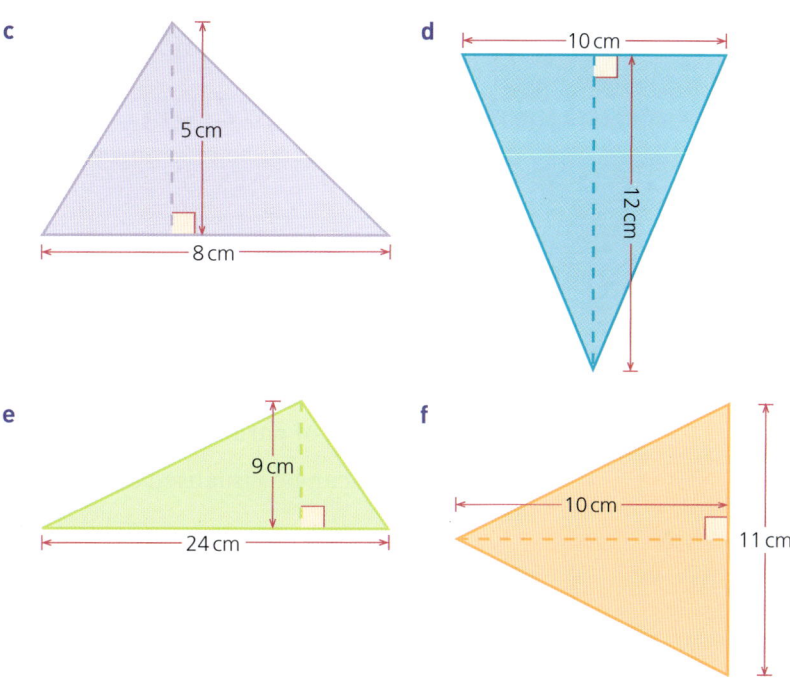

2   Calculate the areas of the shapes below:

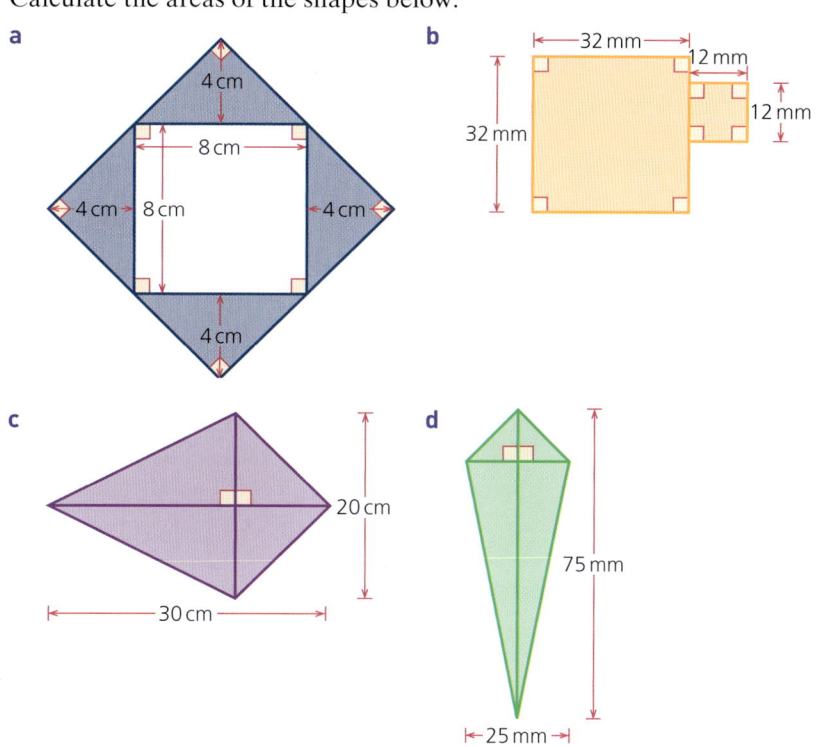

27 PERIMETER, AREA AND VOLUME

The area of a parallelogram and a trapezium

A **parallelogram** can be rearranged to form a rectangle as shown below:

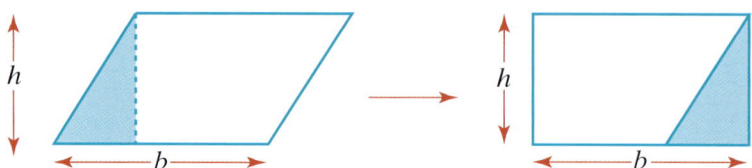

Therefore: **area of parallelogram** = base length × perpendicular height.

A **trapezium** can be visualised as being split into two triangles as shown:

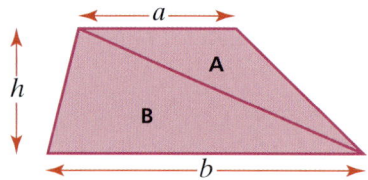

Area of triangle A = $\frac{1}{2} \times a \times h$

Area of triangle B = $\frac{1}{2} \times b \times h$

Area of the trapezium

= area of triangle A + area of triangle B

= $\frac{1}{2} ah + \frac{1}{2} bh$

= $\frac{1}{2} h(a + b)$

➡ Worked examples

a   Calculate the area of the parallelogram.

Area = base length × perpendicular height

= 8 × 6

= 48 cm²

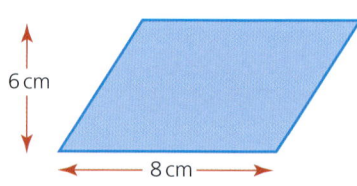

The area of a parallelogram and a trapezium

b Calculate the shaded area in the shape (below).

Area of rectangle = 12 × 8
= 96 cm²

Area of trapezium = $\frac{1}{2}$ × 5(3 + 5)
= 2.5 × 8
= 20 cm²

Shaded area = 96 − 20
= 76 cm²

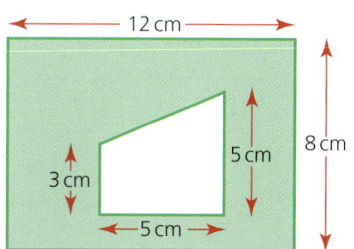

Exercise 27.2 Find the area of each of the following shapes:

1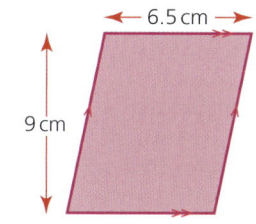

2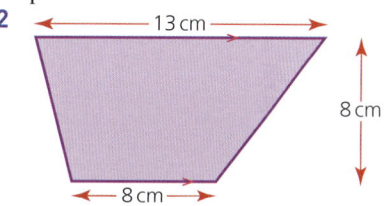

3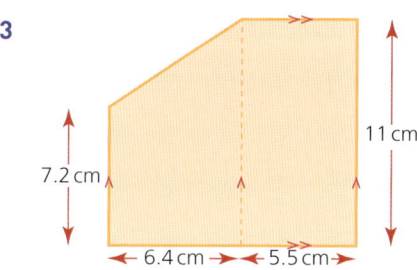

4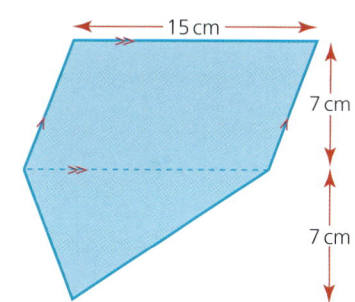

Exercise 27.3

1 Calculate *a*.

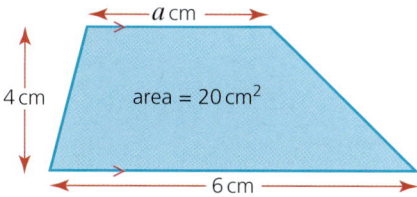

2 If the areas of this trapezium and parallelogram are equal, calculate *x*.

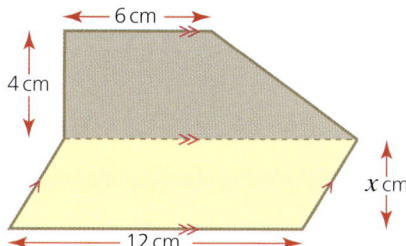

359

27 PERIMETER, AREA AND VOLUME

Exercise 27.3 (cont)

3 The end view of a house is as shown in the diagram (below). If the door has a width and height of 0.75 m and 2 m respectively, calculate the area of brickwork.

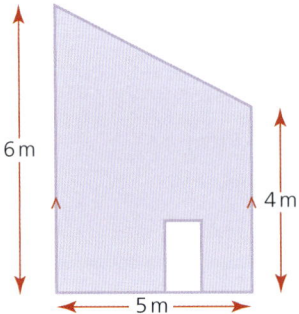

4 A garden in the shape of a trapezium is split into three parts: a flower bed in the shape of a triangle, a flower bed in the shape of a parallelogram and a section of grass in the shape of a trapezium, as shown (right). The area of the grass is two and a half times the total area of flower beds. Calculate:

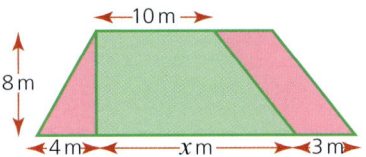

   a the area of each flower bed,
   b the area of grass,
   c the value of x.

The circumference and area of a circle

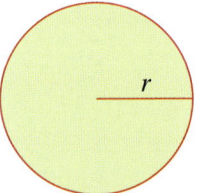 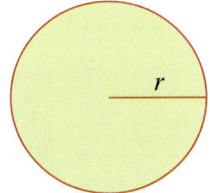

The circumference is $2\pi r$.   The area is πr^2.
   $C = 2\pi r$ $A = \pi r^2$

Note

You should always use the π button on your calculator unless a question says otherwise.

→ **Worked examples**

a Calculate the circumference of this circle, giving your answer to 3 s.f.

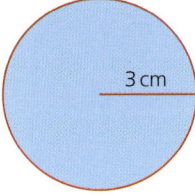

   $C = 2\pi r$
   $ = 2\pi \times 3 = 18.8$

   The circumference is 18.8 cm.

The circumference and area of a circle

Note: The answer 18.8 cm is only correct to 3 s.f. and therefore only an approximation. An **exact** answer involves leaving the answer in terms of π.

i.e. $C = 2\pi r$
$= 2\pi \times 3$
$= 6\pi$ cm

b If the circumference of this circle is 12 cm, calculate the radius, giving your answer
   i to 3 s.f.
   ii in terms of π

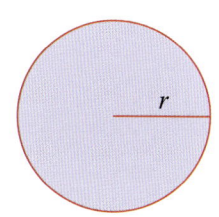

   i $C = 2\pi r$

$$r = \frac{12}{2\pi}$$
$$= 1.91$$

The radius is 1.91 cm.

   ii $r = \dfrac{C}{2\pi} = \dfrac{12}{2\pi}$

$$= \frac{6}{\pi} \text{ cm}$$

c Calculate the area of this circle, giving your answer
   i to 3 s.f.
   ii in exact form

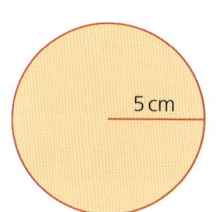

   i $A = \pi r^2$
      $= \pi \times 5^2 = 78.5$

The area is 78.5 cm².
   ii $A = \pi r^2$
      $= \pi \times 5^2$
      $= 25\pi$ cm²

d The area of a circle is 34 cm², calculate the radius, giving your answer
   i to 3 s.f.
   ii in terms of π
   i $A = \pi r^2$

$$r = \sqrt{\frac{A}{\pi}}$$
$$r = \sqrt{\frac{34}{\pi}} = 3.29$$

The radius is 3.29 cm.

   ii $r = \sqrt{\dfrac{A}{\pi}} = \sqrt{\dfrac{34}{\pi}}$ cm

27 PERIMETER, AREA AND VOLUME

Exercise 27.4

1. Calculate the circumference of each circle, giving your answer to 3 s.f.

   a 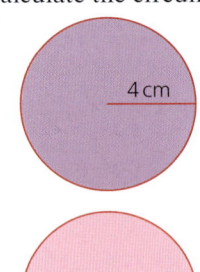   b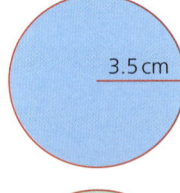

   c 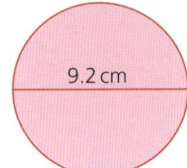   d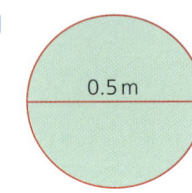

2. Calculate the area of each of the circles in Question 1. Give your answers to 3 s.f.

3. Calculate the radius of a circle when the circumference is:
   a 15 cm   b π cm
   c 4 m   d 8 mm

4. Calculate the diameter of a circle when the area is:
   a 16 cm²   b 9π cm²
   c 8.2 m²   d 14.6 mm²

Exercise 27.5

1. The wheel of a car has an outer radius of 25 cm. Calculate:
   a how far the car has travelled after one complete turn of the wheel,
   b how many times the wheel turns for a journey of 1 km.

2. The back wheel of a wheelchair has a diameter of 60 cm. Calculate how far a the wheelchair has travelled after it has rotated 100 times.

3. A circular ring has a cross-section as shown. If the outer radius is 22 mm and the inner radius 20 mm, calculate the cross-sectional area of the ring. Give your answer in terms of π.

   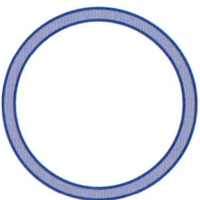

4. Four circles are drawn in a line and enclosed by a rectangle as shown (below). If the radius of each circle is 3 cm, calculate the unshaded area within the rectangle. Give your answer in exact form.

The surface area of a cuboid and a cylinder

5 A garden is made up of a rectangular patch of grass and two semicircular vegetable patches. If the dimensions of the rectangular patch are 16 m (length) and 8 m (width) respectively, calculate in exact form:
  a the perimeter of the garden,
  b the total area of the garden.

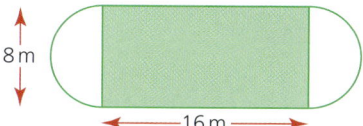

The surface area of a cuboid and a cylinder

To calculate the surface area of a **cuboid**, start by looking at its individual faces. These are either squares or rectangles. The surface area of a cuboid is the sum of the areas of its faces.

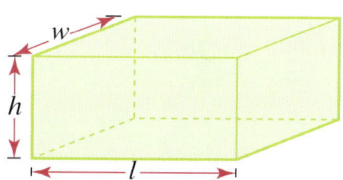

Area of top = wl Area of bottom = wl
Area of front = lh Area of back = lh
Area of one side = wh Area of other side = wh
Total surface area
  = $2wl + 2lh + 2wh$
  = $2(wl + lh + wh)$

For the surface area of a **cylinder**, it is best to visualise the net of the solid: it is made up of one rectangular piece and two circular pieces.

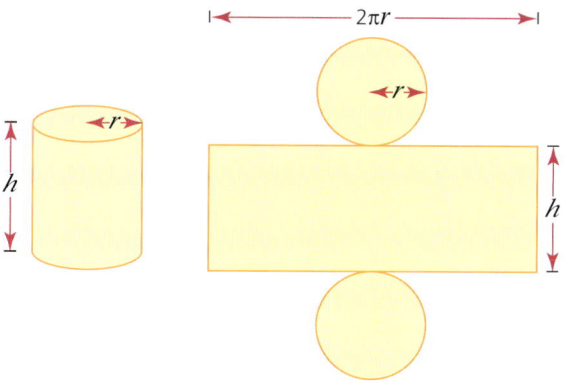

Area of circular pieces = $2 \times \pi r^2$
Area of rectangular piece = $2\pi r \times h$
Total surface area = $2\pi r^2 + 2\pi rh$
          = $2\pi r(r + h)$

27 PERIMETER, AREA AND VOLUME

Worked examples

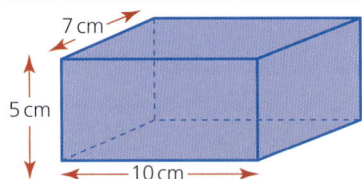

a Calculate the surface area of the cuboid shown.

Total area of top and bottom = $2 \times 7 \times 10 = 140\,\text{cm}^2$

Total area of front and back = $2 \times 5 \times 10 = 100\,\text{cm}^2$

Total area of both sides = $2 \times 5 \times 7 = 70\,\text{cm}^2$

Total surface area = $310\,\text{cm}^2$

b If the height of a cylinder is 7 cm and the radius of its circular top is 3 cm, calculate its surface area.

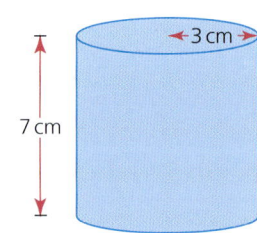

Total surface area = $2\pi r(r + h)$

$= 2\pi \times 3 \times (3 + 7)$

$= 6\pi \times 10$

$= 60\pi$

$= 188\,\text{cm}^2$ (3 s.f.)

The total surface area is $188\,\text{cm}^2$.

If the answer is left in terms of π, it is an exact answer. The answer of 188 cm² is only an approximation correct to three significant figures.

Exercise 27.6

1 Calculate the surface area of each of the following cuboids:
  a $l = 12\,\text{cm}$, $w = 10\,\text{cm}$, $h = 5\,\text{cm}$
  b $l = 4\,\text{cm}$, $w = 6\,\text{cm}$, $h = 8\,\text{cm}$
  c $l = 4.2\,\text{cm}$, $w = 7.1\,\text{cm}$, $h = 3.9\,\text{cm}$
  d $l = 5.2\,\text{cm}$, $w = 2.1\,\text{cm}$, $h = 0.8\,\text{cm}$

2 Calculate the height of each of the following cuboids:
  a $l = 5\,\text{cm}$, $w = 6\,\text{cm}$, surface area = $104\,\text{cm}^2$
  b $l = 2\,\text{cm}$, $w = 8\,\text{cm}$, surface area = $112\,\text{cm}^2$
  c $l = 3.5\,\text{cm}$, $w = 4\,\text{cm}$, surface area = $118\,\text{cm}^2$
  d $l = 4.2\,\text{cm}$, $w = 10\,\text{cm}$, surface area = $226\,\text{cm}^2$

3 Calculate the surface area of each of the following cylinders.
  i Give your answer in terms of π.
  ii Give your answer correct to 3 s.f.
    a $r = 2\,\text{cm}, h = 6\,\text{cm}$ **b** $r = 4\,\text{cm}, h = 7\,\text{cm}$
    c $r = 3.5\,\text{cm}, h = 9.2\,\text{cm}$ **d** $r = 0.8\,\text{cm}, h = 4.3\,\text{cm}$

4 Calculate the height of each of the following cylinders. Give your answers to 1 d.p.
  a $r = 2.0\,\text{cm}$, surface area = $40\,\text{cm}^2$
  b $r = 3.5\,\text{cm}$, surface area = $88\,\text{cm}^2$
  c $r = 5.5\,\text{cm}$, surface area = $250\,\text{cm}^2$
  d $r = 3.0\,\text{cm}$, surface area = $189\,\text{cm}^2$

Exercise 27.7

1. Two cubes are placed next to each other. The length of each of the edges of the larger cube is 4 cm. If the ratio of their surface areas is 1:4, calculate:
   a the surface area of the small cube,
   b the length of an edge of the small cube.

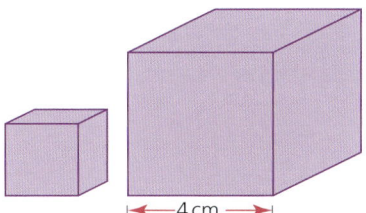

2. A cube and a cylinder have the same surface area. If the cube has an edge length of 6 cm and the cylinder a radius of 2 cm, calculate:
   a the surface area of the cube,
   b the height of the cylinder.

3. Two cylinders have the same surface area.
   The shorter of the two has a radius of 3 cm and a height of 2 cm, and the taller cylinder has a radius of 1 cm. Calculate:

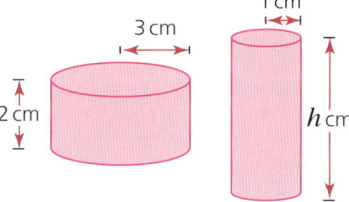

   a the surface area of one of the cylinders in terms of π,
   b the height of the taller cylinder.

4. Two cuboids have the same surface area. The dimensions of one of them are: length = 3 cm, width = 4 cm and height = 2 cm.
   Calculate the height of the other cuboid if its length is 1 cm and its width is 4 cm.

The volume and surface area of a prism

A prism is any three-dimensional object which has a constant cross-sectional area. Below are a few examples of some of the more common types of prism.

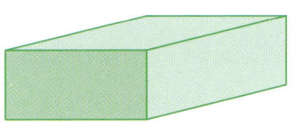
Rectangular prism (cuboid)

Circular prism (cylinder)

Triangular prism

27 PERIMETER, AREA AND VOLUME

When each of the shapes is cut parallel to the shaded face, the cross-section is constant and the shape is therefore classified as a prism.

Volume of a prism = area of cross-section × length

Surface area of a prism = sum of the areas of each of its faces

→ Worked examples

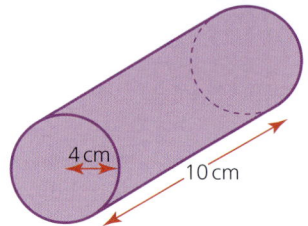

a Calculate the volume of the cylinder in the diagram.

Volume = cross-sectional area × length

$= \pi \times 4^2 \times 10$

Volume = 503 cm³ (3 s.f.)

As an exact value, the volume would be left as 160π cm³

b Calculate the
   i   volume and
   ii surface area of the 'L' shaped prism in the diagram.

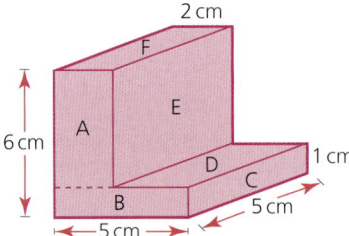

The cross-sectional area can be split into two rectangles:

i Area of rectangle A = 5 × 2
                      = 10 cm²
   Area of rectangle B = 5 × 1
                      = 5 cm²

Total cross-sectional area = (10 cm² + 5 cm²) = 15 cm²

Volume of prism = 15 × 5
               = 75 cm³

The volume and surface area of a prism

ii Area of rectangle A = 5 × 2 = 10 cm²
    Area of rectangle B = 5 × 1 = 5 cm²
    Area of rectangle C = 5 × 1 = 5 cm²
    Area of rectangle D = 3 × 5 = 15 cm²
    Area of rectangle E = 5 × 5 = 25 cm²
    Area of rectangle F = 2 × 5 = 10 cm²
    Area of back is the same as A + B = 15 cm²
    Area of left face is the same as C + E = 30 cm²
    Area of base = 5 × 5 = 25 cm²
    Total surface area = 10 + 5 + 5 + 15 + 25 + 10 + 15 + 30 + 25 = 140 cm²

Exercise 27.8

1 Calculate the volume of each of the following cuboids, where w, l and h represent the width, length and height respectively.
   a $w = 2$ cm, $l = 3$ cm, $h = 4$ cm
   b $w = 6$ cm, $l = 1$ cm, $h = 3$ cm
   c $w = 6$ cm, $l = 23$ mm, $h = 2$ cm
   d $w = 42$ mm, $l = 3$ cm, $h = 0.007$ m

2 Calculate the volume of each of the following cylinders, where r represents the radius of the circular face and h the height of the cylinder.
   a $r = 4$ cm, $h = 9$ cm
   b $r = 3.5$ cm, $h = 7.2$ cm
   c $r = 25$ mm, $h = 10$ cm
   d $r = 0.3$ cm, $h = 17$ mm

3 Calculate the volume and total surface area of each of the following right-angled triangular prisms.

   a   b

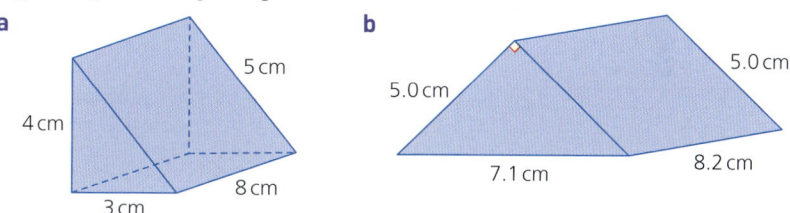

4 The diagram below shows the net of a right-angled triangular prism.

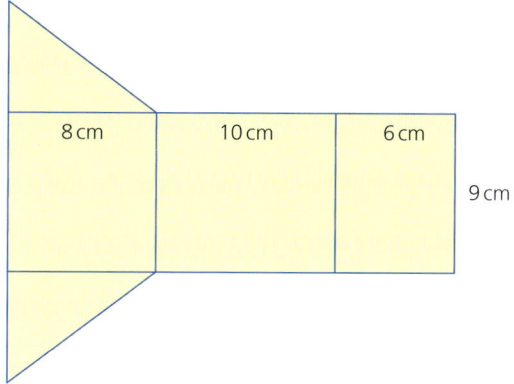

367

27 PERIMETER, AREA AND VOLUME

Exercise 27.8 (cont)

Calculate:
a The surface area of the prism.
b The volume of the prism.

5 Calculate the volume of each of the following prisms. All dimensions are given in centimetres.

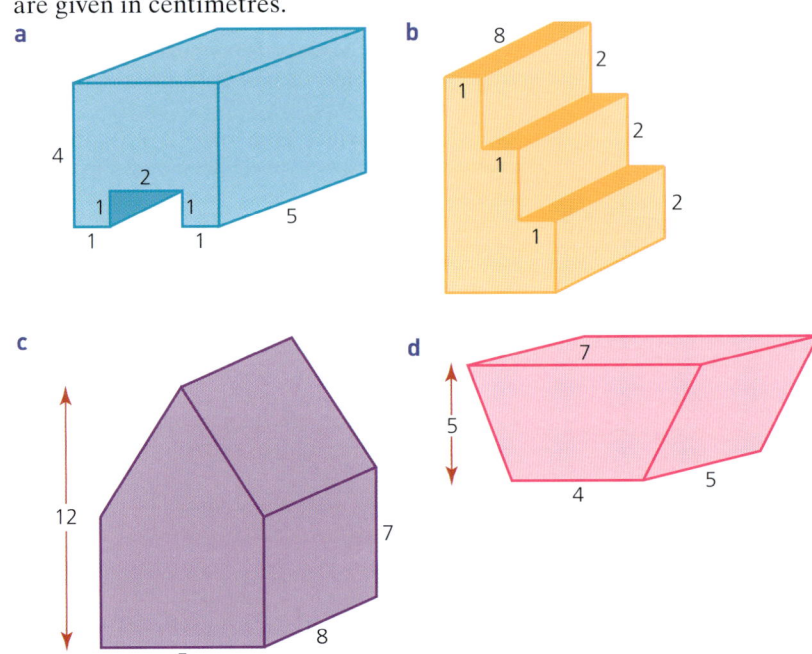

Exercise 27.9

1 The diagram shows a plan view of a cylinder inside a box the shape of a cube. If the radius of the cylinder is 8 cm, calculate the percentage volume of the cube not occupied by the cylinder.

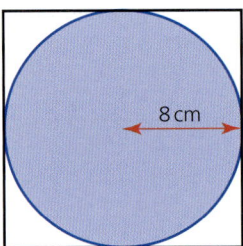

2 A chocolate bar is made in the shape of a triangular prism. The triangular face of the prism is equilateral and has an edge length of 4 cm and a perpendicular height of 3.5 cm. The manufacturer also sells these in special packs of six bars arranged as a hexagonal prism. If the prisms are 20 cm long, calculate:

Arc length

  a the cross-sectional area of the pack,
  b the volume of the pack.

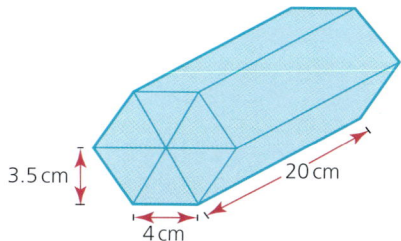

3 A cuboid and a cylinder have the same volume. The radius and height of the cylinder are 2.5 cm and 8 cm respectively. If the length and width of the cuboid are each 5 cm, calculate its height to 1 d.p.

4 A section of steel pipe is shown in the diagram. The inner radius is 35 cm and the outer radius 36 cm.

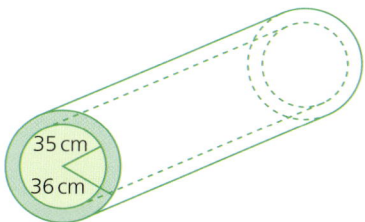

Calculate the volume of steel used in making the pipe if it has a length of 130 m. Give your answer in terms of π.

Arc length

An **arc** is part of the circumference of a circle between two radii.

Its length is proportional to the size of the angle x between the two radii. The length of the arc as a fraction of the circumference of the whole circle is therefore equal to the fraction that x is of $360°$.

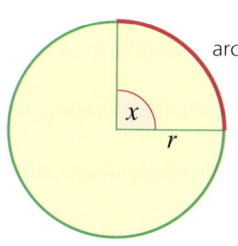

Arc length = $\frac{x}{360} \times 2\pi r$

→ Worked examples

a Find the length of the minor arc in the circle.
   i   Give your answer to 3 s.f.
       Arc length = $\frac{80}{360} \times 2 \times \pi \times 6$
       = 8.38 cm
   ii Give your answer in terms of π
       Arc length = $\frac{80}{360} \times 2 \times \pi \times 6$
       = $\frac{8}{3} \pi$ cm

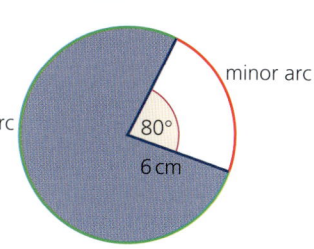

27 PERIMETER, AREA AND VOLUME

> **Note**
>
> The Core syllabus only requires calculations involving factors of 360.

b In the circle the length of the minor arc is 12.4 cm and the radius is 7 cm.
  i Calculate the angle x.
  Arc length $= \frac{x}{360} \times 2\pi r$

  $12.4 = \frac{x}{360} \times 2 \times \pi \times 7$

  $\frac{12.4 \times 360}{2 \times \pi \times 7} = x$

  $= 101.5°$ (1 d.p.)

  ii Calculate the length of the major arc.
  $C = 2\pi r$
  $= 2 \times \pi \times 7 = 44.0$ cm (3 s.f.)
  Major arc = circumference − minor arc = (44.0 − 12.4) = 31.6 cm

Exercise 27.10

1 For each of the following, give the length of the arc to 3 s.f. O is the centre of the circle.

a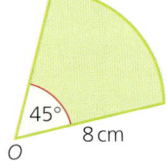

b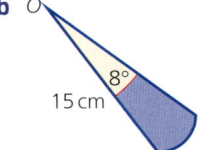

c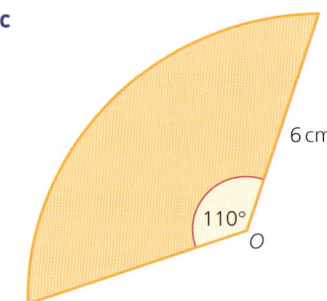

d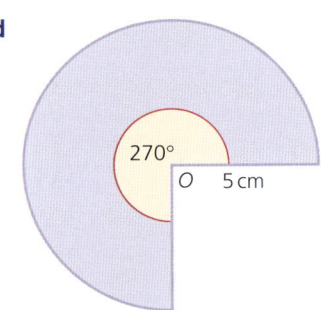

2 A sector is the region of a circle enclosed by two radii and an arc. Calculate the angle x for each of the following sectors. The radius r and arc length a are given in each case.
  a $r = 14$ cm, $a = 8$ cm
  b $r = 4$ cm, $a = 16$ cm
  c $r = 7.5$ cm, $a = 7.5$ cm
  d $r = 6.8$ cm, $a = 13.6$ cm

3 Calculate the radius r for each of the following sectors. The angle x and arc length a are given in each case.
  a $x = 75°$, $a = 16$ cm
  b $x = 300°$, $a = 24$ cm
  c $x = 20°$, $a = 6.5$ cm
  d $x = 243°$, $a = 17$ cm

The area of a sector

Exercise 27.11

1 Calculate the perimeter of each of these shapes. Give your answers in exact form.

   a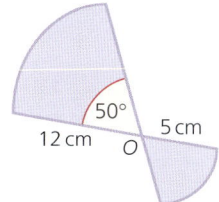

   b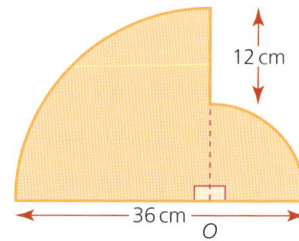

2 A shape is made from two sectors arranged in such a way that they share the same centre.
   The radius of the smaller sector is 7 cm and the radius of the larger sector is 10 cm. If the angle at the centre of the smaller sector is 30° and the arc length of the larger sector is 12 cm, calculate:
   a   the arc length of the smaller sector,
   b   the total perimeter of the two sectors,
   c   the angle at the centre of the larger sector.

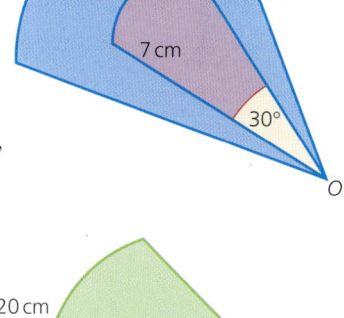

3 For the diagram (right), calculate:
   a   the radius of the smaller sector,
   b   the perimeter of the shape,
   c   the angle x.

The area of a sector

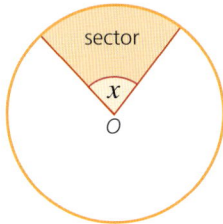

A **sector** is the region of a circle enclosed by two radii and an arc. Its area is proportional to the size of the angle x between the two radii. The area of the sector as a fraction of the area of the whole circle is therefore equal to the fraction that x is of 360°.

Area of sector $= \frac{x}{360} \times \pi r^2$

27 PERIMETER, AREA AND VOLUME

> **Worked examples**

a Calculate the area of the sector, giving your answer
   i to 3 s.f.   ii in terms of π

   i Area = $\frac{45}{360} \times \pi \times 12^2$

      = 56.5 cm²

   ii Area = 18π cm²

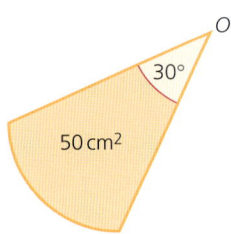

b Calculate the radius of the sector, giving your answer to 3 s.f.

   Area = $\frac{x}{360} \times \pi r^2$

   $50 = \frac{30}{360} \times \pi \times r^2$

   $\frac{50 \times 360}{30\pi} = r^2$

   r = 13.8
   The radius is 13.8 cm.

Exercise 27.12

1 Calculate the area of each of the following sectors, using the values of the angle x and radius r in each case.
   a $x = 60°$, $r = 8$ cm
   b $x = 120°$, $r = 14$ cm
   c $x = 2°$, $r = 18$ cm
   d $x = 320°$, $r = 4$ cm

2 Calculate the radius for each of the following sectors, using the values of the angle x and the area A in each case.
   a $x = 40°$, $A = 120$ cm²
   b $x = 12°$, $A = 42$ cm²
   c $x = 150°$, $A = 4$ cm²
   d $x = 300°$, $A = 400$ cm²

3 Calculate the value of the angle x, to the nearest degree, for each of the following sectors, using the values of r and A in each case.
   a $r = 12$ cm, $A = 60$ cm²
   b $r = 26$ cm, $A = 0.02$ m²
   c $r = 0.32$ m, $A = 180$ cm²
   d $r = 38$ mm, $A = 16$ cm²

Exercise 27.13

1 A rotating sprinkler is placed in one corner of a garden (below). If it has a reach of 8 m and rotates through an angle of 30°, calculate the area of garden not being watered. Give your answer in terms of π.

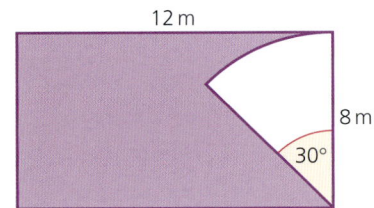

The area of a sector

2 Two sectors AOB and COD share the same centre O. The area of AOB is three times the area of COD. Calculate:
  a the area of sector AOB,
  b the area of sector COD,
  c the radius r cm of sector COD.

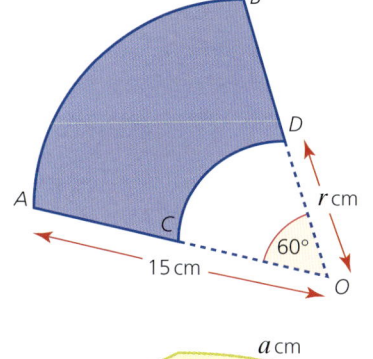

3 A circular cake is cut. One of the slices is shown. Calculate:
  a the length a cm of the arc,
  b the total surface area of all the sides of the slice,
  c the volume of the slice.

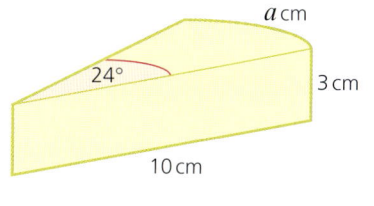

4 The diagram shows a plan view of four tiles in the shape of sectors placed in the bottom of a box. C is the midpoint of the arc AB and intersects the chord AB at point D. If the length OB is 10 cm, calculate:
  a the length OD,
  b the length CD,
  c the area of the sector AOB,
  d the length and width of the box,
  e the area of the base of the box not covered by the tiles.

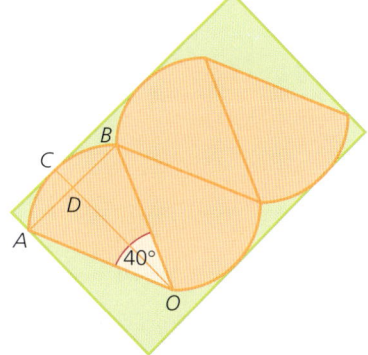

5 The tiles in Question 4 are repackaged and are now placed in a box, the base of which is a parallelogram. Given that C and F are the midpoints of arcs AB and OG respectively, calculate:
  a the angle OCF,
  b the length CE,
  c the length of the sides of the box,
  d the area of the base of the box not covered by the tiles.

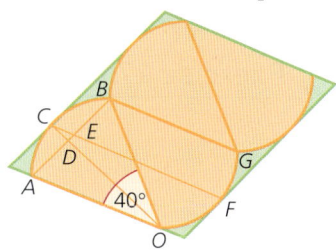

27 PERIMETER, AREA AND VOLUME

You should know how to use this formula for the volume of a sphere, but you do not need to memorise it.

The volume of a sphere

Volume of sphere = $\frac{4}{3}\pi r^3$

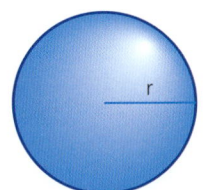

Worked examples

a Calculate the volume of the sphere, giving your answer:
  i to 3 s.f.   **ii** in terms of π

  i Volume of sphere = $\frac{4}{3}\pi r^3$
  $= \frac{4}{3} \times \pi \times 3^3$
  $= 113.1$
  The volume is 113 cm³.

  ii Volume of sphere = $\frac{4}{3}\pi \times 3^3$
  $= 36\pi$ cm³

b Given that the volume of a sphere is 150 cm³, calculate its radius to 3 s.f.
  $V = \frac{4}{3}\pi r^3$
  $r^3 = \frac{3V}{4\pi}$
  $r^3 = \frac{3 \times 150}{4 \times \pi}$
  $r = \sqrt[3]{35.8} = 3.30$
  The radius is 3.30 cm.

Exercise 27.14

1 Calculate the volume of each of the following spheres. The radius *r* is given in each case.
  a $r = 6$ cm   **b** $r = 9.5$ cm
  c $r = 8.2$ cm   **d** $r = 0.7$ cm

2 Calculate the radius of each of the following spheres. Give your answers in centimetres and to 1 d.p. The volume *V* is given in each case.
  a $V = 130$ cm³   **b** $V = 720$ cm³
  c $V = 0.2$ m³   **d** $V = 1000$ mm³

Exercise 27.15

1 Given that sphere B has twice the volume of sphere A, calculate the radius of sphere B. Give your answer to 1 d.p.

 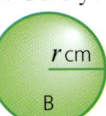

The surface area of a sphere

Note

A hemisphere is half a sphere.

2 Calculate the volume of material used to make the hemispherical bowl if the inner radius of the bowl is 5 cm and its outer radius 5.5 cm. Give your answer in terms of π.

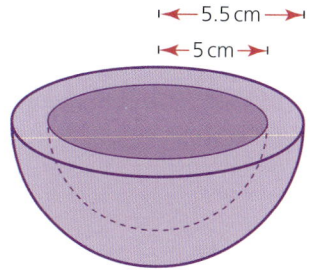

3 The volume of the material used to make the sphere and hemispherical bowl are the same. Given that the radius of the sphere is 7 cm and the inner radius of the bowl is 10 cm, calculate, to 1 d.p., the outer radius r cm of the bowl.

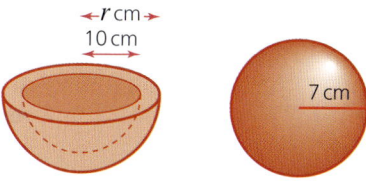

4 A steel ball is melted down to make eight smaller identical balls. If the radius of the original steel ball was 20 cm, calculate the radius of each of the smaller balls.

5 A steel ball of volume 600 cm³ is melted down and made into three smaller balls A, B and C. If the volumes of A, B and C are in the ratio 7:5:3, calculate to 1 d.p. the radius of each of A, B and C.

6 The cylinder and sphere shown have the same radius and the same height. Calculate the ratio of their volumes, giving your answer in the form volume of cylinder : volume of sphere.

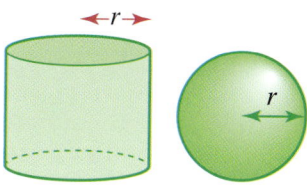

You should know how to use this formula for the surface area of a sphere, but you do not need to memorise it.

The surface area of a sphere

Surface area of sphere = $4\pi r^2$

Exercise 27.16

1 Calculate the surface area of each of the following spheres when:
   a $r = 6$ cm   b $r = 4.5$ cm
   c $r = 12.25$ cm d $r = \frac{1}{\sqrt{\pi}}$ cm

2 Calculate the radius of each of the following spheres, given the surface area in each case.
   a $A = 50$ cm²   b $A = 16.5$ cm²
   c $A = 120$ mm²   d $A = \pi$ cm²

27 PERIMETER, AREA AND VOLUME

3 Sphere A has a radius of 8 cm and sphere B has a radius of 16 cm. Calculate the ratio of their surface areas in the form $1:n$.

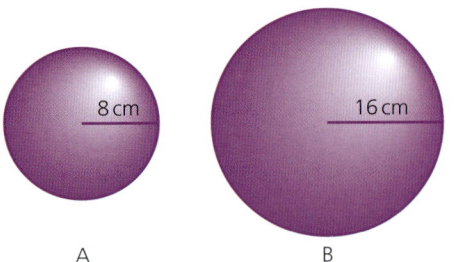

4 A hemisphere of diameter 10 cm is attached to a cylinder of equal diameter as shown. If the total length of the shape is 20 cm, calculate the surface area of the whole shape.

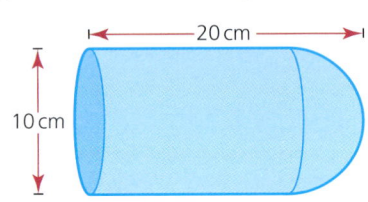

5 A sphere and a cylinder both have the same surface area and the same height of 16 cm. Calculate the radius of the cylinder.

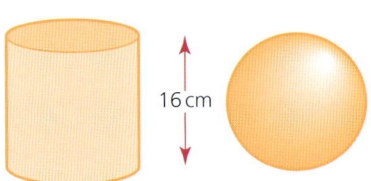

The volume of a pyramid

A pyramid is a three-dimensional shape in which each of its faces must be plane. A pyramid has a polygon for its base and the other faces are triangles with a common vertex, known as the **apex**. Its individual name is taken from the shape of the base.

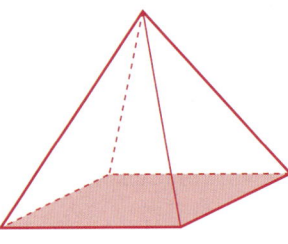

Square-based pyramid

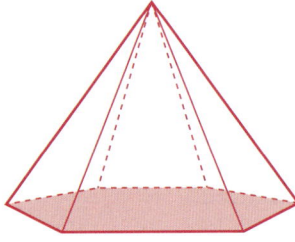

Hexagonal-based pyramid

Volume of any pyramid
$= \frac{1}{3} \times$ area of base $\times$ perpendicular height

You should know how to use this formula for the volume of a pyramid, but you do not need to memorise it.

With a pyramid, if a cut is made parallel to the base and the top part of the pyramid removed, the shape that is left is known as a '**frustum**' or a '**truncated pyramid**'.

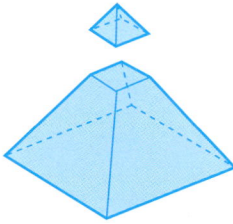

The volume of a pyramid

> **Worked examples**

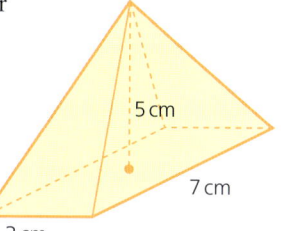

a A rectangular-based pyramid has a perpendicular height of 5 cm and base dimensions as shown. Calculate the volume of the pyramid.

Volume = $\frac{1}{3}$ × base area × height

= $\frac{1}{3}$ × 3 × 7 × 5 = 35

The volume is 35 cm³.

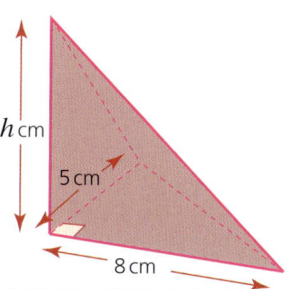

b The pyramid shown has a volume of 60 cm³. Calculate its perpendicular height h cm.

Volume = $\frac{1}{3}$ × base area × height

Height = $\frac{3 \times \text{volume}}{\text{base area}}$

$h = \dfrac{3 \times 60}{\frac{1}{2} \times 8 \times 5}$

$h = 9$

The height is 9 cm.

Exercise 27.17

Find the volume of each of the following pyramids:

1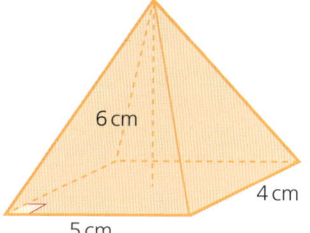

2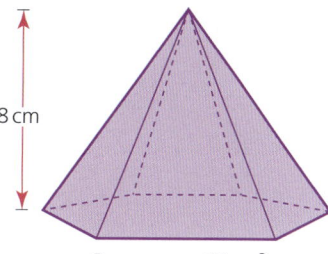

3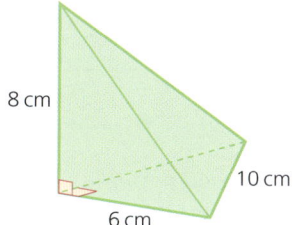

4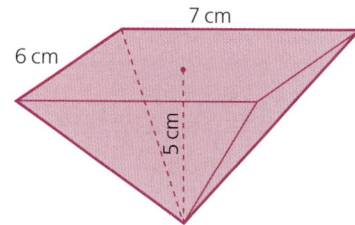

27 PERIMETER, AREA AND VOLUME

Exercise 27.18

1. Calculate the perpendicular height h cm for the pyramid, given that it has a volume of 168 cm³.

   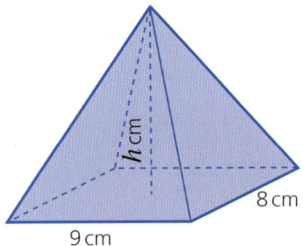

2. Calculate the length of the edge marked x cm, given that the volume of the pyramid is 14 cm³.

   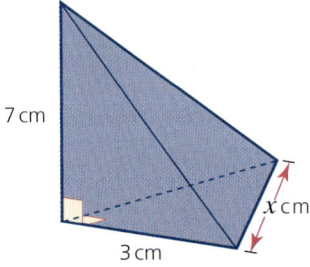

3. The top of a triangular-based pyramid is cut off. The cut is made parallel to the base. If the vertical height of the top is 6 cm, calculate:
   a. the height of the a truncated pyramid,
   b. the volume of the small pyramid,
   c. the volume of the original pyramid.

   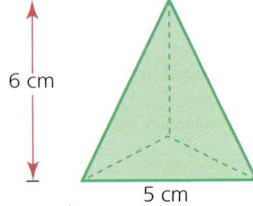
   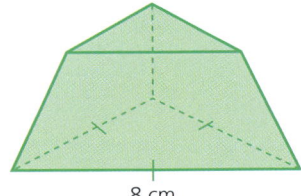

4. The top of a square-based pyramid (right) is cut off. The cut is made parallel to the base. If the base of the smaller pyramid has a side length of 3 cm and the vertical height of the truncated pyramid is 6 cm, calculate:
   a. the height of the original pyramid,
   b. the volume of the original pyramid,
   c. the volume of the truncated pyramid.

   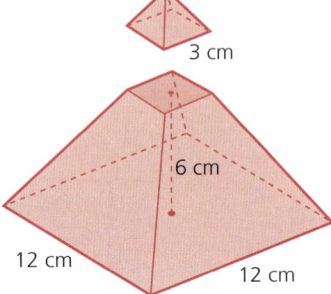

The surface area of a pyramid

The surface area of a pyramid is found simply by adding together the areas of all of its faces.

Exercise 27.19

1. Calculate the surface area of a regular tetrahedron with edge length 2 cm.

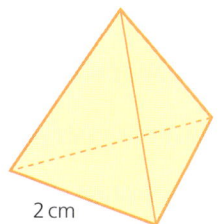

2. The rectangular-based pyramid shown has a sloping edge length of 12 cm. Calculate its surface area.

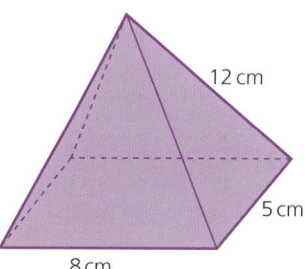

3. Two square-based pyramids are glued together as shown (right). Given that all the triangular faces are congruent, calculate the surface area of the whole shape.

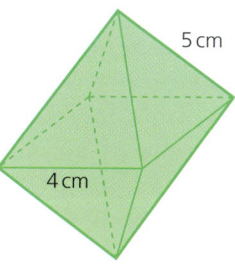

4. The two pyramids shown below have the same surface area.

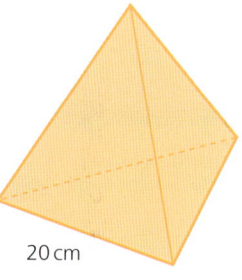 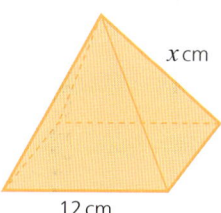

Calculate:
   a. the surface area of the regular tetrahedron,
   b. the area of one of the triangular faces on the square-based pyramid,
   c. the value of x.

5. Calculate the surface area of the frustum shown. Assume that all the sloping faces are congruent.

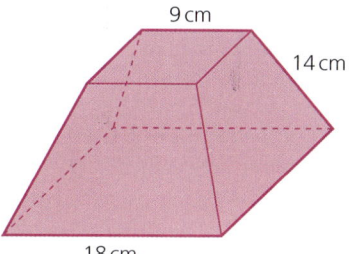

379

27 PERIMETER, AREA AND VOLUME

Exercise 27.19 (cont)

Note

To help you answer this question, you can refer to the section on Pythagoras' theorem applied to three dimensions, in Chapter 30.

6 The diagram below shows the net of a square-based pyramid, with dimensions as shown.

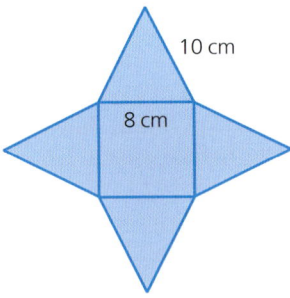

Calculate, giving your answers to 3 s.f.:
a the total surface area of the pyramid,
b the total volume of the pyramid.

The volume of a cone

A cone is a pyramid with a circular base. The formula for its volume is therefore the same as for any other pyramid.

Volume = $\frac{1}{3}$ × base area × height

$= \frac{1}{3}\pi r^2 h$

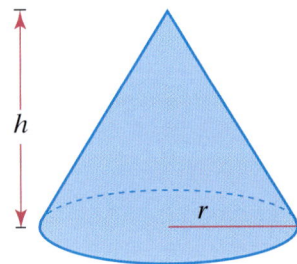

You should know how to use the formula for the volume of a cone, but you do not need to memorise it.

➡ Worked examples

a Calculate the **volume of the cone**.

Volume = $\frac{1}{3}\pi r^2 h$

$= \frac{1}{3} \times \pi \times 4^2 \times 8$

$= 134.0$ (1 d.p.)

The volume is 134 cm³ (3 s.f.).

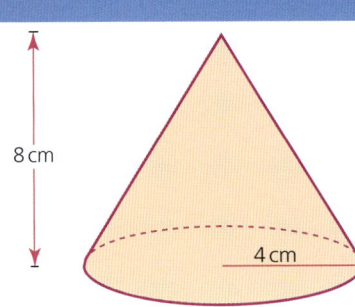

b The sector below is assembled to form a cone as shown.
  i Calculate, in terms of π, the base circumference of the cone.

  The base circumference of the cone is equal to the arc length of the sector.
  The radius of the sector is equal to the slant height of the cone (i.e. 12 cm).

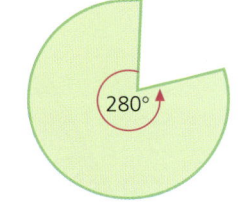
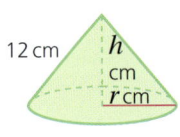

The volume of a cone

Sector arc length $= \frac{\phi}{360} \times 2\pi r$

$\qquad = \frac{280}{360} \times 2\pi \times 12 = \frac{56}{3}\pi$

So the base circumference is $\frac{56}{3}\pi$ cm.

ii Calculate, in exact form, the base radius of the cone.

The base of a cone is circular, therefore:

$C = 2\pi r$

$r = \frac{C}{2\pi}$

$= \frac{\frac{56}{3}\pi}{2\pi} = \frac{56}{6} = \frac{28}{3}$

So the radius is $\frac{28}{3}$ cm.

iii Calculate the exact height of the cone.

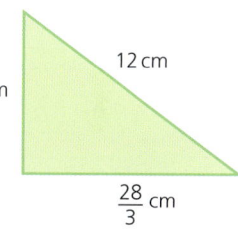

The vertical height of the cone can be calculated using Pythagoras' theorem on the right-angled triangle enclosed by the base radius, vertical height and the sloping face, as shown (right).

Note that the length of the sloping face is equal to the radius of the sector.

$12^2 = h^2 + \left(\frac{28}{3}\right)^2$

$h^2 = 12^2 - \left(\frac{28}{3}\right)^2$

$h^2 = \frac{512}{9}$

$h = \frac{\sqrt{512}}{3} = \frac{16\sqrt{2}}{3}$

Therefore the vertical height is $\frac{16\sqrt{2}}{3}$ cm.

iv Calculate the volume of the cone, leaving your answer both in terms of π and to 3 s.f.

Volume $= \frac{1}{3}\pi r^2 h$

$\qquad = \frac{1}{3}\pi \times \left(\frac{28}{3}\right)^2 \times \frac{16\sqrt{2}}{3}$

$\qquad = \frac{12544\sqrt{2}}{81}\pi$ cm^3

$\qquad = 688$ cm^3

> **Note**
>
> In the worked examples, the previous answer was used to calculate the next stage of the question. By using exact values each time, you will avoid introducing rounding errors into the calculation.

381

27 PERIMETER, AREA AND VOLUME

Exercise 27.20

1 Calculate the volume of each of the following cones. Use the values for the base radius r and the vertical height h given in each case.
   a $r = 3$ cm, $h = 6$ cm
   b $r = 6$ cm, $h = 7$ cm
   c $r = 8$ mm, $h = 2$ cm
   d $r = 6$ cm, $h = 44$ mm

2 Calculate the base radius of each of the following cones. Use the values for the volume V and the vertical height h given in each case.
   a $V = 600$ cm³, $h = 12$ cm
   b $V = 225$ cm³, $h = 18$ mm
   c $V = 1400$ mm³, $h = 2$ cm
   d $V = 0.04$ m³, $h = 145$ mm

3 The base circumference C and the length of the sloping face l are given for each of the following cones. Calculate
   i the base radius,
   ii the vertical height,
   iii the volume in each case.
   Give all answers to 3 s.f.
   a $C = 50$ cm, $l = 15$ cm   b $C = 100$ cm, $l = 18$ cm
   c $C = 0.4$ m, $l = 75$ mm   d $C = 240$ mm, $l = 6$ cm

Exercise 27.21

1 The two cones A and B shown below have the same volume. Using the dimensions shown and given that the base circumference of cone B is 60 cm, calculate the height h cm.

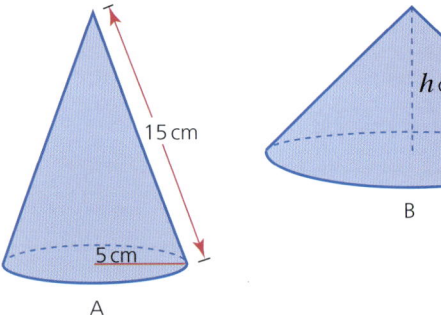

2 A cone is placed inside a cuboid as shown. If the base diameter of the cone is 12 cm and the height of the cuboid is 16 cm, calculate the volume of the cuboid not occupied by the cone.

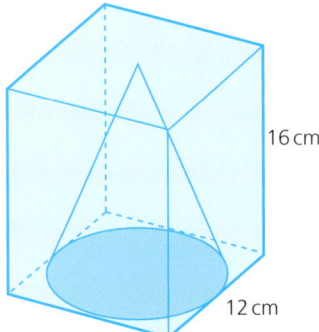

382

The volume of a cone

3 The sector shown is assembled to form a cone. Calculate the following, giving your answers to parts **a**, **b** and **c** in exact form:
  a the base circumference of the cone,
  b the base radius of the cone,
  c the vertical height of the cone,
  d the volume of the cone. Give your answer correct to 3 s.f.

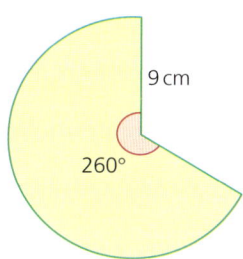

4 Two similar sectors are assembled into cones (below). Calculate the ratio of their volumes.

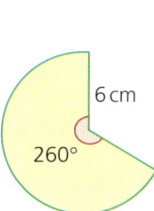

Exercise 27.22

1 An ice cream consists of a hemisphere and a cone (right). Calculate, in exact form, its total volume.

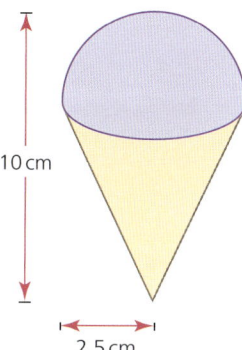

2 A cone is placed on top of a cylinder. Using the dimensions given (right), calculate the total volume of the shape.

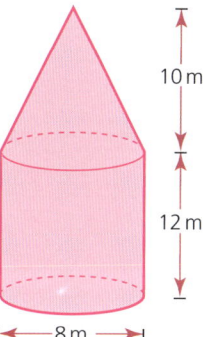

27 PERIMETER, AREA AND VOLUME

Exercise 27.22 (cont)

3 Two identical truncated cones are placed end to end as shown. Calculate the total volume of the shape.

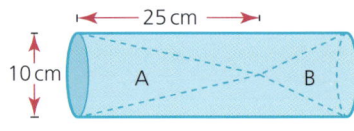

4 Two cones A and B are placed either end of a cylindrical tube as shown. Given that the volumes of A and B are in the ratio 2:1, calculate:
   a the volume of cone A,
   b the height of cone B,
   c the volume of the cylinder.

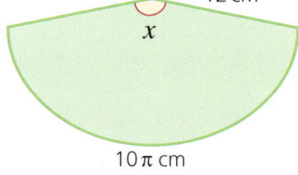

The surface area of a cone

The surface area of a cone comprises the area of the circular base and the area of the curved face. The area of the curved face is equal to the area of the sector from which it is formed.

→ Worked examples

Calculate the total surface area of the cone shown

Surface area of base = $\pi r^2 = 25\pi \text{ cm}^2$

The curved surface area can best be visualised if drawn as a sector as shown in the diagram:

The radius of the sector is equivalent to the slant height of the cone. The curved perimeter of the sector is equivalent to the base circumference of the cone.

$$\frac{x}{360} = \frac{10\pi}{24\pi}$$

Therefore $x = 150°$

Area of sector = $\frac{150}{360} \times \pi \times 12^2 = 60\pi \text{ cm}^2$

Total surface area = $60\pi + 25\pi$
$= 85\pi$
$= 267$ (3 s.f.)

The total surface area is 267 cm^2.

The area of the sector was calculated to be $60\pi \text{ cm}^2$. This is therefore also the area of the curved surface of the cone.

The curved surface of a cone can also be calculated using the formula Area $=\pi rl$, where r represents the radius of the circular base and l represents the slant length of the cone.

In the example above, the curved surface area = $\pi \times 5 \times 12$
$= 60\pi \text{ cm}^2$

You should know how to use the formula for the curved surface of a cone, i.e. the area of the sector, but you do not need to memorise it.

The surface area of a cone

Exercise 27.23

1 Calculate the surface area of each of the following cones (you may use the formula to work out the area of the curved surface of each cone):

a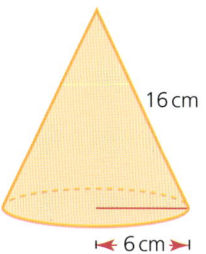

b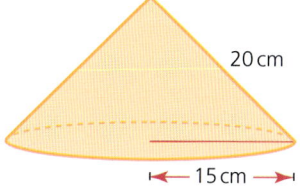

2 Two cones with the same base radius are stuck together as shown. Calculate the surface area of the shape.

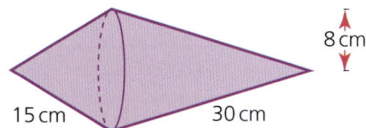

Student assessment 1

1 Calculate the area of the shape below.

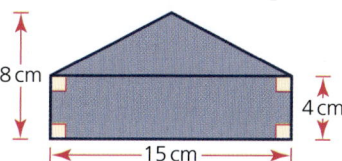

2 Calculate the circumference and area of each of the following circles. Give your answers to 3 s.f.

a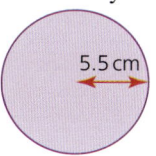

b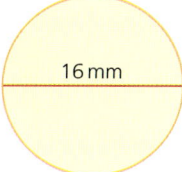

3 A semicircular shape is cut out of the side of a rectangle as shown. Calculate the shaded area, giving your answer in exact form.

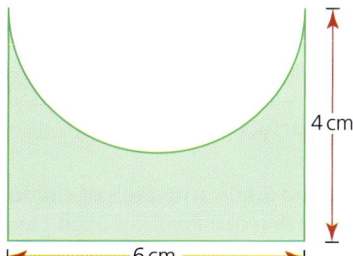

27 PERIMETER, AREA AND VOLUME

4 For the diagram (right), calculate the area of:
   a the semicircle,
   b the trapezium,
   c the whole shape.

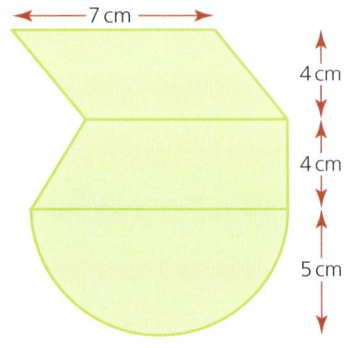

5 A cylindrical tube has an inner diameter of 6 cm, an outer diameter of 7 cm and a length of 15 cm. Calculate the following to 3 s.f.:
   a the surface area of the shaded end,
   b the inside surface area of the tube,
   c the total surface area of the tube.

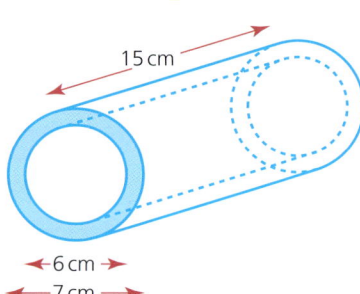

6 Calculate the volume of each of the following cylinders. Give your answers in terms of π.

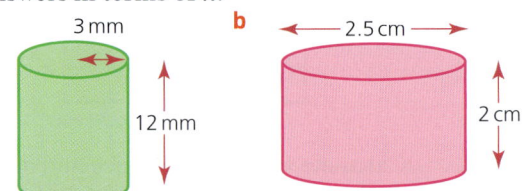

Student assessment 2

1 Calculate the arc length of each of the following sectors. The angle x and radius r are given in each case.
   a $x = 45°$ b $x = 150°$
       $r = 15$ cm $r = 13.5$ cm

2 Calculate the area of the sector shown, giving your answer in exact form.

3 A sphere has a radius of 6.5 cm. Calculate to 3 s.f.:
   a its total surface area,
   b its volume.

4 Calculate the angle x in each of the following sectors. The radius r and arc length a are given in each case.
   a $r = 20$ mm   b $r = 9$ cm
       $a = 95$ mm $a = 9$ mm

Student assessment 3

1. A ball is placed inside a box into which it will fit tightly. If the radius of the ball is 10 cm, calculate the percentage volume of the box not occupied by the ball.

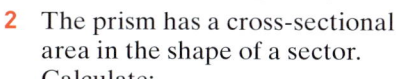

2. The prism has a cross-sectional area in the shape of a sector. Calculate:
   a the radius r cm,
   b the cross-sectional area of the prism,
   c the total surface area of the prism,
   d the volume of the prism.

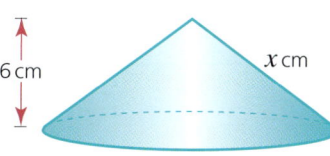

3. The cone and sphere shown (below) have the same volume.

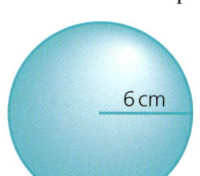

If the radius of the sphere and the height of the cone are both 6 cm, calculate each of the following. Give your answers in exact form:
   a the volume of the sphere,
   b the base radius of the cone,
   c the slant height x cm,
   d the surface area of the cone.

4. The top of a cone is cut off and a cylindrical hole is drilled out of the remaining frustum as shown (right).
   Calculate:
   a the height of the original cone,
   b the volume of the original cone,
   c the volume of the solid frustum,
   d the volume of the cylindrical hole,
   e the volume of the remaining part of the frustum.

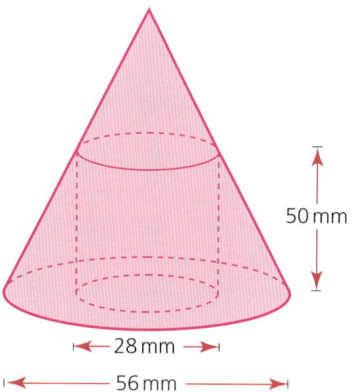

5 Mathematical investigations and ICT 5

Metal trays

A rectangular sheet of metal measures 30 cm by 40 cm.

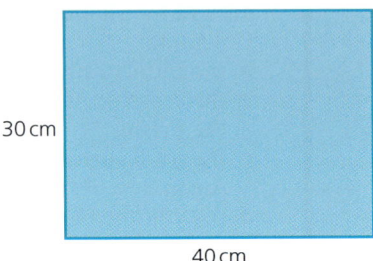

The sheet has squares of equal size cut from each corner. It is then folded to form a metal tray as shown.

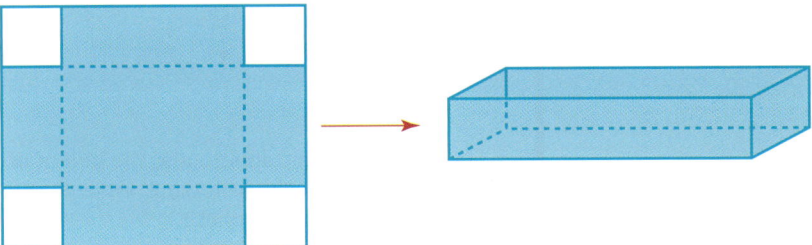

1. **a** Calculate the length, width and height of the tray if a square of side length 1 cm is cut from each corner of the sheet of metal.
   b Calculate the volume of this tray.
2. **a** Calculate the length, width and height of the tray if a square of side length 2 cm is cut from each corner of the sheet of metal.
   b Calculate the volume of this tray.
3. Using a spreadsheet if necessary, investigate the relationship between the volume of the tray and the size of the square cut from each corner. Enter your results in an ordered table.
4. Calculate, to 1 d.p., the side length of the square that produces the tray with the greatest volume.
5. State the greatest volume to the nearest whole number.

Tennis balls

Tennis balls are spherical and have a radius of 3.3 cm.

A manufacturer wishes to make a cuboidal container with a lid that holds 12 tennis balls. The container is to be made of cardboard. The manufacturer wishes to use as little cardboard as possible.

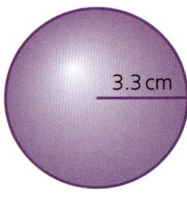

ICT activity

1 Sketch some of the different containers that might be considered.
2 For each container, calculate the total area of cardboard used and therefore decide on the most economical design.

The manufacturer now considers the possibility of using other flat-faced containers.

3 Sketch some of the different containers that might be considered.
4 Investigate the different amounts of cardboard used for each design.
5 Which type of container would you recommend to the manufacturer?

ICT activity

In this activity you will be using a spreadsheet to investigate the maximum possible volume of a cone constructed from a sector of fixed radius.

Circles of radius 10 cm are cut from paper and used to construct cones. Different sized sectors are cut from the circles and then arranged to form a cone, e.g.

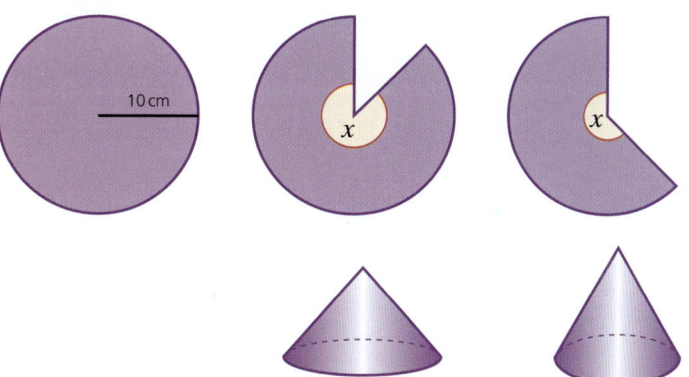

1 Using a spreadsheet like the one below, calculate the maximum possible volume, for a cone constructed from one of these circles:

	A	B	C	D	E	F
1	Angle of sector (θ)	Sector arc length (cm)	Base circumference of cone (cm)	Base radius of cone (cm)	Vertical height of cone (cm)	Volume of cone (cm³)
2	5	0.873	0.873	0.139	9.999	0.202
3	10	1.745	1.745	0.278	9.996	0.808
4	15	2.618	2.618	0.417	9.991	1.816
5	20					
6	25					
7	30					
8	Continue to 355°	Enter formulae here to calculate the results for each column				

2 Plot a graph to show how the volume changes as x increases. Comment on your graph.

TOPIC 6

Trigonometry

Contents

Chapter 28   Bearings (E4.3)
Chapter 29   Trigonometry (E6.1, E6.2, E6.3, E6.4)
Chapter 30   Further trigonometry (E6.2, E6.5)

Learning objectives

E6.1 Pythagoras' theorem
Know and use Pythagoras' theorem.

E6.2 Right-angled triangles
1. Know and use the sine, cosine and tangent ratios for acute angles in calculations involving sides and angles of a right-angled triangle.
2. Solve problems in two dimensions using Pythagoras' theorem and trigonometry.
3. Know that the perpendicular distance from a point to a line is the shortest distance to the line.
4. Carry out calculations involving angles of elevation and depression.

E6.3 Exact trigonometric values
Know the exact values of:
1. $\sin x$ and $\cos x$ for $x = 0°, 30°, 45°, 60°$ and $90°$
2. $\tan x$ for $x = 0°, 30°, 45°, 60°$.

E6.4 Trigonometric functions
1. Recognise, sketch and interpret the following graphs for $0° \leqslant x \leqslant 360°$:
   - $y = \sin x$
   - $y = \cos x$
   - $y = \tan x$.
2. Solve trigonometric equations involving $\sin x$, $\cos x$ or $\tan x$, for $0° \leqslant x \leqslant 360°$.

E6.5 Non-right-angled triangles
1. Use the sine and cosine rules in calculations involving lengths and angles for any triangle.
2. Use the formula area of triangle = $\frac{1}{2} ab \sin C$.

E6.6 Pythagoras' theorem and trigonometry in 3D
Carry out calculations and solve problems in three dimensions using Pythagoras' theorem and trigonometry, including calculating the angle between a line and a plane.

The Swiss

Leonhard Euler

Leonhard Euler (1707–1783)

Euler, like Newton before him, was the greatest mathematician of his generation. He studied all areas of mathematics and continued to work hard after he had gone blind.

As a young man, Euler discovered and proved:

the sum of the infinite series $\sum_{n=1}^{\infty} \left(\frac{1}{n^2}\right) = \frac{\pi^2}{6}$

(i.e.) $\frac{1}{1^2} + \frac{1}{2^2} + \frac{1}{3^2} + \ldots + \frac{1}{n^2} = \frac{\pi^2}{6}$

This brought him to the attention of other mathematicians.

Euler made discoveries in many areas of mathematics, especially calculus and trigonometry. He also developed the ideas of Newton and Leibniz.

Euler worked on graph theory and functions and was the first to prove several theorems in geometry. He studied relationships between a triangle's height, midpoint and circumscribing and inscribing circles, and he also discovered an expression for the volume of a tetrahedron (a triangular pyramid) in terms of its sides.

He also worked on number theory and found the largest prime number known at the time.

Some of the most important constant symbols in mathematics, π, e and i (the square root of −1), were introduced by Euler.

The Bernoulli family

The Bernoullis were a family of Swiss merchants who were friends of Euler. The two brothers, Johann and Jacob, were very gifted mathematicians and scientists, as were their children and grandchildren. They made discoveries in calculus, trigonometry and probability theory in mathematics. In science, they worked on astronomy, magnetism, mechanics, thermodynamics and more.

Unfortunately, many members of the Bernoulli family were not pleasant people. The older members of the family were jealous of each other's successes and often stole the work of their sons and grandsons and pretended that it was their own.

28 Bearings

> **Note**
> All diagrams are not drawn to scale.

Bearings

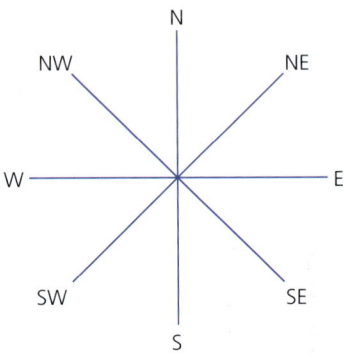
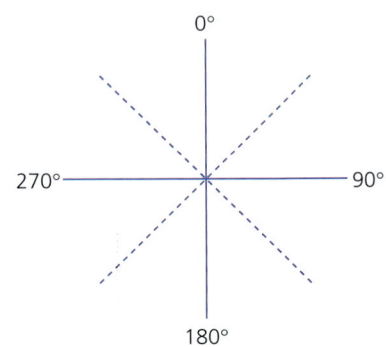

In the days when sailing ships travelled the oceans of the world, compass **bearings** like the ones in the diagram above were used.

As the need for more accurate direction arose, extra points were added to N, S, E, W, NE, SE, SW and NW. Midway between North and North East was North North East, and midway between North East and East was East North East, and so on. This gave 16 points of the compass. This was later extended even further, eventually to 64 points.

As the speed of travel increased, a new system was required. The new system was the **three-figure bearing** system. North was given the bearing zero. 360° in a clockwise direction was one full rotation.

Back bearings

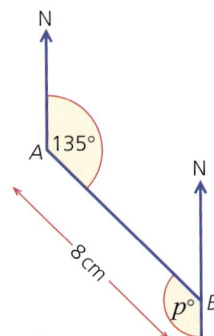

The bearing of B from A is 135° and the distance from A to B is 8 cm, as shown (left). The bearing of A from B is called the **back bearing**.

Since the two North lines are parallel:

$p = 135°$ (alternate angles), so the back bearing is $(180 + 135)°$.

That is, 315°.

(There are a number of methods of solving this type of problem.)

Back bearings

→ Worked example

The bearing of B from A is 245°.
What is the bearing of A from B?

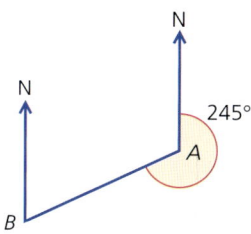 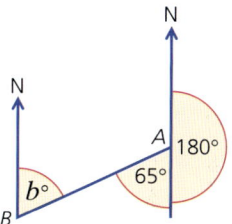

Since the two North lines are parallel:
$b = 65°$ (alternate angles), so the bearing is $(245 - 180)°$. That is, 065°.

Exercise 28.1

1 Draw a diagram to show the following compass bearings and journey. Use a scale of 1 cm : 1 km. North can be taken to be a line vertically up the page.
Start at point A. Travel a distance of 7 km on a bearing of 135° to point B. From B, travel 12 km on a bearing of 250° to point C. Measure the distance and bearing of A from C.

2 Given the following bearings of point B from point A, draw diagrams and use them to calculate the bearings of A from B.
  a bearing 163°   b bearing 214°

3 Given the following bearings of point D from point C, draw diagrams and use them to calculate the bearings of C from D.
  a bearing 300°   b bearing 282°

Student assessment 1

1 A climber gets to the top of Mont Blanc. He can see in the distance a number of ski resorts. He uses his map to find the bearing and distance of the resorts, and records them as shown below:

  Val d'Isère 30 km   bearing 082°
  Les Arcs 40 km   bearing 135°
  La Plagne 45 km   bearing 205°
  Méribel 35 km   bearing 320°

Choose an appropriate scale and draw a diagram to show the position of each resort. What are the distance and bearing of the following?
  a Val d'Isère from La Plagne.   b Méribel from Les Arcs.

2 A coastal radar station picks up a distress call from a ship. It is 50 km away on a bearing of 345°. The radar station contacts a lifeboat at sea which is 20 km away on a bearing of 220°.
Make a scale drawing and use it to find the distance and bearing of the ship from the lifeboat.

3 An aircraft is seen on radar at airport A. The aircraft is 210 km away from the airport on a bearing of 065°. The aircraft is diverted to airport B, which is 130 km away from A on a bearing of 215°. Choose an appropriate scale and make a scale drawing to find the distance and bearing of airport B from the aircraft.

29 Trigonometry

> **Note**
>
> All diagrams are not drawn to scale.

In this chapter, unless instructed otherwise, give your answers exactly or correct to three significant figures as appropriate. Answers in degrees should be given to one decimal place.

There are three basic trigonometric ratios: sine, cosine and tangent.

Each of these relates an angle of a right-angled triangle to a ratio of the lengths of two of its sides.

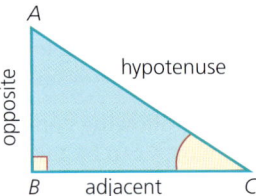

The sides of the triangle have names, two of which are dependent on their position in relation to a specific angle.

The longest side (always opposite the right angle) is called the **hypotenuse**. The side opposite the angle is called the **opposite** side and the side next to the angle is called the **adjacent** side.

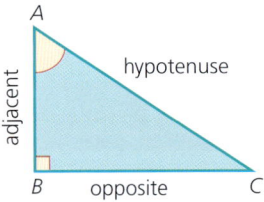

Note that, when the chosen angle is at A, the sides labelled opposite and adjacent change.

Tangent

$$\tan C = \frac{\text{length of opposite side}}{\text{length of adjacent side}}$$

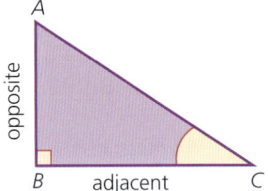

Tangent

→ Worked examples

a Calculate the size of angle BAC in each of the triangles.

i $\tan x° = \dfrac{\text{opposite}}{\text{adjacent}} = \dfrac{4}{5}$

$x = \tan^{-1}\left(\dfrac{4}{5}\right)$

$x = 38.7$ (1 d.p.)

angle $BAC = 38.7°$

ii $\tan x° = \dfrac{8}{3}$

$x = \tan^{-1}\left(\dfrac{8}{3}\right)$

$x = 69.4$ (1 d.p.)

angle $BAC = 69.4°$

Answers involving angles should be given to 1 d.p. unless stated otherwise.

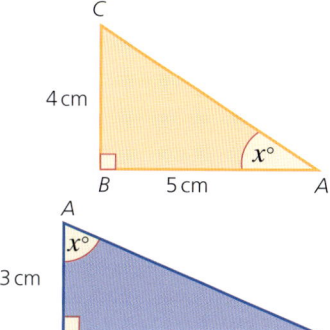

b Calculate the length of the opposite side QR.

$\tan 42° = \dfrac{p}{6}$

$6 \times \tan 42° = p$

$p = 5.40$ (3 s.f.)

$QR = 5.40$ cm (3 s.f.)

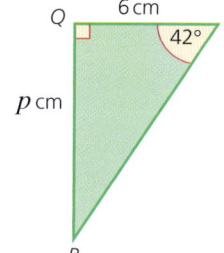

c Calculate the length of the adjacent side XY.

$\tan 35° = \dfrac{6}{z}$

$z \times \tan 35° = 6$

$z = \dfrac{6}{\tan 35°}$

$z = 8.57$ (3 s.f.)

$XY = 8.57$ cm (3 s.f.)

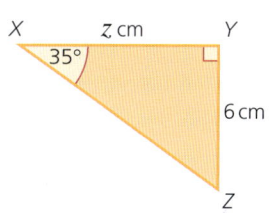

Exercise 29.1

Calculate the length of the side marked x cm in each of the diagrams in Questions 1 and 2.

1 a 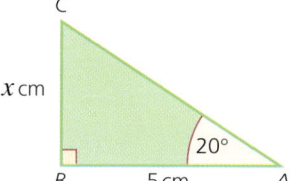 b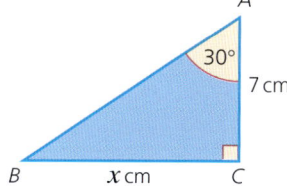

29 TRIGONOMETRY

Exercise 29.1 (cont)

c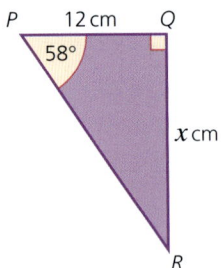

d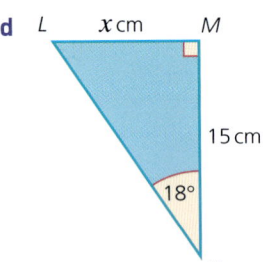

e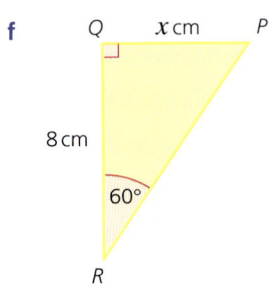

f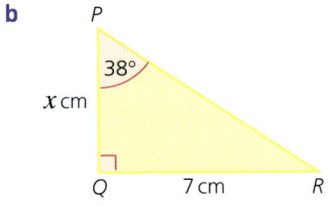

2 a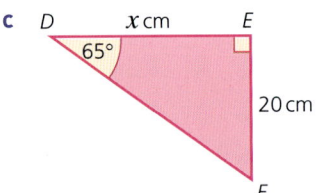

b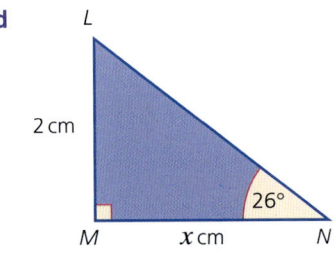

c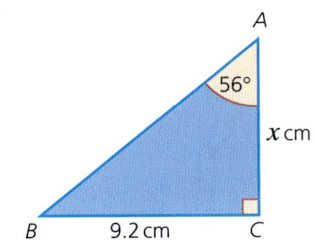

Sine

3 Calculate the size of the angle marked $x°$ in each of the following diagrams.

a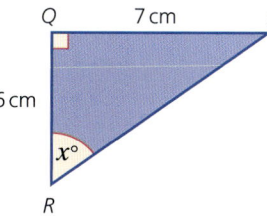

b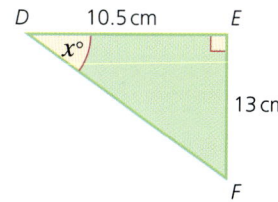

c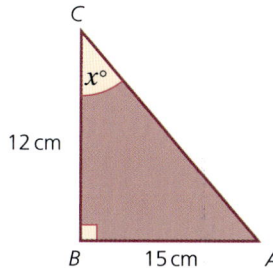

d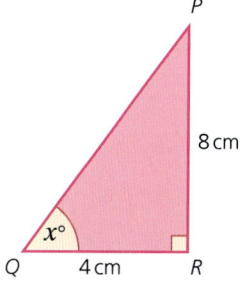

e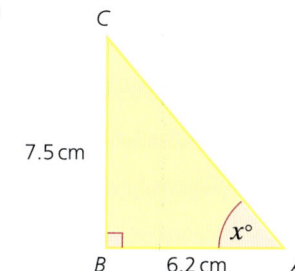

f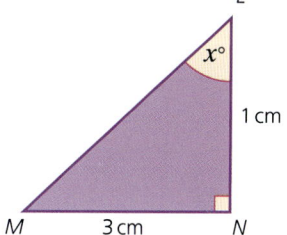

Sine

$$\sin N = \frac{\text{length of opposite side}}{\text{length of hypotenuse}}$$

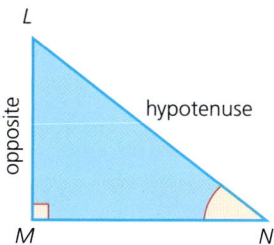

29 TRIGONOMETRY

> **Worked examples**

a Calculate the size of angle *BAC*.

$$\sin x = \frac{\text{opposite}}{\text{hypotenuse}} = \frac{7}{12}$$

$$x = \sin^{-1}\left(\frac{7}{12}\right)$$

$$x = 35.7 \text{ (1 d.p.)}$$

angle $BAC = 35.7°$ (1 d.p.)

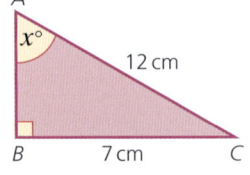

b Calculate the length of the hypotenuse *PR*.

$$\sin 18° = \frac{11}{q}$$

$$q \times \sin 18° = 11$$

$$q = \frac{11}{\sin 18°}$$

$$q = 35.6 \text{ (3 s.f.)}$$

$$PR = 35.6 \text{ cm (3 s.f.)}$$

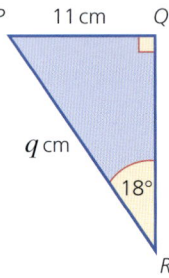

Exercise 29.2

1 Calculate the length of the marked side in each of the following diagrams.

a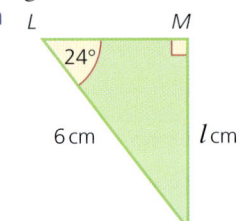

b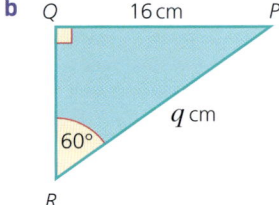

c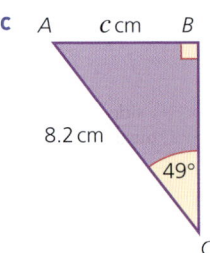

d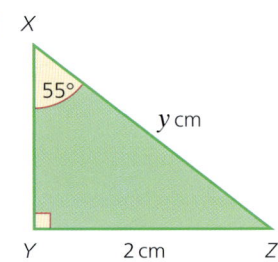

e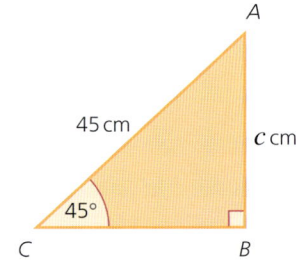

f

Cosine

2 Calculate the size of the angle marked *x* in each of the following diagrams.

a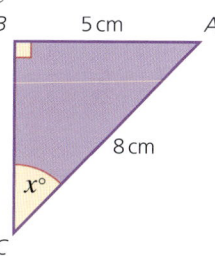

b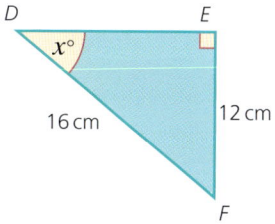

c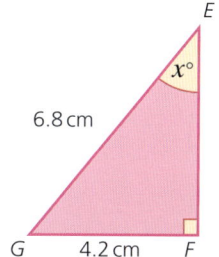

d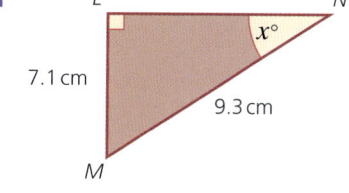

e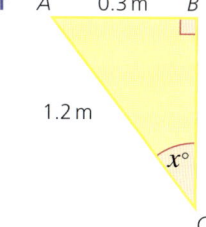

f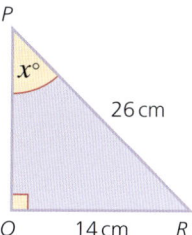

Cosine

$$\cos Z = \frac{\text{length of adjacent side}}{\text{length of hypotenuse}}$$

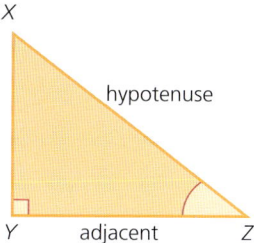

29 TRIGONOMETRY

> **Worked examples**

a Calculate the length XY.

$$\cos 62° = \frac{\text{adjacent}}{\text{hypotenuse}} = \frac{z}{20}$$

$$z = 20 \times \cos 62°$$

$$z = 9.39 \text{ (3 s.f.)}$$

$$XY = 9.39 \text{ cm (3 s.f.)}$$

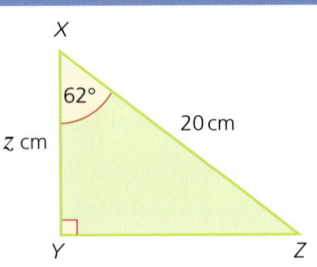

b Calculate the size of angle ABC.

$$\cos x = \frac{5.3}{12}$$

$$x = \cos^{-1}\left(\frac{5.3}{12}\right)$$

$$x = 63.8 \text{ (1 d.p.)}$$

angle $ABC = 63.8°$ (1 d.p.)

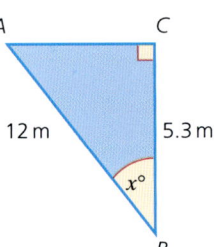

Exercise 29.3

1 Calculate the marked side or angle in each of the following diagrams.

a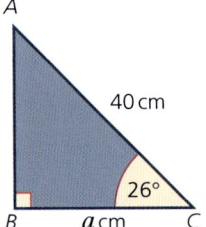

b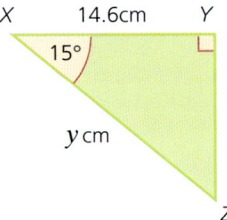

c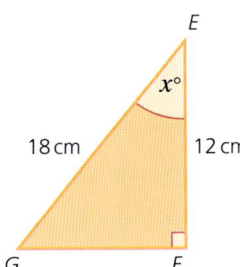

d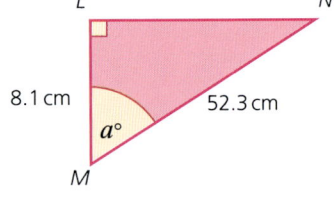

e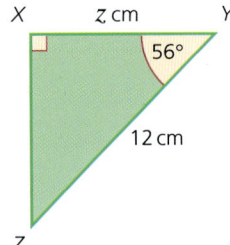

f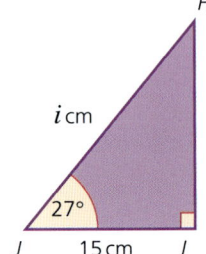

Pythagoras' theorem

g 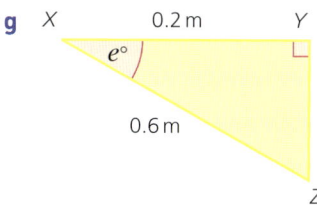   h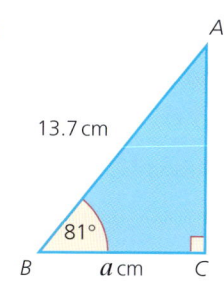

Pythagoras' theorem

Pythagoras' theorem states the relationship between the lengths of the three sides of a right-angled triangle.

Pythagoras' theorem states that:
$a^2 = b^2 + c^2$

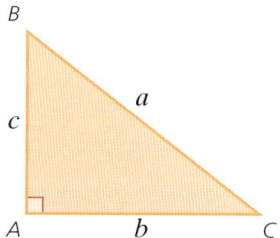

> ## Worked examples

a Calculate the length of the side BC.
Using Pythagoras:
$a^2 = b^2 + c^2$
$a^2 = 8^2 + 6^2$
$a^2 = 64 + 36 = 100$
$a = \sqrt{100}$
$a = 10$
$BC = 10\,\text{m}$

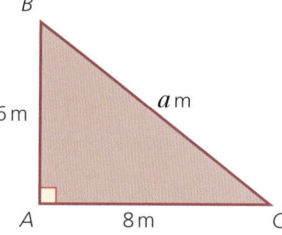

b Calculate the length of the side AC.
Using Pythagoras:
$a^2 = b^2 + c^2$
$12^2 = b^2 + 5^2$
$b^2 = 144 - 25 = 119$
$b = \sqrt{119}$
$b = 10.9$ (3 s.f.)
$AC = 10.9\,\text{m}$ (3 s.f.)

Note that if the answer is left in surd form as $\sqrt{119}$, this is known as leaving your answer in exact form. However, leaving an answer as a surd is in the Extended syllabus only.

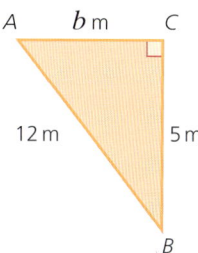

29 TRIGONOMETRY

Exercise 29.4 In each of the diagrams in Questions 1 and 2, use Pythagoras' theorem to calculate the length of the marked side. Where the answer is not an integer, leave it in exact form.

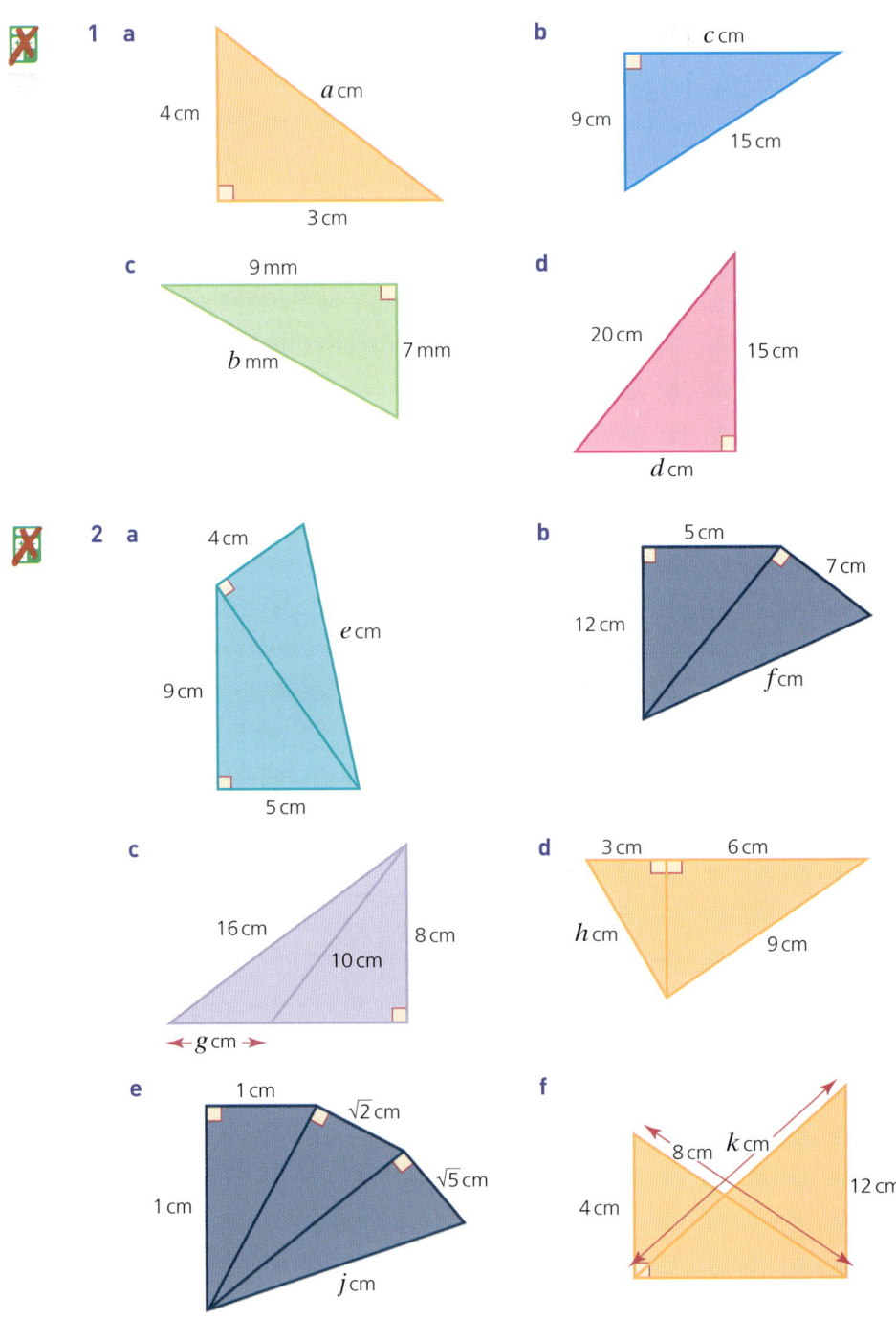

Pythagoras' theorem

3 Villages A, B and C lie on the edge of the Namib Desert. Village A is 30 km due north of village C. Village B is 65 km due east of A. Calculate the shortest distance between villages C and B, giving your answer to the nearest 0.1 km.

4 Town X is 54 km due west of town Y. The shortest distance between town Y and town Z is 86 km. If town Z is due south of X, calculate the distance between X and Z, giving your answer to the nearest kilometre.

5 Village B is on a bearing of 135° and at a distance of 40 km from village A, as shown. Village C is on a bearing of 225° and a distance of 62 km from village A.
   a Show that triangle ABC is right-angled.
   b Calculate the distance from B to C, giving your answer to the nearest 0.1 km.

6 Two boats set off from X at the same time (below). Boat A sets off on a bearing of 325° and with a velocity of 14 km/h. Boat B sets off on a bearing of 235° with a velocity of 18 km/h. Calculate the distance between the boats after they have been travelling for 2.5 hours. Give your answer to the nearest metre.

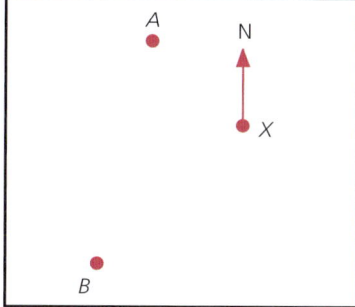

7 A boat sets off on a trip from S. It heads towards B, a point 6 km away and due north. At B it changes direction and heads towards point C, 6 km away from and due east of B. At C it changes direction once again and heads on a bearing of 135° towards D, which is 13 km from C.
   a Calculate the distance between S and C to the nearest 0.1 km.
   b Calculate the distance the boat will have to travel if it is to return to S from D.

29 TRIGONOMETRY

Exercise 29.4 (cont)

8 Two trees are standing on flat ground.

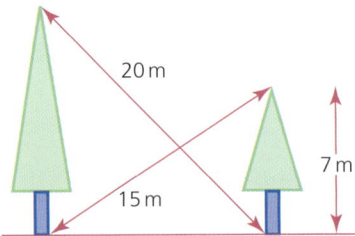

The height of the smaller tree is 7 m. The distance between the top of the smaller tree and the base of the taller tree is 15 m.
The distance between the top of the taller tree and the base of the smaller tree is 20 m.
a Calculate the horizontal distance between the two trees.
b Calculate the height of the taller tree.

Exercise 29.5

1 By using Pythagoras' theorem, trigonometry or both, calculate the marked value in each of the following diagrams. In each case give your answer to 1 d.p.

a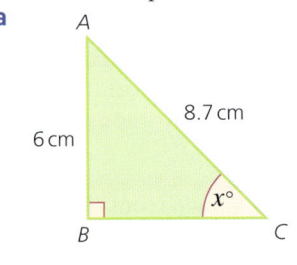

b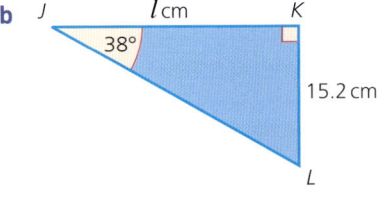

c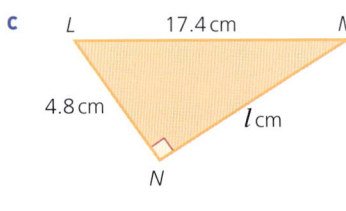

d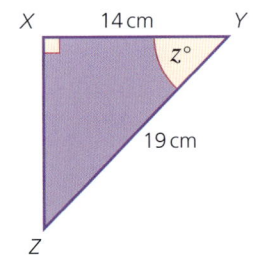

2 A sailing boat sets off from a point X and heads towards Y, a point 17 km north. At point Y, it changes direction and heads towards point Z, a point 12 km away on a bearing of 090°. Once at Z, the crew want to sail back to X. Calculate:
   a the distance ZX,
   b the bearing of X from Z.

3 An aeroplane sets off from G on a bearing of 024° towards H, a point 250 km away. At H, it changes course and heads towards J, which is 180 km away on a bearing of 055°.
   a How far is H to the north of G?
   b How far is H to the east of G?
   c How far is J to the north of H?

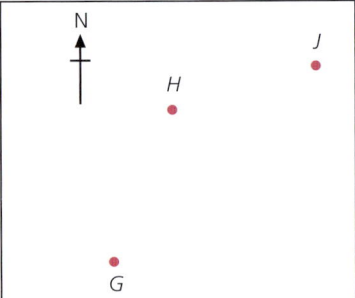

d How far is J to the east of H?
   e What is the shortest distance between G and J?
   f What is the bearing of G from J?

4 $PQRS$ is a quadrilateral. The sides RS and QR are the same length. The sides QP and RS are parallel. Calculate:
   a angle SQR,
   b angle PSQ,
   c length PQ,
   d length PS,
   e the area of $PQRS$.

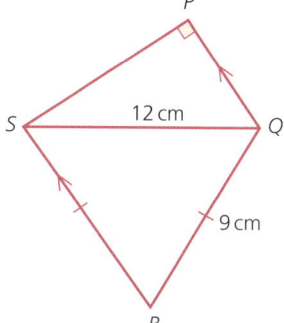

Angles of elevation and depression

The **angle of elevation** is the angle above the horizontal through which a line of view is raised. The **angle of depression** is the angle below the horizontal through which a line of view is lowered.

➡ Worked examples

a The base of a tower is 60 m away from a point X on the ground. If the angle of elevation of the top of the tower from X is 40°, calculate the height of the tower.

Give your answer to the nearest metre.

$\tan 40° = \dfrac{h}{60}$

$h = 60 \times \tan 40° = 50.3$

The height is 50 m to the nearest metre.

b An aeroplane receives a signal from a point X on the ground. If the angle of depression of point X from the aeroplane is 30°, calculate the height at which the aeroplane is flying.

Give your answer to the nearest 0.1 km.

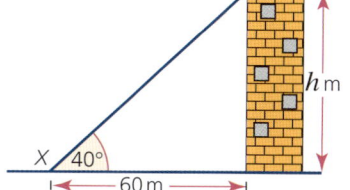

$\sin 30° = \dfrac{h}{6}$

$h = 6 \times \sin 30° = 3$

The height is 3 km.

29 TRIGONOMETRY

Exercise 29.6

1. *A* and *B* are two villages. If the horizontal distance between them is 12 km and the vertical distance between them is 2 km, calculate:
   a the shortest distance between the two villages,
   b the angle of elevation of *B* from *A*.

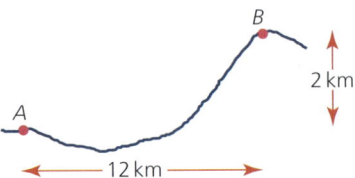

2. *X* and *Y* are two towns. If the horizontal distance between them is 10 km and the angle of depression of *Y* from *X* is 7°, calculate:
   a the shortest distance between the two towns,
   b the vertical height between the two towns.

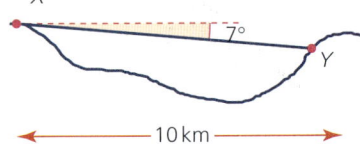

3. A girl standing on a hill at *A* can see a small boat at a point *B* on a lake.

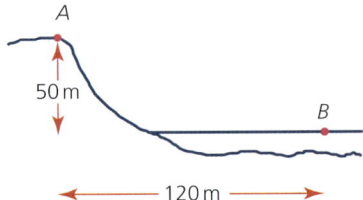

   If the girl is at a height of 50 m above *B* and at a horizontal distance of 120 m away from *B*, calculate:
   a the angle of depression of the boat from the girl,
   b the shortest distance between the girl and the boat.

4. Two hot air balloons are 1 km apart in the air. If the angle of elevation of the higher from the lower balloon is 20°, calculate, giving your answers to the nearest metre:
   a the vertical height between the two balloons,
   b the horizontal distance between the two balloons.

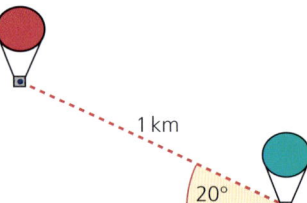

5. A boy *X* can be seen by two of his friends, *Y* and *Z*, who are swimming in the sea.

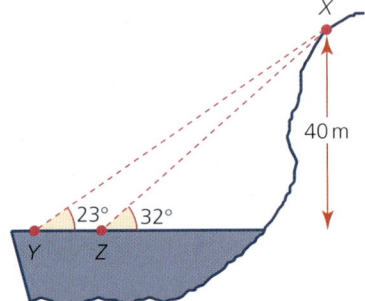

Angles of elevation and depression

If the angle of elevation of X from Y is 23° and from Z is 32°, and the height of X above Y and Z is 40 m, calculate:
a the horizontal distance between X and Z,
b the horizontal distance between Y and Z.
Note: XYZ is a vertical plane

6 A plane is flying at an altitude of 6 km directly over the line AB. It spots two boats, A and B, on the sea. If the angles of depression of A and B from the plane are 60° and 30° respectively, calculate the horizontal distance between A and B.

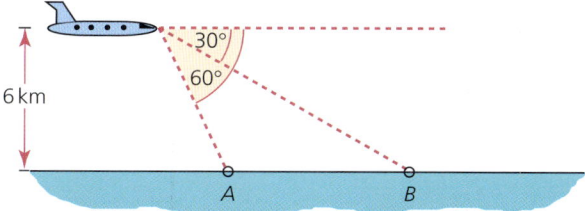

7 Two planes are flying directly above each other. A person standing at P can see both of them. The horizontal distance between the two planes and the person is 2 km. If the angles of elevation of the planes from the person are 65° and 75°, calculate:
a the altitude at which the higher plane is flying,
b the vertical distance between the two planes.

> **Note**
> You may find a sketch of this information will help you solve the problem.

8 Three villages, A, B and C, can see each other across a valley. The horizontal distance between A and B is 8 km, and the horizontal distance between B and C is 12 km. The angle of depression of B from A is 20° and the angle of elevation of C from B is 30°. Calculate, giving all answers to 1 d.p.:
a the vertical height between A and B,
b the vertical height between B and C,
c the angle of elevation of C from A,
d the shortest distance between A and C.
Note: A, B and C are in the same vertical plane.

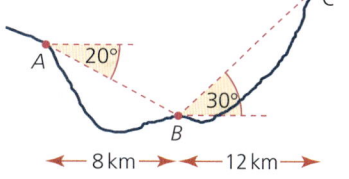

9 Using binoculars, three people, P, Q and R, can see each other across a valley. The horizontal distance between P and Q is 6.8 km and the horizontal distance between Q and R is 10 km. If the shortest distance between P and Q is 7 km and the angle of depression of Q from R is 15°, calculate, giving appropriate answers:
a the vertical height between Q and R,
b the vertical height between P and R,
c the angle of elevation of R from P,
d the shortest distance between P and R.
Note: P, Q and R are in the same vertical plane.

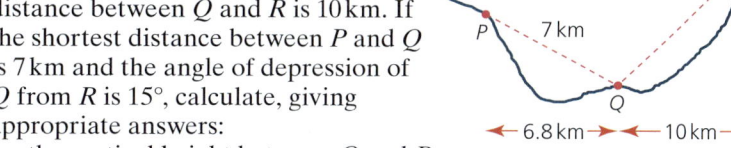

29 TRIGONOMETRY

Exercise 29.6 (cont)

10 Two people, A and B, are standing either side of a transmission mast. A is 130 m away from the mast and B is 200 m away.

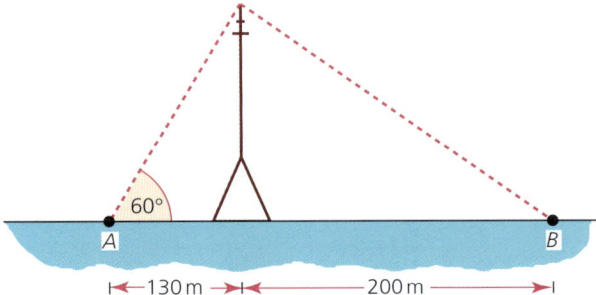

If the angle of elevation of the top of the mast from A is 60°, calculate:
a the height of the mast to the nearest metre,
b the angle of elevation of the top of the mast from B.

11 Two trees are standing on flat ground. The angle of elevation of their tops from a point X on the ground is 40°. If the horizontal distance between X and the small tree is 8 m and the distance between the tops of the two trees is 20 m, calculate:
a the height of the small tree,
b the height of the tall tree,
c the horizontal distance between the trees.

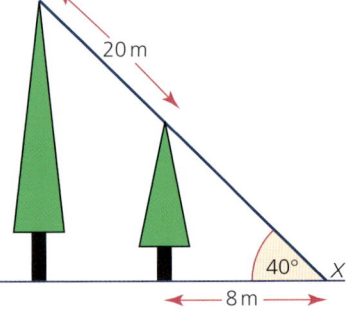

Special angles and their trigonometric ratios

So far, most of the angles you have worked with have required the use of a calculator in order to calculate their sine, cosine or tangent. However, some angles produce exact values and a calculator is both unnecessary and unhelpful when exact solutions are required.

There are a number of angles which have 'neat' trigonometric ratios, for example 0°, 30°, 45°, 60° and 90°. Their trigonometric ratios are derived below.

Consider the right-angled isosceles triangle ABC.

Let the perpendicular sides AC and BC each have a length of 1 unit.

As triangle ABC is isosceles, angle ABC = angle CAB = 45°.

Special angles and their trigonometric ratios

Using Pythagoras' theorem, AB can also be calculated:

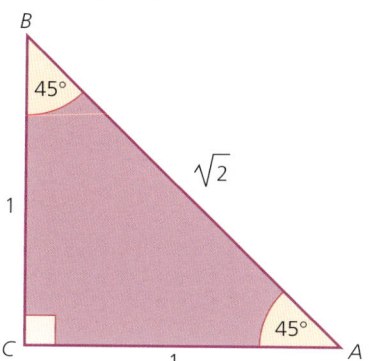

$(AB)^2 = (AC)^2 + (BC)^2$

$(AB)^2 = 1^2 + 1^2 = 2$

$AB = \sqrt{2}$

From the triangle, it can be deduced that $\sin 45° = \frac{1}{\sqrt{2}}$.

When rationalised, this can be written as $\sin 45° = \frac{\sqrt{2}}{2}$.

Similarly, $\cos 45° = \frac{1}{\sqrt{2}} = \frac{\sqrt{2}}{2}$

Therefore $\sin 45° = \cos 45°$

$\tan 45° = \frac{1}{1} = 1$

Consider also the equilateral triangle XYZ (below) in which each of its sides has a length of 2 units.

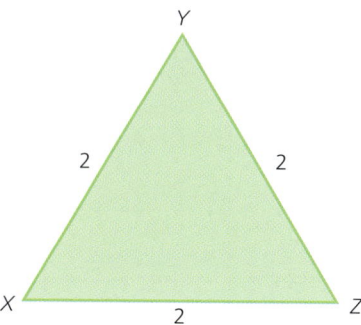

If a vertical line is dropped from the vertex Y until it meets the base XZ at P, the triangle is bisected. Consider now the right-angled triangle XYP.

29 TRIGONOMETRY

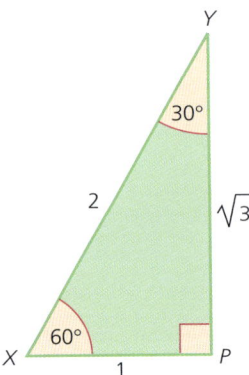

Angle $XYP = 30°$ as it is half of angle XYZ.

$XP = 1$ unit length as it is half of XZ.

The length YP can be calculated using Pythagoras' theorem:

$$(XY)^2 = (XP)^2 + (YP)^2$$
$$(YP)^2 = (XY)^2 - (XP)^2 = 2^2 - 1^2 = 3$$
$$YP = \sqrt{3}$$

Therefore from this triangle the following trigonometric ratios can be deduced:

$\sin 30° = \frac{1}{2}$ $\qquad \cos 30° = \frac{\sqrt{3}}{2}$ $\qquad \tan 30° = \frac{1}{\sqrt{3}} = \frac{\sqrt{3}}{3}$

$\sin 60° = \frac{\sqrt{3}}{2}$ $\qquad \cos 60° = \frac{1}{2}$ $\qquad \tan 60° = \frac{\sqrt{3}}{1} = \sqrt{3}$

These results and those obtained from the trigonometric graphs shown on the next page are summarised in the table below:

Angle (°)	sin (°)	cos (°)	tan (°)
0°	0	1	0
30°	$\frac{1}{2}$	$\frac{\sqrt{3}}{2}$	$\frac{1}{\sqrt{3}} = \frac{\sqrt{3}}{3}$
45°	$\frac{1}{\sqrt{2}} = \frac{\sqrt{2}}{2}$	$\frac{1}{\sqrt{2}} = \frac{\sqrt{2}}{2}$	1
60°	$\frac{\sqrt{3}}{2}$	$\frac{1}{2}$	$\sqrt{3}$
90°	1	0	—

There are other angles which have the same trigonometric ratios as those shown in the table. The following section explains why, using a unit circle, i.e. a circle with a radius of 1 unit.

Graphs of trigonometric functions

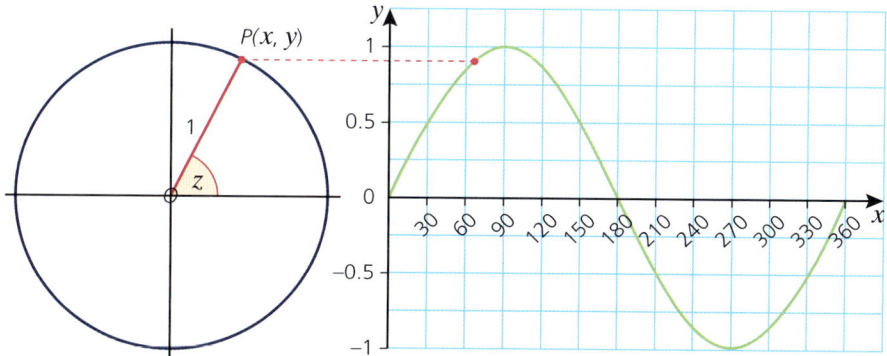

In the diagram above, *P* is a point on the circumference of a circle with centre at *O* and a radius of 1 unit. *P* has coordinates (*x*, *y*). As the position of *P* changes, then so does the angle *z*.

$\sin z = \frac{y}{1} = y$, i.e. the sine of the angle *z* is represented by the *y*-coordinate of *P*.

The graph therefore shows the different values of sin *z* as *z* varies. A more complete diagram is shown below. Note that the angle *z* is measured anticlockwise from the positive *x*-axis.

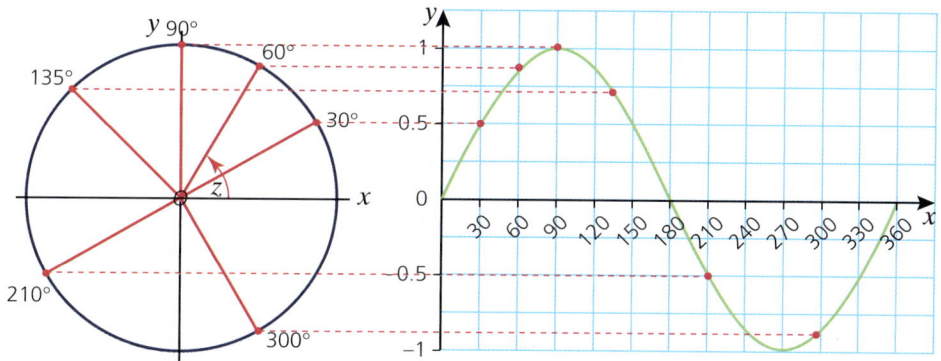

The graph of *y* = sin *x* has:
- » a period of 360° (i.e. it repeats itself every 360°)
- » a maximum value of +1
- » a minimum value of –1
- » symmetry, e.g. sin *z* = sin (180 – *z*).

Similar diagrams and graphs can be constructed for cos *z* and tan *z*.

From the unit circle, it can be deduced that $\cos z = \frac{x}{1} = x$,

i.e. the cosine of the angle *z* is represented by the *x*-coordinate of *P*. Since cos *z* = *x*, to be able to compare the graphs, the axes should be rotated through 90° as shown.

29 TRIGONOMETRY

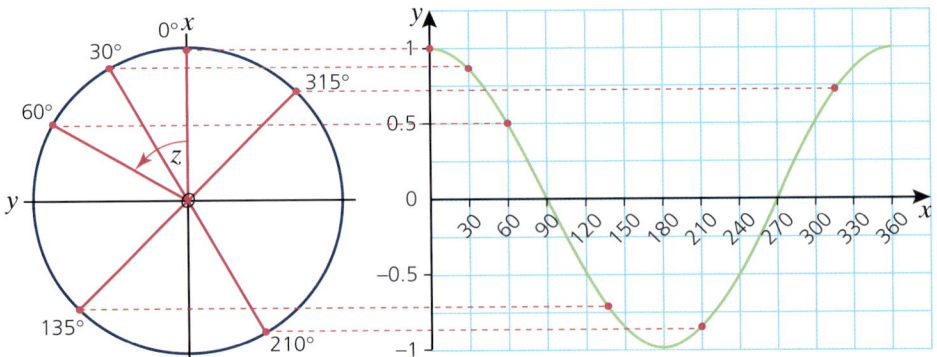

The properties of the cosine curve are similar to those of the sine curve. It has:
- a period of 360°
- a maximum value of +1
- a minimum value of −1
- symmetry, e.g. cos z = cos (360 − z).

The cosine curve is a translation of the sine curve of −90°, i.e. cos z = sin (z + 90).

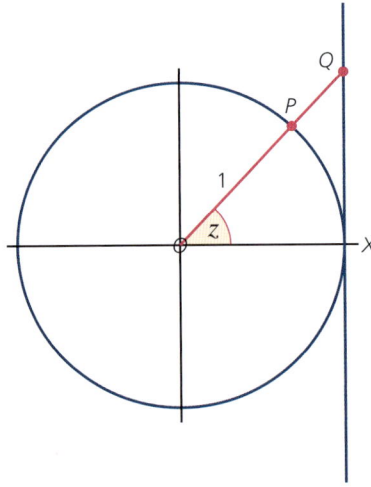

From the unit circle it can be deduced that tan $z = \frac{y}{x}$.

In order to compare all the graphs, a tangent to the unit circle is drawn at (1, 0). OP is extended to meet the tangent at Q as shown.

As OX = 1 (radius of the unit circle), tan $z = \frac{QX}{OX} = QX$.

i.e. tan z is equal to the y-coordinate of Q.

Graphs of trigonometric functions

The graph of tan z is therefore shown below:

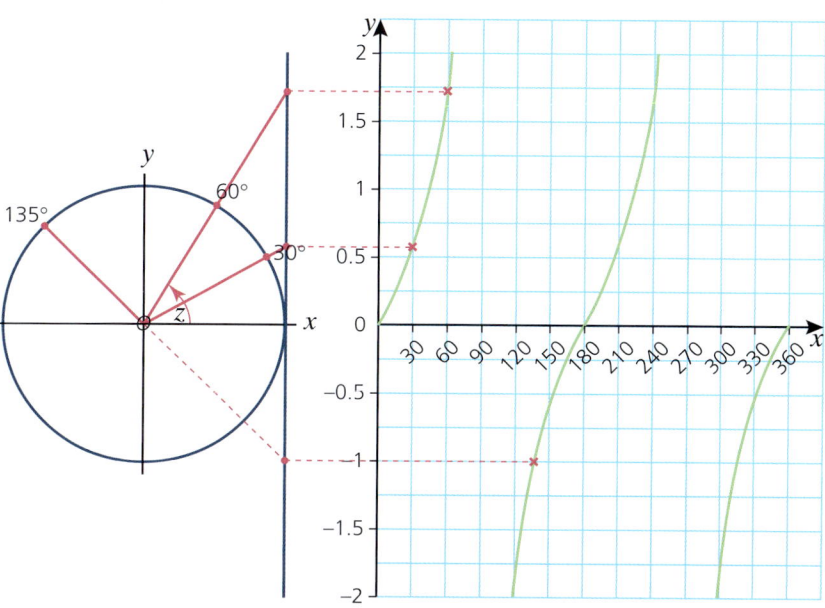

The graph of tan z has:
» a period of 180°
» no maximum or minimum value
» symmetry
» asymptotes at 90° and 270°.

➜ Worked examples

a   sin 30° = 0.5. Which other angle between 0° and 360° has a sine of 0.5?

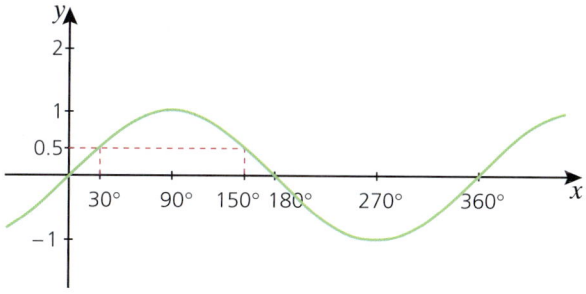

From the graph above it can be seen that sin 150° = 0.5.
Also sin x = sin (180° − x); therefore sin 30° = sin (180° − 30) = sin 150°.

29 TRIGONOMETRY

b cos 60° = 0.5. Which other angle between 0° and 360° has a cosine of 0.5?

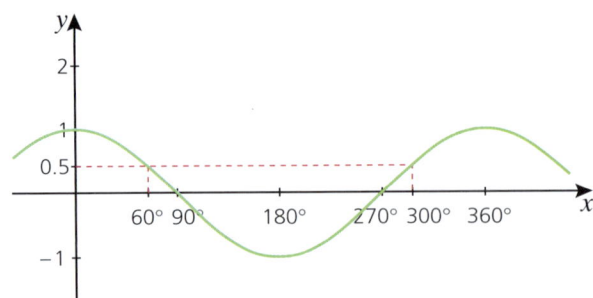

From the graph above it can be seen that cos 300° = 0.5.

c The cosine of which angle between 0° and 180° is equal to the negative of cos 50°?

Cos 50° has the same magnitude but different sign to cos 130° because of the symmetrical properties of the cosine curve.
Therefore cos 130° = −cos 50°

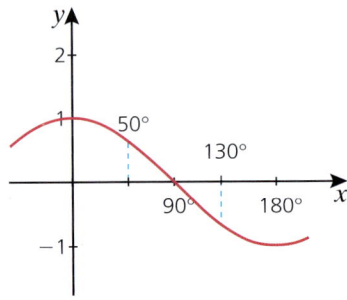

Exercise 29.7

1 Write each of the following in terms of the sine of another angle between 0° and 360°.
   a sin 60°   **b** sin 80°   **c** sin 115°
   d sin 200°   **e** sin 300°   **f** sin 265°

2 Write each of the following in terms of the sine of another angle between 0° and 360°.
   a sin 35°   **b** sin 50°   **c** sin 30°
   d sin 248°   **e** sin 304°   **f** sin 327°

3 Find the two angles between 0° and 360° which have the following sine. Give each angle to the nearest degree.
   a 0.33   **b** 0.99   **c** 0.09
   d $-\frac{1}{2}$ **e** $-\frac{\sqrt{3}}{2}$ **f** $-\frac{1}{\sqrt{2}}$

4 Find the two angles between 0° and 360° which have the following sine. Give each angle to the nearest degree.
   a 0.94   **b** 0.16   **c** 0.80
   d −0.56   **e** −0.28   **f** −0.33

Solving trigonometric equations

Exercise 29.8

1. Write each of the following in terms of the cosine of another angle between 0° and 360°.
   a cos 20°   b cos 85°   c cos 32°
   d cos 95°   e cos 147°   f cos 106°

2. Write each of the following in terms of the cosine of another angle between 0° and 360°.
   a cos 98°   b cos 144°   c cos 160°
   d cos 183°   e cos 211°   f cos 234°

3. Write each of the following in terms of the cosine of another angle between 0° and 180°.
   a −cos 100°   b −cos 90°   c −cos 110°
   d −cos 45°   e −cos 122°   f −cos 25°

4. The cosine of which acute angle has the same value as:
   a cos 125°   b cos 107°   c −cos 120°
   d −cos 98°   e −cos 92°   f −cos 110°?

5. Explain with reference to a right-angled triangle why the tangent of 90° is undefined.

Solving trigonometric equations

Knowledge of the graphs of the trigonometric functions enables us to solve trigonometric equations.

Worked examples

This ratio is one of the special angles covered earlier. Therefore, a calculator is not really needed to work out the size of A.

a Angle A is an obtuse angle. If $\sin A = \frac{\sqrt{3}}{2}$, calculate the size of A.

$$\sin A = \frac{\sqrt{3}}{2}$$

$$A = \sin^{-1}\left(\frac{\sqrt{3}}{2}\right)$$

$$A = 60°.$$

However, the question states that A is an obtuse angle (i.e. $90° < A < 180°$), therefore $A \neq 60°$. Because of the symmetry properties of the sine curve, it can be deduced that $\sin 60° = \sin 120°$ as shown.

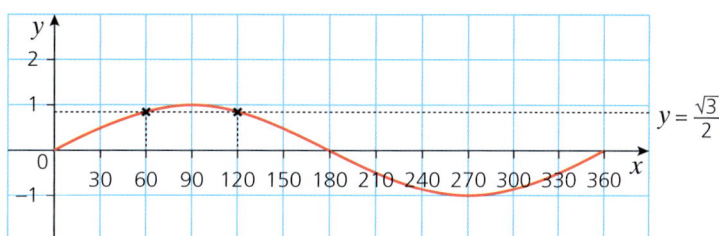

Therefore $A = 120°$.

29 TRIGONOMETRY

This ratio is also of a special angle. A calculator is therefore not needed.

b If $\tan x = \frac{1}{\sqrt{3}}$, calculate the possible values for x in the range $0 \leqslant x \leqslant 360$.

$\tan x = \frac{1}{\sqrt{3}}$

$x = \tan^{-1}\left(\frac{1}{\sqrt{3}}\right) = 30°$

But the graph of $y = \tan x$ has a period of $180°$. Therefore another solution in the range would be $30° + 180° = 210°$.

Therefore $x = 30°, 210°$.

Exercise 29.9

*Try and do parts **b**, **c** and **d** without a calculator.*

1 Solve each of the following equations, giving all the solutions in the range $0 \leqslant x \leqslant 360$.

a $\sin x = \frac{1}{4}$ b $\cos x = \frac{1}{\sqrt{2}}$ c $\sin x = -\frac{1}{2}$

d $\tan x = -\sqrt{3}$ e $5\cos x + 1 = 2$ f $\frac{1}{2}\tan x + 2 = 1$

2 In the triangle below, $\tan x = \frac{3}{4}$

Deduce, without a calculator, the value of:
a $\sin x$
b $\cos x$

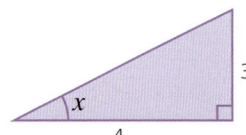

3 In the triangle below, $\sin x = \frac{3}{10}$

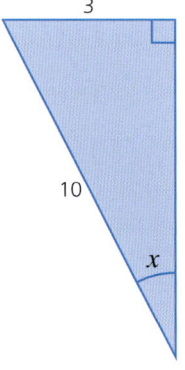

Deduce, without a calculator, the exact value of:
a $\cos x$
b $\tan x$

4 By using the triangle below as an aid, explain why the solution to the equation $\sin x = \cos x$ occurs when $x = 45°$.

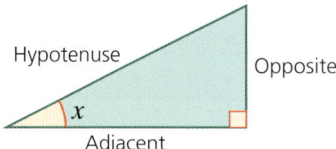

Student assessment 1

1 Calculate the length of the side marked x cm in each of the following.

a

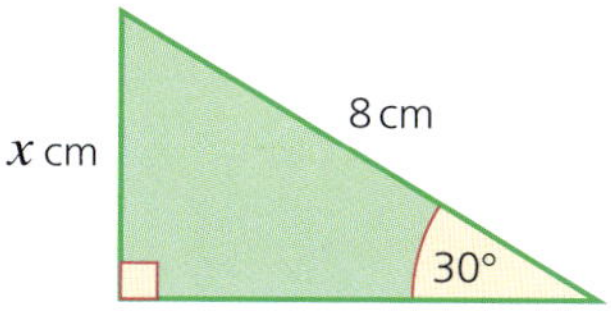
8 cm
x cm
30°

b

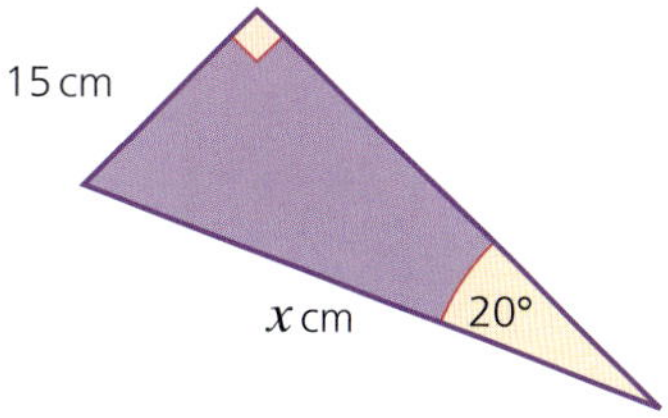
15 cm
x cm
20°

c

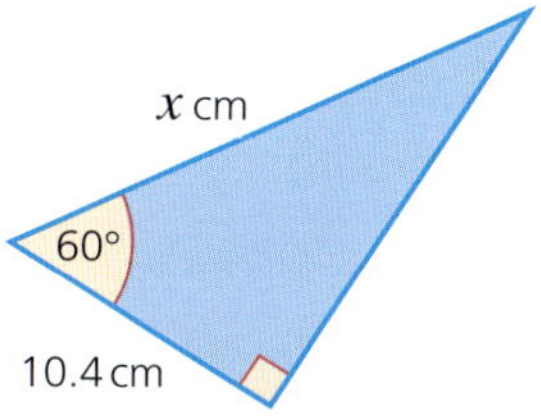
x cm
60°
10.4 cm

d

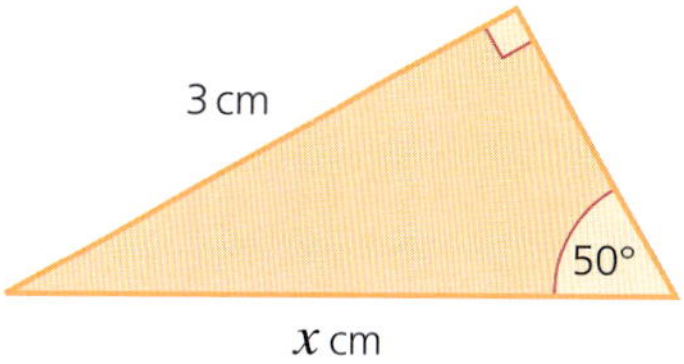
3 cm
50°
x cm

2 Calculate the size of the angle marked z in each of the following.

a

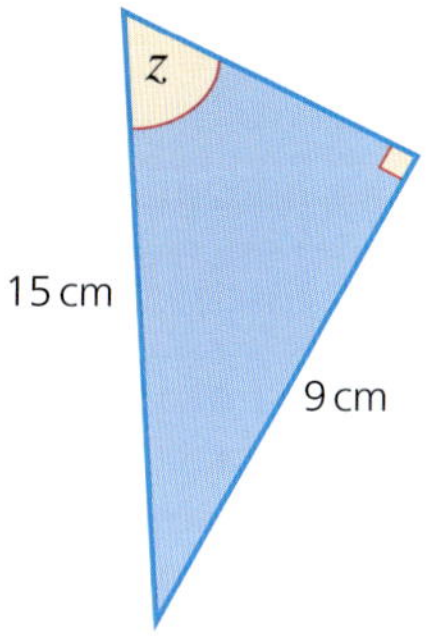
z
15 cm
9 cm

b

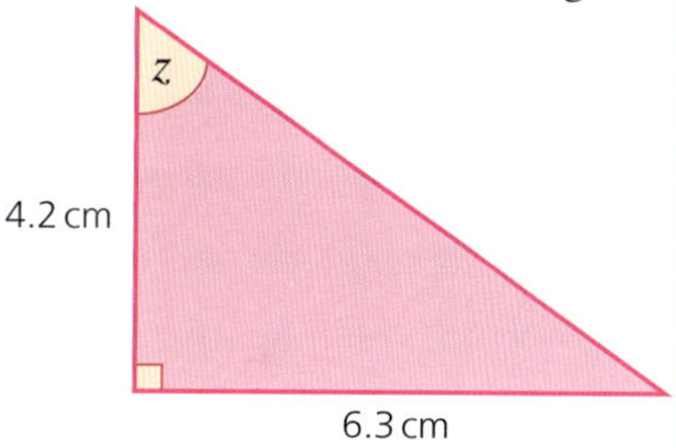
z
4.2 cm
6.3 cm

c

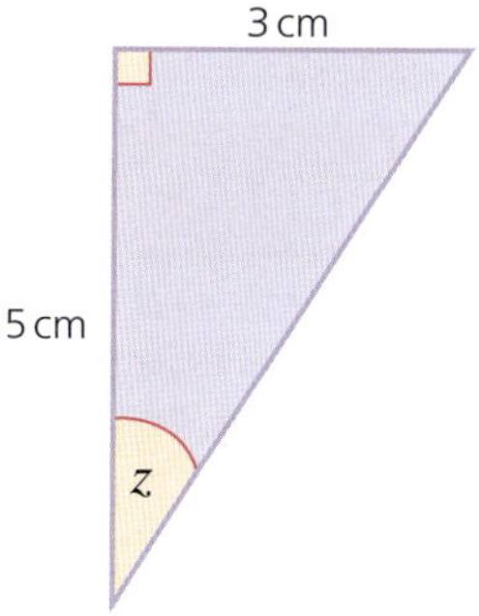
3 cm
5 cm
z

d

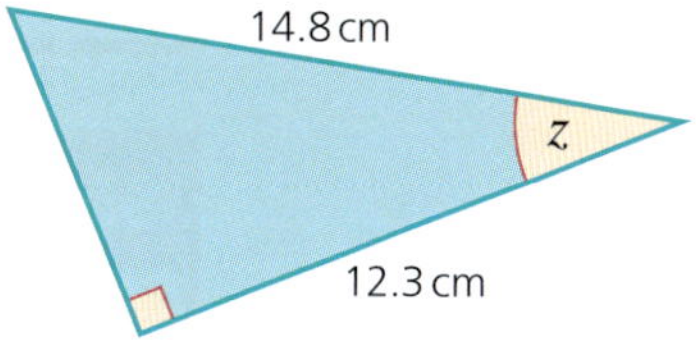
14.8 cm
z
12.3 cm

3 Calculate the length of the side marked q cm in each of the following.

a

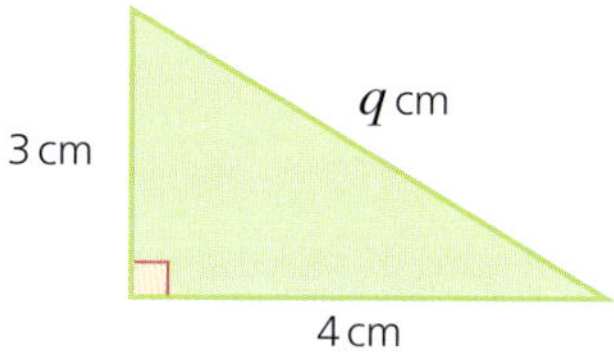
q cm
3 cm
4 cm

29 TRIGONOMETRY

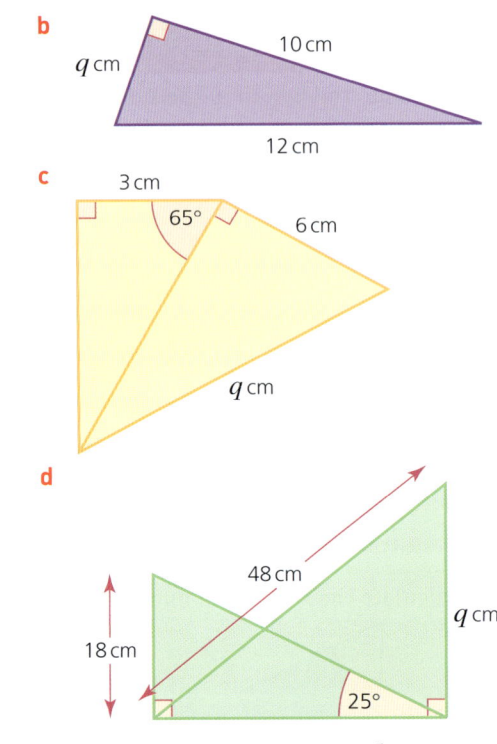

b

c

d

4 In the triangle below, $\cos x = \frac{5}{6}$.

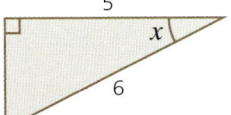

Showing your working clearly, deduce the exact value of:
a $\sin x$
b $\tan x$

Student assessment 2

1 A map shows three towns A, B and C. Town A is due north of C. Town B is due east of A. The distance AC is 75 km and the bearing of C from B is 245°. Calculate, giving your answers to the nearest 100 m:

a the distance AB,
b the distance BC.

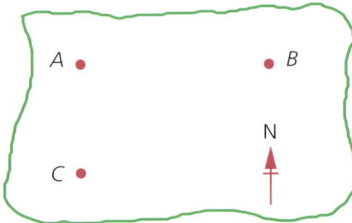

2 Two boats X and Y, sailing in a race, are shown in the diagram. Boat X is 145 m due north of a buoy B. Boat Y is due east of buoy B. Boats X and Y are 320 m apart. Calculate:

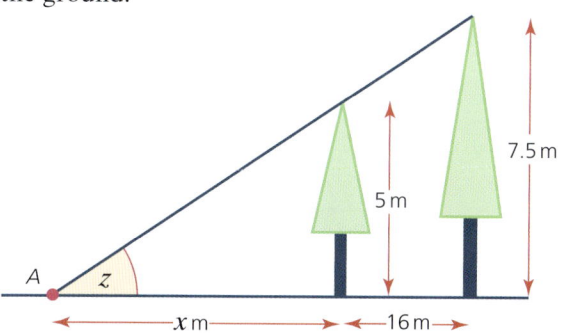

a the distance BY,
b the bearing of Y from X,
c the bearing of X from Y.

3 Two trees stand 16 m apart. Their tops make an angle z at point A on the ground.

a Express z in terms of the height of the shorter tree and its distance x metres from point A.
b Express z in terms of the height of the taller tree and its distance from A.
c Form an equation in terms of x.
d Calculate the value of x.
e Calculate the value z.

4 Two hawks P and Q are flying vertically above one another. Hawk Q is 250 m above hawk P. They both spot a snake at R. Using the information given, calculate:
a the height of P above the ground,
b the distance between P and R,
c the distance between Q and R.

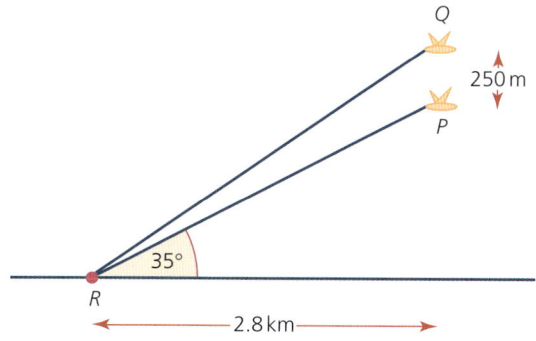

5 Solve the following trigonometric equations, giving all the solutions in the range $0 \leqslant x \leqslant 360$.

a $\sin x = \dfrac{2}{5}$
b $\tan x = -\dfrac{\sqrt{3}}{3}$
c $\cos x = -0.1$
d $\sin x = 1$

29 TRIGONOMETRY

Student assessment 3

1. Explain, with the aid of a graph, why the equation $\cos x = \frac{3}{2}$ has no solutions.

2. The cosine of which other angle between 0 and 180° has the same value as:
   a $\cos 128°$
   b $-\cos 80°$?

3. A circle of radius 3 cm, centre at O, is shown on the axes below.

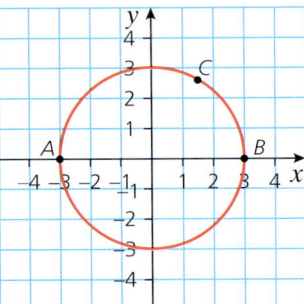

   The points A and B lie where the circumference of the circle intersects the x-axis. Point C is free to move on the circumference of the circle.
   a Deduce, justifying your answer, the size of angle ACB.
   b If $BC = 3$ cm, calculate the possible coordinates of point C, giving your answers in exact form.

4. The Great Pyramid at Giza is 146 m high. Two people A and B are looking at the top of the pyramid. The angle of elevation of the top of the pyramid from B is 12°. The distance between A and B is 25 m.

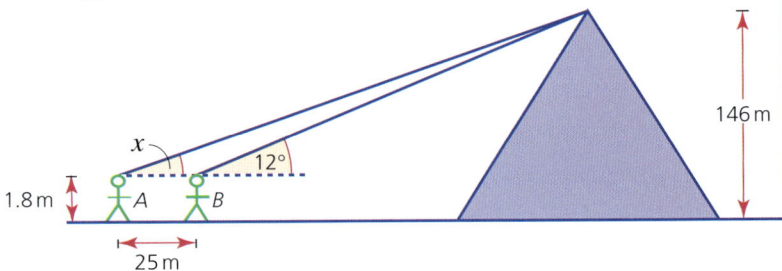

   If both A and B are 1.8 m tall, calculate:
   a the distance on the ground from B to the centre of the base of the pyramid,
   b the angle of elevation x of the top of the pyramid from A,
   c the distance between A and the top of the pyramid.
   Note: A, B and the top of the pyramid are in the same vertical plane.

5 Two hot air balloons A and B are travelling in the same horizontal direction as shown in the diagram below. A is travelling at $2\,\text{m/s}$ and B at $3\,\text{m/s}$. Their heights above the ground are $1.6\,\text{km}$ and $1\,\text{km}$, respectively.

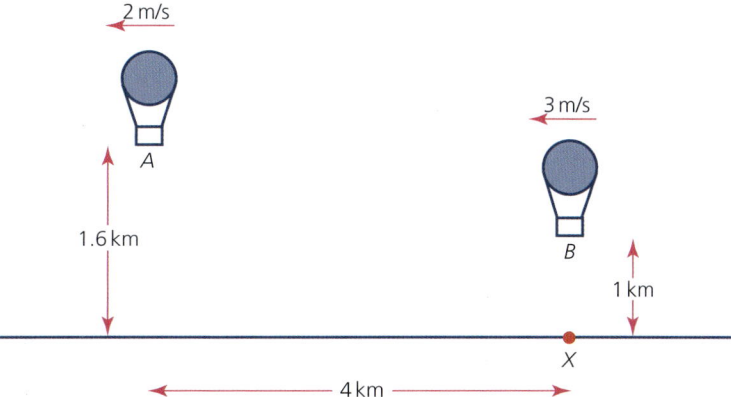

At midday, their horizontal distance apart is $4\,\text{km}$ and balloon B is directly above a point X on the ground.
Calculate:
a the angle of elevation of A from X at midday,
b the angle of depression of B from A at midday,
c their horizontal distance apart at 12 30,
d the angle of elevation of B from X at 12 30,
e the angle of elevation of A from B at 12 30,
f how much closer A and B are at 12 30 compared with midday.

6 a On one diagram, plot the graph of $y = \sin x°$ and the graph of $y = \cos x°$, for $0° \leqslant x° \leqslant 180°$.
b Use your graph to find the angles for which $\sin x° = \cos x°$.

30 Further trigonometry

The sine rule

With right-angled triangles, we can use the basic trigonometric ratios of sine, cosine and tangent. The **sine rule** is a relationship which can be used with non-right-angled triangles.

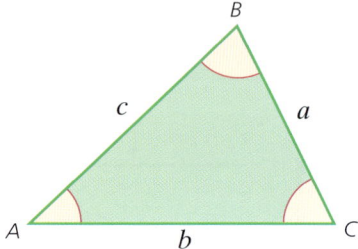

The sine rule states that:

$$\frac{a}{\sin A} = \frac{b}{\sin B} = \frac{c}{\sin C}$$

You should know this formula, but you do not need to memorise it.

or alternatively

$$\frac{\sin A}{a} = \frac{\sin B}{b} = \frac{\sin C}{c}$$

➡ Worked examples

a Calculate the length of side BC.
Using the sine rule:

$$\frac{a}{\sin A} = \frac{b}{\sin B}$$

$$\frac{a}{\sin 40°} = \frac{6}{\sin 30°}$$

$$a = \frac{6 \times \sin 40°}{\sin 30°}$$

$a = 7.7$ (1 d.p.)

$BC = 7.7$ cm

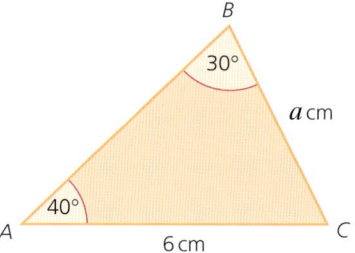

b Calculate the size of angle C.
Using the sine rule:

$$\frac{\sin A}{a} = \frac{\sin C}{c}$$

$$\sin C = \frac{6.5 \times \sin 60°}{6}$$

$C = \sin^{-1}(0.94)$

$C = 69.8°$ (1 d.p.)

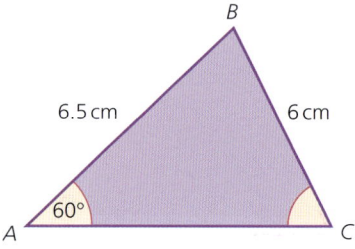

The sine rule

In the worked example **b** above, $C = 69.8°$ is not the only possible solution from the information given, as the question does not state that the angle is acute and often, diagrams are not drawn to scale.

The triangle ABC_1 is as shown in **b** above and therefore, angle $AC_1B = 69.8°$ as calculated.

However, if a circle of radius 6 cm and centre at B is drawn, then it can be seen to intersect the base of the triangle in another place, C_2.

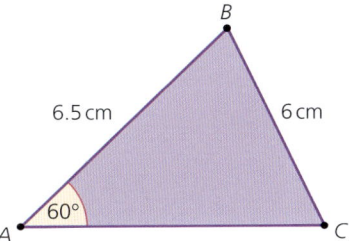

Therefore, triangle ABC_2 is another possible triangle.

But angle AC_2B is not the same size as angle AC_1B.

It can be calculated in the same way as before, however. Using our knowledge of the sine curve, it can be calculated that angle AC_2B is $180 - 69.8 = 110.2°$

This is known as the **ambiguous case** of the sine rule as there are two possible answers.

Exercise 30.1

1 Calculate the length of the side marked x in each of the following.

a
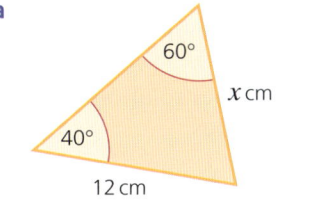

b
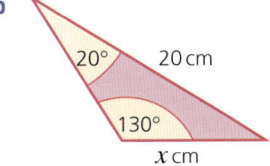

c
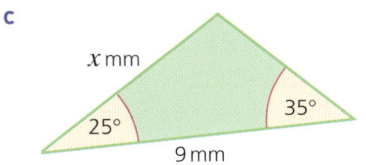

d
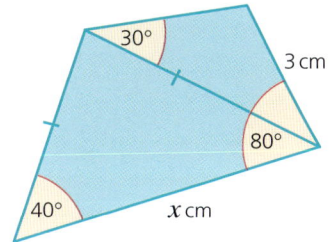

30 FURTHER TRIGONOMETRY

Exercise 30.1 (cont)

2 Calculate the size of the angle marked x in each of the following. If two values for the angle are possible, give both values.

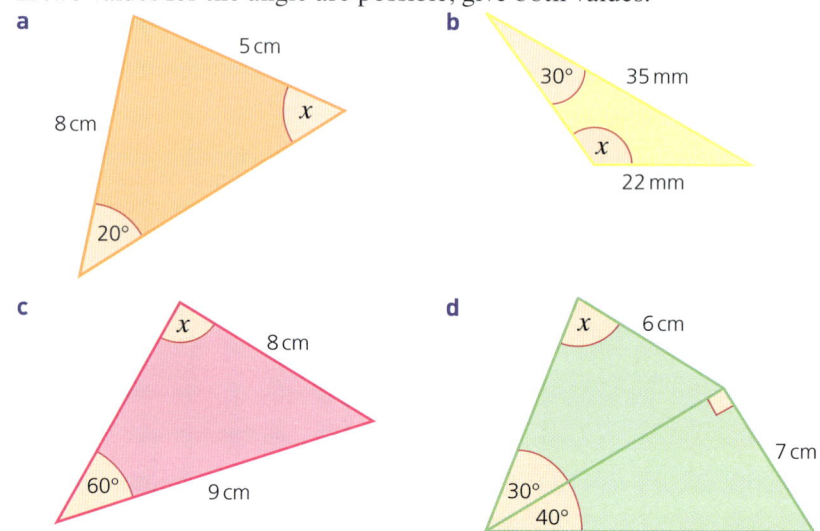

3 Triangle ABC has the following dimensions:
$AC = 10$ cm, $AB = 8$ cm and angle $ACB = 20°$.
  a Calculate the two possible values for angle CBA.
  b Sketch and label the two possible shapes for triangle ABC.

4 Triangle PQR has the following dimensions:
$PQ = 6$ cm, $PR = 4$ cm and angle $PQR = 40°$.
  a Calculate the two possible values for angle QRP.
  b Sketch and label the two possible shapes for triangle PQR.

The cosine rule

The **cosine rule** is another relationship which can be used with non-right-angled triangles.

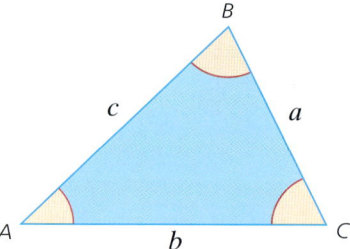

The cosine rule states that:

$$a^2 = b^2 + c^2 - 2bc \cos A$$

You should know this formula, but you do not need to memorise it.

The cosine rule

→ Worked examples

a Calculate the length of the side BC.
Using the cosine rule:
$$a^2 = b^2 + c^2 - 2bc \cos A$$
$$a^2 = 9^2 + 7^2 - (2 \times 9 \times 7 \times \cos 50°)$$
$$= 81 + 49 - (126 \times \cos 50°) = 49.0$$
$$a = \sqrt{49.0}$$
$$a = 7.00 \text{ (3 s.f.)}$$
$$BC = 7.00 \text{ cm (3 s.f.)}$$

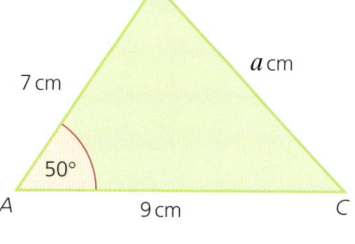

b Calculate the size of angle A.
Using the cosine rule:
$$a^2 = b^2 + c^2 - 2bc \cos A$$
Rearranging the equation gives:
$$\cos A = \frac{b^2 + c^2 - a^2}{2bc}$$
$$\cos A = \frac{15^2 + 12^2 - 20^2}{2 \times 15 \times 12} = -0.086$$
$$A = \cos^{-1}(-0.086)$$
$$A = 94.9° \text{ (1 d.p.)}$$

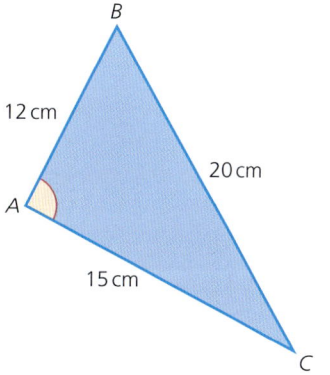

Exercise 30.2

1 Calculate the length of the side marked x in each of the following.

a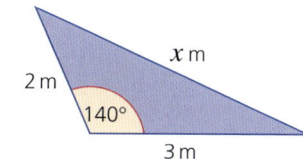

b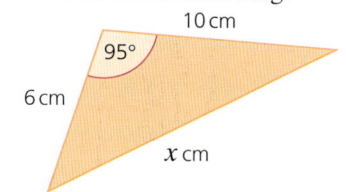

c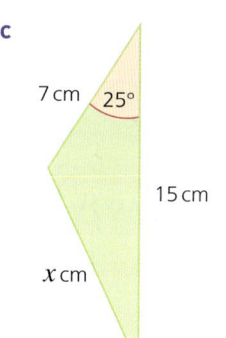

d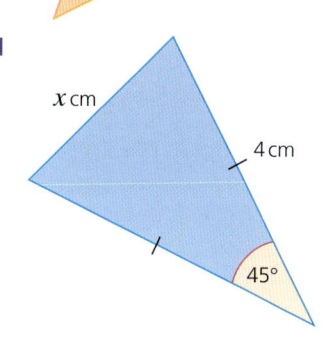

30 FURTHER TRIGONOMETRY

Exercise 30.2 (cont)

e
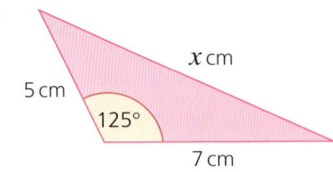

2 Calculate the angle marked x in each of the following.

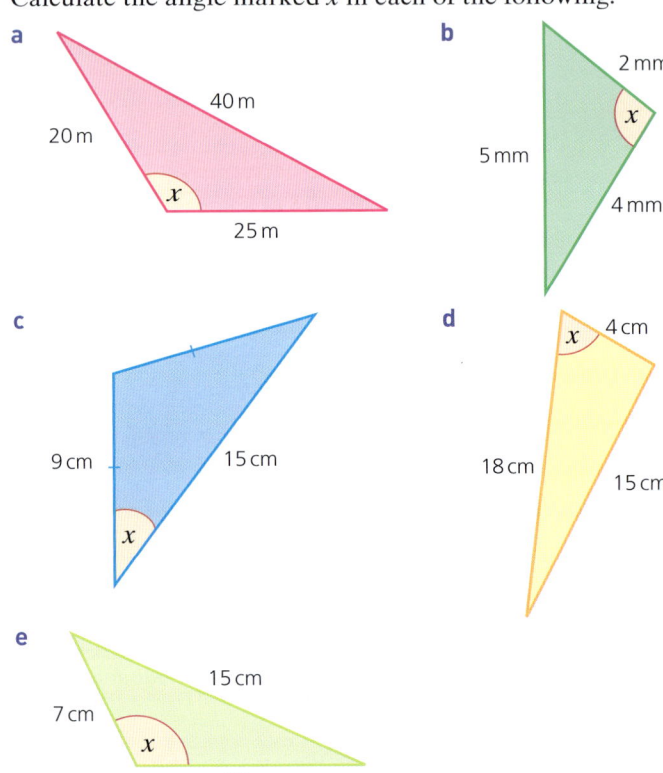

Exercise 30.3

1 Four players, W, X, Y and Z, are on a rugby pitch. The diagram shows a plan view of their relative positions.
Calculate:
  a the distance between players X and Z,
  b angle ZWX,
  c angle WZX,
  d angle YZX,
  e the distance between players W and Y.

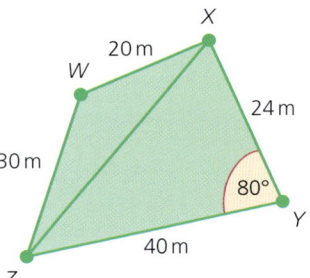

The area of a triangle

2. Three yachts, A, B and C, are racing off Cape Comorin in India. Their relative positions are shown (below).

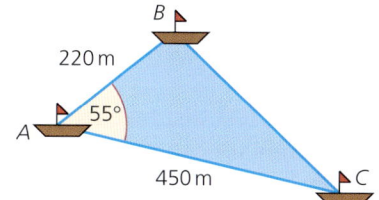

Calculate the distance between B and C to the nearest 10 m.

3. There are two trees standing on one side of a river bank. On the opposite side, a boy is standing at X.

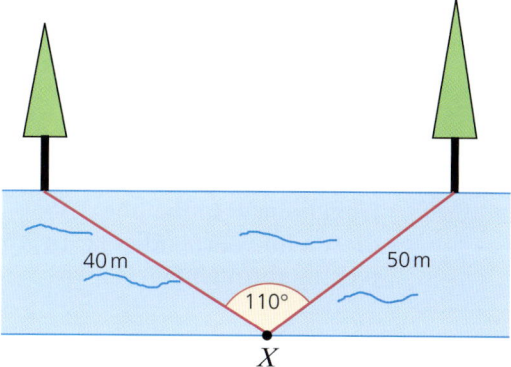

Using the information given, calculate the distance between the two trees.

The area of a triangle

Area $= \frac{1}{2}bh$

Also:

$\sin C = \frac{h}{a}$

Rearranging:

$h = a \sin C$

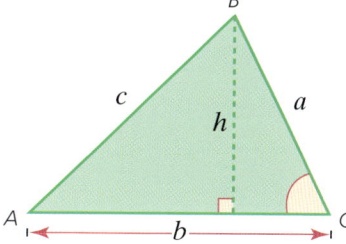

Therefore

→ Area $= \frac{1}{2}ab\sin C$

You should know this formula, but you do not need to memorise it.

30 FURTHER TRIGONOMETRY

Shortest distance from a point to a line

The height of a triangle is measured perpendicular to the base of the triangle, as shown above.

In general, the shortest distance from a point to a line is the distance measured perpendicular to the line and passing through the point.

e.g.

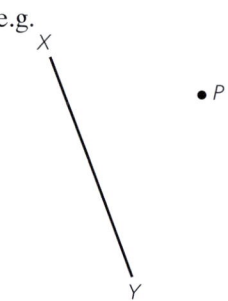
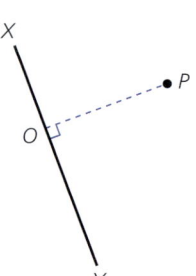

To calculate the shortest distance from point P to the line XY, draw a line perpendicular to XY passing through P.

The shortest distance from P to the line XY is therefore the distance OP.

Exercise 30.4

1 Calculate the area of the following triangles.

a
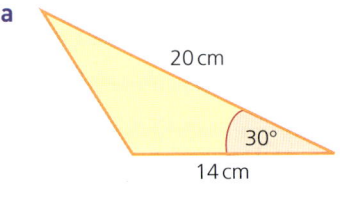

b
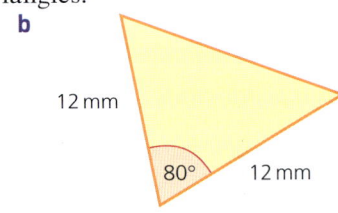

c

d
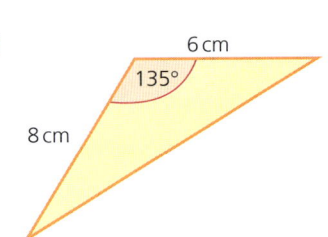

Shortest distance from a point to a line

2 Calculate the value of x in each of the following.

a 12 cm, area = 40 cm², 16 cm, $x°$

b 9 cm, 160°, area = 20 cm², x cm

c x cm, 60°, area = 150 cm², 15 cm

d $x°$, 14 cm, 8 cm, area = 50 cm²

3 A straight stretch of coast AB is shown below. A ship S at sea is 1.1 km from A.

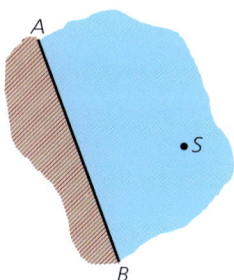

If angle $SAB = 25°$, calculate the shortest distance between the ship and the coast, giving your answer to the nearest metre.

4 The four corners, A, B, C and D, of a large nature reserve form a quadrilateral and are shown below.
$AC = 70$ km, $AD = 45$ km and $AB = 60$ km.

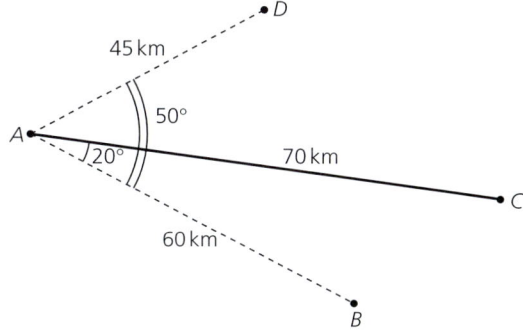

If angle $BAC = 20°$ and angle $BAD = 50°$, calculate the area of the nature reserve, giving your answer to the nearest square kilometre.

30 FURTHER TRIGONOMETRY

Exercise 30.4 (cont)

5 *ABCD* is a school playing field. The following lengths are known:
$OA = 83\,m$,
$OB = 122\,m$,
$OC = 106\,m$,
$OD = 78\,m$.
Calculate the area of the school playing field to the nearest $100\,m^2$.

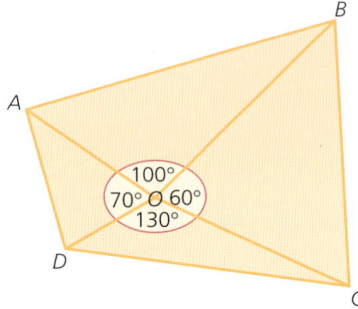

6 The roof of a garage has a slanting length of 3 m and makes an angle of 120° at its vertex. The height of the walls of the garage is 4 m and its depth is 9 m.

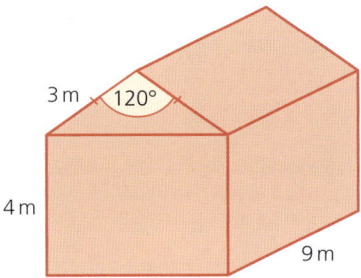

Calculate:
a the cross-sectional area of the roof,
b the volume occupied by the whole garage.

Trigonometry in three dimensions

Worked examples

The diagram (below) shows a cube of edge length 3 cm.

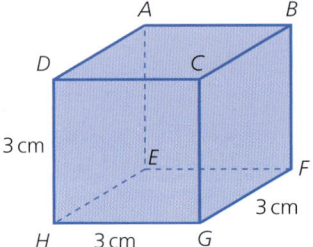

a Calculate the length EG.

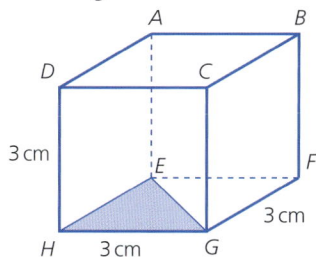 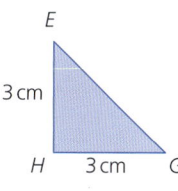

Triangle EHG (above) is right-angled. Use Pythagoras' theorem to calculate the length EG.

$(EG)^2 = (EH)^2 + (HG)^2$
$(EG)^2 = 3^2 + 3^2 = 18$
$EG = \sqrt{18}\,\text{cm} = 4.24\,\text{cm}$ (3 s.f.)

b Calculate the length AG.

Triangle AEG (below) is right-angled. Use Pythagoras' theorem to calculate the length AG.

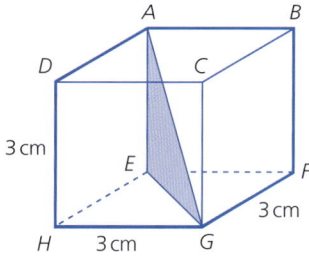 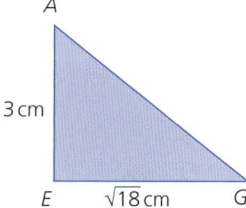

$(AG)^2 = (AE)^2 + (EG)^2$
$(AG)^2 = 3^2 + (\sqrt{18})^2$
$(AG)^2 = 9 + 18$
$AG = \sqrt{27}\,\text{cm} = 5.20\,\text{cm}$ (3 s.f.)

c Calculate the angle EGA.

To calculate angle EGA we use the triangle EGA:

$\tan G = \dfrac{3}{\sqrt{18}}$

$G = 35.3°$ (1 d.p.)

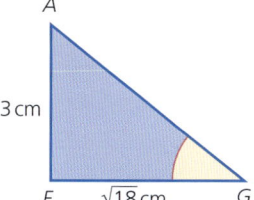

30 FURTHER TRIGONOMETRY

Exercise 30.5

1 a Calculate the length *HF*.
   b Calculate the length *HB*.
   c Calculate the angle *BHG*.

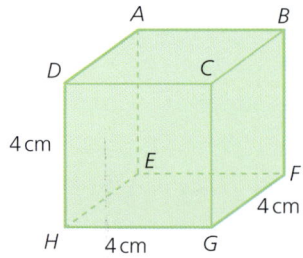

2 a Calculate the length *CA*.
   b Calculate the length *CE*.
   c Calculate the angle *ACE*.

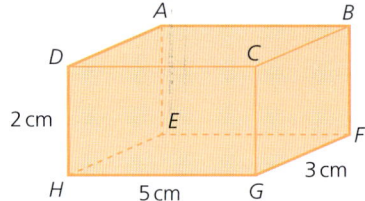

3 a Calculate the length *EG*.
   b Calculate the length *AG*.
   c Calculate the angle *AGE*.

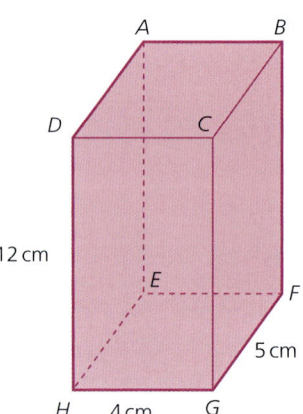

4 a Calculate the angle *BCE*.
   b Calculate the angle *GFH*.

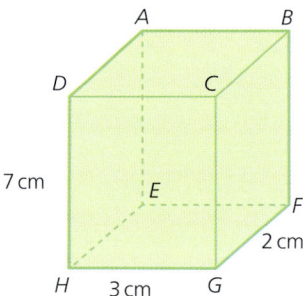

Trigonometry in three dimensions

5 The diagram shows a right pyramid where A is vertically above X.
  a i Calculate the length DB.
  ii Calculate the angle DAX.
  b i Calculate the angle CED.
  ii Calculate the angle DBA.

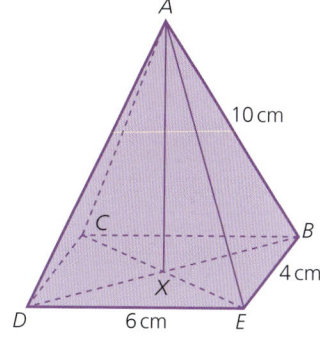

6 The diagram shows a right pyramid where A is vertically above X.
  a i Calculate the length CE.
  ii Calculate the angle CAX.
  b i Calculate the angle BDE.
  ii Calculate the angle ADB.

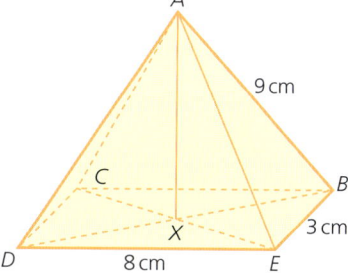

7 In this cone the angle YXZ = 60°. Calculate:
  a the length XY,
  b the length YZ,
  c the circumference of the base.

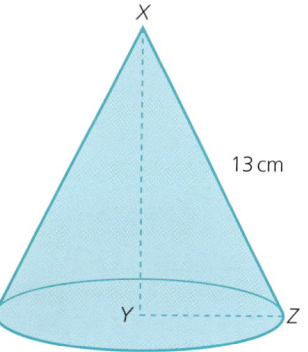

8 In this cone the angle XZY = 40°. Calculate:
  a the length XZ,
  b the length XY.

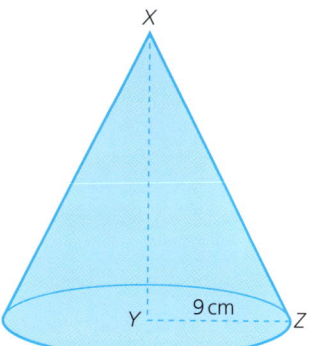

433

30 FURTHER TRIGONOMETRY

Exercise 30.5 (cont)

9 One corner of this cuboid has been sliced off along the plane QTU. $WU = 4$ cm.
   a Calculate the length of the three sides of the triangle QTU.
   b Calculate the three angles Q, T and U in triangle QTU.
   c Calculate the area of triangle QTU.

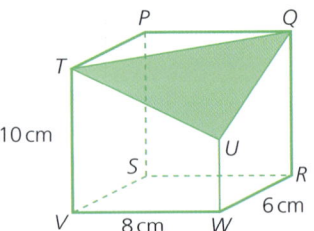

The angle between a line and a plane

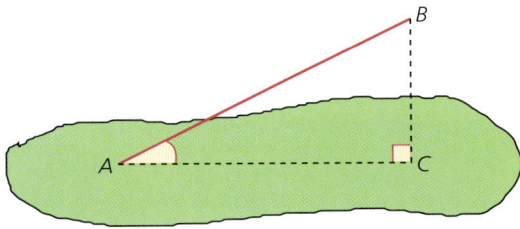

To calculate the size of the angle between the line AB and the shaded plane, drop a perpendicular from B. It meets the shaded plane at C. Then join AC.

The angle between the lines AB and AC represents the angle between the line AB and the shaded plane.

The line AC is the projection of the line AB on the shaded plane.

➡ Worked examples

a Calculate the length EC.

First use Pythagoras' theorem to calculate the length EG:

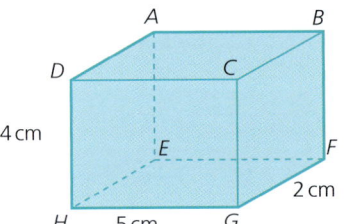

The angle between a line and a plane

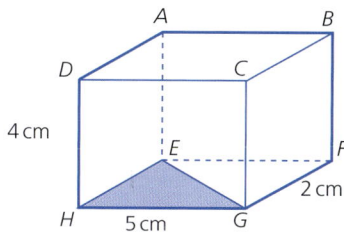 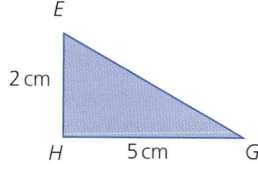

$(EG)^2 = (EH)^2 + (HG)^2$

$(EG)^2 = 2^2 + 5^2$

$(EG)^2 = 29$

$EG = \sqrt{29}$ cm

Now use Pythagoras' theorem to calculate CE:

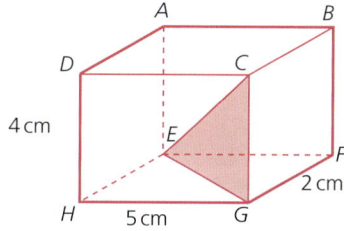 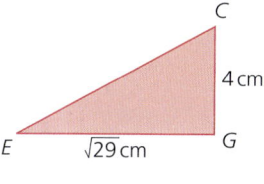

$(EC)^2 = (EG)^2 + (CG)^2$

$(EC)^2 = (\sqrt{29})^2 + 4^2$

$(EC)^2 = 29 + 16$

$EC = \sqrt{45}$ cm $= 6.71$ cm (3 s.f.)

b Calculate the angle between the line CE and the plane $ADHE$.

To calculate the angle between the line CE and the plane $ADHE$ use the right-angled triangle CED and calculate the angle CED.

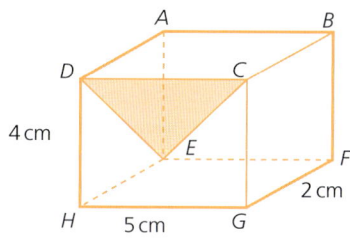 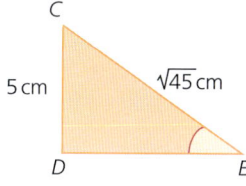

$\sin E = \dfrac{CD}{CE} = \dfrac{5}{\sqrt{45}}$

$E = \sin^{-1} \dfrac{5}{\sqrt{45}} = 48.2°$ (1 d.p.)

435

30 FURTHER TRIGONOMETRY

Exercise 30.6

1 Name the projection of each line onto the given plane:
   a *TR* onto *RSWV*
   b *TR* onto *PQUT*
   c *SU* onto *PQRS*
   d *SU* onto *TUVW*
   e *PV* onto *QRVU*
   f *PV* onto *RSWV*

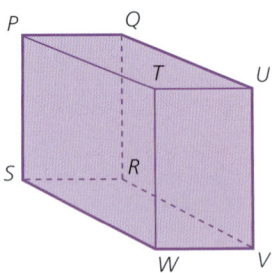

2 Name the projection of each line onto the given plane:
   a *KM* onto *IJNM*
   b *KM* onto *JKON*
   c *KM* onto *HIML*
   d *IO* onto *HLOK*
   e *IO* onto *JKON*
   f *IO* onto *LMNO*

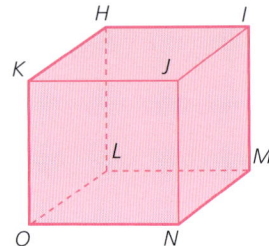

3 Name the angle between the given line and plane:
   a *PT* and *PQRS*
   b *PU* and *PQRS*
   c *SV* and *PSWT*
   d *RT* and *TUVW*
   e *SU* and *QRVU*
   f *PV* and *PSWT*

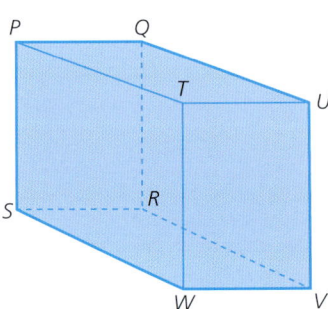

4 a Calculate the length *BH*.
   b Calculate the angle between the line *BH* and the plane *EFGH*.

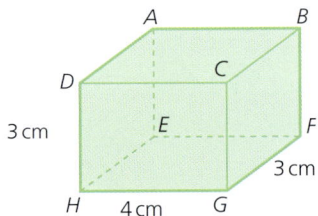

5 a Calculate the length *AG*.
   b Calculate the angle between the line *AG* and the plane *EFGH*.
   c Calculate the angle between the line *AG* and the plane *ADHE*.

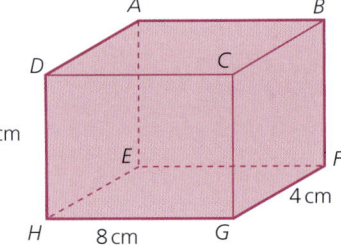

The angle between a line and a plane

6 The diagram shows a right pyramid where A is vertically above X.
  a Calculate the length BD.
  b Calculate the angle between AB and $CBED$.

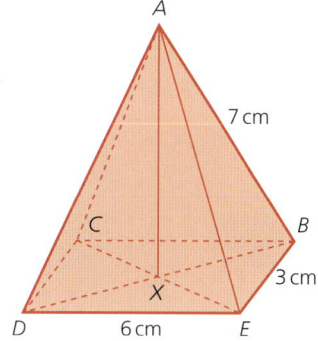

7 The diagram shows a right pyramid where U is vertically above X.
  a Calculate the length WY.
  b Calculate the length UX.
  c Calculate the angle between UX and UZY.

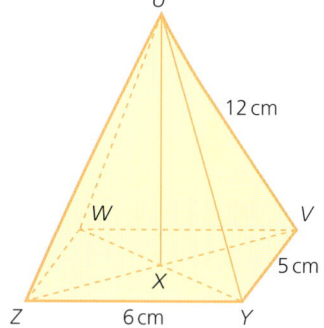

8 $ABCD$ and $EFGH$ are square faces lying parallel to each other. Calculate:
  a the length DB,
  b the length HF,
  c the vertical height of the object,
  d the angle DH makes with the plane $ABCD$.

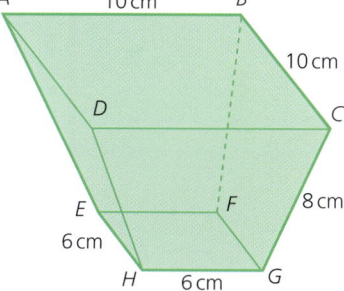

9 $ABCD$ and $EFGH$ are square faces lying parallel to each other. Calculate:
  a the length AC,
  b the length EG,
  c the vertical height of the object,
  d the angle CG makes with the plane $EFGH$.

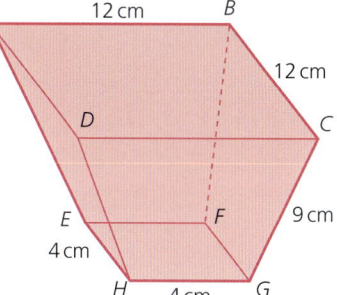

30 FURTHER TRIGONOMETRY

Student assessment 1

1. The triangle *PQR* is shown right.
   a. Calculate the two possible values for angle *PRQ*.
   b. Calculate the shortest distance possible of *R* from side *PQ*.

   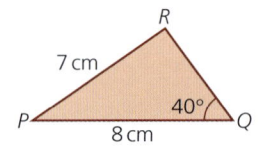

2. Calculate the size of the obtuse angle marked *x* in the triangle (right).

   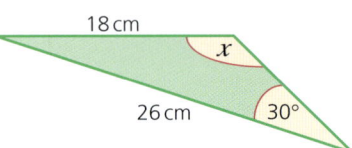

3. The area of the triangle is 10.5 cm^2.
   a. Calculate the value of sin *x*.
   b. If *x* is an obtuse angle, calculate the value of *x*.

   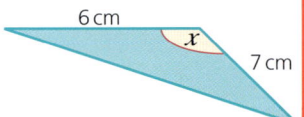

4. For the cuboid, calculate:
   a. the length *EG*,
   b. the length *EC*,
   c. angle *BEC*.

   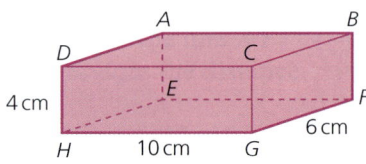

5. For the quadrilateral (right), calculate:
   a. the length *JL*,
   b. angle *KJL*,
   c. the length *JM*,
   d. the area of *JKLM*.

   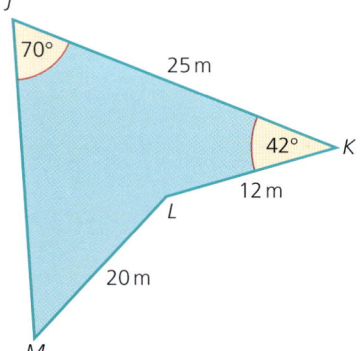

6. For the square-based right pyramid, calculate:
   a. the length *BD*,
   b. angle *ABD*,
   c. the area of triangle *ABD*,
   d. the vertical height of the pyramid.

   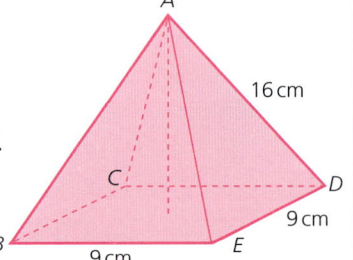

The angle between a line and a plane

Student assessment 2

1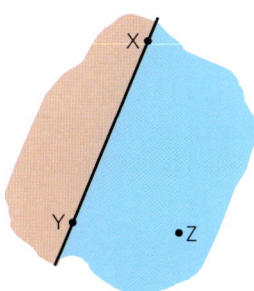

Two points *X* and *Y* are 1000 m apart on a straight piece of coastline. A buoy *Z* is out at sea. *XZ* = 850 m and *YZ* = 625 m. Imani wishes to swim from the coast to the buoy.
  a What is the shortest distance she will have to swim? Give your answer to 3 s.f.
  b How far along the coast from *Y* will she have to set off in order to swim the least distance?

2 Using the triangular prism, calculate:
  a the length *AD*,
  b the length *AC*,
  c the angle *AC* makes with the plane *CDEF*,
  d the angle *AC* makes with the plane *ABFE*.

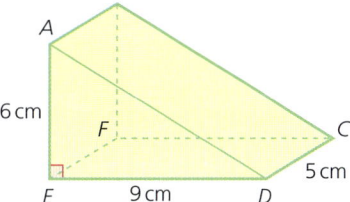

3 For the triangle, calculate:
  a the length *PS*,
  b angle *QRS*,
  c the length *SR*.

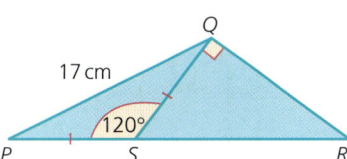

4 The cuboid (right) has one of its corners removed to leave a flat triangle *BDC*.
Calculate:
  a length *DC*,
  b length *BC*,
  c length *DB*,
  d angle *CBD*,
  e the area of triangle *BDC*,
  f the angle *AC* makes with the plane *AEHD*.

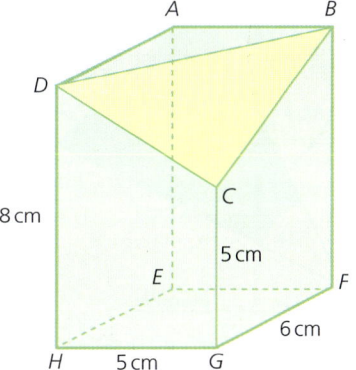

6 Mathematical investigations and ICT 6

Numbered balls

The balls below start with the number 25 and then subsequent numbered balls are added according to a rule. The process stops when ball number 1 is added.

1. Express in words the rule for generating the sequence of numbered balls.
2. What is the longest sequence of balls starting with a number less than 100?
3. Is there a strategy for generating a long sequence?
4. Use your rule to state the longest sequence of balls starting with a number less than 1000.
5. Extend the investigation by having a different term-to-term rule.

Towers of Hanoi

This investigation is based on an old Vietnamese legend. The legend is as follows:

At the beginning of time a temple was created by the Gods. Inside the temple stood three giant rods. On one of these rods, 64 gold discs, all of different diameters, were stacked in descending order of size, i.e. the largest at the bottom rising to the smallest at the top. Priests at the temple were responsible for moving the discs onto the remaining two rods until all 64 discs were stacked in the same order on one of the other rods. When this task was completed, time would cease and the world would come to an end.

The discs, however, could only be moved according to certain rules. These were:
» Only one disc could be moved at a time.
» A disc could only be placed on top of a larger one.

The diagram (left) shows the smallest number of moves required to transfer three discs from the rod on the left to the rod on the right.

With three discs, the smallest number of moves is seven.
1. What is the smallest number of moves needed for two discs?
2. What is the smallest number of moves needed for four discs?
3. Investigate the smallest number of moves needed to move different numbers of discs.

4 Display the results of your investigation in an ordered table.
5 Describe any patterns you see in your results.
6 Predict, from your results, the smallest number of moves needed to move ten discs.
7 Determine a formula for the smallest number of moves for n discs.
8 Assume the priests have been transferring the discs at the rate of one per second and assume the Earth is approximately 4.54 billion years old (4.54×10^9 years).

According to the legend, is the world coming to an end soon? Justify your answer with relevant calculations.

ICT activity

In this activity you will need to use a graphics calculator to investigate the relationship between different trigonometric ratios.

1 a Using the calculator, plot the graph of $y = \sin x$ for $0° \leqslant x \leqslant 360°$. The graph should look similar to the one shown below:

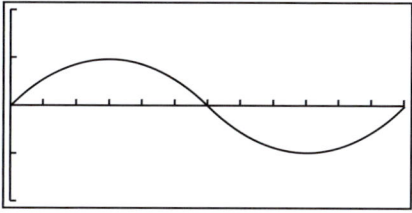

b Using the equation solving facility, evaluate the following:
i sin 70°
ii sin 125°
iii sin 300°
c Referring to the graph, explain why $\sin x = 0.7$ has two solutions between 0° and 360°.
d Use the graph to solve the equation $\sin x = 0.5$.
2 a On the same axes as before, plot $y = \cos x$.
b How many solutions are there to the equation $\sin x = \cos x$ between 0° and 360°?
c What is the solution to the equation $\sin x = \cos x$ between 180° and 270°?
3 By plotting appropriate graphs, solve the following for $0° \leqslant x \leqslant 360°$.
a $\sin x = \tan x$
b $\cos x = \tan x$

TOPIC 7

Vectors and transformations

Contents

Chapter 31 Vectors (E7.2, E7.3, E7.4)
Chapter 32 Transformations (E7.1)

Learning objectives

E7.1 Transformations
Recognise, describe and draw the following transformations:
1. Reflection of a shape in a straight line.
2. Rotation of a shape about a centre through multiples of 90°.
3. Enlargement of a shape from a centre by a scale factor.
4. Translation of a shape by a vector $\begin{pmatrix} x \\ y \end{pmatrix}$.

E7.2 Vectors in two dimensions
1. Describe a translation using a vector represented by $\begin{pmatrix} x \\ y \end{pmatrix}$, $\overrightarrow{AB}$ or **a**.
2. Add and subtract vectors.
3. Multiply a vector by a scalar.

E7.3 Magnitude of a vector
Calculate the magnitude of a vector $\begin{pmatrix} x \\ y \end{pmatrix}$ as $\sqrt{x^2 + y^2}$.

E7.4 Vector geometry
1. Represent vectors by directed line segments.
2. Use position vectors.
3. Use the sum and difference of two or more vectors to express given vectors in terms of two coplanar vectors.
4. Use vectors to reason and to solve geometric problems.

The Italians

Leonardo Pisano (known today as Fibonacci) introduced new methods of arithmetic to Europe from the Hindus, Persians and Arabs. He discovered the sequence 1, 1, 2, 3, 5, 8, 13, … , which is now called the Fibonacci sequence, and some of its occurrences in nature. He also brought the decimal system, algebra and the 'lattice' method of multiplication to Europe. Fibonacci has been called the 'most talented mathematician of the middle ages'. Many books say that he brought Islamic mathematics to Europe, but in Fibonacci's own introduction to *Liber Abaci*, he credits the Hindus.

Fibonacci (1170–1250)

The Renaissance began in Italy. Art, architecture, music and the sciences flourished. Girolamo Cardano (1501–1576) wrote his great mathematical book *Ars Magna* (Great Art) in which he showed, among much algebra that was new, calculations involving the solutions to cubic equations. He published this book, the first algebra book in Latin, to great acclaim. However, although he continued to study mathematics, no other work of his was ever published.

31 Vectors

Translations

A **translation** (a sliding movement) can be described using **column vectors**. A column vector describes the movement of the object in both the *x* direction and the *y* direction.

> ### Worked examples

Define **a** and **b** in the diagram (left) using column vectors.

$$\mathbf{a} = \begin{pmatrix} 2 \\ 2 \end{pmatrix} \quad \mathbf{b} = \begin{pmatrix} -2 \\ 1 \end{pmatrix}$$

Note: When you represent **vectors** by single letters, i.e. **a**, in handwritten work, you should write them as a̰.

If $\mathbf{a} = \begin{pmatrix} 2 \\ 5 \end{pmatrix}$ and $\mathbf{b} = \begin{pmatrix} -3 \\ -2 \end{pmatrix}$, they can be represented diagrammatically.

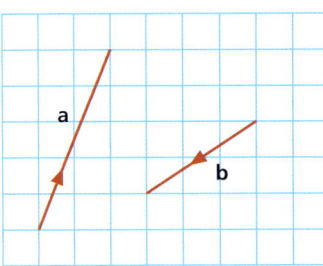

The diagrammatic representation of −**a** and −**b** is shown below.

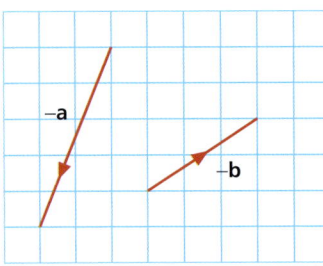

It can be seen from the diagram above that $-\mathbf{a} = \begin{pmatrix} -2 \\ -5 \end{pmatrix}$ and $-\mathbf{b} = \begin{pmatrix} 3 \\ 2 \end{pmatrix}$.

Translations can also be named using letters to represent the start and end point, with an arrow above the letters showing the direction of the translation.

Translations

→ Worked examples

Note

The notation $\vec{AB}$ or **a** for vectors is only required for the Extended syllabus.

a Describe the translation from A to B in the diagram in terms of a column vector.

$$\vec{AB} = \begin{pmatrix} 1 \\ 3 \end{pmatrix}$$

i.e. 1 unit in the x direction, 3 units in the y direction.

b Describe $\vec{BC}, \vec{CD}$ and $\vec{DA}$ in terms of column vectors.

$$\vec{BC} = \begin{pmatrix} 2 \\ 0 \end{pmatrix} \quad \vec{CD} = \begin{pmatrix} 0 \\ -2 \end{pmatrix} \quad \vec{DA} = \begin{pmatrix} -3 \\ -1 \end{pmatrix}$$

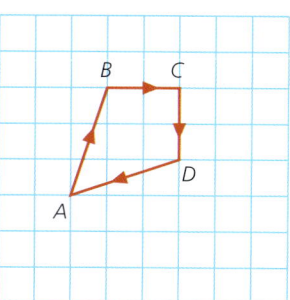

Exercise 31.1

1 Describe each translation using a column vector.
   a **a**   b **b**   c **c**   d **d**   e **e**
   f **−b**   g **−c**   h **−d**   i **−a**

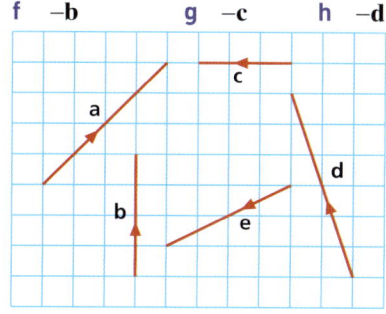

2 Draw and label the following vectors on a square grid.

   a $\mathbf{a} = \begin{pmatrix} 2 \\ 4 \end{pmatrix}$ b $\mathbf{b} = \begin{pmatrix} -3 \\ 6 \end{pmatrix}$ c $\mathbf{c} = \begin{pmatrix} 3 \\ -5 \end{pmatrix}$

   d $\mathbf{d} = \begin{pmatrix} -4 \\ -3 \end{pmatrix}$ e $\mathbf{e} = \begin{pmatrix} 0 \\ -6 \end{pmatrix}$ f $\mathbf{f} = \begin{pmatrix} -5 \\ 0 \end{pmatrix}$

   g **g** = −**c**   h **h** = −**b**   i **i** = −**f**

3 Describe each translation using a column vector.
   a $\vec{AB}$ b $\vec{BC}$ c $\vec{CD}$ d $\vec{DE}$ e $\vec{EA}$
   f $\vec{AE}$ g $\vec{DA}$ h $\vec{CA}$ i $\vec{DB}$

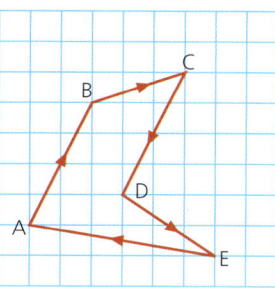

445

31 VECTORS

Addition and subtraction of vectors

Vectors can be added together and represented diagrammatically as shown.

The translation represented by **a** followed by **b** can be written as a single transformation **a** + **b**:

i.e. $\begin{pmatrix} 2 \\ 5 \end{pmatrix} + \begin{pmatrix} -3 \\ -2 \end{pmatrix} = \begin{pmatrix} -1 \\ 3 \end{pmatrix}$

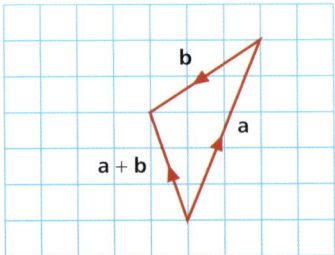

Worked examples

$\mathbf{a} = \begin{pmatrix} 2 \\ 5 \end{pmatrix} \quad \mathbf{b} = \begin{pmatrix} -3 \\ -2 \end{pmatrix}$

a Draw a diagram to represent **a** − **b**, where **a** − **b** = (**a**) + (−**b**).

b Calculate the vector represented by **a** − **b**.

$\begin{pmatrix} 2 \\ 5 \end{pmatrix} - \begin{pmatrix} -3 \\ -2 \end{pmatrix} = \begin{pmatrix} 5 \\ 7 \end{pmatrix}$

Exercise 31.2 In the following questions,

$\mathbf{a} = \begin{pmatrix} 3 \\ 4 \end{pmatrix} \quad \mathbf{b} = \begin{pmatrix} -2 \\ 1 \end{pmatrix} \quad \mathbf{c} = \begin{pmatrix} -4 \\ -3 \end{pmatrix} \quad \mathbf{d} = \begin{pmatrix} 3 \\ -2 \end{pmatrix}$

1 Draw vector diagrams to represent the following.
   a **a** + **b**   b **b** + **a**   c **a** + **d**
   d **d** + **a**   e **b** + **c**   f **c** + **b**

2 What conclusions can you draw from your answers to Question 1 above?

3 Draw vector diagrams to represent the following.
   a **b** − **c**   b **d** − **a**   c −**a** − **c**
   d **a** + **c** − **b** e **d** − **c** − **b** f −**c** + **b** + **d**

4 Represent each of the vectors in Question 3 by a single column vector.

Multiplying a vector by a scalar

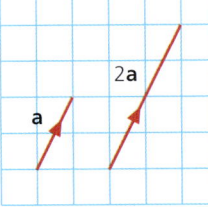

Look at the two vectors in the diagram.

$\mathbf{a} = \begin{pmatrix} 1 \\ 2 \end{pmatrix} \qquad 2\mathbf{a} = 2\begin{pmatrix} 1 \\ 2 \end{pmatrix} = \begin{pmatrix} 2 \\ 4 \end{pmatrix}$

Multiplying a vector by a scalar

> **Worked example**

If $\mathbf{a} = \begin{pmatrix} 2 \\ -4 \end{pmatrix}$, express the vectors **b**, **c**, **d** and **e** in terms of **a**.

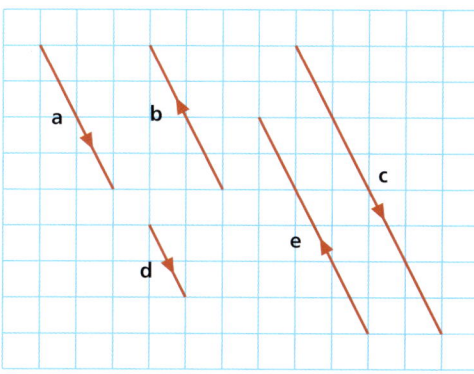

$\mathbf{b} = -\mathbf{a} \quad \mathbf{c} = 2\mathbf{a} \quad \mathbf{d} = \tfrac{1}{2}\mathbf{a} \quad \mathbf{e} = -\tfrac{3}{2}\mathbf{a}$

Exercise 31.3

1 $\mathbf{a} = \begin{pmatrix} 1 \\ 4 \end{pmatrix} \qquad \mathbf{b} = \begin{pmatrix} -4 \\ -2 \end{pmatrix} \qquad \mathbf{c} = \begin{pmatrix} -4 \\ 6 \end{pmatrix}$

Express the following vectors in terms of either **a**, **b** or **c**.

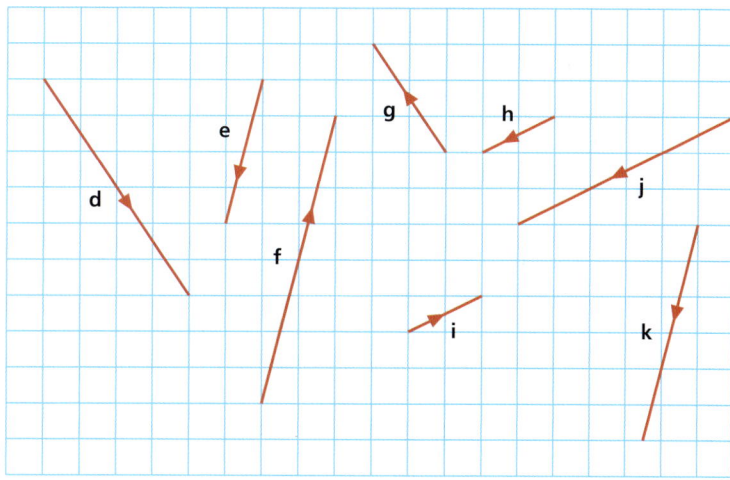

2 $\mathbf{a} = \begin{pmatrix} 2 \\ 3 \end{pmatrix} \qquad \mathbf{b} = \begin{pmatrix} -4 \\ -1 \end{pmatrix} \qquad \mathbf{c} = \begin{pmatrix} -2 \\ 4 \end{pmatrix}$

Represent each of the following as a single column vector.
  a 2**a**   **b** 3**b**   **c** −**c**   **d** **a** + **b**   **e** **b** − **c**
  f 3**c** − **a**   **g** 2**b** − **a**   **h** $\tfrac{1}{2}$(**a** − **b**)   **i** 2**a** − 3**c**

31 VECTORS

Exercise 31.3 (cont)

3 $\mathbf{a} = \begin{pmatrix} -2 \\ 3 \end{pmatrix}$ $\mathbf{b} = \begin{pmatrix} 0 \\ -3 \end{pmatrix}$ $\mathbf{c} = \begin{pmatrix} 4 \\ -1 \end{pmatrix}$

Express each of the following vectors in terms of **a**, **b** and **c**.

a $\begin{pmatrix} -4 \\ 6 \end{pmatrix}$ b $\begin{pmatrix} 0 \\ 3 \end{pmatrix}$ c $\begin{pmatrix} 4 \\ -4 \end{pmatrix}$

d $\begin{pmatrix} -2 \\ 6 \end{pmatrix}$ e $\begin{pmatrix} 8 \\ -2 \end{pmatrix}$ f $\begin{pmatrix} 10 \\ -5 \end{pmatrix}$

The magnitude of a vector

The **magnitude** or size of a vector is represented by its length, i.e. the longer the length, the greater the magnitude. The magnitude of a vector **a** or $\overrightarrow{AB}$ is denoted by $|\mathbf{a}|$ or $|\overrightarrow{AB}|$ respectively and is calculated using Pythagoras' theorem.

Worked examples

$\mathbf{a} = \begin{pmatrix} 3 \\ 4 \end{pmatrix}$ $\overrightarrow{BC} = \begin{pmatrix} -6 \\ 8 \end{pmatrix}$

a Represent both of the above vectors diagrammatically.

b i Calculate $|\mathbf{a}|$.

$|\mathbf{a}| = \sqrt{(3^2 + 4^2)}$

$= \sqrt{25} = 5$

ii Calculate $|\overrightarrow{BC}|$.

$|\overrightarrow{BC}| = \sqrt{(-6)^2 + 8^2}$

$= \sqrt{100} = 10$

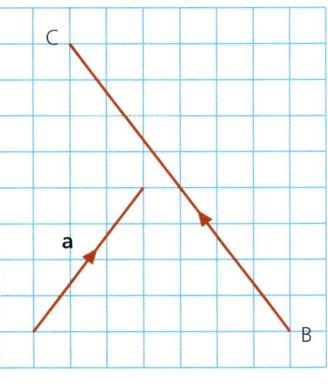

Exercise 31.4

1 Calculate the magnitude of the vectors shown below. Give your answers correct to 1 d.p. where appropriate.

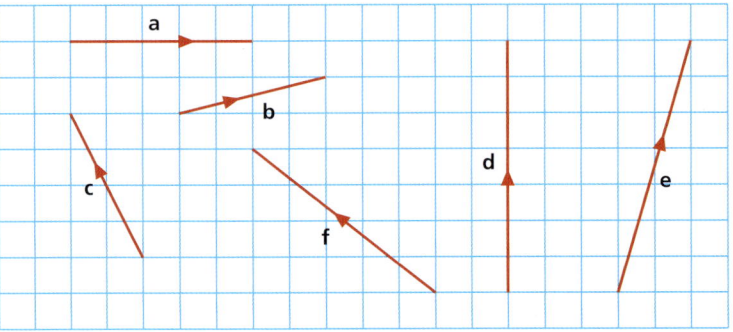

2 Calculate the magnitude of the following vectors, giving your answers to 1 d.p. where appropriate.

a $\overrightarrow{AB} = \begin{pmatrix} 0 \\ 4 \end{pmatrix}$ b $\overrightarrow{BC} = \begin{pmatrix} 2 \\ 5 \end{pmatrix}$ c $\overrightarrow{CD} = \begin{pmatrix} -4 \\ -6 \end{pmatrix}$

d $\overrightarrow{DE} = \begin{pmatrix} -5 \\ 12 \end{pmatrix}$ e $2\overrightarrow{AB}$ f $2\overrightarrow{CD}$

3 $\mathbf{a} = \begin{pmatrix} 4 \\ -3 \end{pmatrix}$ $\mathbf{b} = \begin{pmatrix} -5 \\ 7 \end{pmatrix}$ $\mathbf{c} = \begin{pmatrix} -1 \\ -8 \end{pmatrix}$

Calculate the magnitude of the following, giving your answers to 1 d.p.
a **a + b**   b **2a − b**   c **b − c**
d **2c + 3b**   e **2b − 3a**   f **a + 2b − c**

Position vectors

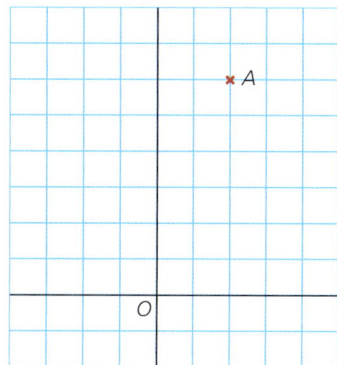

Sometimes a vector is fixed in position relative to a specific point.

In the diagram, the position vector of A relative to O is $\begin{pmatrix} 2 \\ 6 \end{pmatrix}$.

Exercise 31.5   Give the position vectors of A, B, C, D, E, F, G and H relative to O in the diagram (below).

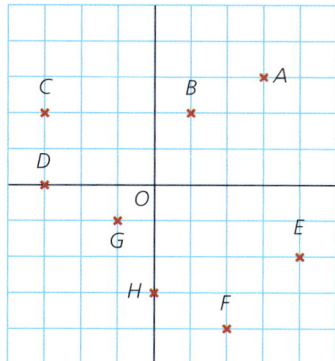

31 VECTORS

Vector geometry

In general, vectors are not fixed in position. If a vector **a** has a specific magnitude and direction, then any other vector with the same magnitude and direction as **a** can also be labelled **a**.

If $\mathbf{a} = \begin{pmatrix} 3 \\ 2 \end{pmatrix}$ then all the vectors shown in the diagram can also be labelled **a**, as they all have the same magnitude and direction.

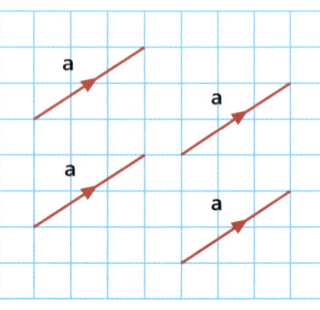

This property of vectors can be used to solve problems in vector geometry.

Worked examples

a Name a vector equal to $\overrightarrow{AD}$.
  $\overrightarrow{BC} = \overrightarrow{AD}$

b Write $\overrightarrow{BD}$ in terms of $\overrightarrow{BE}$.
  $\overrightarrow{BD} = 2\overrightarrow{BE}$

c Express $\overrightarrow{CD}$ in terms of $\overrightarrow{AB}$.
  $\overrightarrow{CD} = \overrightarrow{BA} = -\overrightarrow{AB}$

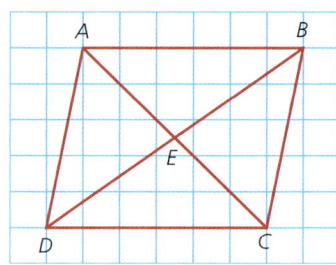

Exercise 31.6

1 If $\overrightarrow{AG} = \mathbf{a}$ and $\overrightarrow{AE} = \mathbf{b}$, express the following in terms of **a** and **b**.
  a $\overrightarrow{EI}$ b $\overrightarrow{HC}$ c $\overrightarrow{FC}$
  d $\overrightarrow{DE}$ e $\overrightarrow{GH}$ f $\overrightarrow{CD}$
  g $\overrightarrow{AI}$ h $\overrightarrow{GE}$ i $\overrightarrow{FD}$

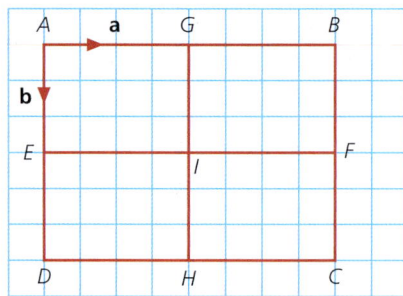

2 If $\overrightarrow{LP} = \mathbf{a}$ and $\overrightarrow{LR} = \mathbf{b}$, express the following in terms of **a** and **b**.
  a $\overrightarrow{LM}$ b $\overrightarrow{PQ}$ c $\overrightarrow{PR}$
  d $\overrightarrow{MQ}$ e $\overrightarrow{MP}$ f $\overrightarrow{NP}$

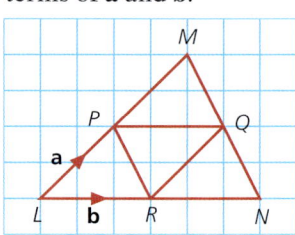

Vector geometry

3 *ABCDEF* is a regular hexagon.

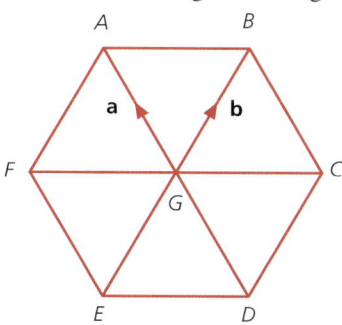

If $\vec{GA} = \mathbf{a}$ and $\vec{GB} = \mathbf{b}$, express the following in terms of **a** and **b**.

a $\vec{AD}$ b $\vec{FE}$ c $\vec{DC}$
d $\vec{AB}$ e $\vec{FC}$ f $\vec{EC}$
g $\vec{BE}$ h $\vec{FD}$ i $\vec{AE}$

4 If $\vec{AB} = \mathbf{a}$ and $\vec{AG} = \mathbf{b}$, express the following in terms of **a** and **b**.

a $\vec{AF}$ b $\vec{AM}$ c $\vec{FM}$
d $\vec{FO}$ e $\vec{EI}$ f $\vec{KF}$
g $\vec{CN}$ h $\vec{AN}$ i $\vec{DN}$

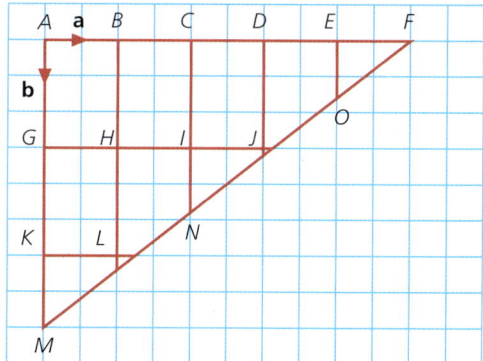

Exercise 31.7

1 *T* is the midpoint of the line *PS* and *R* divides the line *QS* in the ratio 1:3.
$\vec{PT} = \mathbf{a}$ and $\vec{PQ} = \mathbf{b}$.

a Express each of the following in terms of **a** and **b**.
   i $\vec{PS}$
   ii $\vec{QS}$
   iii $\vec{PR}$

b Show that $\vec{RT} = \frac{1}{4}(2\mathbf{a} - 3\mathbf{b})$.

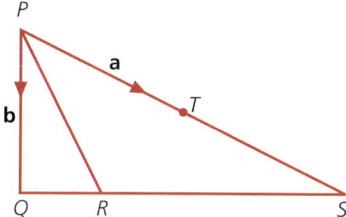

31 VECTORS

Exercise 31.7 (cont)

2 $\vec{PM} = 3\vec{LP}$ and $\vec{QN} = 3\vec{LQ}$
   Prove that:
   a the line PQ is parallel to the line MN,
   b the line MN is four times the length of the line PQ.

3 $PQRS$ is a parallelogram. The point T divides the line PQ in the ratio $1:3$, and U, V and W are the midpoints of SR, PS and QR respectively.

   $\vec{PT} = \mathbf{a}$ and $\vec{PV} = \mathbf{b}$.

   a Express each of the following in terms of **a** and **b**.
   i $\vec{PQ}$
   ii $\vec{SU}$
   iii $\vec{PU}$
   iv $\vec{VX}$
   b Show that $\vec{XR} = \frac{1}{2}(5\mathbf{a} + 2\mathbf{b})$.

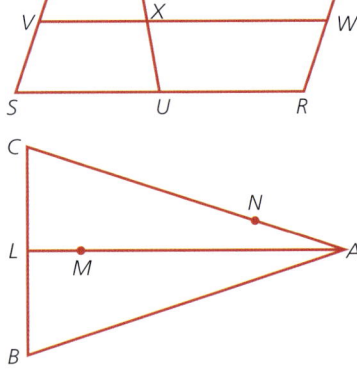

4 ABC is an isosceles triangle. L is the midpoint of BC. M divides the line LA in the ratio $1:5$, and N divides AC in the ratio $2:5$.
   a $\vec{BC} = \mathbf{p}$ and $\vec{BA} = \mathbf{q}$. Express the following in terms of **p** and **q**.
   i $\vec{LA}$ ii $\vec{AN}$
   b Show that $\vec{MN} = \frac{1}{84}(46\mathbf{q} - 11\mathbf{p})$.

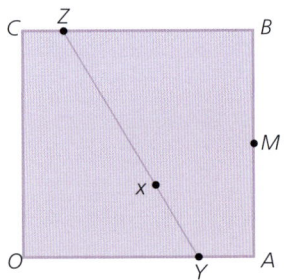

Note
Points which lie on the same line are collinear.

5 A square $OABC$ is shown opposite. Point Y divides OA in the ratio $5:3$. Point M is the midpoint of AB. Point Z divides CB in the ratio $1:7$. If $\vec{OA} = \mathbf{a}$ and if $\vec{OC} = \mathbf{c}$, prove that O, X and M are collinear.

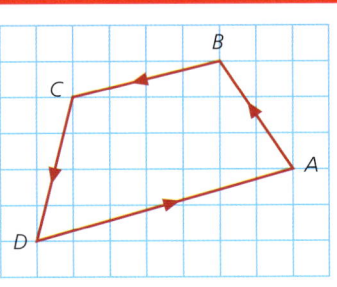

Student assessment 1

1 Using the diagram, describe the following translations using column vectors.
   a $\vec{AB}$ b $\vec{DA}$ c $\vec{CA}$

452

Vector geometry

2 Describe each of the translations shown using column vectors.

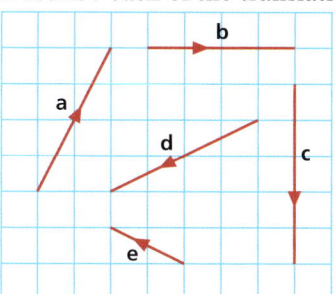

3 Using the vectors in Question 2, draw diagrams to represent:
   a $\mathbf{a} + \mathbf{b}$ **b** $\mathbf{e} - \mathbf{d}$ **c** $\mathbf{c} - \mathbf{e}$ **d** $2\mathbf{e} + \mathbf{b}$

4 In the following, $\mathbf{a} = \begin{pmatrix} 2 \\ 6 \end{pmatrix}$ $\mathbf{b} = \begin{pmatrix} -3 \\ -1 \end{pmatrix}$ $\mathbf{c} = \begin{pmatrix} -2 \\ 4 \end{pmatrix}$

Calculate:
   a $\mathbf{a} + \mathbf{b}$ **b** $\mathbf{c} - \mathbf{b}$ **c** $2\mathbf{a} + \mathbf{b}$ **d** $3\mathbf{c} - 2\mathbf{b}$

Student assessment 2

1 a Calculate the magnitude of the vector $\overrightarrow{AB}$ shown in the diagram.

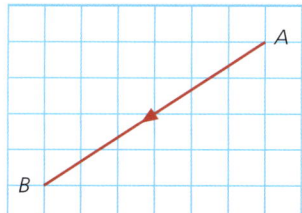

  b Calculate the magnitude of the following vectors.

$\mathbf{a} = \begin{pmatrix} 2 \\ 9 \end{pmatrix}$ $\mathbf{b} = \begin{pmatrix} -7 \\ -4 \end{pmatrix}$ $\mathbf{c} = \begin{pmatrix} -5 \\ 12 \end{pmatrix}$

2 $\mathbf{p} = \begin{pmatrix} 3 \\ 2 \end{pmatrix}$ $\mathbf{q} = \begin{pmatrix} -4 \\ 1 \end{pmatrix}$ $\mathbf{r} = \begin{pmatrix} 3 \\ -4 \end{pmatrix}$

Calculate the magnitude of the following, giving your answers to 3 s.f.
   a $3\mathbf{p} - 2\mathbf{q}$ **b** $\tfrac{1}{2}\mathbf{r} + \mathbf{q}$

3 Give the position vectors of A, B, C, D and E relative to O for the diagram below.

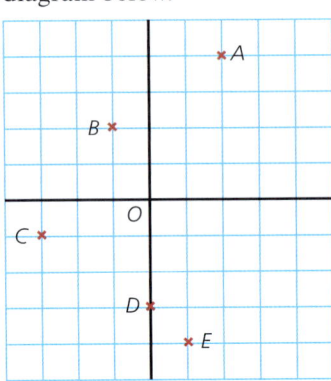

4 a Name another vector equal to $\overrightarrow{DE}$ in the diagram.
   b Express $\overrightarrow{DF}$ in terms of $\overrightarrow{BC}$.
   c Express $\overrightarrow{CF}$ in terms of $\overrightarrow{DE}$.

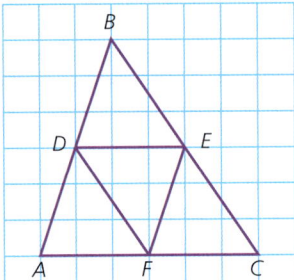

Student assessment 3

1 In the triangle PQR, the point S divides the line PQ in the ratio $1:3$, and T divides the line RQ in the ratio $3:2$.
$\overrightarrow{PR} = \mathbf{a}$ and $\overrightarrow{PQ} = \mathbf{b}$.

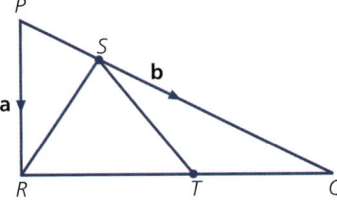

   a Express the following in terms of **a** and **b**.
      i $\overrightarrow{PS}$ ii $\overrightarrow{SR}$ iii $\overrightarrow{TQ}$
   b Show that $\overrightarrow{ST} = \frac{1}{20}(8\mathbf{a} + 7\mathbf{b})$.

2 In the triangle ABC, the point D divides the line AB in the ratio $1:3$, and E divides the line AC also in the ratio $1:3$.

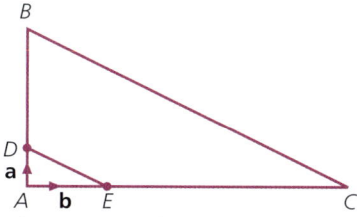

If $\overrightarrow{AD} = \mathbf{a}$ and $\overrightarrow{AE} = \mathbf{b}$, prove that:
a $\overrightarrow{BC} = 4\overrightarrow{DE}$,
b $BCED$ is a trapezium.

3 The parallelogram $ABCD$ shows the points P and Q dividing each of the lines AD and DC in the ratio $1:4$ respectively.

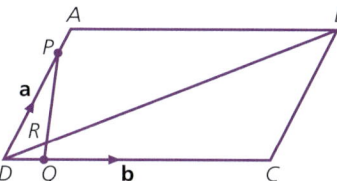

a If $\overrightarrow{DA} = \mathbf{a}$ and $\overrightarrow{DC} = \mathbf{b}$, express the following in terms of $\mathbf{a}$ and $\mathbf{b}$.
  i $\overrightarrow{AC}$ **ii** $\overrightarrow{CB}$ **iii** $\overrightarrow{DB}$
b i Find the ratio in which R divides DB.
  ii Find the ratio in which R divides PQ.

32 Transformations

An object undergoing a transformation changes in either position or shape. In its simplest form this change can occur because of a **reflection**, **rotation**, **translation** or **enlargement**. When an object undergoes a transformation, then its new position or shape is known as the **image**.

Reflection

When an object is reflected, it undergoes a 'flip' movement about a dashed (broken) line known as the **mirror line**, as shown in the diagram:

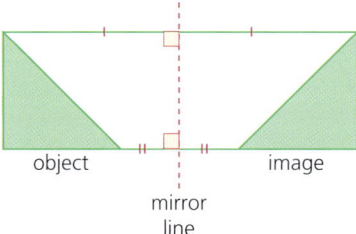

A point on the object and its equivalent point on the image are equidistant from the mirror line. This distance is measured at right angles to the mirror line. The line joining the point to its image is perpendicular to the mirror line.

The position of the mirror line is essential when describing a reflection. Sometimes, its equation as well as its position will be required.

➔ Worked examples

a Find the equation of the mirror line in the reflection given in the diagram (below).

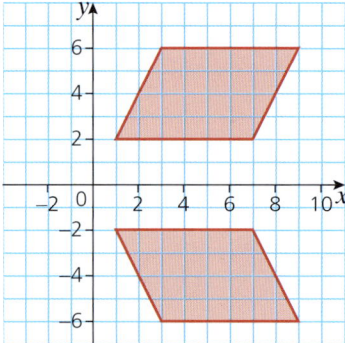

Here, the mirror line is the x-axis. The equation of the mirror line is therefore $y = 0$.

Reflection

b A **reflection** is shown below.
  i Draw the position of the mirror line.

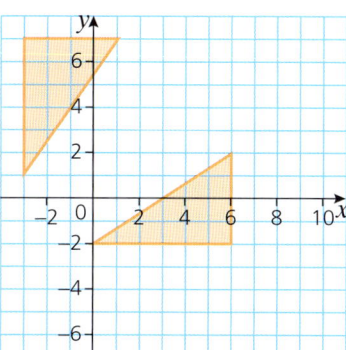 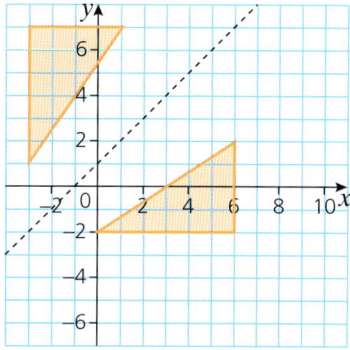

  ii Give the equation of the mirror line.
      Equation of mirror line: $y = x + 1$.

Exercise 32.1

Copy each of the following diagrams, then:
a draw the position of the mirror line(s),
b give the equation of the mirror line(s).

1
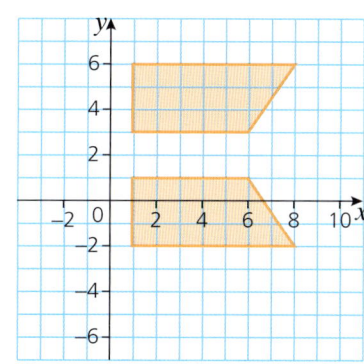

2
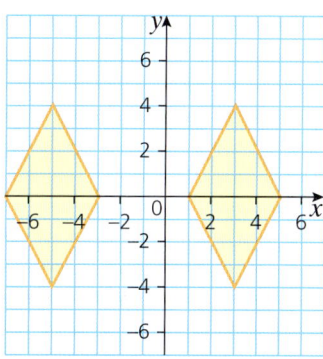

3
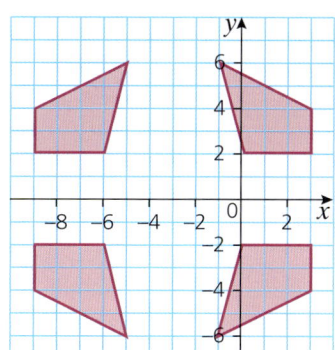

4
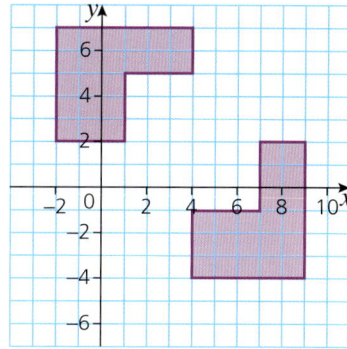

32 TRANSFORMATIONS

Exercise 32.1 (cont)

5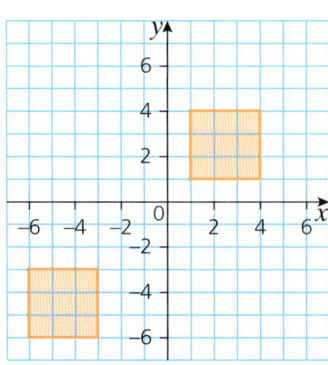

6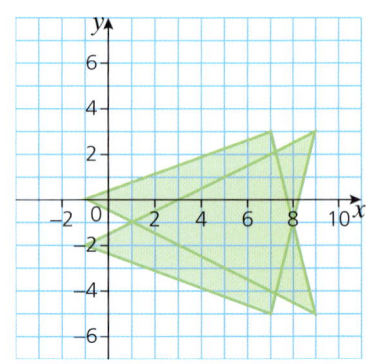

7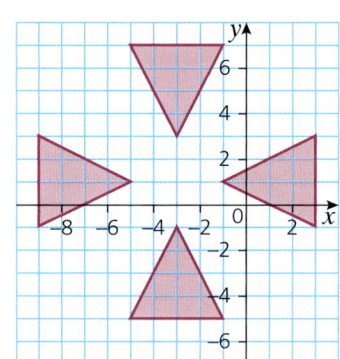

8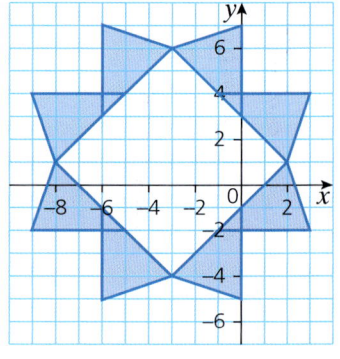

Exercise 32.2 In Questions 1 and 2, copy each diagram four times and reflect the object in each of the lines given.

1 a $x = 2$
  b $y = 0$
  c $y = x$
  d $y = -x$

2 a $x = -1$
  b $y = -x - 1$
  c $y = x + 2$
  d $x = 0$

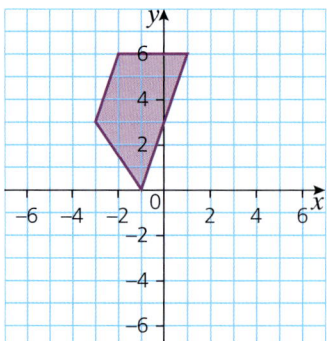

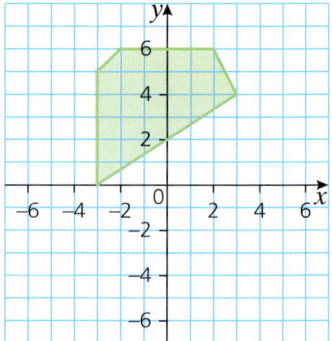

3 Copy the diagram (below), and reflect the triangles in the following lines:

$x = 1$ and $y = -3$.

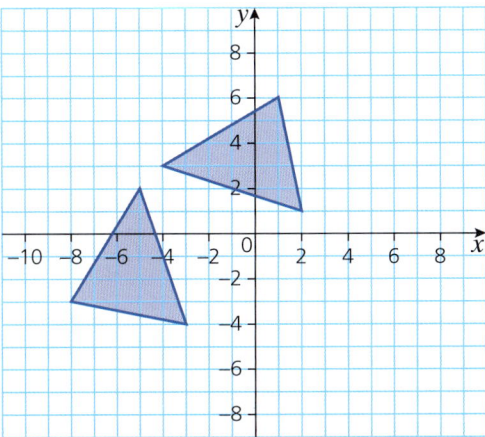

Rotation

When an object is rotated, it undergoes a 'turning' movement about a specific point known as the **centre of rotation**. When describing a rotation, it is necessary to identify not only the position of the centre of rotation, but also the angle and direction of the turn, as shown in the diagram:

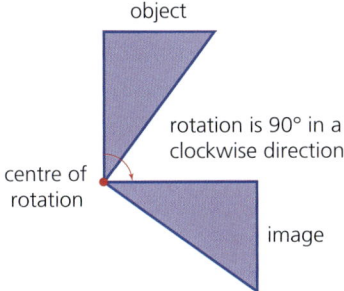

32 TRANSFORMATIONS

Exercise 32.3 In the following, the object and centre of rotation have both been given. Copy each diagram and draw the object's image under the stated rotation about the marked point.

1
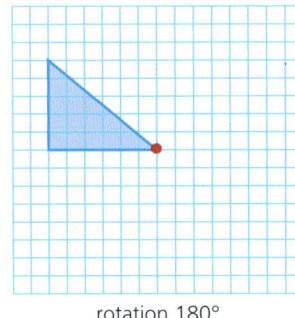
rotation 180°

2
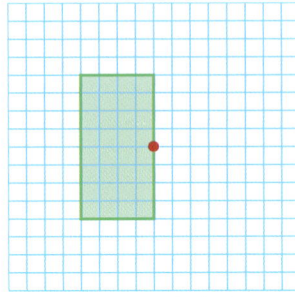
rotation 90° clockwise

3
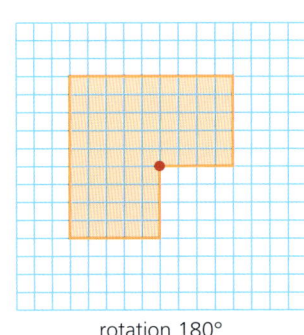
rotation 180°

4
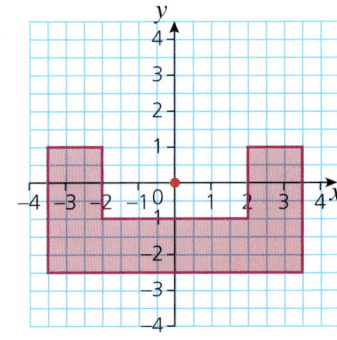
rotation 90° anticlockwise about (0, 0)

5
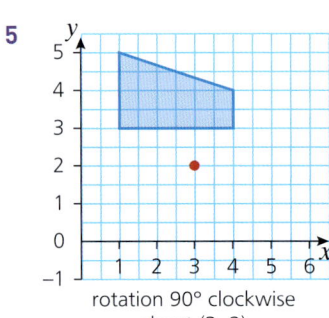
rotation 90° clockwise about (3, 2)

6
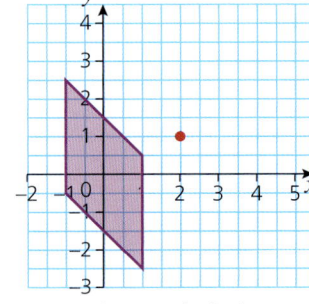
rotation 90° clockwise about (2, 1)

Rotation

Exercise 32.4

In the following, the object (unshaded) and image (shaded) have been drawn.

Copy each diagram and:
a mark the centre of rotation,
b calculate the angle and direction of rotation.

1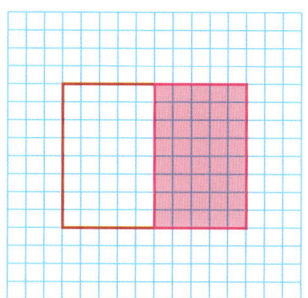

2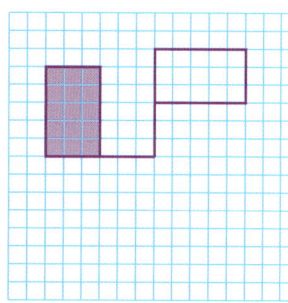

3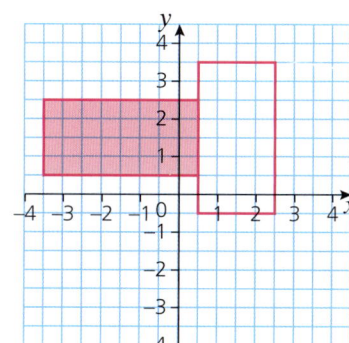

4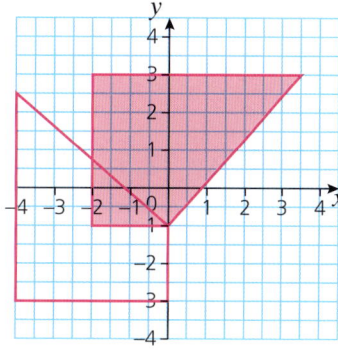

5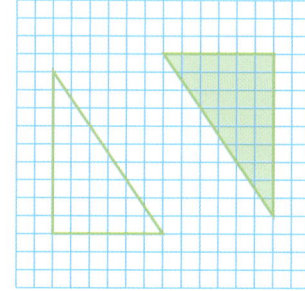

6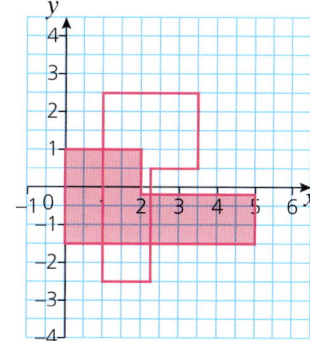

Note

To describe a rotation, three pieces of information need to be given. These are the centre of rotation, the angle of rotation and the direction of rotation.

32 TRANSFORMATIONS

Translation

When an object is translated, it undergoes a 'straight sliding' movement. When describing a translation, it is necessary to give the translation vector. As no rotation is involved, each point on the object moves in the same way to its corresponding point on the image, e.g.

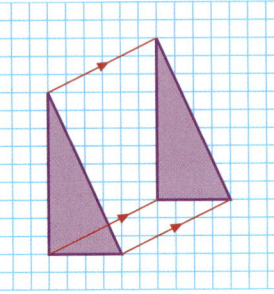 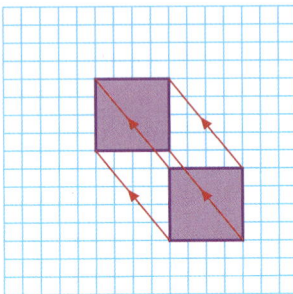

Vector = $\begin{pmatrix} 6 \\ 3 \end{pmatrix}$ Vector = $\begin{pmatrix} -4 \\ 5 \end{pmatrix}$

Exercise 32.5 In the following diagrams, object A has been translated to each of images B and C.

Give the translation vectors in each case.

1
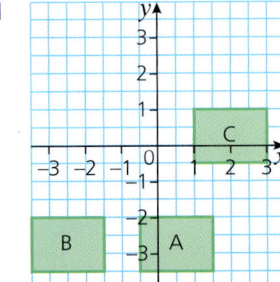

2
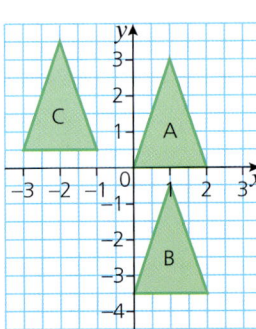

3
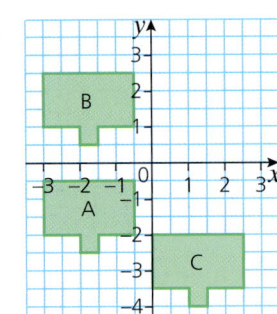

4
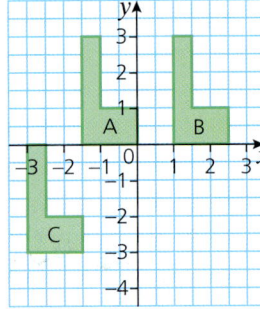

Translation

Exercise 32.6
a Copy each of the following diagrams and draw the object.
b Translate the object by the vector given in each case and draw the image in its position.
(Note that a bigger grid than the one shown may be needed.)

1
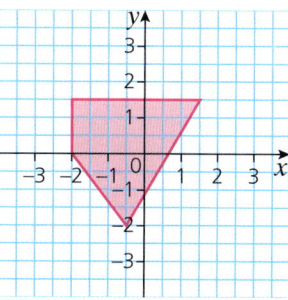

Vector = $\begin{pmatrix} 3 \\ 5 \end{pmatrix}$

2
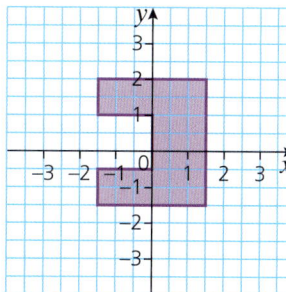

Vector = $\begin{pmatrix} 5 \\ -4 \end{pmatrix}$

3
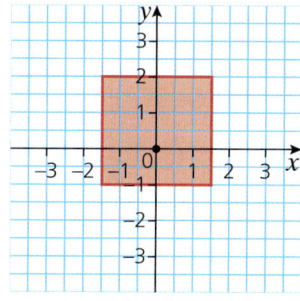

Vector = $\begin{pmatrix} -4 \\ 6 \end{pmatrix}$

4
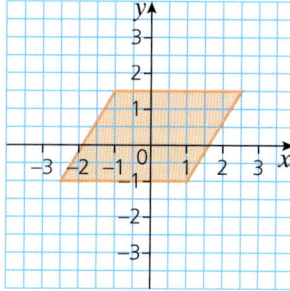

Vector = $\begin{pmatrix} -2 \\ -5 \end{pmatrix}$

5
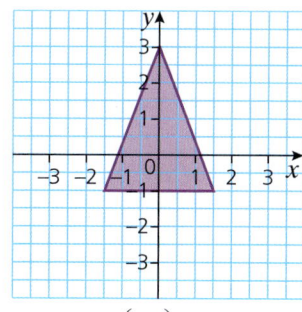

Vector = $\begin{pmatrix} -6 \\ 0 \end{pmatrix}$

6
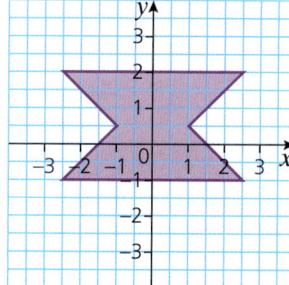

Vector = $\begin{pmatrix} 0 \\ -1 \end{pmatrix}$

32 TRANSFORMATIONS

Enlargement

When an object is enlarged, the result is an image which is mathematically similar to the object but of a different size. The image can be either larger or smaller than the original object. When describing an enlargement, two pieces of information need to be given, the position of the **centre of enlargement** and the **scale factor of enlargement**.

→ Worked examples

a   In the diagram below, triangle ABC is enlarged to form triangle $A'B'C'$.

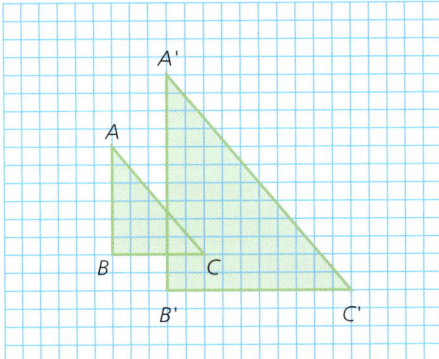

i   Find the centre of enlargement.

The centre of enlargement is found by joining corresponding points on the object and image with a straight line. These lines are then extended until they meet. The point at which they meet is the centre of enlargement O.

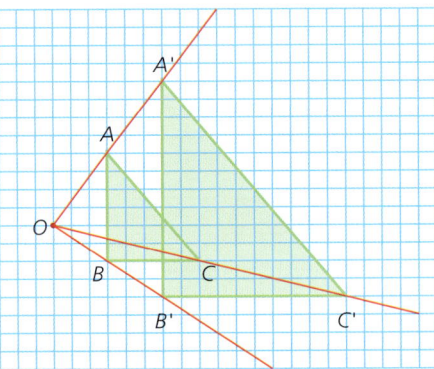

ii   Calculate the scale factor of enlargement.

The scale factor of enlargement can be calculated in one of two ways. From the diagram above it can be seen that the distance OA' is twice the distance OA. Similarly OC' and OB' are both twice OC and OB respectively, hence the scale factor of enlargement is 2.

Enlargement

Alternatively, the scale factor can be found by considering the ratio of the length of a side on the image to the length of the corresponding side on the object, i.e.

$$\frac{A'B'}{AB} = \frac{12}{6} = 2$$

Hence the scale factor of enlargement is 2.

b In the diagram below, the rectangle $ABCD$ undergoes a transformation to form rectangle $A'B'C'D'$.

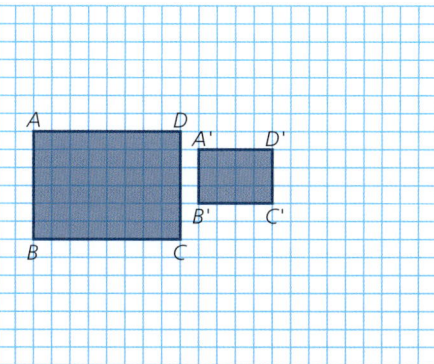

i Find the centre of enlargement.
By joining corresponding points on both the object and the image, the centre of enlargement is found at O.

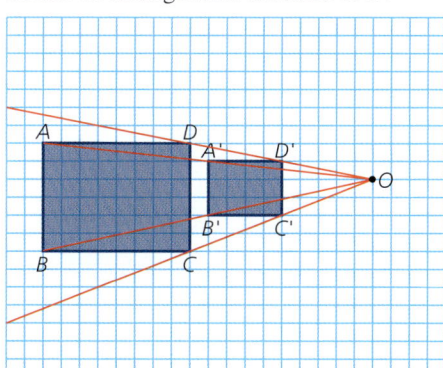

ii Calculate the scale factor of enlargement.

The scale factor of enlargement = $\frac{A'B'}{AB} = \frac{3}{6} = \frac{1}{2}$

Note

If the scale factor of enlargement is greater than 1, then the image is larger than the object. If the scale factor lies between 0 and 1, then the resulting image is smaller than the object. In these cases, although the image is smaller than the object, the transformation is still known as an enlargement.

32 TRANSFORMATIONS

Exercise 32.7 Copy the following diagrams and find:
  a the centre of enlargement,
  b the scale factor of enlargement.

1
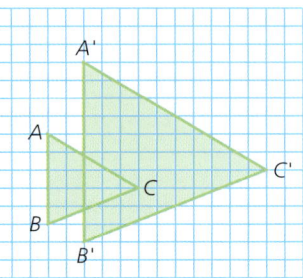

2
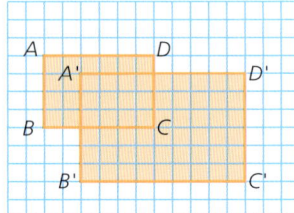

3
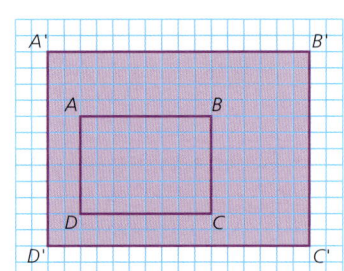

4
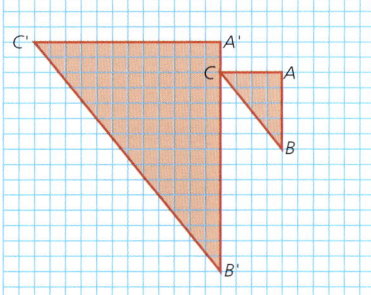

5
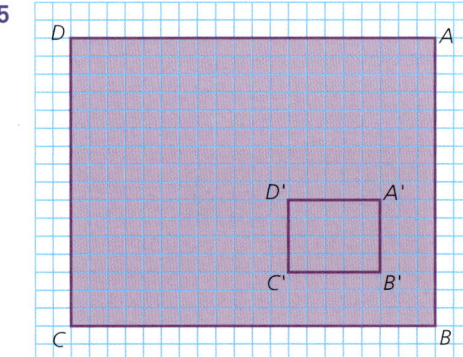

Exercise 32.8 Copy the following diagrams and enlarge the objects by the scale factor given and from the centre of enlargement shown.
(Grids larger than those shown may be needed.)

1
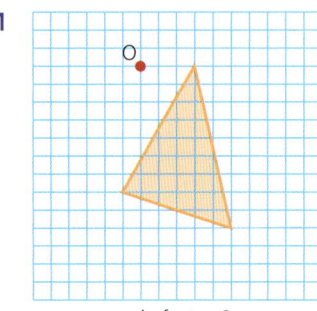

scale factor 2

2
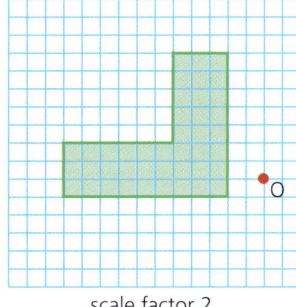

scale factor 2

Negative enlargement

3
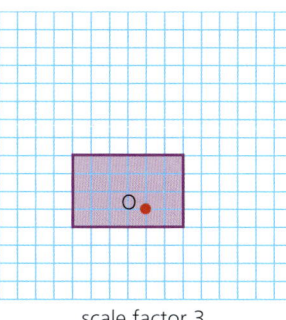
scale factor 3

4
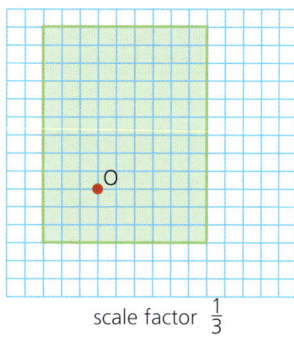
scale factor $\frac{1}{3}$

Negative enlargement

The diagram below shows an example of **negative enlargement**.

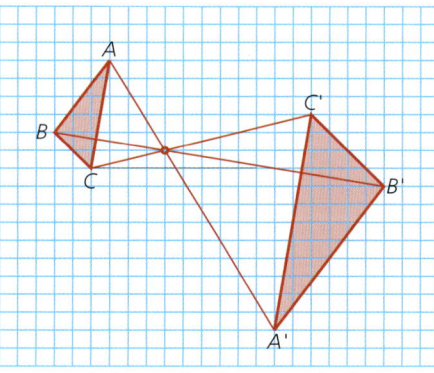

scale factor of enlargement is −2

With negative enlargement each point and its image are on opposite sides of the centre of enlargement. The scale factor of enlargement is calculated in the same way, remembering, however, to write a '−' sign before the number.

Exercise 32.9

1 Copy the following diagram and then calculate the scale factor of enlargement and show the position of the centre of enlargement.

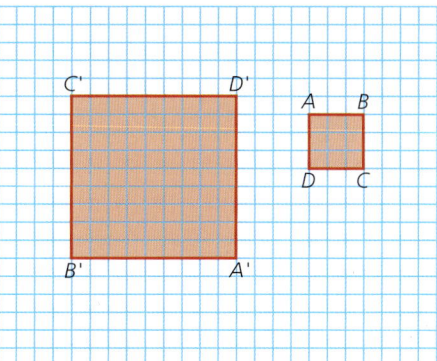

32 TRANSFORMATIONS

Exercise 32.9 (cont)

2   The scale factor of enlargement and centre of enlargement are both given. Copy and complete the diagram.

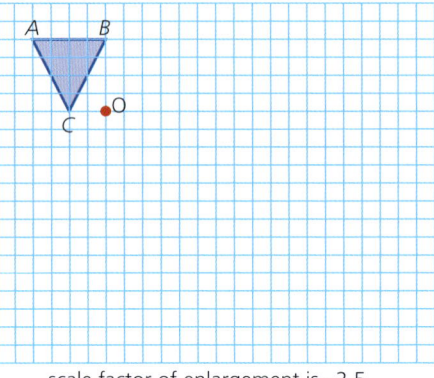

scale factor of enlargement is −2.5

3   The scale factor of enlargement and centre of enlargement are both given. Copy and complete the diagram.

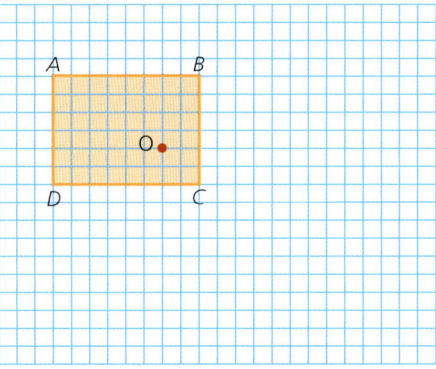

scale factor of enlargement is −2

4   Copy the following diagram and then calculate the scale factor of enlargement and show the position of the centre of enlargement.

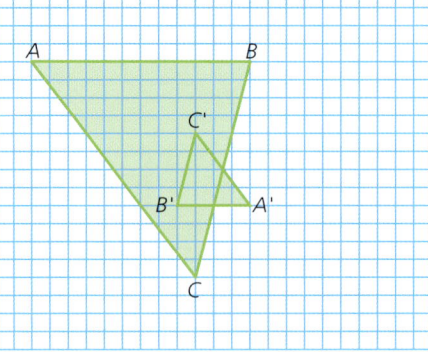

5 An object and part of its image under enlargement are given in the diagram below. Copy the diagram and complete the image. Also find the centre of enlargement and calculate the scale factor of enlargement.

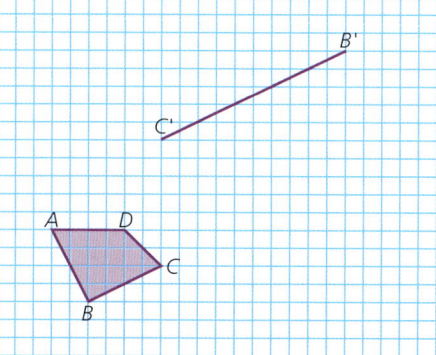

6 In the diagram below, part of an object in the shape of a quadrilateral and its image under enlargement are drawn. Copy and complete the diagram. Also find the centre of enlargement and calculate the scale factor of enlargement.

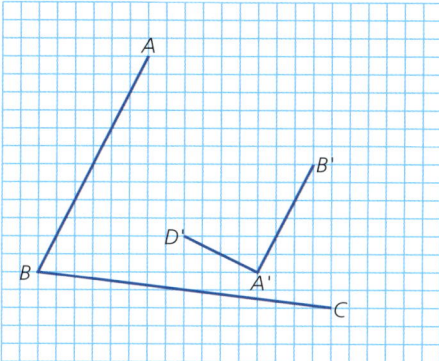

Combinations of transformations

An object may not just undergo one type of transformation. It can undergo a succession of different transformations.

32 TRANSFORMATIONS

 Worked example

A triangle ABC maps onto $A'B'C'$ after an enlargement of scale factor 3 from the centre of enlargement $(0, 7)$. $A'B'C'$ is then mapped onto $A"B"C"$ by a reflection in the line $x = 1$.

a Draw and label the image $A'B'C'$.
b Draw and label the image $A"B"C"$.

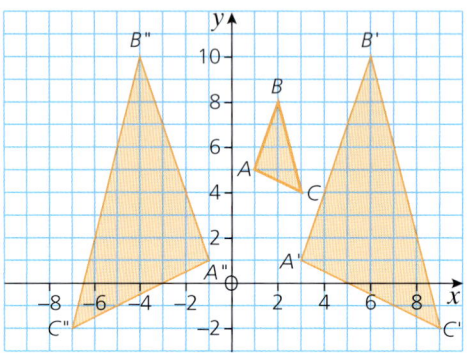

Exercise 32.10

For each of the following questions, copy the diagram.

After each transformation, draw the image on the same grid and label it clearly.

1 The square $ABCD$ is mapped onto $A'B'C'D'$ by a reflection in the line $y = 3$. $A'B'C'D'$ then maps onto $A"B"C"D"$ as a result of a 90° rotation in a clockwise direction about the point $(-2, 5)$.

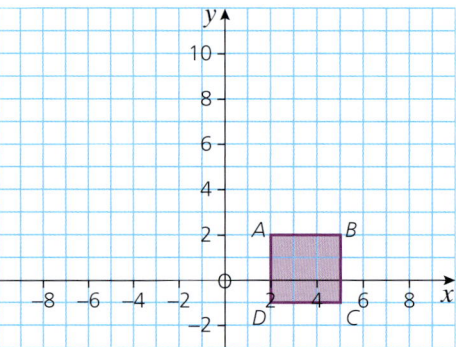

2 The rectangle $ABCD$ is mapped onto $A'B'C'D'$ by an enlargement of scale factor -2 with its centre at $(0, 5)$. $A'B'C'D'$ then maps onto $A''B''C''D''$ as a result of a reflection in the line $y = -x + 7$.

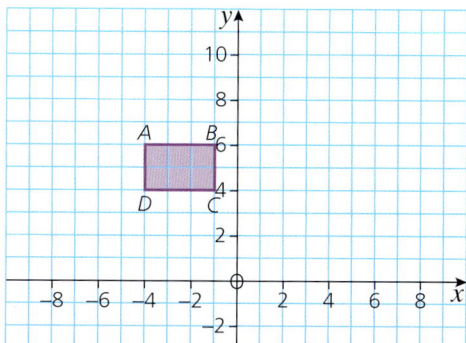

Student assessment 1

1 Write down the column vector of the translation which maps:
   a rectangle A to rectangle B,
   b rectangle B to rectangle C.

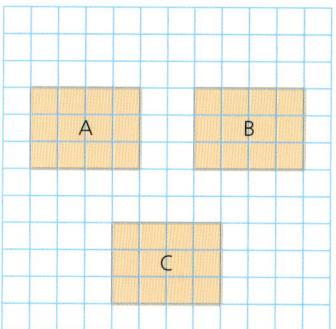

2 Enlarge the triangle below by scale factor 2 and from the centre of enlargement O.

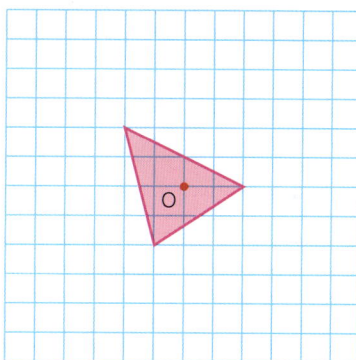

3 Copy the diagram below, which shows an object and its reflected image.
   a Draw on your diagram the position of the mirror line.
   b Find the equation of the mirror line.

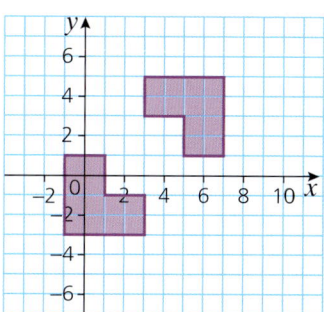

4 The triangle ABC is mapped onto triangle $A'B'C'$ by a rotation (below).
   a Find the coordinates of the centre of rotation.
   b Give the angle and direction of rotation.

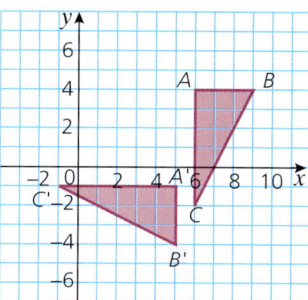

32 TRANSFORMATIONS

Student assessment 2

1 Write down the column vector of the translation which maps:
   a triangle A to triangle B,
   b triangle B to triangle C.

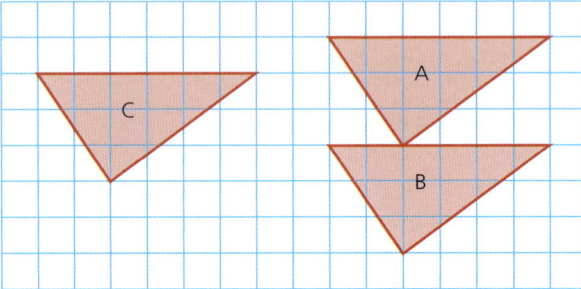

2 Enlarge the rectangle below by scale factor 1.5 and from the centre of enlargement O.

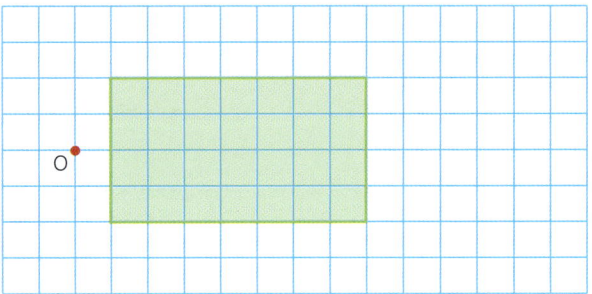

3 Copy the diagram below.

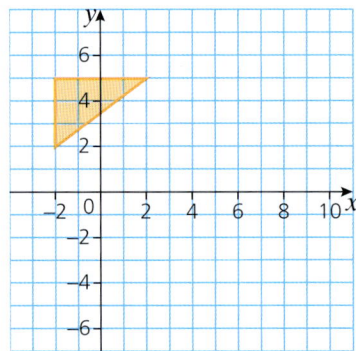

   a Draw in the mirror line with equation $y = x - 1$.
   b Reflect the object in the mirror line.

4 Square $ABCD$ is mapped onto square $A'B'C'D'$. Square $A'B'C'D'$ is subsequently mapped onto square $A''B''C''D''$.

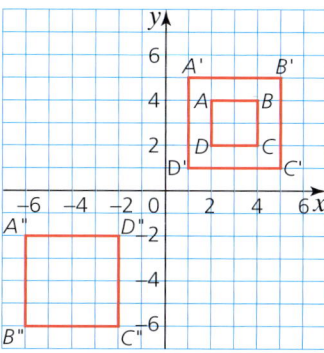

a Describe fully the transformation which maps $ABCD$ onto $A'B'C'D'$.
b Describe fully the transformation which maps $A'B'C'D'$ onto $A''B''C''D''$.

Student assessment 3

1 An object $ABCD$ and its image $A'B'C'D'$ are shown below.
   a Find the position of the centre of enlargement.
   b Calculate the scale factor of enlargement.

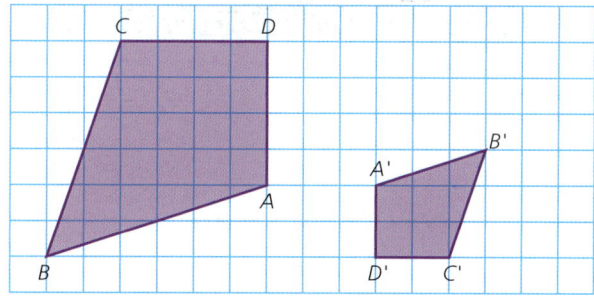

2 The square $ABCD$ is mapped onto $A'B'C'D'$. $A'B'C'D'$ is subsequently mapped onto $A''B''C''D''$.

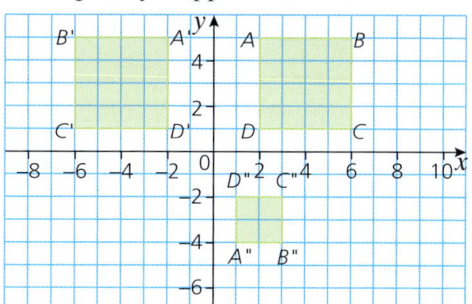

a Describe in full the transformation which maps $ABCD$ onto $A'B'C'D'$.
  b Describe in full the transformation which maps $A'B'C'D'$ onto $A"B"C"D"$.

3 An object $WXYZ$ and its image $W'X'Y'Z'$ are shown below.
  a Find the position of the centre of enlargement.
  b Calculate the scale factor of enlargement.

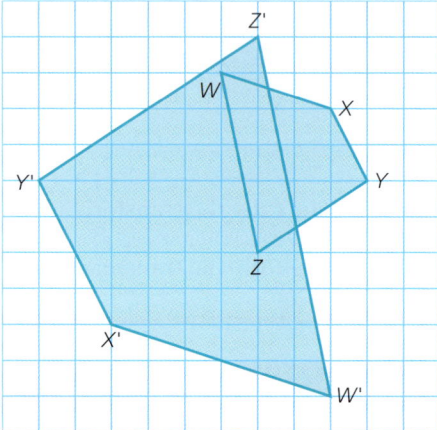

4 Triangle ABC is mapped onto $A'B'C'$. $A'B'C'$ is subsequently mapped onto $A"B"C"$.
  a Describe in full the transformation which maps ABC onto $A'B'C'$.
  b Describe in full the transformation which maps $A'B'C'$ onto $A"B"C"$.

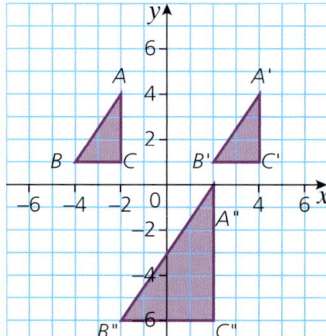

7 Mathematical investigations and ICT 7

A painted cube

A $3 \times 3 \times 3$ cm cube is painted on the outside as shown in the left-hand diagram below:

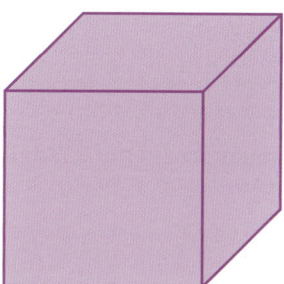 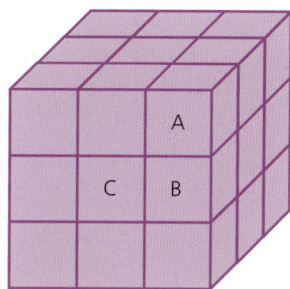

The large cube is then cut up into 27 smaller cubes, each $1 \text{ cm} \times 1 \text{ cm} \times 1 \text{ cm}$ as shown on the right.

$1 \times 1 \times 1$ cm cubes with 3 painted faces are labelled type A.

$1 \times 1 \times 1$ cm cubes with 2 painted faces are labelled type B.

$1 \times 1 \times 1$ cm cubes with 1 face painted are labelled type C.

$1 \times 1 \times 1$ cm cubes with no faces painted are labelled type D.

1 a How many of the 27 cubes are type A?
   b How many of the 27 cubes are type B?
   c How many of the 27 cubes are type C?
   d How many of the 27 cubes are type D?
2 Consider a $4 \times 4 \times 4$ cm cube cut into $1 \times 1 \times 1$ cm cubes. How many of the cubes are type A, B, C and D?
3 How many type A, B, C and D cubes are there when a $10 \times 10 \times 10$ cm cube is cut into $1 \times 1 \times 1$ cm cubes?
4 Generalise for the number of type A, B, C and D cubes in an $n \times n \times n$ cube.
5 Generalise for the number of type A, B, C and D cubes in a cuboid l cm long, w cm wide and h cm high.

MATHEMATICAL INVESTIGATIONS AND ICT 7

Triangle count

The diagram below shows an isosceles triangle with a vertical line drawn from its apex to its base.

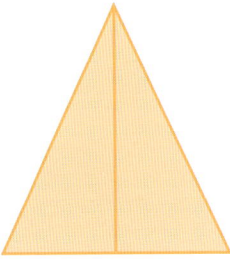

There is a total of 3 triangles in this diagram.

If a horizontal line is drawn across the triangle, it will look as shown:

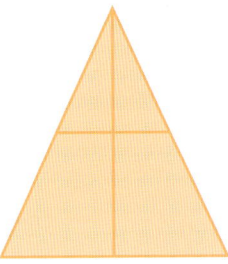

There is a total of 6 triangles in this diagram.

When one more horizontal line is added, the number of triangles increases further:

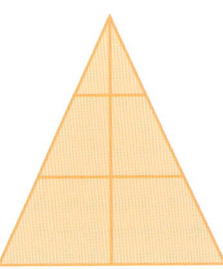

1 Calculate the total number of triangles in the diagram above with the two inner horizontal lines.
2 Investigate the relationship between the total number of triangles (t) and the number of inner horizontal lines (h). Enter your results in an ordered table.
3 Write an algebraic rule linking the total number of triangles and the number of inner horizontal lines.

The triangle (right) has two lines drawn from the apex to the base.

There is a total of six triangles in this diagram.

If a horizontal line is drawn through this triangle, the number of triangles increases as shown:

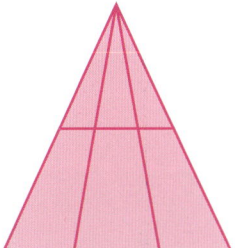

4 Calculate the total number of triangles in the diagram above with two lines from the vertex and one inner horizontal line.
5 Investigate the relationship between the total number of triangles (t) and the number of inner horizontal lines (h) when two lines are drawn from the apex. Enter your results in an ordered table.
6 Write an algebraic rule linking the total number of triangles and the number of inner horizontal lines.

ICT activity

Using Autograph or another appropriate software package, prepare a help sheet for your revision that demonstrates the addition, subtraction and multiplication of vectors.

An example is shown below:

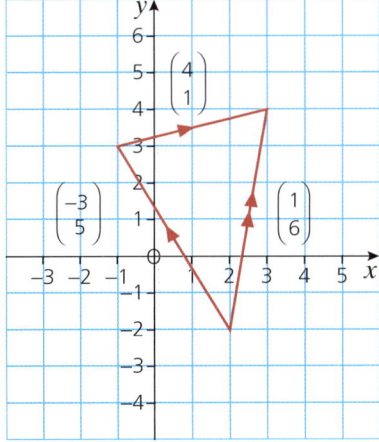

Vector addition:

$$\begin{pmatrix} -3 \\ 5 \end{pmatrix} + \begin{pmatrix} 4 \\ 1 \end{pmatrix} = \begin{pmatrix} 1 \\ 6 \end{pmatrix}$$

TOPIC 8
Probability

Contents

Chapter 33 Probability (E8.1, E8.2, E8.3)
Chapter 34 Further probability (E8.3, E8.4)

Learning objectives

E8.1 Introduction to probability
1. Understand and use the probability scale from 0 to 1.
2. Understand and use probability notation.
3. Calculate the probability of a single event.
4. Understand that the probability of an event not occurring = 1 – the probability of the event occurring.

E8.2 Relative and expected frequencies
1. Understand relative frequency as an estimate of probability.
2. Calculate expected frequencies.

E8.3 Probability of combined events
Calculate the probability of combined events using, where appropriate:
- Sample space diagrams
- Venn diagrams
- Tree diagrams.

E8.4 Conditional probability
Calculate conditional probability using Venn diagrams, tree diagrams and tables.

Order and chaos

Blaise Pascal and Pierre de Fermat (known for his last theorem) corresponded about problems connected to games of chance.

Although Newton and Galileo had had some thoughts on the subject, this is accepted as the beginning of the study of what is now called probability. Later, in 1657, Christiaan Huygens wrote the first book on the subject, entitled *The Value of all Chances in Games of Fortune*.

In 1821, Carl Friedrich Gauss (1777–1855) worked on normal distribution.

At the start of the nineteenth century, the French mathematician Pierre-Simon Laplace was convinced of the existence of a Newtonian universe. In other words, if you knew the position and velocities of all the particles in the universe, you would be able to predict the future because their movement would be predetermined by scientific laws. However, quantum mechanics has since shown that this is not true. Chaos theory is at the centre of understanding these limits.

Blaise Pascal (1623–1662)

33 Probability

Probability is the study of chance, or the likelihood of an event happening. However, because probability is based on chance, what theory predicts does not necessarily happen in practice.

A **favourable outcome** refers to the event in question actually happening. The **total number of possible outcomes** refers to all the different types of outcome one can get in a particular situation. In general:

$$\text{Probability of an event} = \frac{\text{number of favourable outcomes}}{\text{total number of equally likely outcomes}}$$

If the probability = 0, the event is impossible.

If the probability = 1, the event is certain to happen.

If an event can either happen or not happen then:

Probability of the event not occurring
= 1 − the probability of the event occurring.

→ Worked examples

Although probabilities are written here as fractions, they could also be expressed as decimals or percentages.

a A fair spinner numbered 1 - 6 is spun. Calculate the probability of getting a six.

Number of favourable outcomes = 1   (i.e. getting a 6)
Total number of possible outcomes = 6   (i.e. getting a 1, 2, 3, 4, 5 or 6)
Probability of getting a six = $\frac{1}{6}$

Probability of not getting a six = $1 - \frac{1}{6} = \frac{5}{6}$

b A fair spinner numbered 1 - 6 is spun. Calculate the probability of getting an even number.

Number of favourable outcomes = 3   (i.e. getting a 2, 4 or 6)
Total number of possible outcomes = 6   (i.e. getting a 1, 2, 3, 4, 5 or 6)
Probability of getting an even number = $\frac{3}{6} = \frac{1}{2}$

c Thirty students are asked to choose their favourite subject out of Maths, English and Art. The results are shown in the table below.

	Maths	English	Art
Girls	7	4	5
Boys	5	3	6

A student is chosen at random.
i   What is the probability that it is a girl?
    Total number of girls is 16.
    Probability of choosing a girl is $\frac{16}{30} = \frac{8}{15}$.
ii What is the probability that it is a boy whose favourite subject is Art?
    Number of boys whose favourite subject is Art is 6.
    Probability is therefore $\frac{6}{30} = \frac{1}{5}$.
iii What is the probability of **not** choosing a girl whose favourite subject is English?
    There are two ways of approaching this:
    Method 1:
    Total number of students who are not girls whose favourite subject is English is $7 + 5 + 5 + 3 + 6 = 26$.
    Therefore probability is $\frac{26}{30} = \frac{13}{15}$.
    Method 2:
    Total number of girls whose favourite subject is English is 4.
    Probability of choosing a girl whose favourite subject is English is $\frac{4}{30}$.
    Therefore the probability of **not** choosing a girl whose favourite subject is English is:
    $1 - \frac{4}{30} = \frac{26}{30} = \frac{13}{15}$

The likelihood of an event such as 'you will play sport tomorrow' will vary from person to person. Therefore, the probability of the event is not constant. However, the probability of some events, such as the result of spinning a coin or spinner, can be found by experiment or calculation.

A probability scale goes from 0 to 1.

Exercise 33.1

1 Copy the probability scale above.
   Mark on the probability scale the probability that:
   a a day chosen at random is a Saturday,
   b a coin will show tails when spun,
   c the sun will rise tomorrow,
   d a woman will run 100 metres in under 10 seconds,
   e the next car you see will be silver.

2 Express your answers to Question 1 as fractions, decimals and percentages.

33 PROBABILITY

Exercise 33.2

1. Calculate the theoretical probability, when spinning a fair 1–6 spinner, of getting each of the following:
   - **a** a score of 1
   - **b** a score of 2, 3, 4, 5 or 6
   - **c** an odd number
   - **d** a score less than 6
   - **e** a score of 7
   - **f** a score less than 7

2. **a** Calculate the probability of:
   - **i** being born on a Wednesday,
   - **ii** not being born on a Wednesday.

   b Explain the result of adding the answers to **a i** and **ii** together.

3. 250 balls are numbered from 1 to 250 and placed in a box. A ball is picked at random. Find the probability of picking a ball with:
   - **a** the number 1
   - **b** an even number
   - **c** a three-digit number
   - **d** a number less than 300

4. In a class there are 25 girls and 15 boys. The teacher takes in all of their books in a random order. Calculate the probability that the teacher will:
   - **a** mark a book belonging to a girl first,
   - **b** mark a book belonging to a boy first.

5. Twenty-six tiles, each printed with one different letter of the alphabet, are put into a bag. If one tile is taken out at random, calculate the probability that it is:
   - **a** an A or P
   - **b** a vowel
   - **c** a consonant
   - **d** an X, Y or Z
   - **e** a letter in your first name.

6. A boy was late for school 5 times in the previous 30 school days. If tomorrow is a school day, calculate the probability that he will arrive late.

7. **a** Three red, 10 white, 5 blue and 2 green counters are put into a bag. If one is picked at random, calculate the probability that it is:
   - **i** a green counter
   - **ii** a blue counter.

   b If the first counter taken out is green and it is not put back into the bag, calculate the probability that the second counter picked is:
   - **i** a green counter
   - **ii** a red counter.

8. A circular spinner with an arrow has the numbers 0 to 36 equally spaced around its edge. Assuming that it is unbiased, calculate the probability on spinning the arrow of getting:
   - **a** the number 5
   - **b** not 5
   - **c** an odd number
   - **d** zero
   - **e** a number greater than 15
   - **f** a multiple of 3
   - **g** a multiple of 3 or 5
   - **h** a prime number.

9. The letters R, C and A can be combined in several different ways.
   - **a** Write the letters in as many different orders as possible.

   If a computer writes these three letters at random, calculate the probability that:
   - **b** the letters will be written in alphabetical order,
   - **c** the letter R is written before both the letters A and C,
   - **d** the letter C is written after the letter A,
   - **e** the computer will spell the word CART if the letter T is added.

10. A normal pack of playing cards contains 52 cards. These are made up of four suits (hearts, diamonds, clubs and spades). Each suit consists of 13 cards. These are labelled Ace, 2, 3, 4, 5, 6, 7, 8, 9, 10, Jack, Queen and King. The hearts and diamonds are red; the clubs and spades are black.

Probability

If a card is picked at random from a normal pack of cards, calculate the probability of picking:
- **a** a heart
- **b** not a heart
- **c** a 4
- **d** a red King
- **e** a Jack, Queen or King
- **f** the Ace of spades
- **g** an even-numbered card
- **h** a 7 or a club.

 **Exercise 33.3**

1 Zuri conducts a survey on the types of vehicle that pass her house. The results are shown below.

Vehicle type	Car	Lorry	Van	Bicycle	Motorbike	Other
Frequency	28	6	20	48	32	6

- **a** How many vehicles passed Zuri's house?
- **b** A vehicle is chosen at random from the results. Calculate the probability that it is:
  - **i** a car
  - **ii** a lorry
  - **iii** not a van.

2 In a class, data is collected about whether each student is right-handed or left-handed. The results are shown below.

	Left-handed	Right-handed
Boys	2	12
Girls	3	15

- **a** How many students are in the class?
- **b** A student is chosen at random. Calculate the probability that the student is:
  - **i** a girl
  - **ii** left-handed
  - **iii** a right-handed boy
  - **iv** not a right-handed boy.

3 A library keeps a record of the books that are borrowed during one day. The results are shown in the chart below.

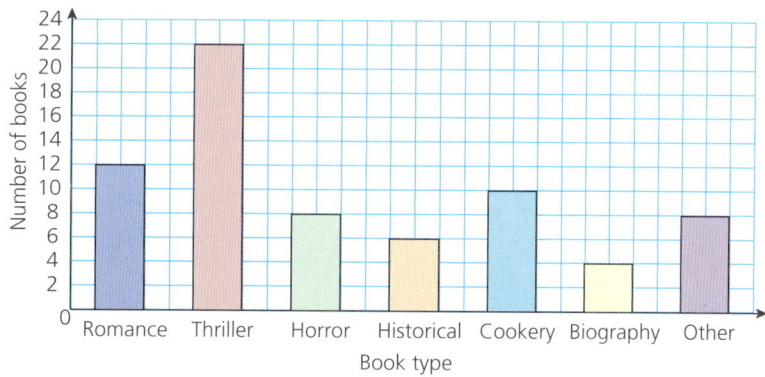

- **a** How many books were borrowed that day?
- **b** A book is chosen at random from the ones borrowed. Calculate the probability that it is:
  - **i** a thriller
  - **ii** a horror or a romance
  - **iii** not a horror or a romance
  - **iv** not a biography.

33 PROBABILITY

Venn diagrams

In Chapter 10, we saw how Venn diagrams are used to display information written as sets. Venn diagrams are also a good way of representing information when carrying out probability calculations.

Set notation symbols include:

ξ universal set

{ } set

∪ union of

∩ intersection of

→ Worked examples

Cards numbered 1–15 are arranged in two sets:
$A = \{$multiples of 3$\}$ and $B = \{$multiples of 5$\}$.

a Represent this information in a Venn diagram.

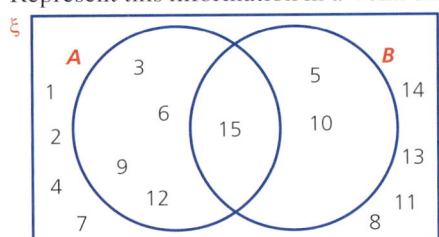

b A card is picked at random. Calculate the probability that it is from set A. (i.e. $P(A)$).

From the diagram it can be seen that the number of elements in set A is five. (i.e $n(A) = 5$)

Therefore $P(A) = \frac{5}{15} = \frac{1}{3}$

Probability notation is only required for the Extended syllabus.

c If a card is picked at random, calculate the probability that it is not from set A. (i.e. $P(A')$).

As $P(A) = \frac{1}{3}$,

then $P(A') = 1 - \frac{1}{3} = \frac{2}{3}$

Note

A' means not A

d Calculate $P(A \cap B)$.

This represents the probability of picking a card belonging to both set A and set B.

$n(A \cap B) = 1$

Therefore $P(A \cap B) = \frac{1}{15}$

Venn diagrams

Exercise 33.4

1. In a class of 30 students, all students study at least one language: Spanish (*S*) or Mandarin Chinese (*M*). 16 students study only Spanish and 8 students study only Mandarin.

   a Copy and complete the Venn diagram for the number of students studying each language.

   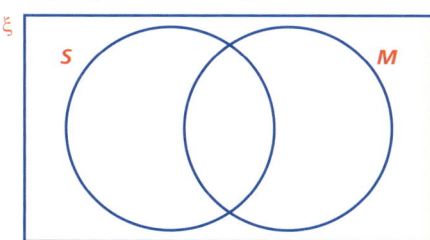

   b If a student is chosen at random, calculate:

   i $P(S)$
   ii $P(M)$
   iii $P(S \cap M)$
   iv $P(S \cup M)$

2. One hundred students take part in a school sports event. 70 students take part in track events (*T*), 62 students take part in field events (*F*) and 8 students take part in neither.

   a Construct a Venn diagram to show this information.
   b If a student is chosen at random, calculate:

   i $P(T)$
   ii $P(T \cap F)$
   iii $P(F')$
   iv $P(T \cup F)'$

3. Sixty students were asked where they had travelled to in the last twelve months. The three most popular destinations were Singapore (*S*), Dubai (*D*) and Great Britain (*B*). The number of students travelling to each destination is shown in the Venn diagram.

   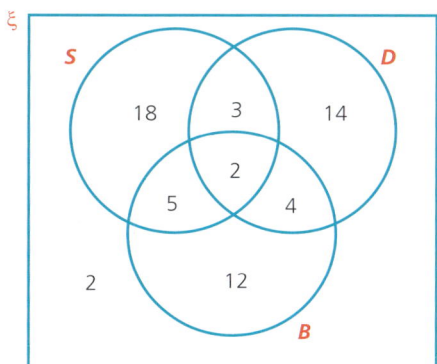

   If a student is chosen at random, calculate the probability of the following:

   a $P(S)$
   b $P(B \cap D)$
   c $P(S \cup D)$
   d $P(S \cap D \cap B)$
   e $P(B')$
   f $P(S \cap B \cap D')$

33 PROBABILITY

Exercise 33.4 (cont)

4 Cards numbered 1–20 are arranged in three sets as follows:
A = {multiples of 2}
B = {multiples of 3}
C = {multiples of 4}

a Copy and complete the Venn diagram:

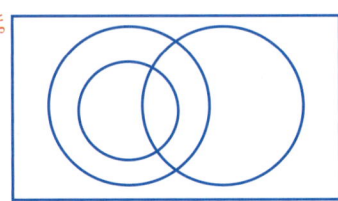

b If a number is chosen at random, calculate each of the following probabilities:

i $P(A)$
ii $P(A \cap B)$
iii $P(B \cap C)$
iv $P(B \cup C)$
v $P(A \cup B)'$
vi $P(A \cup B \cup C')$

Relative frequency

A football referee always used a special coin to toss for ends. She noticed that out of the last twenty matches the coin had come down heads far more often than tails. She wanted to know if the coin was fair, that is, if it was as likely to come down heads as tails.

She decided to do a simple experiment by spinning the coin lots of times. Her results are shown below.

Number of trials	Number of heads	Relative frequency
100	40	0.4
200	90	0.45
300	142	0.47…
400	210	0.525
500	260	0.52
600	290	0.48…
700	345	0.49…
800	404	0.505
900	451	0.50…
1000	499	0.499

The **relative frequency** = $\dfrac{\text{number of successful trials}}{\text{total number of trials}}$

In the 'long run', that is after many trials, did the coin appear to be fair?

Notice that the greater the number of trials, the better the estimated probability or relative frequency is likely to be. The key idea is that increasing the number of trials gives a better estimate of the probability and the closer the result obtained by experiment will be to that obtained by calculation.

Relative frequency

> ### Worked examples

a There is a group of 250 people in a hall. A girl calculates that the probability of randomly picking someone that she knows from the group is 0.032. Calculate the number of people in the group that the girl knows.

$$\text{Probability} = \frac{\text{number of favourable results } (F)}{\text{number of possible results}}$$

$$0.032 = \frac{F}{250}$$

$250 \times 0.032 = F$ so $8 = F$

The girl knows 8 people in the group.

b A boy enters 8 cakes into a baking competition. His mother knows how many cakes have been entered into the competition in total and tells her son that he has a probability of 0.016 of winning the first prize (assuming all the cakes have an equal chance). How many cakes were entered into the competition?

$$\text{Probability} = \frac{\text{number of favourable results}}{\text{number of possible results } (T)}$$

$$0.016 = \frac{8}{T}$$

$$T = \frac{8}{0.016} = 500$$

So, 500 cakes were entered into the competition.

Exercise 33.5

1 Mikki calculates that she has a probability of 0.004 of winning the first prize in a photography competition if the selection is made at random. If 500 photographs are entered into the competition, how many photographs did Mikki enter?

2 The probability of getting any particular number on a spinner game is given as 0.04. How many numbers are there on the spinner?

3 A bag contains 7 red counters, 5 blue, 3 green and 1 yellow. If one counter is drawn, what is the probability that it is:
   a yellow   b red   c blue or green
   d red, blue or green   e not blue?

4 Luca collects marbles. He has the following colours in a bag: 28 red, 14 blue, 25 yellow, 17 green and 6 purple. If he draws one marble from the bag, what is the probability that it is:
   a red   b blue   c yellow or blue
   d purple   e not purple?

5 The probability of Hanane drawing a marble of one of the following colours from another bag of marbles is:
   blue 0.25   red 0.2   yellow 0.15   green 0.35   white 0.05
   If there are 49 green marbles, how many of each other colour does she have in her bag?

6 There are six red sweets in a bag. If the probability of randomly picking a red sweet is 0.02, calculate the number of sweets in the bag.

7 The probability of getting a bad egg in a batch of 400 is 0.035. How many bad eggs are there likely to be in a batch?

8 A sports arena has 25 000 seats, some of which are VIP seats. For a charity event, all the seats are allocated randomly. The probability of getting a VIP seat is 0.008. How many VIP seats are there?

9 The probability of Harts Utd winning 4–0 is 0.05. How many times are they likely to win by this score in a season of 40 matches?

33 PROBABILITY

Student assessment 1

1. What is the probability of spinning the following numbers with a fair 1–6 spinner?
   a a 2
   b not a 2
   c less than 5
   d a 7

2. If you have a normal pack of 52 cards, what is the probability of drawing:
   a a diamond
   b a 6
   c a black card
   d a picture card
   e a card less than 5?

3. 250 coins, one of which is gold, are placed in a bag. What is the probability of getting the gold coin if I take, without looking, the following numbers of coins?
   a 1
   b 5
   c 20
   d 75
   e 250

4. A bag contains 11 blue, 8 red, 6 white, 5 green and 10 yellow counters. If one counter is taken from the bag, what is the probability that it is:
   a blue
   b green
   c yellow
   d not red?

5. The probability of drawing a red, blue or green marble from a bag containing 320 marbles is:
   red 0.5   blue 0.3   green 0.2
   How many marbles of each colour are there?

6. In a small town there are a number of sports clubs. The clubs have 750 members in total. The table below shows the types of sports club and the number of members each has.

	Tennis	Football	Golf	Hockey	Athletics
Men	30	110	40	15	10
Women	15	25	20	45	30
Boys	10	200	5	10	40
Girls	20	35	0	30	60

   A sports club member is chosen at random from the town. Calculate the probability that the member is:
   a a man
   b a girl
   c a woman who does athletics
   d a boy who plays football
   e not a boy who plays football
   f not a golf player
   g a male who plays hockey.

7. A 1–6 spinner is thought to be biased. In order to test it, Monique spins it 12 times and gets the following results:

Number	1	2	3	4	5	6
Frequency	2	2	2	2	2	2

   Jas decides to test the same spinner and spins it 60 times. The table below shows her results:

Number	1	2	3	4	5	6
Frequency	3	3	47	3	2	2

   a Which results are likely to be more reliable? Justify your answer.
   b What conclusion can you make about whether the spinner is biased?

8 In a school, students can study science as the individual subjects of Physics, Chemistry and Biology. Each student must study at least one of the subjects. The following Venn diagram gives the number of students studying each subject:

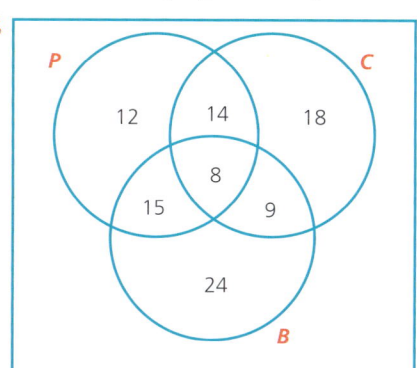

If a student is chosen at random, calculate each of the following probabilities:
a $P(C)$
b $P(P \cap B)$
c $P(P \cap C \cap B)$
d $P(B \cap C \cap P')$
e $P(P \cup C)$

Student assessment 2

1 An octagonal spinner has the numbers 1 to 8 on it as shown.

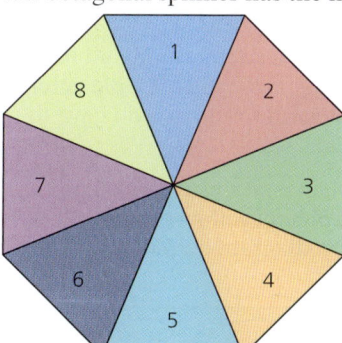

What is the probability of spinning:
a a 7
b not a 7
c a factor of 12
d a 9?

2 A game requires the use of all the playing cards in a normal pack from 6 to King inclusive.
a How many cards are used in the game?
b What is the probability of choosing:
   i   a 6   ii   a picture
   iii a club   iv   a prime number
   v   an 8 or a spade?

3 180 students in a school are offered a chance to attend a football match for free. If the students are chosen at random, what is the chance of being picked to go if the following numbers of tickets are available?
a 1   b 9   c 15   d 40   e 180

33 PROBABILITY

4 A bag contains 11 white, 9 blue, 7 green and 5 red counters. What is the probability that a single counter drawn will be:
   a blue   **b** red or green   **c** not white?

5 The probability of drawing a red, blue or green marble from a bag containing 320 marbles is:
   red 0.4   blue 0.25   green 0.35
   If there are no other colours in the bag, how many marbles of each colour are there?

6 Students in a class conduct a survey to see how many friends they have on a social media website. The results were grouped and are shown in the pie chart below.

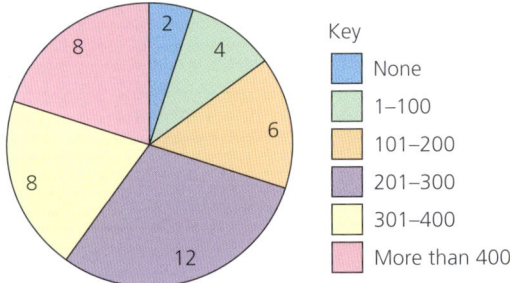
Number of friends on social media website

A student is chosen at random. What is the probability that she:
   a has 101–200 friends on the social media website
   b uses the social media website
   c has more than 200 friends on the website?

7 a If I enter a competition and have a 0.00002 probability of winning, how many people entered the competition?
   b What assumption do you have to make in order to answer part **a**?

8 A large bag contains coloured discs. The discs are completely red (R), completely yellow (Y) or half red and half yellow. The Venn diagram below shows the probability of picking each type of disc:

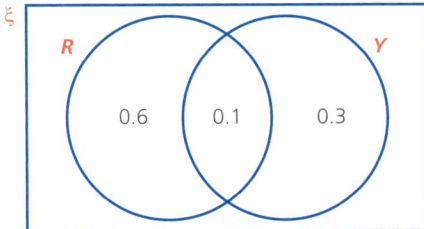

If there are 120 discs that are coloured yellow (either fully or partly), calculate:
   a the number of discs coloured completely red,
   b the total number of discs in the bag.

34 Further probability

Combined events

Combined events look at the probability of two or more events.

→ Worked examples

a Two coins are tossed. Show in a **two-way table** all the possible outcomes.

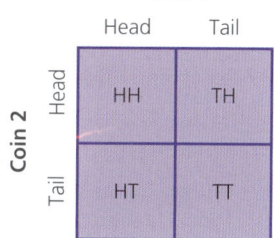

A two-way table is a simple form of a sample space diagram.

b Calculate the probability of getting two heads.
All four outcomes are equally likely: therefore, the probability of getting HH is $\frac{1}{4}$.

c Calculate the probability of getting a head and a tail in any order.
The probability of getting a head and a tail in any order, i.e. HT or TH, is $\frac{2}{4} = \frac{1}{2}$.

Exercise 34.1

1 a Two fair, tetrahedral dice are rolled. If each is numbered 1–4, draw a two-way table to show all the possible outcomes.
   b What is the probability that both dice show the same number?
   c What is the probability that the number on one dice is double the number on the other?
   d What is the probability that the sum of both numbers is prime?

2 Two fair dice are rolled. Copy and complete the diagram to show all the possible combinations.
   What is the probability of getting:
   a a double 3,
   b any double,
   c a total score of 11,
   d a total score of 7,
   e an even number on both dice,
   f an even number on at least one dice,
   g a total of 6 or a double,
   h scores which differ by 3,
   i a total which is either a multiple of 2 or 5?

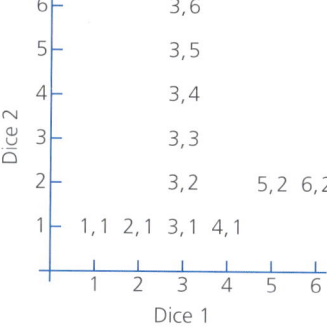

34 FURTHER PROBABILITY

Tree diagrams

When more than two combined events are being considered, then two-way tables cannot be used and another method of representing information diagrammatically is needed. Tree diagrams are a good way of doing this.

→ Worked examples

a If a coin is flipped three times, show all the possible outcomes on a tree diagram, writing each of the probabilities at the side of the branches.

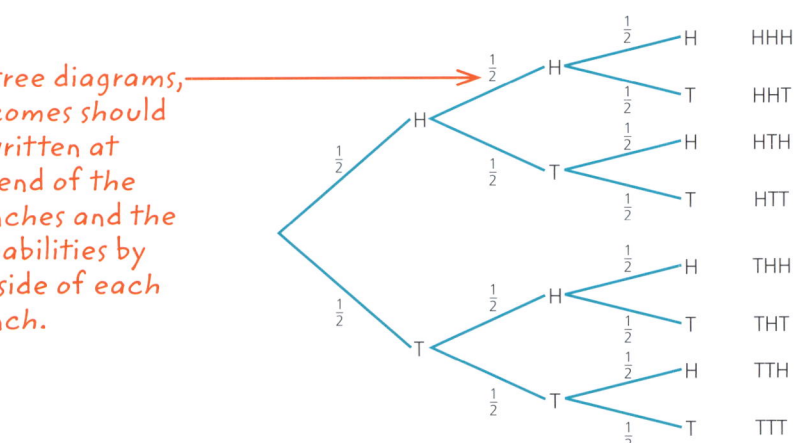

On tree diagrams, outcomes should be written at the end of the branches and the probabilities by the side of each branch.

b What is the probability of getting three heads?
   There are eight equally likely outcomes, therefore the probability of getting HHH is $\frac{1}{8}$.

c What is the probability of getting two heads and one tail in any order?
   The successful outcomes are HHT, HTH, THH.
   Therefore the probability is $\frac{3}{8}$.

d What is the probability of getting at least one head?
   This refers to any outcome with either one, two or three heads, i.e. all of them *except* TTT.
   Therefore the probability is $\frac{7}{8}$.

e What is the probability of getting no heads?
   The only successful outcome for this event is TTT.
   Therefore the probability is $\frac{1}{8}$.

Tree diagrams

 Exercise 34.2

1. **a** A computer uses the numbers 1, 2 or 3 at random to make three-digit numbers. Assuming that a number can be repeated, show on a tree diagram all the possible combinations that the computer can print.
   b Calculate the probability of getting:
   i the number 131,
   ii an even number,
   iii a multiple of 11,
   iv a multiple of 3,
   v a multiple of 2 or 3.

2. **a** A cat has four kittens. Draw a tree diagram to show all the possible combinations of males and females. [assume P (male) = P (female)]
   b Calculate the probability of getting:
   i all female,
   ii two females and two males,
   iii at least one female,
   iv more females than males.

3. **a** A netball team plays three matches. In each match the team is equally likely to win, lose or draw. Draw a tree diagram to show all the possible outcomes over the three matches.
   b Calculate the probability that the team:
   i wins all three matches,
   ii wins more times than loses,
   iii loses at least one match,
   iv either draws or loses all three matches.
   c Explain why it is not very realistic to assume that the outcomes are equally likely in this case.

4. A spinner is split into quarters as shown.

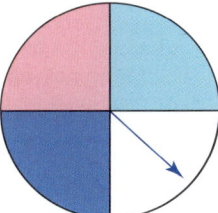

   a If it is spun twice, draw a probability tree showing all the possible outcomes.
   b Calculate the probability of getting:
   i two dark blues,
   ii two blues of either shade,
   iii a pink and a white in any order.

In each of the cases considered so far, all of the outcomes have been assumed to be equally likely. However, this need not be the case.

493

34 FURTHER PROBABILITY

> **Worked example**

In winter, the probability that it rains on any one day is $\frac{5}{7}$.

a Using a tree diagram, show all the possible combinations for two consecutive days.

b Write each of the probabilities by the sides of the branches.

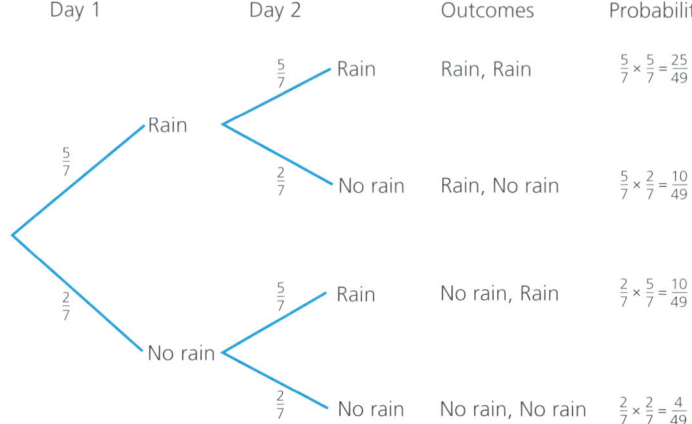

Note how the probability of each outcome is arrived at by multiplying the probabilities of the branches.

c Calculate the probability that it will rain on both days.
$P(R,R) = \frac{5}{7} \times \frac{5}{7} = \frac{25}{49}$

d Calculate the probability that it will rain on the first but not the second day.
$P(R,NR) = \frac{5}{7} \times \frac{2}{7} = \frac{10}{49}$

e Calculate the probability that it will rain on at least one day.
The outcomes which satisfy this event are (R, R) (R, NR) and (NR, R).
Therefore the probability is $\frac{25}{49} + \frac{10}{49} + \frac{10}{49} = \frac{45}{49}$.

Exercise 34.3

1 A board game involves players rolling a dice. However, before a player can start, each player needs to roll a 6.
  a Copy and complete the tree diagram below showing all the possible combinations for the first two rolls of the dice.

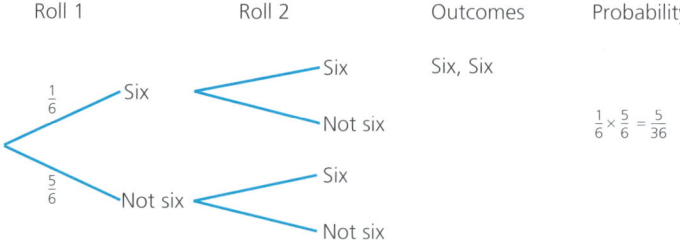

Exercise 34.3 (cont)

b Calculate the probability of the following:
  i getting a 6 on the first roll,
  ii starting within the first two rolls,
  iii starting on the second roll,
  iv not starting within the first three rolls,
  v starting within the first three rolls.
c If you add the answers to **b iv** and **v** what do you notice? Explain.

2 In Italy $\frac{3}{5}$ of the cars are foreign made. By drawing a tree diagram and writing the probabilities next to each of the branches, calculate the following probabilities:
  a the next two cars to pass a particular spot are both Italian,
  b two of the next three cars are foreign,
  c at least one of the next three cars is Italian.

3 The probability that a morning bus arrives on time is 65%.
  a Draw a tree diagram showing all the possible outcomes for three consecutive mornings.
  b Label your tree diagram and use it to calculate the probability that:
    i the bus is on time on all three mornings,
    ii the bus is late the first two mornings,
    iii the bus is on time two out of the three mornings,
    iv the bus is on time at least twice.

4 A normal pack of 52 cards is shuffled and three cards are picked at random. Draw a tree diagram to help calculate the probability of picking:
  a two clubs first, **b** three clubs,
  c no clubs, **d** at least one club.

5 Light bulbs are packaged in cartons of three. 10% of the bulbs are found to be faulty. Calculate the probability of finding two faulty bulbs in a single carton.

6 A volleyball team has a 0.25 chance of losing a game. Calculate the probability of the team achieving:
  a two consecutive wins,
  b three consecutive wins,
  c 10 consecutive wins.

7 A bowl of fruit contains one kiwi fruit, one banana, two mangos and two lychees. Two pieces of fruit are chosen at random and eaten.
  a Draw a probability tree showing all the possible combinations of the two pieces of fruit eaten.
  b Use your tree diagram to calculate the probability that:
    i both the pieces of fruit eaten are mangos,
    ii a kiwi fruit and a banana are eaten,
    iii at least one lychee is eaten.

8 A class has n number of girls and n number of boys. Two students are chosen at random.
  a Draw a tree diagram to show all the possible outcomes, labelling the probability of each branch in terms of n, where appropriate.
  b Show that the probability of two girls being chosen is $\frac{n-1}{2(2n-1)}$.

9 A bag of candies contains n red candies and $n+3$ yellow candies. A child takes two candies from the bag at random.
  a Draw a tree diagram to show all the possible outcomes and label the probability of each branch in terms of n.
  b Calculate the probability that the child picks two yellow candies.

34 FURTHER PROBABILITY

Conditional probability

So far, all the probability considered has been based on random events with no other information given. However, sometimes more information is known and, as a result, the probability of the event happening changes. This is an example of **conditional probability**.

→ Worked examples

a The table shows the number of boys and girls studying Maths and Art in a school.

	Maths	Art	Total
Boys	26	14	40
Girls	34	12	46
Total	60	26	86

i A student is chosen at random. Calculate the probability that they study Maths.
60 students out of a total of 86 study Maths.
Therefore, the probability that they study Maths is $\frac{60}{86} = \frac{30}{43}$.

ii A student is chosen at random. Calculate the probability that they study Maths **given** that the student is a girl.
Here more information has been given than in part i. We already know that the student chosen is a girl, therefore, the student studying Maths is only being chosen from a group of 46 students (i.e. the total number of girls). Therefore, the probability of choosing a student who studies Maths who is a girl is $\frac{34}{46} = \frac{17}{23}$.
In general, the notation used for conditional probability is $P(A|B)$. This states the probability of event A happening given that event B has happened.
In the example ii above, if the event of a student studying Maths is M and the event of choosing a girl is G, then the probability of choosing a student who studies Maths who is a girl is written as follows:
$P(M|G) = \frac{34}{46}$

b The numbers 1–15 are arranged into two sets of numbers where O = {odd numbers} and P = {prime numbers}.
i Represent the numbers in a Venn diagram.

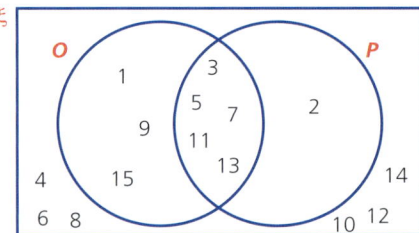

ii A number is chosen at random. Calculate the probability that it is prime.
$P(P) = \frac{6}{15} = \frac{2}{5}$

> **Note**
>
> Although the notation $P(A|B)$ is not explicitly required for this course, it has been included here as a more efficient way of describing conditional probability.

Conditional probability

iii A number is chosen at random. Calculate the probability that it is prime **given** that it is odd, i.e. calculate $P(P|O)$.

We already know that the number is odd, therefore the total number of possible outcomes is only eight.

Therefore $P(P|O) = \frac{5}{8}$.

c A bag contains ten red beads (R) and five black beads (B). A bead is taken out of the bag at random, its colour is noted and then it is **not** placed back in the bag.

i Draw a tree diagram to show the possible outcomes and the probabilities of the first two beads removed from the bag.

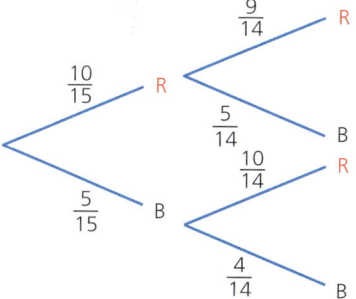

ii Calculate the probability that the first two beads are red.

$P(RR) = \frac{10}{15} \times \frac{9}{14} = \frac{3}{7}$

iii Calculate the probability that the second bead is black given that the first bead chosen is red.

As the first bead is already known to be red, then the bottom half of the tree diagram can be ignored as it involves a black bead having been chosen first.

The top half of the tree diagram has a total probability of $\frac{2}{3}$

$\left(\text{i.e. } \frac{10}{15} \times \frac{9}{14} + \frac{10}{15} \times \frac{5}{14} = \frac{2}{3} \right)$

The probability of getting a red bead followed by a black bead when considering only the top half of the tree diagram is

$\dfrac{\frac{10}{15} \times \frac{5}{14}}{\frac{2}{3}} = \frac{5}{14}$.

Another way to consider this is to note that, as it is already known that the first bead is red, we are already at the stage indicated in the diagram below:

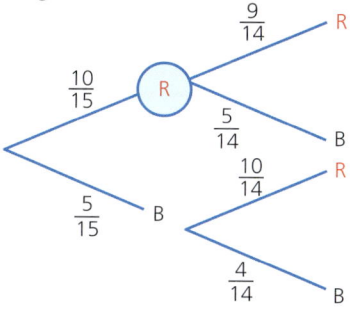

The probability of picking a black bead after this stage is therefore $\frac{5}{14}$.

34 FURTHER PROBABILITY

 Exercise 34.4

1. Each of the boys and girls in a class are asked how many brothers and sisters they have. The results are shown in the table:

	Number of brothers and sisters					
	0	1	2	3	≥4	Total
Boys	2	3	3	2	0	10
Girls	3	1	6	3	1	14
Total	5	4	9	5	1	24

   If a student is picked at random, calculate the probability:
   a that it is a girl,
   b that it is a boy with two or more brothers or sisters,
   c that they have no brothers or sisters given that it is a girl,
   d that it is a boy given that they have four or more brothers or sisters.

2. The Venn diagram shows the type of vehicle owned by a group of 60 people, where:
   C = {car}, M = {motorbike} and B = {bicycle}.

   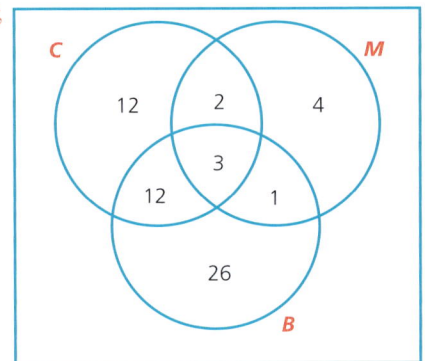

   If a person is picked at random, calculate the probability that they own the following vehicle type.
   a $P(C)$ b $P(M \cup B)$
   c $P(C \cap B)$ d $P(B|C)$
   e $P(C|B)$ f $P(C \cap M|B)$
   g $P(M|C')$ h $P(M'|B)$

3. Over the course of a season, a volleyball player records how often he is selected to play for his team and whether or not his team wins. At the end of the season, he analyses the results and finds that the probability of being selected was 0.8. If he was selected, the probability of the team winning was 0.65; if he wasn't selected, the probability of the team winning was 0.45.
   a Draw a tree diagram to represent this information.
   b If the team played 100 matches during the season, calculate how many they won.
   c Given that he played in a match, calculate the probability that the team won.
   d Given that the team won a match, calculate the probability that he played.

Conditional probability

4 The numbers 1–15 are arranged into three sets as follows:
   A = {odd numbers}, B = {prime numbers} and C = {multiples of two}.
   a Copy and complete the Venn diagram, labelling each circle and then placing each number in the correct region.

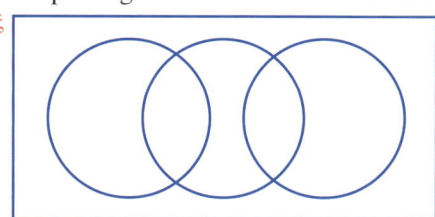

   b If a number is picked at random, calculate the following probabilities:
      i $P(B)$ ii $P(B \cap C)$
      iii $P(A \cap C)$ iv $P(A|B)$
      v $P(B|C)$ vi $P(C|B)$
      vii $P(B|B \cap C)$
   c Explain why two of the circles do not overlap each other.

Student assessment 1

1 Two normal and fair dice are rolled and their scores added together.
   a Using a two-way table, show all the possible scores that can be achieved.
   b Using your two-way table, calculate the probability of getting:
      i   a score of 12,   ii   a score of 7,
      iii a score less   iv   a score of 7
           than 4,   or more.
   c Two dice are rolled 180 times. In theory, how many times would you expect to get a total score of 6?

2 A spinner is numbered as shown.

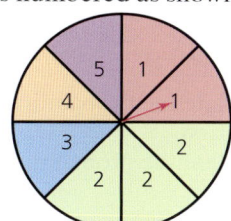

   a If it is spun once, calculate the probability of getting:
      i   a 1,   ii   a 2.
   b If it is spun twice, calculate the probability of getting:
      i   a 2 followed by a 4,
      ii   a 2 and a 4 in any order,
      iii at least one 1,
      iv   at least one 2.

3 Two spinners are coloured as shown (below).

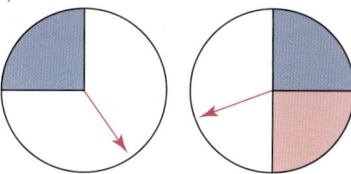

   a They are both spun. Draw and label a tree diagram showing all the possible outcomes.
   b Using your tree diagram, calculate the probability of getting:
      i   two blues,   ii   two whites,
      iii a white and   iv   at least one
           a pink,   white.

4 Two spinners are labelled as shown:

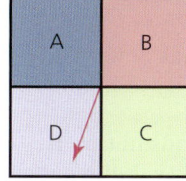 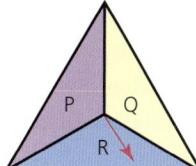

   Calculate the probability of getting:
   a A and P,
   b A or B and R,
   c C but not Q.

34 FURTHER PROBABILITY

5 A vending machine accepts $1 and $2 coins. The probability of a $2 coin being rejected is 0.2. The probability of a $1 coin being rejected is 0.1.
A sandwich costing $3 is bought. Calculate the probability of getting a sandwich first time if:
  a one of each coin is used,
  b three $1 coins are used.

6 A biased coin is flipped three times. On each occasion, the probability of getting a head is 0.6.
  a Draw a tree diagram to show all the possible outcomes after three flips. Label each branch clearly with the probability of each outcome.
  b Using your tree diagram, calculate the probability of getting:
    i three heads,
    ii three tails,
    iii at least two heads.

7 A ball enters a chute at X.

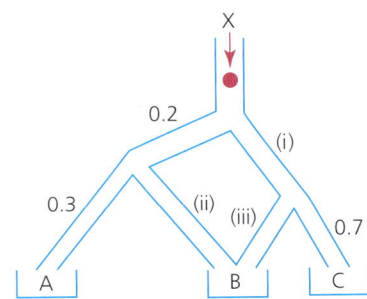

  a What are the probabilities of the ball going down each of the chutes labelled (i), (ii) and (iii)?
  b Calculate the probability of the ball landing in:
    i tray A,   **ii** tray C,
    iii tray B.
  c Given that the ball goes down chute (i), calculate the probability of it landing in tray B.
  d Given that the ball lands in tray B, calculate the probability that it came down chute (iii).

8 A fish breeder keeps three types of fish: A, B and C. Each type can be categorised into two sizes, small (S) and large (L). The probabilities of each are given in the tree diagram:

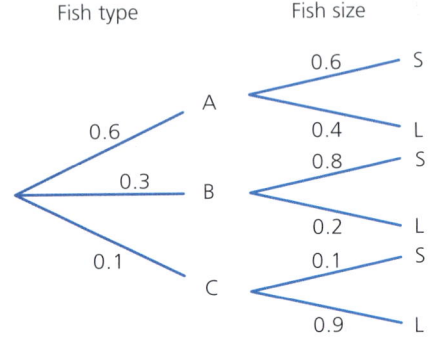

  a A fish is chosen at random. Calculate the probability that:
    i it is a large fish of type B,
    ii it is a large fish,
    iii it is a large fish given that it is of type B.
  b If a fish chosen at random is small, calculate the probability of it being of type:
    i A,   **ii** B,   **iii** C.

9 A football team decide to analyse their match results. They look at their probability of winning depending on whether they scored first in the game or not. The results are presented in the table:

	Win	Lose	Draw
Score first	0.38	0.12	0.18
Don't score first	0.02	0.22	0.08

  a Calculate the probability that the team score first.
  b Calculate the probability that the team draw.
  c Given that the team score first, calculate the probability that they draw.
  d Given that the team lose, calculate the probability that they didn't score first.

8 Mathematical investigations and ICT 8

Probability drop

A game involves dropping a red marble down a chute. On hitting a triangle divider, the marble can bounce either left or right. On completing the drop, the marble lands in one of the trays along the bottom. The trays are numbered from left to right. Different sizes of game exist, the four smallest versions are shown below:

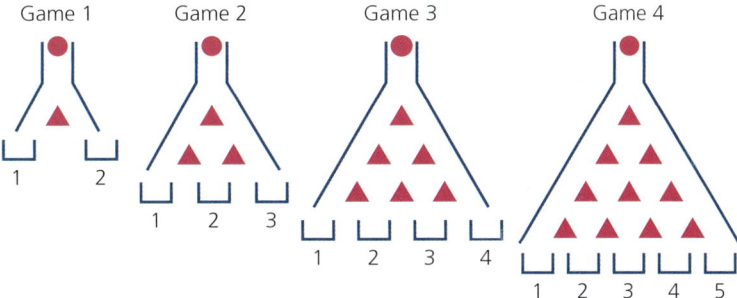

To land in tray 2 in the second game above, the ball can travel in one of two ways. These are: Left – Right or Right – Left.

This can be abbreviated to LR or RL.

1. State the different routes the marble can take to land in each of the trays in the third game.
2. State the different routes the marble can take to land in each of the trays in the fourth game.
3. State, giving reasons, the probability of a marble landing in tray 1 in the fourth game.
4. State, giving reasons, the probability of a marble landing in each of the other trays in the fourth game.
5. Investigate the probability of the marble landing in each of the different trays in larger games.
6. Using your findings from your investigation, predict the probability of a marble landing in tray 7 in the tenth game (11 trays at the bottom).
7. Investigate the links between this game and the sequence of numbers generated in Pascal's triangle.

MATHEMATICAL INVESTIGATIONS AND ICT 8

Dice sum

Two ordinary dice are rolled and their scores added together. Below is an incomplete table showing the possible outcomes:

		\multicolumn{6}{c}{Dice 1}

		1	2	3	4	5	6
Dice 2	1	2			5		
	2						
	3				7		
	4				8		
	5				9	10	11
	6						12

1. Copy and complete the table to show all possible outcomes.
2. How many possible outcomes are there?
3. What is the most likely total when two dice are rolled?
4. What is the probability of getting a total score of 4?
5. What is the probability of getting the most likely total?
6. How many times more likely is a total score of 5 compared with a total score of 2?

Now consider rolling two four-sided dice, each numbered 1–4. Their scores are also added together.

7. Draw a table to show all the possible outcomes when the two four-sided dice are rolled and their scores added together.
8. How many possible outcomes are there?
9. What is the most likely total?
10. What is the probability of getting the most likely total?
11. Investigate the number of possible outcomes, the most likely total and its probability when two identical dice are rolled together and their scores are added, i.e. consider eight-sided dice, ten-sided dice, etc.
12. Consider two m-sided dice rolled together and their scores added.
    a What is the total number of outcomes in terms of m?
    b What is the most likely total, in terms of m?
    c What, in terms of m, is the probability of the most likely total?
13. Consider an m-sided and n-sided dice rolled together, where $m > n$.
    a In terms of m and n, deduce the total number of outcomes.
    b In terms of m and/or n, deduce the most likely total(s).
    c In terms of m and/or n, deduce the probability of getting the most likely total.

ICT activity: Buffon's needle experiment

You will need to use a spreadsheet for this activity.

The French count Le Comte de Buffon devised the following probability experiment.

1. Measure the length of a match (with the head cut off) as accurately as possible.
2. On a sheet of paper, draw a series of straight lines parallel to each other. The distance between each line should be the same as the length of the match.
3. Take ten identical matches and drop them randomly on the paper. Count the number of matches that cross or touch any of the lines.

For example, in the diagram below, the number of matches crossing or touching lines is six.

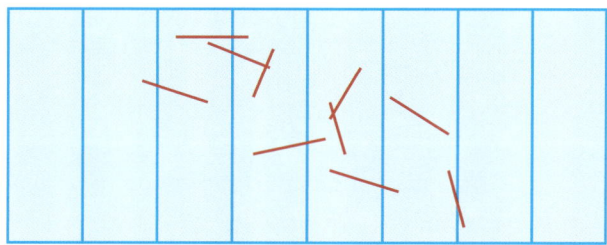

4. Repeat the experiment a further nine times, making a note of your results, so that altogether you have dropped 100 matches.
5. Set up a spreadsheet similar to the one shown below and enter your results in cell B2.

	A	B	C	D	E	F	G	H	I	J	K
1	Number of drops (N)	100	200	300	400	500	600	700	800	900	1000
2	Number of matches crossing/touching lines (n)										
3	Probability of crossing a line ($p = n/N$)										
4	$2/p$										

6. Repeat 100 match drops again, making a total of 200 drops, and enter cumulative results in cell C2.
7. By collating the results of your fellow students, enter the cumulative results of dropping a match 300–1000 times in cells D2–K2 respectively.
8. Using an appropriate formula, get the spreadsheet to complete the calculations in Rows 3 and 4.
9. Use the spreadsheet to plot a line graph of N against $\frac{2}{p}$.
10. What value does $\frac{2}{p}$ appear to get closer to?

503

TOPIC 9

Statistics

Contents

Chapter 35 Mean, median, mode and range (E9.2, E9.3)
Chapter 36 Collecting, displaying and interpreting data (E9.1, E9.2, E9.4, E9.5, E9.7)
Chapter 37 Cumulative frequency (E9.3, E9.6)

Learning objectives

E9.1 Classifying statistical data
Classify and tabulate statistical data.

E9.2 Interpreting statistical data
1. Read, interpret and draw inferences from tables and statistical diagrams.
2. Compare sets of data using tables, graphs and statistical measures.
3. Appreciate restrictions on drawing conclusions from given data.

E9.3 Averages and measures of spread
1. Calculate the mean, median, mode, quartiles, range and interquartile range for individual data and distinguish between the purposes for which these are used.
2. Calculate an estimate of the mean for grouped discrete or continuous data.
3. Identify the modal class from a grouped frequency distribution.

E9.4 Statistical charts and diagrams
Draw and interpret:
(a) bar charts
(b) pie charts
(c) pictograms
(d) stem-and-leaf diagrams
(e) simple frequency distributions.

E9.5 Scatter diagrams
1. Draw and interpret scatter diagrams.
2. Understand what is meant by positive, negative and zero correlation.
3. Draw by eye, interpret and use a straight line of best fit.

E9.6 Cumulative frequency diagrams
1. Draw and interpret cumulative frequency tables and diagrams.
2. Estimate and interpret the median, percentiles, quartiles and interquartile range from cumulative frequency diagrams.

E9.7 Histograms
1. Draw and interpret histograms.
2. Calculate with frequency density.

Statistics in history

The earliest writing on statistics was found in a ninth-century book entitled *Manuscript on Deciphering Cryptographic Messages*, written by the Arab philosopher Al-Kindi (801–873), who lived in Baghdad. In his book, he gave a detailed description of how to use statistics to unlock coded messages.

The *Nuova Cronica*, a fourteenth-century history of Florence by the Italian banker Giovanni Villani, includes much statistical information on population, commerce, trade and education.

Florence Nightingale (1820–1910) was a famous British nurse who treated casualties in the Crimean War (1853–1856). By using statistics she realised that most of the deaths that occurred were not as a result of battle injuries but from preventable illnesses afterwards, such as cholera and typhoid. By understanding these statistics, Florence Nightingale was able to improve the sanitary conditions of the injured soldiers and therefore reduce their mortality rates.

Early statistics served the needs of states, state-istics. By the early nineteenth century, statistics included the collection and analysis of data in general. Today, statistics are widely employed in government, business, and natural and social sciences. The use of modern computers has enabled large-scale statistical computation and has also made possible new methods that are impractical to perform manually.

35 Mean, median, mode and range

Average

'Average' is a word which in general use is taken to mean somewhere in the middle. For example, a woman may describe herself as being of average height. A student may think they are of average ability in Maths. Mathematics is more exact and uses three principal methods to measure average.

- » The **mode** is the value occurring the most often.
- » The **median** is the middle value when all the data is arranged in order of size.
- » The **mean** is found by adding together all the values of the data and then dividing that total by the number of data values.

Spread

It is often useful to know how spread out the data is. It is possible for two sets of data to have the same mean and median but very different spreads.

The simplest measure of spread is the **range**. The range is simply the difference between the largest and smallest values in the data.

Another measure of spread is known as the interquartile range. This is covered in more detail in Chapter 37.

> **Note**
> Interquartile range is not part of the Core syllabus.

Advantages and disadvantages of different averages		
	Advantage	**Disadvantage**
Mean	It includes all data values.	It can be affected by extreme values.
Median	It is not affected by extreme values. It gives a sense of where the 'middle' value is.	As it only considers the middle value, it does not take into account all data values.
Mode	It is not affected by extreme values. The most common result is used.	It does not take into account all data values. It is not useful if there are lots of different data values.

Spread

> **Worked examples**

a i Find the mean, median and mode of the data listed below.
1, 0, 2, 4, 1, 2, 1, 1, 2, 5, 5, 0, 1, 2, 3

Mean = $\frac{1+0+2+4+1+2+1+1+2+5+5+0+1+2+3}{15}$ = 2

Arranging all the data in order of magnitude and then picking out the middle number gives the median:
0, 0, 1, 1, 1, 1, 1, ②, 2, 2, 2, 3, 4, 5, 5
The mode is the number which appears most often.
Therefore the mode is 1.

ii Calculate the range of the data.
Largest value = 5
Smallest value = 0
Therefore the range is 5 − 0 = 5

b i The bar chart (below) shows the score out of 10 achieved by a class in a maths test.
Calculate the mean, median and mode for this data.

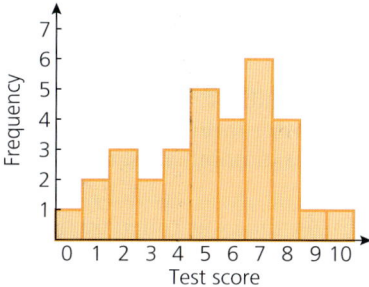

Transferring the results to a **frequency distribution table** gives:

*This frequency table also shows the **frequency distribution** of the test scores.*

Test score	0	1	2	3	4	5	6	7	8	9	10	Total
Frequency	1	2	3	2	3	5	4	6	4	1	1	32
Frequency × score	0	2	6	6	12	25	24	42	32	9	10	168

In the total column we can see the number of students taking the test, i.e. 32, and also the total number of marks obtained by all the students, i.e. 168.

Therefore, the mean score = $\frac{168}{32}$ = 5.25

Arranging all the scores in order gives:
0, 1, 1, 2, 2, 2, 3, 3, 4, 4, 4, 5, 5, 5, 5, ⑤, ⑥, 6, 6, 6, 7, 7, 7, 7, 7, 7, 8, 8, 8, 8, 9, 10

Because there is an even number of students there isn't one middle number. There is, however, a middle pair. The median is $\frac{(5+6)}{2}$ = 5.5.
The mode is 7 as it is the score which occurs most often.

ii Calculate the range of the data.
Largest value = 10   Smallest value = 0
Therefore the range is 10 − 0 = 10.

35 MEAN, MEDIAN, MODE AND RANGE

Exercise 35.1

In Questions 1–5, find the mean, median, mode and range for each set of data.

1. A hockey team plays 15 matches. Below is a list of the numbers of goals scored in these matches.
   1, 0, 2, 4, 0, 1, 1, 1, 2, 5, 3, 0, 1, 2, 2

2. The total scores when two dice are thrown 20 times are:
   7, 4, 5, 7, 3, 2, 8, 6, 8, 7, 6, 5, 11, 9, 7, 3, 8, 7, 6, 5

3. The ages of a group of girls are:
   14 years 3 months, 14 years 5 months,
   13 years 11 months, 14 years 3 months,
   14 years 7 months, 14 years 3 months,
   14 years 1 month

4. The numbers of students present in a class over a three-week period are:
   28, 24, 25, 28, 23, 28, 27, 26, 27, 25, 28, 28, 28, 26, 25

5. An athlete keeps a record of her training times, in seconds, for the 100 m race:
   14.0, 14.3, 14.1, 14.3, 14.2, 14.0, 13.9, 13.8, 13.9, 13.8, 13.8, 13.7, 13.8, 13.8, 13.8

6. The mean mass of the 11 players in a football team is 80.3 kg. The mean mass of the team plus a substitute is 81.2 kg. Calculate the mass of the substitute.

7. After eight matches, a basketball player had scored a mean of 27 points. After three more matches his mean was 29. Calculate the total number of points he scored in the last three games.

Exercise 35.2

1. An ordinary dice was rolled 60 times. The frequency distribution is shown in the table below. Calculate the mean, median, mode and range of the scores.

Score	1	2	3	4	5	6
Frequency	12	11	8	12	7	10

2. Two dice were thrown 100 times. Each time their combined score was recorded. Below is a frequency distribution of the results. Calculate the mean score.

Score	2	3	4	5	6	7	8	9	10	11	12
Frequency	5	6	7	9	14	16	13	11	9	7	3

3. Sixty flowering shrubs of the same variety are planted. At their flowering peak, the number of flowers per shrub is counted and recorded. The frequency distribution is shown in the table below.

Flowers per shrub	0	1	2	3	4	5	6	7	8
Frequency	0	0	0	6	4	6	10	16	18

a. Calculate the mean, median, mode and range of the number of flowers per shrub.
b. Which of the mean, median and mode would be most useful when advertising the shrub to potential buyers?

The mean for grouped data

The mean for grouped data can only be an estimate as the position of the data within a group is not known. An estimate is made by calculating the mid-interval value for a group and then assigning that mid-interval value to all of the data within the group.

➜ Worked example

The history test scores for a group of 40 students are shown in the grouped frequency table below.

Score, S	Frequency	Mid-interval value	Frequency × mid-interval value
$0 \leqslant S < 20$	2	10	20
$20 \leqslant S < 40$	4	30	120
$40 \leqslant S < 60$	14	50	700
$60 \leqslant S < 80$	16	70	1120
$80 \leqslant S < 100$	4	90	360

a Calculate an estimate for the mean test result.

$$\text{Mean} = \frac{20 + 120 + 700 + 1120 + 360}{40} = 58$$

b What is the modal class?
This refers to the class with the greatest frequency, if the class width is constant. Therefore the modal class is $60 \leqslant S < 80$.

Exercise 35.3

1 The heights of 50 basketball players attending a tournament are recorded in the grouped frequency table.

Height (m)	$1.8 < H \leqslant 1.9$	$1.9 < H \leqslant 2.0$	$2.0 < H \leqslant 2.1$	$2.1 < H \leqslant 2.2$	$2.2 < H \leqslant 2.3$	$2.3 < H \leqslant 2.4$
Frequency	2	5	10	22	7	4

a Copy the table and complete it to include the necessary data with which to calculate the mean height of the players.
b Estimate the mean height of the players.
c What is the modal class height of the players?

2 The number of hours of overtime worked by employees at a factory over a period of a month is given in the table (below).

Hours of overtime	0–9	10–19	20–29	30–39	40–49	50–59
Frequency	12	18	22	64	32	20

a Calculate an estimate for the mean number of hours of overtime worked by the employees that month.
b What is the modal class?

35 MEAN, MEDIAN, MODE AND RANGE

Exercise 35.3 (cont)

3 The length of the index finger of 30 students in a class was measured. The results were recorded and are shown in the table below.

Length (cm)	5.0 < L ⩽ 5.5	5.5 < L ⩽ 6.0	6.0 < L ⩽ 6.5	6.5 < L ⩽ 7.0	7.0 < L ⩽ 7.5
Frequency	3	8	10	7	2

a Calculate an estimate for the mean index finger length of the students.
b What is the modal class?

Student assessment 1

1 A javelin thrower keeps a record of her best throws over ten competitions. These are shown in the table below.

Competition	1	2	3	4	5	6	7	8	9	10
Distance (m)	77	75	78	86	92	93	93	93	92	89

Find the mean, median, mode and range of her throws.

2 The bar chart shows the marks out of 10 for a chemistry test taken by a class of students.

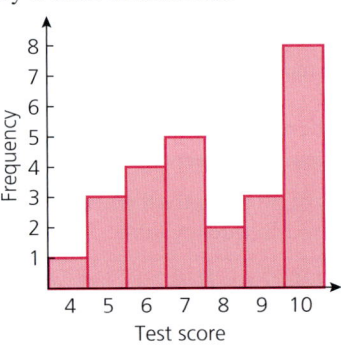

a Calculate the number of students who took the test.
b Calculate for the class:
  i the mean test result,
  ii the median test result,
  iii the modal test result,
  iv the range of the test results.
c The teacher is happy with these results as she says that the average result was 10/10. Another teacher says that the average is only 7.5/10. Which teacher is correct? Give a reason for your answer.

3 The range, mode, median and mean of five positive integers are all equal to 12. Work out one possible set of these five integers.

The mean for grouped data

4 A hundred sacks of coffee with a stated mass of 10 kg are unloaded from a train. The mass of each sack is checked and the results are presented in the table.

Mass (kg)	Frequency
$9.8 \leq M < 9.9$	14
$9.9 \leq M < 10.0$	22
$10.0 \leq M < 10.1$	36
$10.1 \leq M < 10.2$	20
$10.2 \leq M < 10.3$	8

a Calculate an estimate for the mean mass.
b What is the modal class?

36 Collecting, displaying and interpreting data

Tally charts and frequency tables

The number of chocolate buttons in each of twenty packets is:

35 36 38 37 35 36 38 36 37 35
36 36 38 36 35 38 37 38 36 38

The figures can be shown on a tally chart:

Number	Tally	Frequency
35	IIII	4
36	IIII II	7
37	III	3
38	IIII I	6

When the tallies are added up to find the frequency, the chart is usually called a **frequency table**. The information can then be displayed in a variety of ways.

Pictograms

● = 4 packets, ◕ = 3 packets, ◐ = 2 packets, ◔ = 1 packet

Buttons per packet	
35	●
36	● ◕
37	◐
38	● ◐

Stem-and-leaf diagrams

> **Note**
>
> When a bar chart is showing categorical data (such as favourite subjects) rather than numerical data, there are gaps between the bars.

Bar charts

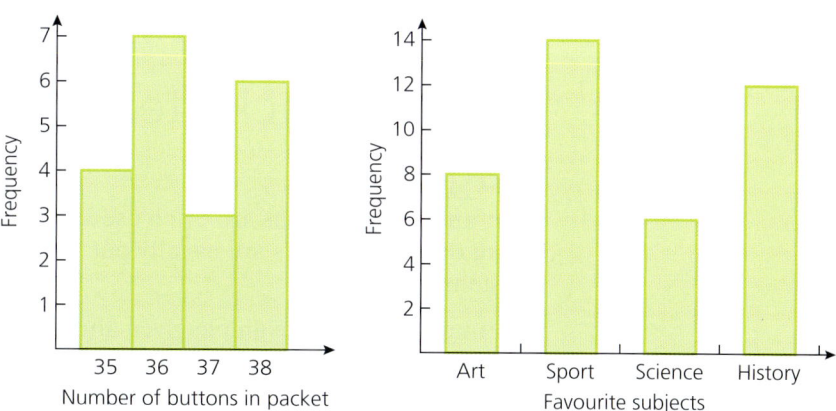

Stem-and-leaf diagrams

Discrete data is data that has a specific, fixed value. A stem-and-leaf diagram can be used to display discrete data in a clear and organised way. It has an advantage over bar charts as the original data can easily be recovered from the diagram.

The ages of people on a coach transferring them from an airport to a ski resort are as follows:

22	24	25	31	33	23	24	26	37	42
40	36	33	24	25	18	20	27	25	33
28	33	35	39	40	48	27	25	24	29

Displaying the data on a stem-and-leaf diagram produces the following graph.

```
1 | 8
2 | 0 2 3 4 4 4 5 5 5 5 6 7 7 8 9
3 | 1 3 3 3 3 5 6 7 9
4 | 0 0 2 8
```

Key 2 | 5 represents 25

In this form the data can be analysed quite quickly:
» The youngest person is 18
» The oldest is 48
» The modal ages are 24, 25 and 33

36 COLLECTING, DISPLAYING AND INTERPRETING DATA

As the data is arranged in order, the median age can also be calculated quickly. The middle people out of 30 will be the 15th and 16th people. In this case the 15th person is 27 years old and the 16th person 28 years old, therefore the median age is 27.5.

Back-to-back stem-and-leaf diagrams

Stem-and-leaf diagrams are often used as an easy way to compare two sets of data. The leaves are usually put 'back-to-back' on either side of the stem.

Continuing from the example given above, consider a second coach from the airport taking people to a golfing holiday. The ages of these people are shown below:

43	46	52	61	65	38	36	28	37	45
69	72	63	55	46	34	35	37	43	48
54	53	47	36	58	63	70	55	63	64

Displaying the two sets of data on a back-to-back stem-and-leaf diagram is shown below:

Golf   **Skiing**

						1	8
					8	2	0 2 3 4 4 4 5 5 5 6 7 7 8 9
8	7	7	6	6	5 4	3	1 3 3 3 5 6 7 9
8	7	6	6	5	3 3	4	0 0 2 8
		8	5	5	4 3 2	5	
9	5	4	3	3	3 1	6	
					2 0	7	

Key: 5 | 3 | 6 represents 35 to the left and 36 to the right

From the back-to-back diagram it is easier to compare the two sets of data. This data shows that the people on the coach going to the golf resort tend to be older than those on the coach to the ski resort.

Grouped frequency tables

If there is a big range in the data, it is easier to group the data in a **grouped frequency table**.

The groups are arranged so that no score can appear in two groups.

The scores for the first round of a golf competition are:

| 71 | 75 | 82 | 96 | 83 | 75 | 76 | 82 | 103 | 85 | 79 | 77 | 83 | 85 | 88 |
| 104 | 76 | 77 | 79 | 83 | 84 | 86 | 88 | 102 | 95 | 96 | 99 | 102 | 75 | 72 |

This data can be grouped as shown:

Score	Frequency
71–75	5
76–80	6
81–85	8
86–90	3
91–95	1
96–100	3
101–105	4
Total	30

Note: It is not possible to score 70.5 or 77.3 at golf. The scores are **discrete**. If the data is **continuous**, for example when measuring time, the intervals can be shown as $0 \leq t < 10$, $10 \leq t < 20$, $20 \leq t < 30$ and so on.

Pie charts

Data can be displayed on a **pie chart** – a circle divided into sectors. The size of the sector is in direct proportion to the frequency of the data. The sector size does not show the actual frequency. The actual frequency can be calculated easily from the size of the sector.

➔ Worked examples

a

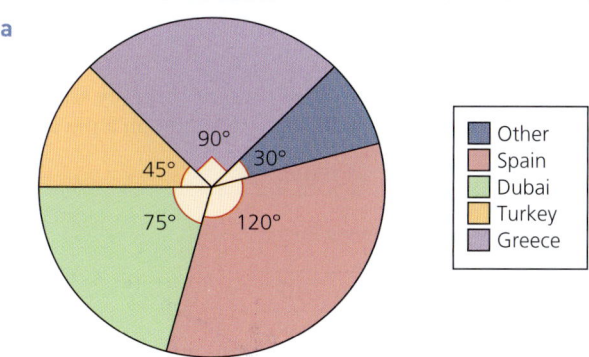

In a survey, 240 people were asked to vote for their favourite holiday destination. The results are shown on the pie chart above. Calculate the actual number of votes for each destination.

The total 240 votes are represented by 360°.

It follows that if 360° represents 240 votes:

There were $240 \times \frac{120}{360}$ votes for Spain so, 80 votes for Spain.

There were $240 \times \frac{75}{360}$ votes for Dubai so, 50 votes for Dubai.

There were $240 \times \frac{45}{360}$ votes for Turkey so, 30 votes for Turkey.

36 COLLECTING, DISPLAYING AND INTERPRETING DATA

There were $240 \times \frac{90}{360}$ votes for Greece so, 60 votes for Greece.

Other destinations received $240 \times \frac{30}{360}$ votes so, 20 votes for other destinations.

Note: It is worth checking your result by adding them:

$80 + 50 + 30 + 60 + 20 = 240$ total votes

b The table below shows what percentage of the money raised during a fundraising campaign that charities, which support different causes, received. If a total of $5 million was raised, how much money did each charitable cause receive?

Charitable cause	Percentage of money
Children and young adults	45%
Disability and mental health	36%
Animal welfare	15%
Others	4%

Children and young adults received $\frac{45}{100} \times \$5$ million
so, $2.25 million.

Disability and mental health received $\frac{36}{100} \times \$5$ million
so, $1.8 million.

Animal welfare received $\frac{15}{100} \times \$5$ million
so, $750 000.

Other charitable causes received $\frac{4}{100} \times \$5$ million
so, $200 000.

Check total:

$2.25 + 1.8 + 0.75 + 0.2 = 5$ (million dollars)

c The table shows the results of a survey among 72 students to find their favourite sport. Display this data on a pie chart.

Sport	Frequency
Football	35
Tennis	14
Volleyball	10
Hockey	6
Basketball	5
Other	2

72 students are represented by 360°, so 1 student is represented by $\frac{360}{72}$ degrees. Therefore, the size of each sector can be calculated:

Football $35 \times \frac{360}{72}$ degrees   i.e. 175°

Tennis $14 \times \frac{360}{72}$ degrees   i.e. 70°

Volleyball $10 \times \frac{360}{72}$ degrees   i.e. 50°

Hockey $6 \times \frac{360}{72}$ degrees   i.e. 30°

Pie charts

Basketball $5 \times \frac{360}{72}$ degrees   i.e. 25°

Other sports $2 \times \frac{360}{72}$ degrees   i.e. 10°

Check total:

175 + 70 + 50 + 30 + 25 + 10 = 360

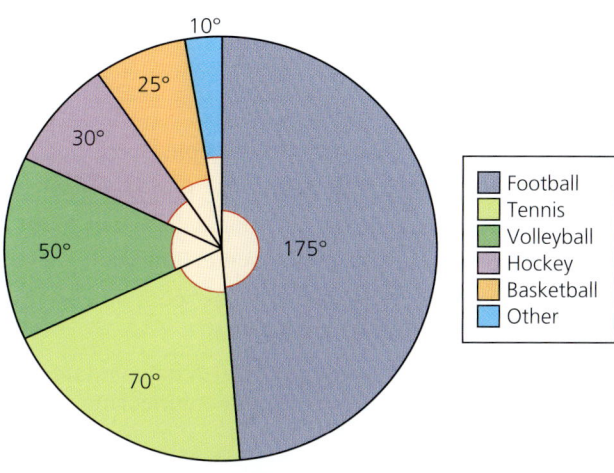

Exercise 36.1

You will need an angle measurer or protractor for this question.

1 The unlabelled pie charts below show how Ayse and her brother, Ahmet, spent one day. Calculate how many hours they spent on each activity. The diagrams are to scale.

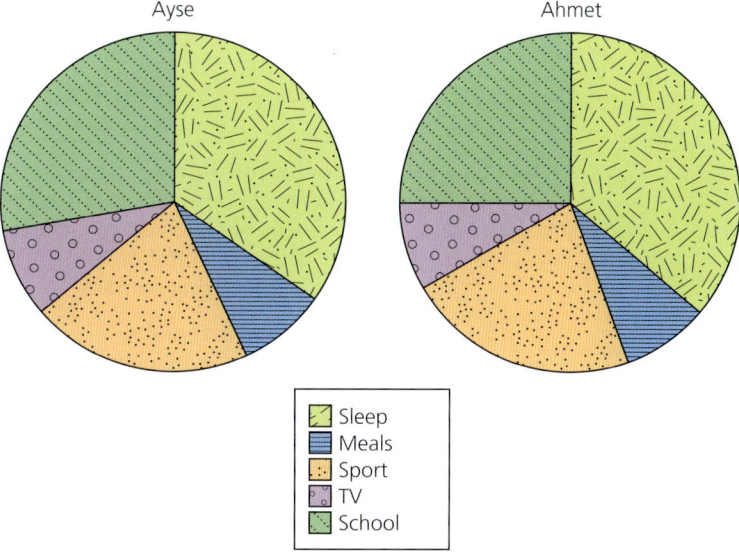

36 COLLECTING, DISPLAYING AND INTERPRETING DATA

Exercise 36.1 (cont)

2 A survey was carried out among a class of 40 students. The question asked was, 'How would you spend a gift of $15?' The results are shown below:

Choice	Frequency
Music	14
Books	6
Clothes	18
Cinema	2

Illustrate these results on a pie chart.

3 A student works during the holidays. He earns a total of $2400. He estimates that the money has been spent as follows: clothes, $\frac{1}{3}$; transport, $\frac{1}{5}$; entertainment, $\frac{1}{4}$. He has saved the rest.

Calculate how much he has spent on each category, and illustrate this information on a pie chart.

4 Two universities in central Asia compared the percentages of people who enrolled on different engineering courses in 2022. The results are shown in the table below.

Engineering course	University A (percentage)	University B (percentage)
Civil	22	38
Mechanical	16	8
Chemical	24	16
Electrical	12	24
Industrial	8	10
Aerospace	18	4

a Illustrate this information on two pie charts, and make two statements that could be supported by the data.
b If 3000 people enrolled on engineering courses at University B in 2022, calculate the number who enrolled on either mechanical or aerospace engineering courses.

5 A village has two sports clubs.
The ages of people in each club are listed below:

Ages in Club 1								
38	8	16	15	18	8	59	12	14	55
14	15	24	67	71	21	23	27	12	48
31	14	70	15	32	9	44	11	46	62

Pie charts

Ages in Club 2									
42	62	10	62	74	18	77	35	38	66
43	71	68	64	66	66	22	48	50	57
60	59	44	57	12					

a Draw a back-to-back stem-and-leaf diagram for the ages of the members of each club.
b For each club, calculate:
  i the age range of their members,
  ii the median age.
c One of the clubs is the golf club, the other is the athletics club. Which club is **likely** to be which? Give a reason for your answer.

6 The birthday months of boys and girls in one class are plotted as a dual bar chart below:

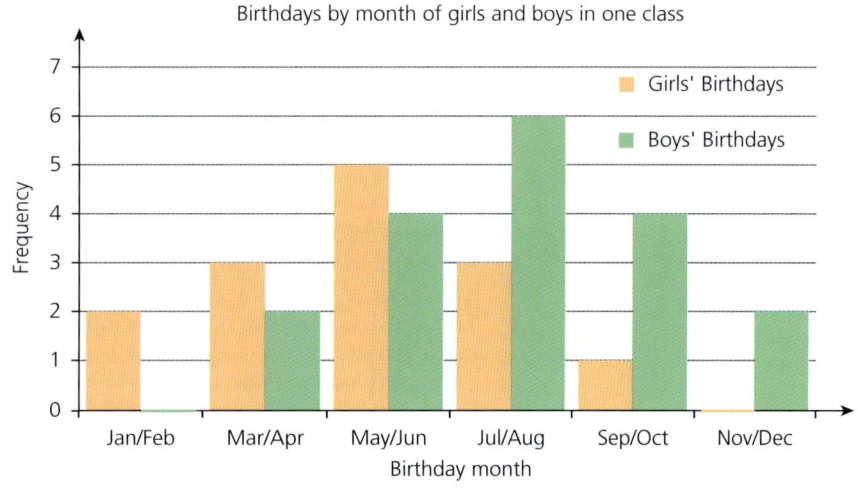

a How many girls are there in the class?
b How many more boys than girls have their birthdays in either July or August?
c Describe the differences in the birthdays of boys and girls in this class.
d Construct a dual bar chart for the birthday months of boys and girls in your own class.

36 COLLECTING, DISPLAYING AND INTERPRETING DATA

Exercise 36.1 (cont)

7 Two fishing boats return to port and the mass of two types of fish caught by each boat is recorded. The masses are shown in the composite (stacked) bar chart below.

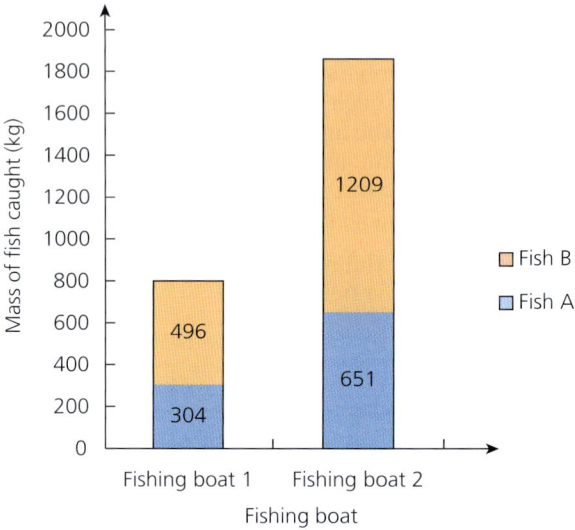

Types of fish caught by two fishing boats

a Which boat caught the greater mass of fish type A?
b Assuming only the two types of fish were caught, which boat's catch had a higher percentage of fish type A? Show your working.
c The above composite bar chart shows the mass of fish in kg on the vertical axis.
   Construct a composite bar chart comparing the catches of both boats, but with percentages on the vertical axis.

Surveys

A survey requires data to be collected, organised, analysed and presented.

A survey may be carried out for interest's sake, for example, to find out how many cars pass your school in an hour. A survey could be carried out to help future planning – information about traffic flow could lead to the building of new roads, or the placing of traffic lights or a pedestrian crossing.

Exercise 36.2

1 Below are ten statements, some of which you may have heard or read before.
   Conduct a survey to collect data which will support or disprove one of the statements. Where possible, use pie charts to illustrate your results.
   a Magazines are full of adverts.
   b If you go to a football match you are lucky to see more than one goal scored.
   c Every other car on the road is white.
   d Most retired people use public transport.
   e Children today do nothing but watch TV.
   f Newspapers have more sport than news in them.

g More people prefer to drink coffee than tea.
h Nobody walks to school any more.
i Nearly everybody has a computer at home.
j Most of what is on TV comes from America.

2 Below are some instructions relating to a washing machine, written in English, French, German, Dutch and Italian.
Analyse the data and write a report. You may wish to comment upon:
  a the length of words in each language,
  b the frequency of letters of the alphabet in different languages.

ENGLISH

ATTENTION

Do not interrupt drying during the programme.

This machine incorporates a temperature safety thermostat which will cut out the heating element in the event of a water blockage or power failure. In the event of this happening, reset the programme before selecting a further drying time. For further instructions, consult the user manual.

FRENCH

ATTENTION

N'interrompez pas le séchage en cours de programme.

Une panne d'électricité ou un manque d'eau momentanés peuvent annuler le programme de séchage en cours. Dans ces cas arrêtez l'appareil, affichez de nouveau le programme et après remettez l'appareil en marche.

Pour d'ultérieures informations, rapportez-vous à la notice d'utilisation.

GERMAN

ACHTUNG

Die Trocknung soll nicht nach Anlaufen des Programms unterbrochen werden.

Ein kurzer Stromausfall bzw. Wassermangel kann das laufende Trocknungsprogramm annullieren. In diesem Falle Gerät ausschalten, Programm wieder einstellen und Gerät wieder einschalten.

Für nähere Angaben beziehen Sie sich auf die Bedienungsanleitung.

ESTONIAN

TÄHELEPANU

Ärge katkestage kuivatamist programmi ajal.

Sellel masinal on temperatuuri turvatermostaat, mis lõikab veeummistuse või voolukatkestuse korral kütteelemendi välja. Juhul kui see peaks juhtuma, lähtestage programm uuesti enne uue kuivamisaja valimist. Täiendavate juhiste saamiseks vaadake kasutusjuhendit.

MALAY

PERHATIAN

Jangan ganggu pengeringan semasa program.

Mesin ini menggabungkan termostat keselamatan suhu yang akan dipotong keluarkan elemen pemanas sekiranya berlaku penyumbatan air atau kegagalan kuasa. Sekiranya ini berlaku, tetapkan semula atur cara sebelum memilih a masa pengeringan selanjutnya. Untuk arahan lanjut, rujuk manual pengguna.

36 COLLECTING, DISPLAYING AND INTERPRETING DATA

Scatter diagrams

Scatter diagrams are particularly useful if we wish to see if there is a **correlation** (relationship) between two sets of data. The two values of data collected represent the coordinates of each point plotted. How the points lie when plotted, indicates the type of relationship between the two sets of data.

➡ Worked example

The heights and masses of 20 children under the age of five were recorded. The heights were recorded in centimetres and the masses in kilograms. The data is shown in a table:

Height	32	34	45	46	52
Mass	5.8	3.8	9.0	4.2	10.1
Height	59	63	64	71	73
Mass	6.2	9.9	16.0	15.8	9.9
Height	86	87	95	96	96
Mass	11.1	16.4	20.9	16.2	14.0
Height	101	108	109	117	121
Mass	19.5	15.9	12.0	19.4	14.3

a Plot a scatter diagram of the above data.

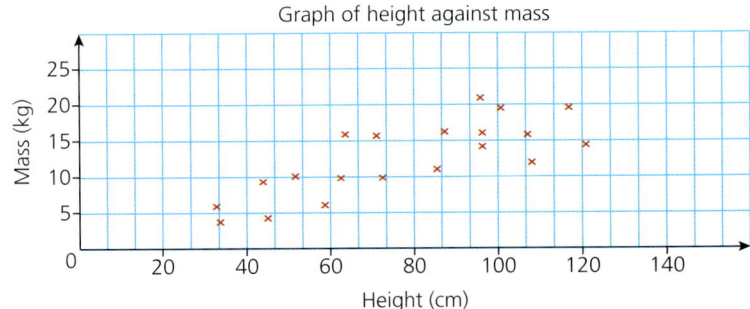

b Comment on any relationship you see.

The points tend to lie in a diagonal direction from bottom left to top right. This suggests that as height increases then, in general, mass increases too. Therefore there is a **positive correlation** between height and mass.

c If another child was measured as having a height of 80 cm, approximately what mass would you expect them to be?

We assume that this child will follow the trend set by the other 20 children. To deduce an approximate value for the mass, we draw a **line of best fit**. This is done by eye and is a solid straight line which passes through the points as closely as possible, as shown.

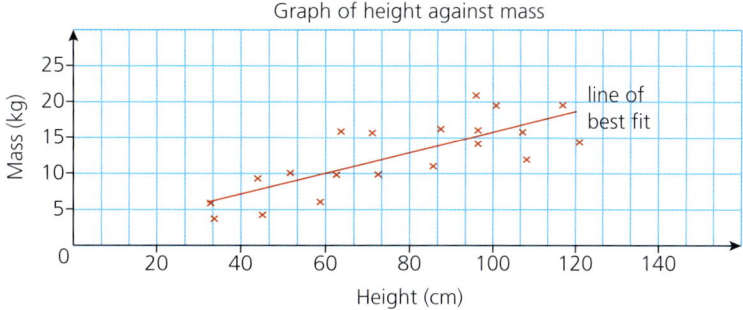

The line of best fit can now be used to give an approximate solution to the question. If a child has a height of 80 cm, you would expect their mass to be in the region of 13 kg.

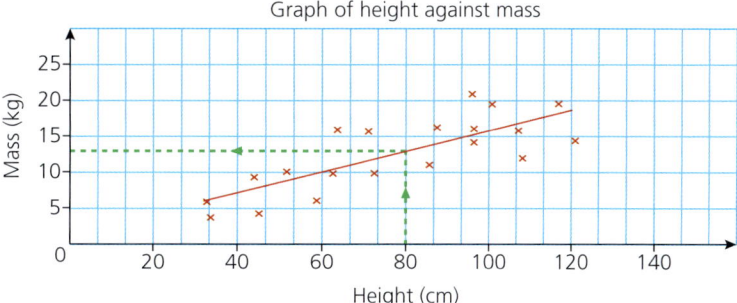

d Someone decides to extend the line of best fit in both directions because they want to make predictions for heights and masses beyond those of the data collected. The graph is shown below.

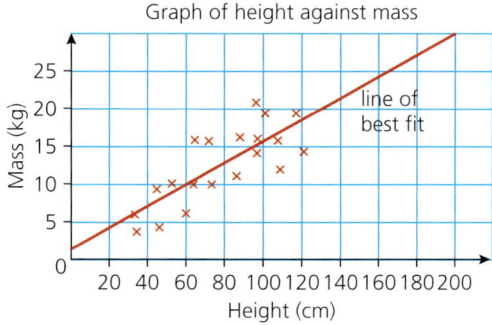

Explain why this should not be done.

It should not be done ecause we cannot assume that the relationship between mass and height continues in the same pattern beyond the collected data.

36 COLLECTING, DISPLAYING AND INTERPRETING DATA

Types of correlation

There are several types of correlation, depending on the arrangement of the points plotted on the scatter diagram.

A **strong positive correlation** between the variables x and y.

The points lie very close to the line of best fit.

As x increases, so does y.

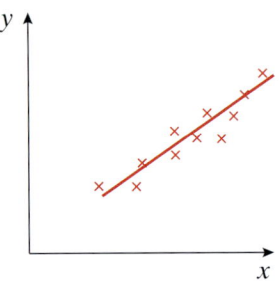

A **weak positive correlation**. Although there is direction to the way the points are lying, they are not tightly packed around the line of best fit.

As x increases, y tends to increase too.

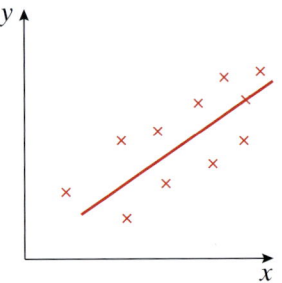

A **strong negative correlation**. The points lie close to the line of best fit. As x increases, y decreases.

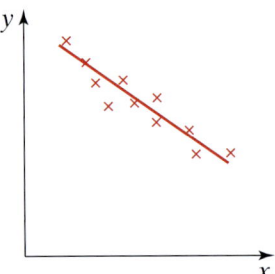

A **weak negative correlation**. The points are not tightly packed around the line of best fit. As x increases, y tends to decrease.

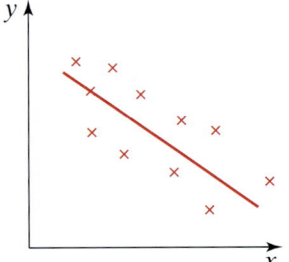

Zero or no correlation. As there is no pattern to the way in which the points are lying, there is no correlation between the variables x and y. As a result there can be no line of best fit.

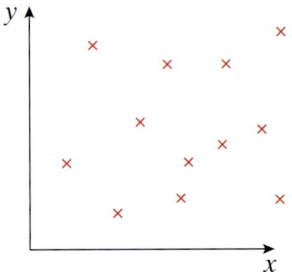

Types of correlation

Exercise 36.3

1. State what type of correlation you might expect, if any, if the following data was collected and plotted on a scatter diagram. Give reasons for your answer.
   a A student's score in a maths exam and their score in a science exam.
   b A student's hair colour and the distance they have to travel to school.
   c The outdoor temperature and the number of cold drinks sold by a shop.
   d The age of a motorcycle and its second-hand selling price.
   e The number of people living in a house and the number of rooms the house has.
   f The number of goals your opponents score and the number of times you win.
   g A child's height and the child's age.
   h A car's engine size and its fuel consumption.

2. A website gives average monthly readings for the number of hours of sunshine and the amount of rainfall in millimetres for several cities in Europe. The table below is a summary for July.

Place	Hours of sunshine	Rainfall (mm)
Athens	12	6
Belgrade	10	61
Copenhagen	8	71
Dubrovnik	12	26
Edinburgh	5	83
Frankfurt	7	70
Geneva	10	64
Helsinki	9	68
Innsbruck	7	134
Krakow	7	111
Lisbon	12	3
Marseilles	11	11
Naples	10	19
Oslo	7	82
Plovdiv	11	37
Reykjavik	6	50
Sofia	10	68
Tallinn	10	68
Valletta	12	0
York	6	62
Zurich	8	136

   a Plot a scatter diagram of the number of hours of sunshine against the amount of rainfall. Use a spreadsheet if possible.
   b What type of correlation, if any, is there between the two variables? Comment on whether this is what you would expect.

525

36 COLLECTING, DISPLAYING AND INTERPRETING DATA

Exercise 36.3 (cont)

3 The United Nations keeps an up-to-date database of statistical information on its member countries. The table below shows some of the information available.

Country	Life expectancy at birth (years, 2005–2010)		Adult illiteracy rate (%, 2009)	Infant mortality rate (per 1000 births, 2005–2010)
	Female	Male		
Australia	84	79	1	5
Barbados	80	74	0.3	10
Brazil	76	69	10	24
Chad	50	47	68.2	130
China	75	71	6.7	23
Colombia	77	69	7.2	19
Cuba	81	77	0.2	5
Democratic Republic of the Congo	55	53	18.9	79
Egypt	72	68	33	35
France	85	78	1	4
Germany	82	77	1	4
India	65	62	34	55
Japan	86	79	1	3
Kenya	55	54	26.4	64
Mexico	79	74	7.2	17
Nepal	67	66	43.5	42
Portugal	82	75	5.1	4
Russian Federation	73	60	0.5	12
Saudi Arabia	75	71	15	19
South Africa	53	50	12	49
United Kingdom	82	77	1	5
United States of America	81	77	1	6

a By plotting a scatter diagram, decide if there is a correlation between the adult illiteracy rate and the infant mortality rate.
b Are your findings in part **a** what you expected? Explain your answer.
c Without plotting a graph, decide if you think there is likely to be a correlation between male and female life expectancy at birth. Explain your reasons.
d Plot a scatter diagram to test if your predictions for part **c** were correct.

Types of correlation

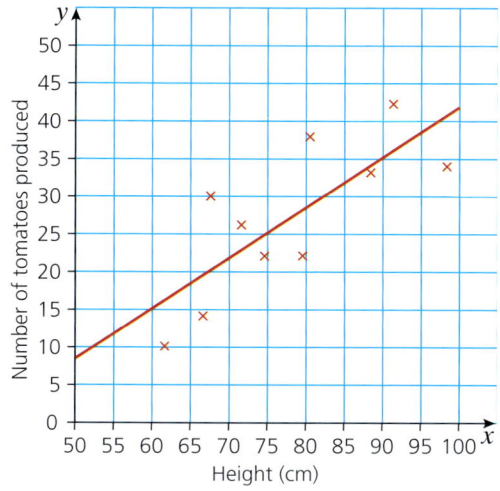

4   Kris plants 10 tomato plants. He wants to see if there is a relationship between the number of tomatoes the plant produces and its height in centimetres.
    The results are presented in the scatter diagram left. The line of best fit is also drawn.
    a   Describe the correlation (if any) between the height of a plant and the number of tomatoes it produced.
    b   Kris has another plant grown in the same conditions as the others. If the height is 85 cm, estimate from the graph the number of tomatoes he can expect it to produce.
    c   Another plant only produces 15 tomatoes. Deduce its height from the graph.
    d   Kris thinks he will be able to make more predictions if the height axis starts at 0 cm rather than 50 cm and if the line of best fit is then extended.
        By re-plotting the data on a new scatter graph and extending the line of best fit, explain whether Kris's idea is correct.

5   The table shows the 15 countries that won the most medals at the 2016 Rio Olympics. In addition, statistics relating to each of those countries' population, wealth, percentage of people with higher education and percentage who are overweight are also given.

Country	Olympic medals			Population (million)	Average wealth per person in 1000's $	% with a Higher Education Qualification	% Adult population that are overweight	
	Gold	Silver	Bronze				Male	Female
USA	46	37	38	322	345	45	73	63
UK	27	23	17	65	289	44	68	59
China	26	18	26	1376	23	10	39	33
Russia	19	18	19	143	10	54	60	55
Germany	17	10	15	81	185	28	64	49
Japan	12	8	21	127	231	50	29	19
France	10	18	14	664	244	34	67	52
S. Korea	9	3	9	50	160	45	38	30
Italy	8	12	8	60	202	18	66	53
Australia	8	11	10	24	376	43	70	58
Holland	8	7	4	17	184	35	63	49
Hungary	8	3	4	10	34	24	67	49
Brazil	7	6	6	208	18	14	55	53
Spain	7	4	6	46	116	35	67	55
Kenya	6	6	1	46	2	11	17	34

36 COLLECTING, DISPLAYING AND INTERPRETING DATA

Exercise 36.3 (cont)

A sports scientist wants to see if there is a correlation between the number of medals a country won and the percentage of overweight people in that country.

To obtain a simple scatter graph, she plots the number of gold medals against the mean percentage of overweight people and adds the line of best fit:

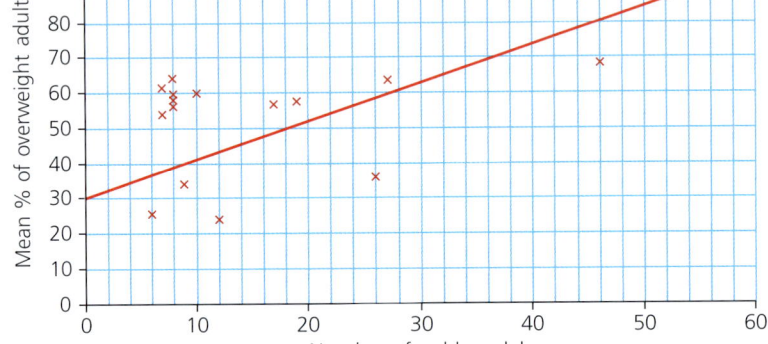

a Describe the type of correlation implied by the above graph.
b The sports scientist states that the graph shows that the more overweight you are the more likely you are to win a gold medal. Give two reasons why this conclusion may not be accurate.
c Analyse the correlation between two other sets of data and comment on whether the results are expected or not. Justify your answer.

Histograms

A **histogram** displays the frequency of either continuous or grouped discrete data in the form of bars. There are several important features of a histogram which distinguish it from a bar chart.

» The bars are joined together.
» The bars can be of varying width.
» The frequency of the data is represented by the area of the bar and not the height (though in the case of bars of equal width, the area is directly proportional to the height of the bar and so the height is usually used as the measure of frequency).

Histograms

Worked example

A zoo keeper measures the length (L cm) of 32 lizards kept in the reptile section of the zoo. Draw a histogram of the data.

Length (cm)	Frequency
$0 < L \leq 10$	0
$10 < L \leq 20$	0
$20 < L \leq 30$	1
$30 < L \leq 40$	2
$40 < L \leq 50$	5
$50 < L \leq 60$	8
$60 < L \leq 70$	7
$70 < L \leq 80$	6
$80 < L \leq 90$	2
$90 < L \leq 100$	1

All the class intervals are the same. As a result, the bars of the histogram will all be of equal width and the frequency can be plotted on the vertical axis. The histogram is shown below.

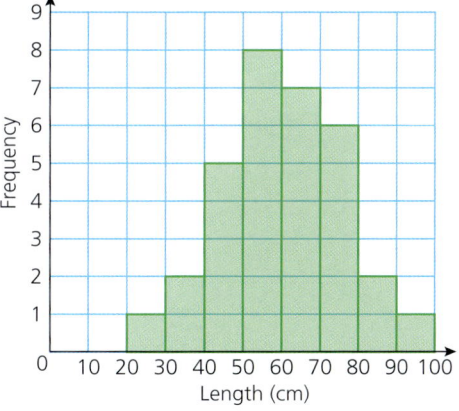

529

36 COLLECTING, DISPLAYING AND INTERPRETING DATA

Exercise 36.4

1. The table (below) shows the distances travelled to school by a class of 30 students. Represent this information on a histogram.

Distance (km)	Frequency
$0 \leq d < 1$	8
$1 \leq d < 2$	5
$2 \leq d < 3$	6
$3 \leq d < 4$	3
$4 \leq d < 5$	4
$5 \leq d < 6$	2
$6 \leq d < 7$	1
$7 \leq d < 8$	1

2. The heights of students in a class were measured. The results are shown in the table (below). Draw a histogram to represent this data.

Height (cm)	Frequency
$145 \leq h < 150$	1
$150 \leq h < 155$	2
$155 \leq h < 160$	4
$160 \leq h < 165$	7
$165 \leq h < 170$	6
$170 \leq h < 175$	3
$175 \leq h < 180$	2
$180 \leq h < 185$	1

Note that both questions in Exercise 36.4 deal with **continuous data**.

So far the work on histograms has only dealt with problems in which the class intervals are of the same width. This, however, need not be the case.

➡ Worked example

The heights of 25 sunflowers were measured and the results recorded in the table (below).

Height (m)	Frequency
$0 \leq h < 1.0$	6
$1.0 \leq h < 1.5$	3
$1.5 \leq h < 2.0$	4
$2.0 \leq h < 2.25$	3
$2.25 \leq h < 2.50$	5
$2.50 \leq h < 2.75$	4

If a histogram were drawn with frequency plotted on the vertical axis, then it could look like the one drawn opposite.

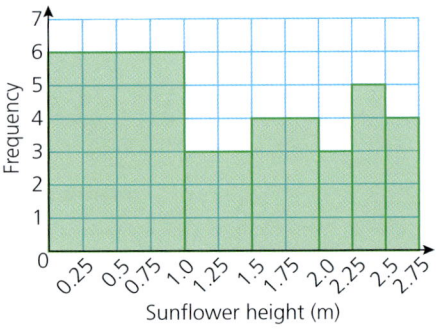

Histograms

This graph is misleading because it leads people to the conclusion that most of the sunflowers were under 1 m, simply because the area of the bar is so great. In fact, only approximately one quarter of the sunflowers were under 1 m.

When class intervals are different it is the area of the bar which represents the frequency, not the height. Instead of frequency being plotted on the vertical axis, **frequency density** is plotted.

$$\text{Frequency density} = \frac{\text{frequency}}{\text{class width}}$$

The results of the sunflower measurements in the example above can therefore be written as:

Height (m)	Frequency	Frequency density
$0 \leqslant h < 1.0$	6	$6 \div 1 = 6$
$1.0 \leqslant h < 1.5$	3	$3 \div 0.5 = 6$
$1.5 \leqslant h < 2.0$	4	$4 \div 0.5 = 8$
$2.0 \leqslant h < 2.25$	3	$3 \div 0.25 = 12$
$2.25 \leqslant h < 2.50$	5	$5 \div 0.25 = 20$
$2.50 \leqslant h < 2.75$	4	$4 \div 0.25 = 16$

The histogram can therefore be redrawn giving a more accurate representation of the data:

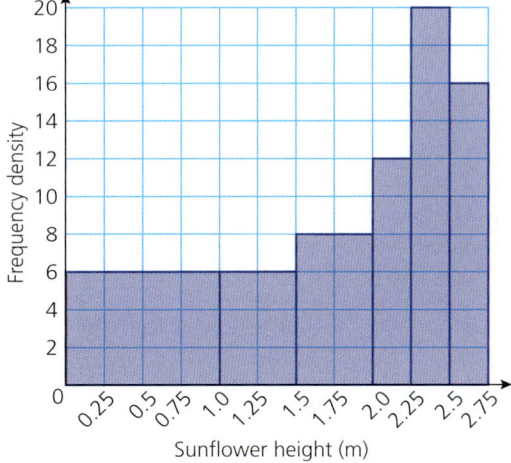

531

36 COLLECTING, DISPLAYING AND INTERPRETING DATA

Exercise 36.5

1. The table below shows the time taken, in minutes, by 40 students to travel to school.

Time (min)	Frequency	Frequency density
$0 \leqslant t < 10$	6	
$10 \leqslant t < 15$	3	
$15 \leqslant t < 20$	13	
$20 \leqslant t < 25$	7	
$25 \leqslant t < 30$	3	
$30 \leqslant t < 40$	4	
$40 \leqslant t < 60$	4	

   a Copy the table and complete it by calculating the frequency density.
   b Represent the information on a histogram.

2. On Sundays, Maria helps her father feed their chickens. Over a period of one year, she kept a record of how long it took. Her results are shown in the table below.

Time (min)	Frequency	Frequency density
$0 \leqslant t < 30$	8	
$30 \leqslant t < 45$	5	
$45 \leqslant t < 60$	8	
$60 \leqslant t < 75$	9	
$75 \leqslant t < 90$	10	
$90 \leqslant t < 120$	12	

   a Copy the table and complete it by calculating the frequency density. Give the answers correct to 1 d.p.
   b Represent the information on a histogram.

3. Frances and Ali did a survey of the ages of the people living in their village. Part of their results are set out in the table below.

Age (years)	Frequency	Frequency density
$0 \leqslant y < 1$	35	
$1 \leqslant y < 5$		12
$5 \leqslant y < 10$		28
$10 \leqslant y < 20$	180	
$20 \leqslant y < 40$	260	
$40 \leqslant y < 60$		14
$60 \leqslant y < 90$	150	

   a Copy the table and complete it by calculating either the frequency or the frequency density.
   b Represent the information on a histogram.

Histograms

4 The table below shows the ages of 150 people, chosen randomly, taking the 06 00 train into a city.

Age (years)	Frequency
$0 \leqslant y < 15$	3
$15 \leqslant y < 20$	25
$20 \leqslant y < 25$	20
$25 \leqslant y < 30$	30
$30 \leqslant y < 40$	32
$40 \leqslant y < 50$	30
$50 \leqslant y < 80$	10

The histogram below shows the results obtained when the same survey was carried out on the 11 00 train.

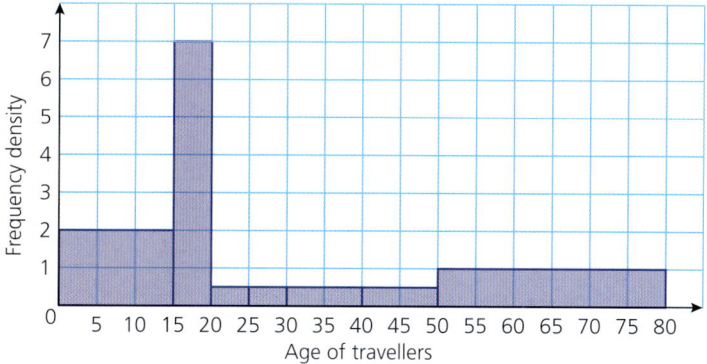

a Draw a histogram for the 06 00 train.
b Compare the two sets of data and give two possible reasons for the differences.

Student assessment 1

1 The table below shows the population (in millions) of the continents. Display this information on a pie chart.

Continent	Asia	Europe	America	Africa	Oceania
Population (millions)	4140	750	920	995	35

2 A department store decides to investigate whether there is a correlation between the number of pairs of gloves it sells and the outside temperature. Over a one-year period the store records, every two weeks, how many pairs of gloves are sold and the mean daytime temperature during the same period. The results are given in the table:

533

36 COLLECTING, DISPLAYING AND INTERPRETING DATA

Mean temperature (°C)	3	6	8	10	10	11	12	14	16	16	17	18	18
Number of pairs of gloves	61	52	49	54	52	48	44	40	51	39	31	43	35
Mean temperature (°C)	19	19	20	21	22	22	24	25	25	26	26	27	28
Number of pairs of gloves	26	17	36	26	46	40	30	25	11	7	3	2	0

a Plot a scatter diagram of mean temperature against number of pairs of gloves.
b What type of correlation is there between the two variables?
c How might this information be useful for the department store in the future?
d The mean daytime temperature during the next two-week period is predicted to be 20°C. Draw a line of best fit on your graph and use it to estimate the number of pairs of gloves the department store can expect to sell.

3 A test in physics is marked out of 40. The scores of a class of 32 students are shown below.

24	27	30	33	26	27	28	39
21	18	16	33	22	38	33	11
16	11	14	23	37	36	38	22
28	11	9	17	28	11	36	34

a Display the data on a stem-and-leaf diagram.
b The teacher says that any student getting less than the average score will have to sit a re-test. How many students will sit the re-test? Justify your answer fully.

4 The grouped frequency table below shows the number of points scored by a school basketball player.

Points	Number of games	Frequency density
$0 \leqslant p < 5$	2	
$5 \leqslant p < 10$	3	
$10 \leqslant p < 15$	8	
$15 \leqslant p < 25$	9	
$25 \leqslant p < 35$	12	
$35 \leqslant p < 50$	3	

a Copy and complete the table by calculating the frequency densities. Give your answers to 1 d.p.
b Draw a histogram to illustrate the data.

37 Cumulative frequency

Cumulative frequency

Calculating the **cumulative frequency** is done by adding up the frequencies as we go along. A cumulative frequency diagram is particularly useful when trying to calculate the median of a large set of data, grouped or continuous data, or when trying to establish how consistent the results in a set of data are.

➜ Worked example

The duration of two different brands of battery, A and B, is tested. 50 batteries of each type are randomly selected and tested in the same way. The duration of each battery is then recorded. The results of the tests are shown in the tables below.

Brand A

Duration (h)	Frequency	Cumulative frequency
$0 \leqslant t < 5$	3	3
$5 \leqslant t < 10$	5	8
$10 \leqslant t < 15$	8	16
$15 \leqslant t < 20$	10	26
$20 \leqslant t < 25$	12	38
$25 \leqslant t < 30$	7	45
$30 \leqslant t < 35$	5	50

Brand B

Duration (h)	Frequency	Cumulative frequency
$0 \leqslant t < 5$	1	1
$5 \leqslant t < 10$	1	2
$10 \leqslant t < 15$	10	12
$15 \leqslant t < 20$	23	35
$20 \leqslant t < 25$	9	44
$25 \leqslant t < 30$	4	48
$30 \leqslant t < 35$	2	50

37 CUMULATIVE FREQUENCY

a Plot a cumulative frequency diagram for each brand of battery.

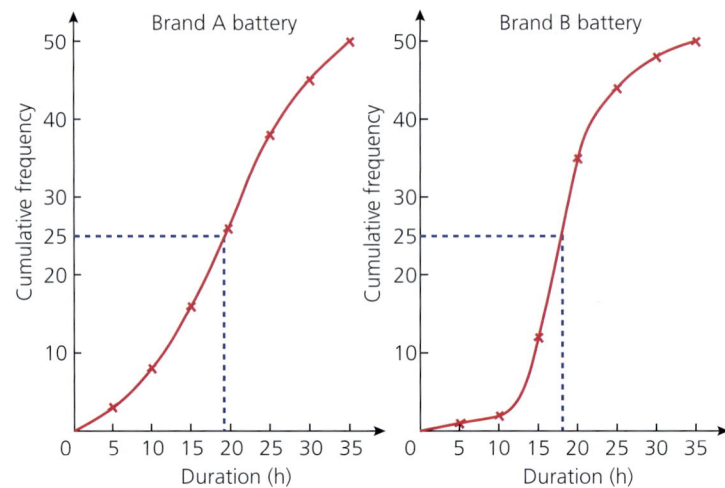

Note that a smooth curve is drawn through the points.

Note

For large data sets the median is at the mid-point of the cumulative frequency.

Both cumulative frequency diagrams are plotted above.

Notice how the points are plotted at the upper boundary of each class interval and *not* at the middle of the interval.

b Calculate the median duration for each brand.

The median value is the value which occurs halfway up the cumulative frequency axis. Therefore:

Median for brand A batteries ≈ 19 h

Median for brand B batteries ≈ 18 h

This tells us that the same number of batteries are still working as have stopped working after approximately 19 h for brand A and approximately 18 h for brand B.

Exercise 37.1

1 Sixty athletes enter a long-distance run. Their finishing times are recorded and are shown in the table below.

Finishing time (h hours)	Frequency	Cumulative freq.
$0 \leqslant h < 0.5$	0	
$0.5 \leqslant h < 1.0$	0	
$1.0 \leqslant h < 1.5$	6	
$1.5 \leqslant h < 2.0$	34	
$2.0 \leqslant h < 2.5$	16	
$2.5 \leqslant h < 3.0$	3	
$3.0 \leqslant h < 3.5$	1	

a Copy the table and calculate the values for the cumulative frequency.
b Draw a cumulative frequency diagram of the results.
c Show how your diagram could be used to find the approximate median finishing time.
d What does the median value tell us?

Quartiles, percentiles and the interquartile range

Exercise 37.1 (cont)

2 Three physics classes take the same test in preparation for their final exam. Their raw scores are shown in the table below.

Class A	12, 21, 24, 30, 33, 36, 42, 45, 53, 53, 57, 59, 61, 62, 74, 88, 92, 93
Class B	48, 53, 54, 59, 61, 62, 67, 78, 85, 96, 98, 99
Class C	10, 22, 36, 42, 44, 68, 72, 74, 75, 83, 86, 89, 93, 96, 97, 99, 99

a Using the class intervals $0 \leqslant x < 20$, $20 \leqslant x < 40$ etc., draw up a grouped frequency and cumulative frequency table for each class.
b Draw a cumulative frequency diagram for each class.
c Show how your diagrams could be used to find the median score for each class.
d What does the median value tell us?

3 The table below shows the heights of students in a class over a three-year period.

Height (cm)	Frequency 2020	Frequency 2021	Frequency 2022
$150 \leqslant h < 155$	6	2	2
$155 \leqslant h < 160$	8	9	6
$160 \leqslant h < 165$	11	10	9
$165 \leqslant h < 170$	4	4	8
$170 \leqslant h < 175$	1	3	2
$175 \leqslant h < 180$	0	2	2
$180 \leqslant h < 185$	0	0	1

a Construct a cumulative frequency table for each year.
b Draw a cumulative frequency diagram for each year.
c Show how your diagrams could be used to find the median height for each year.
d What does the median value tell us?

Quartiles, percentiles and the interquartile range

The cumulative frequency axis can also be represented in terms of **percentiles**. A percentile scale divides the cumulative frequency scale into hundredths. The maximum value of cumulative frequency is found at the 100th percentile. Similarly, the median, being the middle value, is called the 50th percentile. The 25th percentile is known as the **lower quartile**, and the 75th percentile is called the **upper quartile**.

The **range** of a distribution is found by subtracting the lowest value from the highest value. Sometimes this will give a useful result, but often it will not. A better measure of spread is given by looking at the spread of the middle half of the results, i.e. the difference between the upper and lower quartiles. This result is known as the **interquartile range**.

37 CUMULATIVE FREQUENCY

The diagram (below) shows the terms mentioned above.

As the cumulative frequency shows a total of 120, the 90th percentile is calculated from the cumulative frequency as 108, because $120 \times 90 \div 100 = 108$.

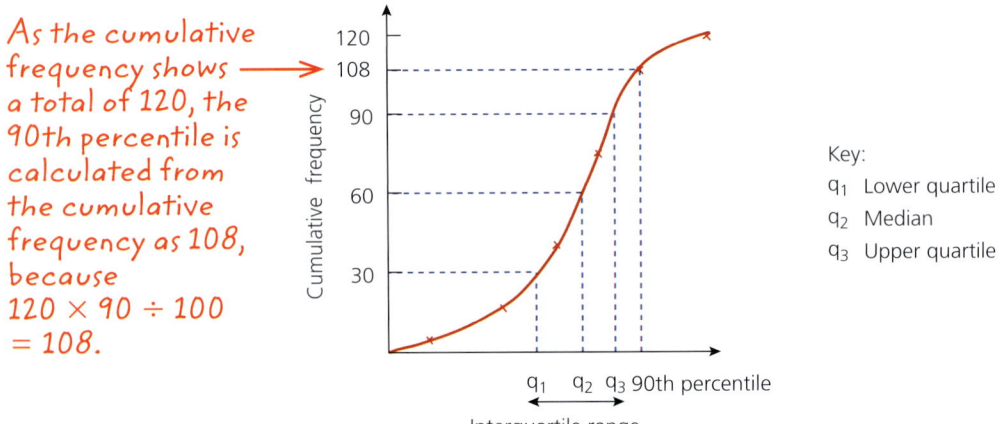

Key:
q_1 Lower quartile
q_2 Median
q_3 Upper quartile

Advantages and disadvantages of different types of range		
	Advantage	**Disadvantage**
Range	The spread takes into account all data values.	It can be affected by extreme values.
Interquartile range	As it only considers the middle 50% of the data, it is not affected by extreme results.	It doesn't take into account all the data values, i.e. it disregards the top and bottom 25%.

→ Worked examples

Consider again the two brands of batteries, A and B, discussed earlier on page 535.

a Using the diagrams, estimate the upper and lower quartiles for each battery.
   Lower quartile of brand A ≈ 13 h
   Upper quartile of brand A ≈ 25 h
   Lower quartile of brand B ≈ 15 h
   Upper quartile of brand B ≈ 21 h

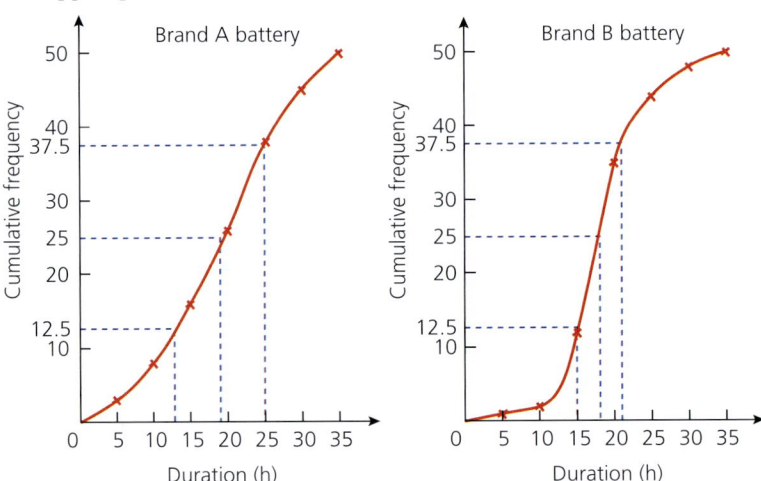

b Calculate the interquartile range for each brand of battery.

Interquartile range of brand A ≈ 12 h

Interquartile range of brand B ≈ 6 h

c Based on these results, how might the manufacturers advertise the two brands of battery?

Brand A: on 'average' the longer-lasting battery.

Brand B: the more reliable battery.

Box-and-whisker plots

The information calculated from a cumulative frequency diagram can be represented in a diagram called a **box-and-whisker plot**.

The typical shape of a box-and-whisker plot is:

Its different components represent different aspects of the data.

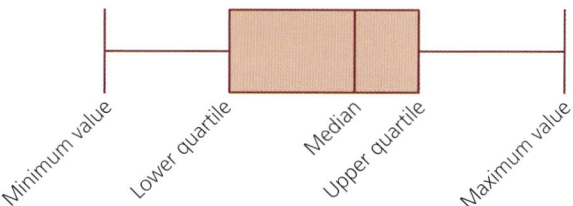

When looking at two sets of data, box-and-whisker plots are an efficient way of comparing them.

→ Worked example

Consider the two battery brands A and B discussed in the previous worked example.
Draw box-and-whisker plots for both brands on the same scale and comment on any similarities and differences.

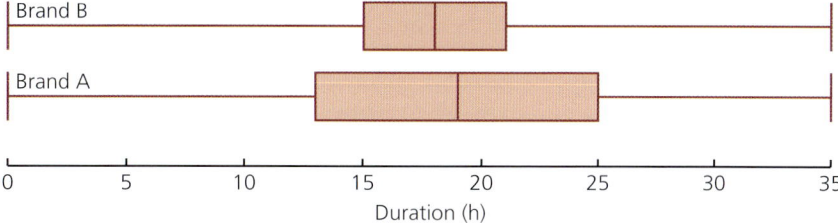

» Both battery brands have the same overall range.
» The median of brand A is slightly bigger than that of brand B.

37 CUMULATIVE FREQUENCY

» The spread of the middle 50% of the data, represented by the rectangular block in the centre and known as the interquartile range, is much smaller for brand B batteries. This implies that the middle 50% are more consistent for brand B than for brand A batteries.

Note: From the tables on page 535, the first group was for a duration of $0 \leqslant t < 5$ hours. For brand A batteries there were three results in that group. It is not known exactly where in the group the three results lie, therefore the minimum value is taken as the lower bound of the group. Similarly, the last group for the data was for a duration of $30 \leqslant t < 35$ hours. There were five brand A batteries in that group. It is not known where those five batteries lie within the group, therefore the maximum value is taken to be the upper bound of the group.

Exercise 37.2

1. Using the results obtained from Question 2 in Exercise 37.1:
   a find the interquartile range of each of the classes taking the mathematics test,
   b analyse your results and write a brief summary comparing the three classes.
2. Using the results obtained from Question 3 in Exercise 37.1:
   a find the interquartile range of the students' heights each year,
   b analyse your results and write a brief summary comparing the three years.
3. Forty students enter a school javelin competition. The distances thrown are recorded below.

Distance thrown (m)	$0 \leqslant d < 20$	$20 \leqslant d < 40$	$40 \leqslant d < 60$	$60 \leqslant d < 80$	$80 \leqslant d < 100$
Frequency	4	9	15	10	2

   a Construct a cumulative frequency table for the above results.
   b Draw a cumulative frequency diagram.
   c If the top 20% of students are considered for the final, estimate (using the cumulative frequency diagram) the qualifying distance.
   d Calculate the interquartile range of the throws.
   e Calculate the median distance thrown.
4. The masses of two different types of orange are compared. Eighty oranges are randomly selected from each type and weighed. The results are shown in the table.

Type A		Type B	
Mass (g)	Frequency	Mass (g)	Frequency
$75 \leqslant m < 100$	4	$75 \leqslant m < 100$	0
$100 \leqslant m < 125$	7	$100 \leqslant m < 125$	16
$125 \leqslant m < 150$	15	$125 \leqslant m < 150$	43
$150 \leqslant m < 175$	32	$150 \leqslant m < 175$	10
$175 \leqslant m < 200$	14	$175 \leqslant m < 200$	7
$200 \leqslant m < 225$	6	$200 \leqslant m < 225$	4
$225 \leqslant m < 250$	2	$225 \leqslant m < 250$	0

   a Construct a cumulative frequency table for each type of orange.
   b Draw a cumulative frequency diagram for each type of orange.

c Calculate the median mass for each type of orange.
d Using your diagrams, estimate for each type of orange:
  i the lower quartile,
  ii the upper quartile,
  iii the interquartile range.
e Write a brief report, comparing the two types of orange. You may wish to draw a box-and-whisker plot to help you.

5 Two competing brands of battery, X and Y, are compared. One hundred batteries of each brand are tested and the duration of each battery is recorded. The results of the tests are shown in the cumulative frequency diagrams below.

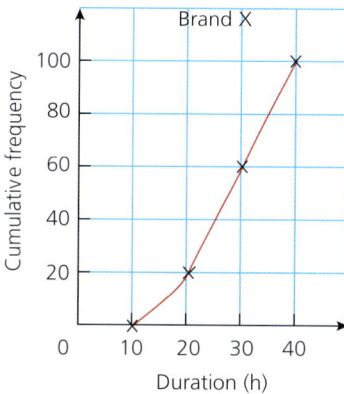
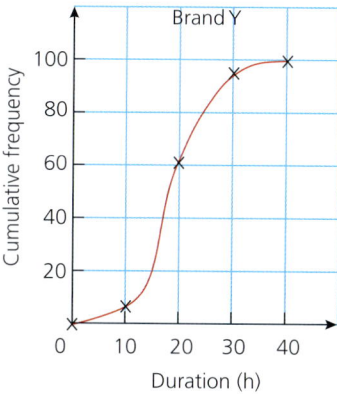

a The manufacturers of brand X claim that on average their batteries will last at least 40% longer than those of brand Y. Showing your method clearly, decide whether this claim is true.
b The manufacturers of brand X also claim that their batteries are more reliable than those of brand Y. Is this claim true? Show your working clearly.

Student assessment 1

1 Thirty students sit a maths exam. Their marks are given as percentages and are shown in the table below.

Mark (%)	$20 \leqslant m < 30$	$30 \leqslant m < 40$	$40 \leqslant m < 50$	$50 \leqslant m < 60$	$60 \leqslant m < 70$	$70 \leqslant m < 80$	$80 \leqslant m < 90$	$90 \leqslant m < 100$
Frequency	2	3	5	7	6	4	2	1

a Construct a cumulative frequency table of the above results.
b Draw a cumulative frequency diagram of the results.
c Using the diagram, estimate a value for:
  i the median,
  ii the upper and lower quartiles,
  iii the interquartile range.

37 CUMULATIVE FREQUENCY

2 Four hundred students sit an exam. Their marks (as percentages) are shown in the table below.

Mark (%)	Frequency	Cumulative frequency
$31 \leq m < 40$	21	
$41 \leq m < 50$	55	
$51 \leq m < 60$	125	
$61 \leq m < 70$	74	
$71 \leq m < 80$	52	
$81 \leq m < 90$	45	
$91 \leq m < 100$	28	

a Copy and complete the above table by calculating the cumulative frequency.
b Draw a cumulative frequency diagram of the results.
c Using the diagram, estimate a value for:
  i the median result,
  ii the upper and lower quartiles,
  iii the interquartile range.

3 Eight hundred students sit an exam. Their marks (as percentages) are shown in the table below.

Mark (%)	Frequency	Cumulative frequency
$1 \leq m < 10$	10	
$11 \leq m < 20$	30	
$21 \leq m < 30$	40	
$31 \leq m < 40$	50	
$41 \leq m < 50$	70	
$51 \leq m < 60$	100	
$61 \leq m < 70$	240	
$71 \leq m < 80$	160	
$81 \leq m < 90$	70	
$91 \leq m < 100$	30	

a Copy and complete the above table by calculating the cumulative frequency.
b Draw a cumulative frequency diagram of the results.
c An A grade is awarded to a student at or above the 80th percentile. What mark is the minimum requirement for an A grade?
d A C grade is awarded to any student between and including the 55th and the 70th percentile. What marks form the lower and upper boundaries of a C grade?
e Calculate the interquartile range for this exam.

4 Identify which graph(s) correspond with each of the following statements.
  a The graph with the highest lower quartile.
  b The graph with the largest interquartile range. Justify your answer.
  c The graphs with the same 84th percentile. Justify your answer.
  d The graphs with the greatest range.

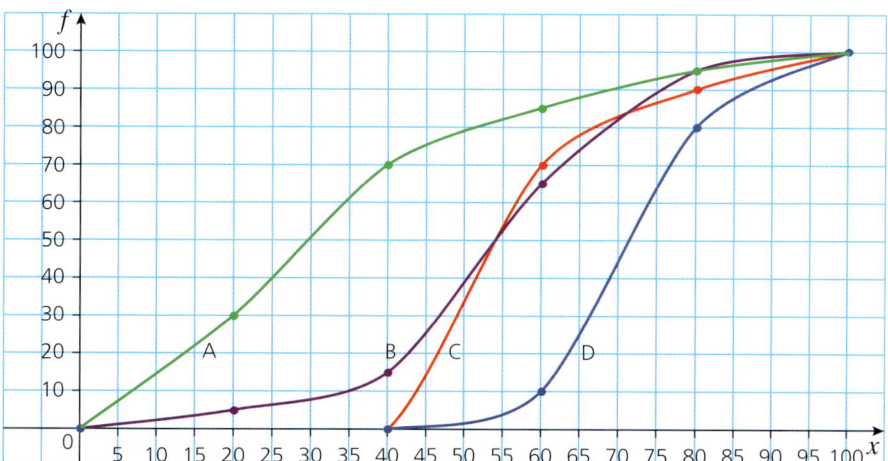

Mathematical investigations and ICT 9

Heights and percentiles

The graphs below show the height charts for males and females from the age of 2 to 20 years in the United States.

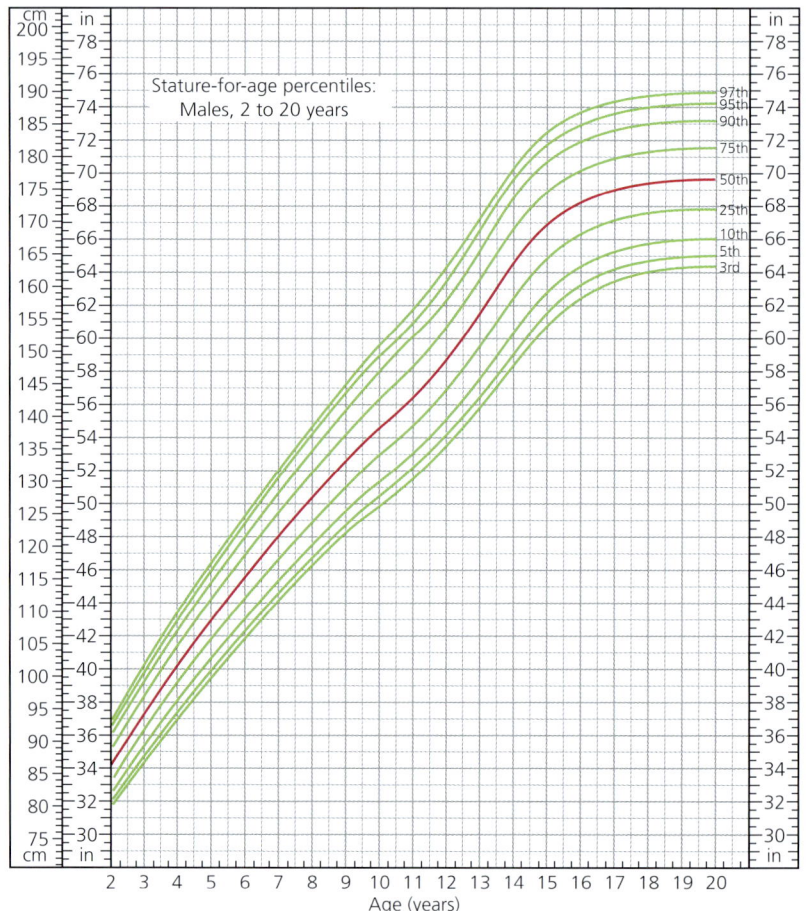

Note: Heights have been given in both centimetres and inches.

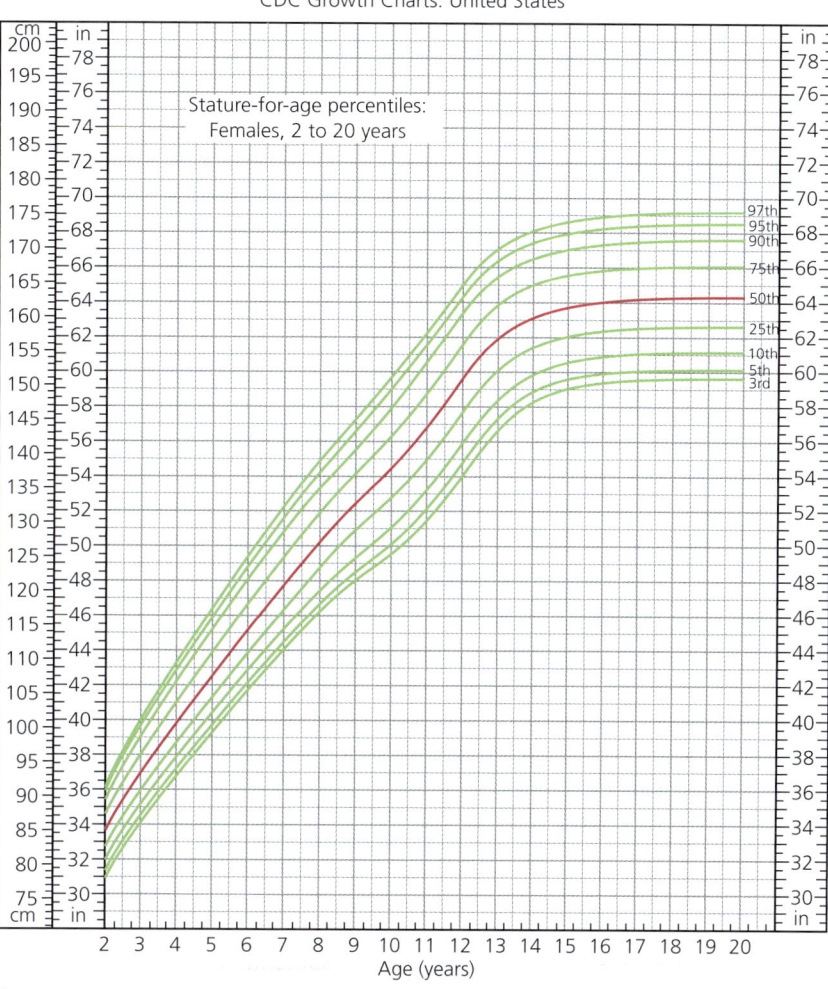

CDC Growth Charts: United States
Stature-for-age percentiles: Females, 2 to 20 years

1. From the graph, find the height corresponding to the 75th percentile for 16-year-old females.
2. Find the height which 75% of 16-year-old males exceed.
3. What is the median height for 12-year-old females?
4. Measure the heights of students in your class. By carrying out appropriate statistical calculations, write a report comparing your data to that shown in the graphs.
5. Would all cultures use the same height charts? Explain your answer.

Reading ages

Depending on their target audience, newspapers, magazines and books have different levels of readability. Some are easy to read and others more difficult.

MATHEMATICAL INVESTIGATIONS AND ICT 9

1. Decide on some factors that you think would affect the readability of a text.
2. Write down the names of two newspapers which you think would have different reading ages. Give reasons for your answer.

There are established formulas for calculating the reading age of different texts. One of these is the Gunning Fog Index. It calculates the reading age as follows:

Reading age = $\frac{2}{5}\left(\frac{A}{n} + \frac{100L}{A}\right)$ where

A = number of words

n = number of sentences

L = number of words with three or more syllables.

3. Select one article from each of the two newspapers you chose in Question 2. Use the Gunning Fog Index to calculate the reading ages for the articles. Do the results support your predictions?
4. Write down some factors which you think may affect the reliability of your results.

ICT activity

In this activity you will be collecting the height data of all the students in your class and then plotting a cumulative frequency diagram of the results.

1. Measure the heights of all the students in your class.
2. Group your data appropriately.
3. Enter your data into graphing software such as Excel or Autograph.
4. Produce a cumulative frequency diagram of the results.
5. From your graph, find:
   a the median height of the students in your class,
   b the interquartile range of the heights.
6. Compare the cumulative frequency diagram from your class with one produced from data collected from another class in a different year group. Comment on any differences or similarities between the two diagrams.

Glossary

= = means is equal to. For example, 3 + 4 = 7.

≠ ≠ means is not equal to. For example, 3 + 4 ≠ 8.

> > means is greater than. For example, 8 > 3 + 4.

< < means is less than. For example, 3 + 4 < 8.

⩾ ⩾ means is greater than or equal to. For example, $x \geq 5$ means x is any number greater than or equal to 5.

⩽ ⩽ means is less than or equal to. For example, $x \leq 5$ means x is any number less than or equal to 5.

∈ ∈ means is an element of. So $e \in S$ means the element e belongs to the set S.

∉ ∉ means is NOT an element of. So $e \notin S$ means the element e does not belong to the set S.

⊆ ⊆ means is a subset of. So $X \subseteq Y$ means X is a subset of Y.

⊄ ⊄ means is NOT a subset of. So $X \not\subseteq Y$ means X is not a subset of Y.

A∩B A∩B means all the elements that belong to BOTH set A and set B. A∩B denotes the elements that are in the intersection of A and B on a Venn diagram.

A∪B The union of sets A and B, A∪B, is all the elements that belong to EITHER set A OR set B OR both sets.

n(A) The number of elements in set A.

A′ The complement of set A.

A⊆B A is a subset of B.

A⊄B A is not a subset of B.

ℰ The universal set, ℰ, for any particular problem is the set which contains all the possible elements for that problem.

∅ The empty set is a set with no elements. It is written as ∅.

12-hour clock 12-hour clock is when the day is split into two halves 'am' and 'pm. The times before 12 noon are written using am and times after 12 noon are written as pm.

24-hour clock 24-hour clock is when the time is given as the number of hours that have passed since midnight. The hours part of the time is given two digits. For example, 01 30 is 1.30 am and 13 30 is 1.30 pm.

accuracy The accuracy of a measurement tells you how close the measurement is to the true value. For example, if you measure a pencil correct to the nearest centimetre, your measurement will be within 0.5 cm of the true measurement.

acute angle An acute angle lies between 0° and 90°.

acute-angled isosceles An acute-angled isosceles triangle has two equal angles and two sides of equal length, and all three angles are less than 90°.

acute-angled triangle In an acute-angled triangle, all three angles are less than 90°.

addition Addition is one of the four operations: addition, subtraction, multiplication and division. It means to find the total or sum of two or more numbers or quantities.

adjacent In a right-angled triangle, the adjacent is the side which is next to the angle.

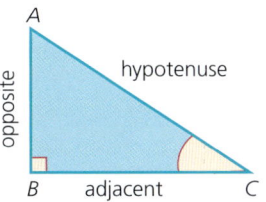

algebraic fraction An algebraic fraction is a fraction with a denominator that is an algebraic expression. For example, $\frac{3}{x}$ or $\frac{1}{x+3}$.

alternate angles Alternate angles are formed when a line crosses a pair of parallel lines. Alternate angles are equal. Look for a Z shape.

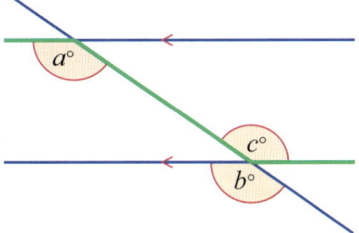

altitude The altitude of a triangle is the perpendicular height.

angle of depression The angle of depression is the angle below the horizontal through which a line of view is lowered.

angle of elevation The angle of elevation is the angle above the horizontal through which a line of view is raised.

angles at a point The angles at a point add up to 360°.

angles on a straight line The angles on a straight line add up to 180°.

apex The apex of a pyramid is the point where the triangular faces of the pyramid meet.

arc An arc is part of the circumference of a circle between two radii. When the angle between the two radii of length r is x, then: arc length $= \frac{x}{360} \times 2\pi r$

area The area of a shape is the amount of surface that it covers. Area is measured in mm², cm², m², km², etc.

area factor When shape A is an enlargement by scale factor k of shape B, then the area factor is k2.

area of a circle The area, A, of a circle of radius r is: $A = \pi r^2$

area of a parallelogram The area, A, of a parallelogram of base length b and perpendicular height h is: $A = bh$

area of a rectangle The area, A, of a rectangle of length l and breadth b is: $A = lb$

area of a trapezium The area, A, of a trapezium is: $A = \frac{1}{2}h(a+b)$

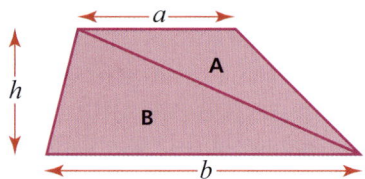

area of a sector The area of a sector is given by: $\frac{\varnothing}{360} \times \pi r^2$

area of a triangle The area, A, of a triangle of base b and perpendicular height h is: $A = \frac{1}{2}bh$

area of any triangle The area of any triangle is given by: area $= \frac{1}{2}ab\sin C$

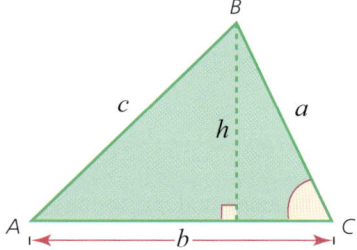

arithmetic sequence An arithmetic sequence is a sequence where the difference between any two terms is a constant.

asymptote An asymptote is a line that a graph tends towards but does not meet. Here the x-axis and y-axis are both asymptotes:

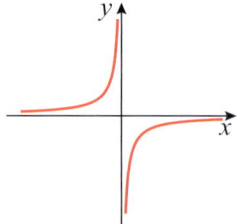

average An average is a measure of the typical value in a data set. There are three measures: mean, mode and median.

average speed average speed $= \frac{\text{total distance}}{\text{total time}}$

back bearing If the bearing of B from A is given, then the back bearing is the bearing of A from B. It is the bearing that takes you from B back to A. The back bearing is in the reverse direction to the original bearing – it represents the direction of the return journey. (*See* bearing)

bar chart A bar chart is a chart that uses rectangular bars to display data. The height of each bar represents the frequency.

base The base of a triangle is one of its sides. Any side can be the base, but the height must be measured perpendicular to the chosen base.

basic pay Basic pay is the fixed pay that an employee is given for working a certain number of hours.

basic week A basic week is the fixed number of hours that an employee is expected to work each week.

bearing A bearing is a direction. It is the angle measured clockwise from North. Bearings are given as 3 figures so, for example, for an angle of 45° the three-figure bearing is 045°.

bisect Bisect means to divide in half.

bonus A bonus is an extra payment that is sometimes added to an employee's basic pay.

breadth The breadth of a rectangle is the measure of its shortest side.

centre of enlargement The centre of enlargement is a specific point about which an object is enlarged.

centre of rotation The centre of rotation is a specific 'pivot' point about which an object is rotated.

chord A chord is a straight line that joins two points on the circumference of a circle.

circumference The circumference is the perimeter of a circle.

circumference of a circle The circumference, C, of a circle of radius r is: $C = 2\pi r$

column vector A column vector describes the movement of an object in both the x direction and the y direction.

common difference The common difference, d, is the difference between one term and the next in an arithmetic sequence.

common ratio The common ratio, r, is the ratio between one term and the next in a geometric sequence.

complement The complement of set A is the set of elements which are in $\mathscr{E}$ but not in A. The complement of A is written as A'.

complementary angle Two angles which add together to total $90°$ are called complementary angles.

composite bar chart A composite bar chart shows the bars stacked on top of each other.

composite function A composite function is when one function is applied to the results of another function. fg(x) means apply function g first, and then apply function f to the result.

compound interest Compound interest is interest that is paid not only on the principal amount, but also on the interest itself. So the amount of interest earned each year increases.

compound measure A compound measure is one made up of two or more other measures.

compound shape A compound shape is a shape that can be split into simpler shapes.

cone A cone is a like a pyramid, but with a circular base.

congruent Congruent shapes are exactly the same shape and size – they are identical.

constant of proportionality When two quantities, x and y, are in direct proportion, $\frac{y}{x} = k$ (or $y = kx$), where k is the constant of proportionality.

construction A construction is an accurate drawing made using a ruler and a pair of compasses.

continuous data Continuous data is numerical data that can take on any value in a certain range. For example, height and weight (mass) are continuous data.

conversion graph A conversion graph is a straight-line graph used to convert one set of units to another.

correlation Correlation is the relationship between two sets of data.

corresponding angles Corresponding angles are formed when a line crosses a pair of parallel lines. Corresponding angles are equal. Look for an F shape.

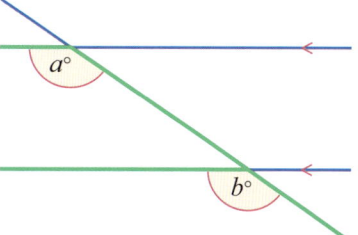

cosine The cosine of an angle, $\cos x$, in a right-angled triangle is the ratio of the adjacent side to the hypotenuse: $\cos x = \frac{\text{length of adjacent side}}{\text{length of hypotenuse}}$

cosine rule The cosine rule is:
$a^2 = b^2 + c^2 - 2bc \cos A$ or $\cos A = \frac{b^2 + c^2 - a^2}{2bc}$.

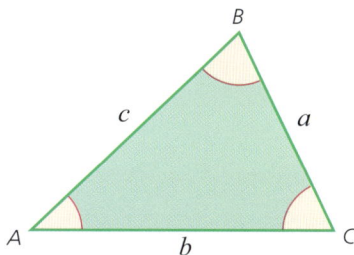

cost price The cost price is the total amount of money that it costs to produce a good or service, before any profit is made.

cube number A cube number is the result when an integer is multiplied by itself twice. The cube numbers are 1, 8, 27, 64, 125, ...

cube root The cube root of a number is the number which when multiplied by itself twice gives the original number. The inverse of cubing is cube rooting. For example, the cube root of 27 is 3 (as $3 \times 3 \times 3 = 27$). The symbol $\sqrt[3]{}$ is used for the cube root of a number, so $\sqrt[3]{27} = 3$.

cubic A cubic function has the form $f(x) = ax^3$

cubic curve A cubic curve is the graph of a cubic function.

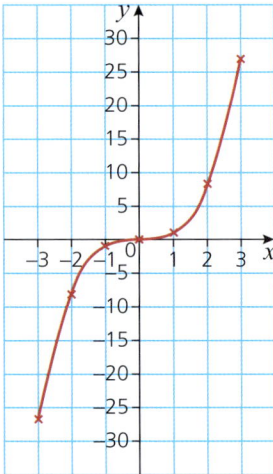

cuboid A cuboid is a prism with a rectangular cross-section.

cumulative frequency The cumulative frequency is the running total of the frequencies in a data set.

cyclic quadrilateral A cyclic quadrilateral is a quadrilateral whose vertices all lie on the circumference of a circle.

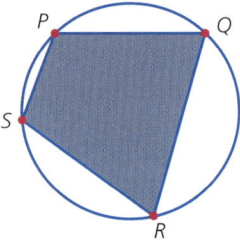

cylinder A cylinder is three-dimensional shape with a constant circular cross-section.

decagon A decagon is a 10-sided polygon.

decimal A decimal is a number with digits after the decimal point. It is a number which is not an integer.

decimal fraction A decimal fraction is a fraction between 0 and 1 in which the denominator is a power of 10 and the numerator is an integer.

decimal place The decimal place is the number of digits after the decimal point. For example, 3.2 has 1 decimal place and 5.678 has 3 decimal places.

denominator The denominator is the bottom line of a fraction; it tells you how many equal parts the whole is divided into. For example, $\frac{3}{8}$ has a denominator of 8, so the 'whole' has been divided into 8 equal parts.

density Density is a measure of the mass of a substance per unit of its volume. It is calculated using the formula: density = $\frac{\text{mass}}{\text{volume}}$

depreciate When the value of something decreases over a period of time, it is said to depreciate.

derivative Differentiating a function produces the derivative (or gradient function).

diameter A diameter is a straight line which passes through the centre of a circle and joins two points on the circumference.

difference of two squares The difference of two squares is an expression in the form $x^2 - y^2$. Note, $x^2 - y^2$ factorises to give $(x + y)(x - y)$.

differentiation Differentiation is the process of finding the derivative or gradient function.

direct proportion Two quantities, x and y, are in direct proportion when the ratio $\frac{y}{x}$ is a constant. So $y = kx$ and the graph of y against x is a straight line passing through the origin. An increase in one quantity causes an increase in the other. For example, when the amount of ingredients is doubled for some cakes, the number of cakes made also doubles.

directed number A directed number is a number that is positive or negative. A number has size (magnitude) and its sign (+ or −) tells you which *direction* to move along a number line from 0 in order to reach that number.

discount An item sold at 10% discount is 10% cheaper than the full selling price.

discrete data Discrete data is numerical data that can only take on certain values, usually whole numbers. For example, the number of peas in a pod is discrete data.

distance between two points The distance, d, between two points (x_1, y_1) and (x_2, y_2) is:

$d = \sqrt{(x_1 - x_2)^2 + (y_1 - y_2)^2}$

division Division is one of the four operations: addition, subtraction, multiplication and division. To divide one number by another means to find how many times one number goes into another number. For example,
$20 \div 5 = 4$
$5 \div 20 = 0.25$

double time Overtime is often paid at a higher rate. When overtime is paid at twice the basic pay, it is called double time.

dual bar chart A dual bar chart shows the bars side by side.

element An object or symbol in a set is called an element.

elevation An elevation is a two-dimensional view of a three-dimensional object. A side elevation is the view from one side of the object and the front elevation is the view from the front.

elimination method The elimination method is a method for solving simultaneous equations. One of the unknowns is eliminated by either adding or subtracting the pair of equations.

enlargement An enlargement changes the size of an object. When a shape is enlarged, the image is mathematically similar to the object but is a different size. Note: the image may be larger or smaller than the original object.

equation An equation says that one expression is equal to another. For example, $6 + 4 = 16 - 6$. When an expression contains an unknown, it can be solved. For example, the solution to the equation $x + 4 = 16 - x$ is $x = 6$.

equation of a straight line The equation of a straight line can be written in the form $y = mx + c$, where m is the gradient of the line and the y-intercept is at $(0, c)$.

equilateral triangle An equilateral triangle has three equal angles (all $60°$) and three sides of equal length.

equivalent fraction Equivalent fractions have the same decimal value. For example, $\frac{3}{5} = 0.6$ and $\frac{9}{15} = 0.6$, so $\frac{3}{5}$ and $\frac{9}{15}$ are equivalent.

estimate / estimation Estimation is a way of working out the approximate answer or estimate to a calculation. The numbers in the calculation are rounded (usually to 1 significant figure) so that the calculation is easier to work out without a calculator. Estimation is useful for checking a calculation.

evaluate Evaluate means to work out the value of something.

expand Expand means to multiply out or remove the brackets.

exponential An exponential expression is in the form a^x where a is a positive constant and x is a variable.

exponential equation An exponential equation has a variable (unknown) as the index. For example, $y = 2^x$ is an exponential equation.

exponential function An exponential function is in the form $f(x) = a^x$, where a is a positive constant.

exponential sequence An exponential sequence is a sequence where there is a common ratio (r) between successive terms. The nth term can be written as $T_n = ar^{n-1}$, where a is the first term.

exterior angle The exterior angle of an n-sided regular polygon $= \frac{360°}{n}$.

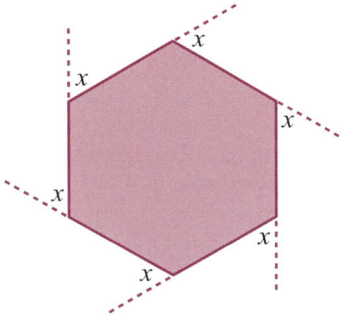

factor A factor of a number divides into that number exactly. For example, the factors of 18 are 1, 2, 3, 6, 9 and 18.

factorise Factorise means to remove common factors and write an equivalent expression using brackets. For example, $2x - 6$ factorises to give $2(x - 3)$.

favourable outcome A favourable outcome refers to the event in question (for example, getting a 6 when a dice is thrown) actually happening.

fraction A fraction represents a part of a whole.

frequency Frequency is the number of times a particular outcome happens.

frequency density frequency density $= \frac{\text{frequency}}{\text{class width}}$

frequency table A frequency table shows the frequency of each data value in a data set.

frustrum A frustrum is the base part of a cone or pyramid when the top of the cone or pyramid is removed.

geometric sequence A geometric sequence is a sequence where the ratio between any two terms is a constant.

gradient Gradient is a measure of how steep a line is. The gradient of the line joining the points (x_1, y_1) and (x_2, y_2) is given by: gradient $= \frac{y_2 - y_1}{x_2 - x_1}$

gradient-intercept form The gradient-intercept form of the equation of a straight line is the form $y = mx + c$, where m is the gradient and c is the y-intercept.

gross earnings Gross earnings are the total earnings *before* all the deductions such as tax, insurance and pension contributions are made.

grouped frequency table A grouped frequency table is a method of displaying a large data set so that it is easier to handle.

height The height of a triangle is the perpendicular distance from its base to its third vertex.

hemisphere A hemisphere is made when a sphere is cut into two congruent halves. It is a half a sphere.

hexagon A hexagon is a 6-sided polygon.

highest common factor The highest common factor (HCF) of two numbers is the greatest integer that divides exactly into both numbers. For example, the highest common factor of 6 and 15 is 3.

histogram A histogram is a chart used to display grouped continuous data as bars. Both axes have continuous scales, and the vertical axis shows the frequency density of each bar.

hyperbola The graph of a reciprocal function in the form $y = \frac{k}{x}$, where k is a constant, is a hyperbola.

hypotenuse The hypotenuse is the longest side of a right-angled triangle.

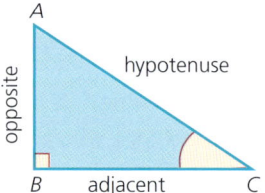

image When an object undergoes a transformation, the resulting position or shape is the image.

improper fraction (or vulgar fraction) In an improper fraction, the numerator is more than the denominator. For example, $\frac{8}{3}$ is an improper fraction.

index The index is the power to which a number is raised. For example, in 4^3 the power (or index) is 3 and so $4^3 = 4 \times 4 \times 4$.

inequality An inequality says that one expression is not equal to a second expression. For example, $x + 2 < 8$ or $7 > 6$

integer An integer is a positive or negative whole number (including zero). The set of integers is $\{..., -3, -2, -1, 0, 1, 2, 3, ...\}$.

interest Interest is the money added by a bank to a sum deposited by a customer. Interest is also the money charged by a bank for a loan to a customer. It can be either simple interest or compound interest (*see* simple interest *and* compound interest).

interior angle The sum of the interior angles of an n-sided polygon is $180(n - 2)°$.

interquartile range interquartile range = upper quartile − lower quartile

intersection The intersection of two sets is the elements that are common to both sets. It is represented by the symbol ∩.

inverse of a function The inverse of a function is its reverse, i.e. it 'undoes' the function's effects. The inverse of the function f(x) is written $f^{-1}(x)$.

inverse proportion Two quantities, x and y, are in inverse proportion when the product of the two quantities xy is constant, i.e. when an increase in one quantity causes a decrease in the second quantity.

irrational number An irrational number is any number (positive or negative) that cannot be written as a fraction. Any decimal which neither terminates nor recurs is irrational. The square root of any number other than square numbers is also irrational. Some examples of irrational numbers are π, $\sqrt{2}$ and $\sqrt{10}$.

irregular polygon An irregular polygon is a polygon that does not have equal sides or equal angles.

isosceles trapezium An isosceles trapezium is a quadrilateral with the following properties:
- one pair of parallel sides
- the other pair of sides are equal in length
- two pairs of equal base angles
- opposite base angles add up to 180°.

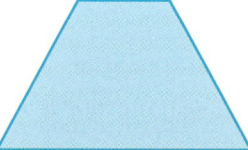

isosceles triangle An isosceles triangle has two equal angles and two sides of equal length.

kite A kite is a quadrilateral with the following properties:
- two pairs of equal sides
- one pair of equal angles
- diagonals which cross at right angles.

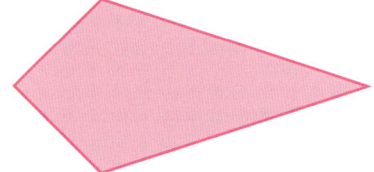

laws of indices The laws of indices are:
- $a^m \times a^n = a^{m+n}$
- $a^m \div a^n$ or $\dfrac{a^m}{a^n} = a^{m-n}$
- $(a^m)^n = a^{mn}$
- $a^1 = a$
- $a^0 = 1$
- $a^{-m} = \dfrac{1}{a^m}$
- $a^{\frac{1}{n}} = \sqrt[n]{a}$
- $a^{\frac{m}{n}} = \sqrt[n]{(a^m)}$ or $\left(\sqrt[n]{(a)}\right)^m$

length The length of a rectangle is the measure of its longest side.

line A line is a one-dimensional object with length but no width. It has infinite length.

line of best fit A line of best fit is a straight line that passes as close as possible to as many points as possible on a scatter diagram.

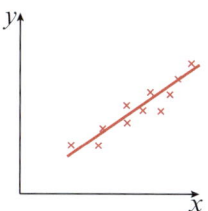

line of symmetry A line of symmetry divides a two-dimensional shape into two congruent (identical) shapes.

line segment A line segment is part of a line.

linear A linear function is a function whose graph is a straight line. It is of the form $f(x) = ax$.

linear equation The graph of a linear equation is a straight line. The highest power of the variable is 1.

linear function The graph of a linear function is a straight line.

linear inequality An inequality says that one expression is not equal to a second expression. For example, x + 2 < 8. In a linear inequality, the highest power of the variable is 1.

local maxima A local maxima of the function f is a stationary point where f(x) reaches a maximum within a given range.

local minima A local minima of the function f is a stationary point where f(x) reaches a minimum within a given range.

loss When an item is sold for less than it cost to make, it is sold at a loss: loss = cost price − selling price

lower bound Measurement is only approximate; the actual value of a measurement could be half the rounded unit above or below the given value. The lower bound is the least possible value that the true measurement could be. For example, the length of a pencil is 15.5 cm to the nearest millimetre. So the lower bound is 15.5 cm − 0.5 mm = 15.45 cm. The actual length, l, of the pencil is greater than or equal to 15.45 cm, so $15.45 \leqslant l$.

lower quartile The lower quartile is the 25th percentile.

lowest common multiple The lowest common multiple (LCM) of two numbers is the lowest integer that is a multiple of both numbers. For example, the lowest common multiple of 6 and 15 is 30.

lowest terms (or simplest form) A fraction is in its lowest terms when the highest common factor of the numerator and denominator is 1. In other words, the fraction cannot be cancelled down any further. For example, $\dfrac{30}{45} = \dfrac{6}{9} = \dfrac{2}{3}$, so $\dfrac{2}{3}$ is a fraction in its lowest terms.

magnitude Magnitude means size.

mean The mean is found by adding together all of the data values and then dividing this total by the number of data values. The mean is one of the three ways to measure an average.

median The median is the middle value when the data set is organised in order of size. The median is one of the three ways to measure an average.

metric units of capacity
1 litre (l) = 1000 millilitres (ml)   1 ml = 1 cm^3

metric units of length
1 kilometre (km) = 1000 metres (m)
1 metre (m) = 100 centimetres (cm)
1 centimetre (cm) = 10 millimetres (mm)

metric units of mass 1 tonne (t) = 1000 kilograms (kg)
1 kilogram (kg) = 1000 grams (g)   1 gram (g)
= 1000 milligrams (mg)

midpoint The midpoint of a line segment AB, where $A(x_1, y_1)$ and $B(x_2, y_2)$ is: $\left(\dfrac{x_1 + x_2}{2}, \dfrac{y_1 + y_2}{2}\right)$

mirror line The mirror line is the line in which an object is reflected.

mixed number A mixed number is made up of a whole number and a proper fraction. For example, $2\dfrac{3}{8}$ is a mixed number.

modal class The modal class is the class or group in a grouped frequency table with highest frequency.

mode The mode is the value occurring most often in a data set. The mode is one of the three ways to measure an average.

multiple The multiple of a number is the result when you multiply that number by a positive integer. For example, the multiples of 6 are 6, 12, 18, 24, 30, …

multiplication Multiplication is one of the four operations: addition, subtraction, multiplication and division. Multiplication is repeated addition, so 3 multiplied by 4 means $3 + 3 + 3 + 3$.

natural number A natural number is a whole number (integer) that is used in counting and starts at zero. The set of natural numbers is $\{0, 1, 2, 3, …\}$

negative correlation Two quantities have negative correlation if, in general, one decreases as the other increases.

negative enlargement In a negative enlargement, the object and image are on opposite sides of the centre of enlargement. The scale factor of enlargement is negative.

negative number A negative number is any number less than 0.

net A net is a two-dimensional shape which can be folded up to form a three-dimensional shape.

net pay Net pay is sometimes called 'take-home' pay. It is the money left *after* all the deductions such as tax, insurance and pension contributions are made.

no correlation Correlation is the relationship between two sets of data. If there is no correlation, then there is no relationship between the two data sets. In a scatter graph showing no correlation, there is no pattern in the plotted points.

numerator The numerator is the top line of a fraction. It represents the number of equal parts of the whole. For example, $\frac{3}{8}$ has a numerator of 3, so there are 3 equal parts and each part is equal to $\frac{1}{8}$ of the 'whole'.

obtuse angle An obtuse angle lies between 90° and 180°.

obtuse-angled triangle In an obtuse-angled triangle, one angle is greater than 90°.

octagon An octagon is an 8-sided polygon.

opposite In a right-angled triangle, the opposite side is the one which is opposite the angle.

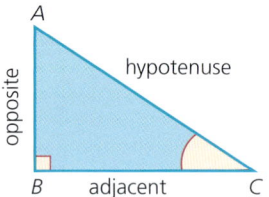

order of operations When a calculation contains a mixture of brackets and/or the operations ($\times$, $\div$, $+$ and $-$), the order that the operations should be carried out in is:
- First work out any …   Brackets and Indices
- … then carry out any …   Multiplication and Division
- … finally   Addition and Subtraction

When a calculation contains operations of equal priority (e.g. $+$ and $-$, or $\times$ and $\div$), work from left to right. For example, $10 - 7 + 2 = 3 + 2 = 5$.

order of rotational symmetry The order of rotational symmetry is the number of times a shape, when rotated about a central point, fits its outline during a complete revolution of 360°.

origin The origin is the point at which the x-axis and the y-axis meet.

overtime Overtime is any hours worked in excess of the basic week.

parabola The graph of a quadratic function is a parabola.

parallel A pair of parallel lines can be continued to infinity in either direction without meeting. Parallel lines have the same gradient.

parallelogram A parallelogram is a quadrilateral with the following properties:
- two pairs of parallel sides
- opposite sides are equal
- opposite angles are equal.

pentagon A pentagon is a 5-sided polygon.

per cent (%) Per cent means parts per 100.

percentage A percentage is the number of parts per 100.

percentage interest (or interest rate) Interest is earned on a fixed percentage of the principal. The interest rate gives the percentage interest earned.

percentage loss percentage loss $= \frac{\text{loss}}{\text{cost price}} \times 100\%$

percentage profit percentage profit $= \frac{\text{profit}}{\text{cost price}} \times 100\%$

percentile The cumulative frequency can be divided into percentiles. The maximum value of the cumulative frequency is the 100th percentile.

perfect square A quadratic equation is called a perfect square if it is in the form $y = x^2 + 2ax + a^2$, where a is a constant. This factorises to give $y = (x + a)^2$.

perimeter The perimeter of a shape is the distance around the outside edge of the shape. Perimeter is measured in mm, cm, m, km, etc.

perimeter of a rectangle The perimeter of a rectangle of length l and breadth b is: $2l + 2b$

perpendicular Two lines are perpendicular if they meet at right angles. The product of the gradients of two perpendicular lines is -1. So if the gradient of a line is m_1, then the gradient of a line perpendicular to it is: $m_2 = -\frac{1}{m_1}$

perpendicular bisector The perpendicular bisector of a line AB is another line which meets AB at right angles and cuts AB exactly in half.

pictogram A pictogram is a chart that uses pictures or symbols to display data.

pie chart A pie chart is a circular chart divided into sectors that is used to display data. The area of each sector is proportional to the frequency.

piece work Piece work is when an employee is paid for the number of articles made (rather than the time spent working).

plan A plan of an object is a scale diagram of the view from above the object, looking directly down on the object.

plane of symmetry A plane of symmetry divides a three-dimensional shape into two congruent (identical) three-dimensional shapes.

point A point is an exact location or position.

polygon A polygon is a two-dimensional shape made up of straight lines.

polynomials A polynomial function is a function such as a quadratic, cubic, etc. that includes only non-negative powers of x.

population density Population density is a measure of the population per unit of area. It is calculated using the formula:

population density $= \frac{\text{population}}{\text{area}}$

positive correlation Two quantities have positive correlation if, in general, one increases as the other increases.

positive number A positive number is any number greater than 0.

power For example, in 4^3 the power is 3 and so $4^3 = 4 \times 4 \times 4$.

pressure Pressure is a compound measurement and is measured in Pascals (Pa) or N/m²:
Pressure $= \frac{\text{Force}}{\text{Area}}$

prime factor A prime factor of a number is any factor of that number that is also a prime. For example, the prime factors of 60 are 2, 3 and 5.

prime number A prime number is a number with exactly two factors: one and itself. The prime numbers are 2, 3, 5, 7, 11, … Note: 1 is not a prime number as it only has one factor.

principal The principal is the amount of money deposited by a customer in a bank account.

prism A prism is a three-dimensional object with a constant cross-sectional area.

probability Probability is the study of chance. The probability of an event happening is a measure of how likely that event is to happen. Probability is given on a scale of 0 (an impossible event) to 1 (a certain event):

probability $= \frac{\text{number of favourable outcomes}}{\text{total number of equally likely outcomes}}$

probability scale A probability scale is a scale that indicates how likely an event is, ranging from impossible to certain.

profit When an item is sold for more than it cost to make, it is sold at a profit: profit = selling price − cost price

proper fraction In a proper fraction, the numerator is less than its denominator. For example, $\frac{3}{8}$ is a proper fraction.

pyramid A pyramid is a three-dimensional shape. It has a polygon for a base and the other faces are triangles which meet at a common vertex, called the apex.

Pythagoras' theorem Pythagoras' theorem states the relationship between the lengths of the three sides of a right-angled triangle. Pythagoras' theorem is: $a^2 = b^2 + c^2$

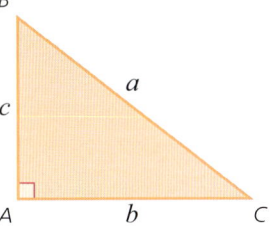

quadratic equation A quadratic equation can be written in the form $y = ax^2 + bx + c$, where a, b and c are constants.

quadratic expression In a quadratic expression, the highest power of any of the terms is 2. A quadratic expression can be written in the form $ax^2 + bx + c$, where a, b and c are constants.

quadratic formula The quadratic formula is used to solve a quadratic equation in the form $ax^2 + bx + c = 0$. The formula is: $x = \frac{-b \pm \sqrt{b^2 - 4ac}}{2a}$

quadratic function A quadratic function is in the form $y = ax^2 + bx + c$, where *a*, *b* and *c* are constants.

quadrilateral A quadrilateral is a 4-sided polygon.

radius A radius is a straight line which joins the centre of a circle to a point on the circumference.

range Range is a measure of the spread of a data set. The range is the difference between the largest and smallest data values.

rate Rate is a ratio of two measurements, usually the second measurement is time. For example, water flows through a pipe at a rate of 1 litre per second or a computer programmer types at a rate of 30 words per minute.

ratio A ratio is the comparison of one quantity with another.

ratio method The ratio method is used to solve problems involving direct proportion by comparing the ratios. For example, a bottling machine fills 500 bottles in 15 minutes. How many bottles will it fill in 90 minutes?

$\frac{x}{90} = \frac{500}{15}$ so $x = \frac{500 \times 90}{15} = 3000$

3000 bottles are filled in 90 minutes.

rational number A rational number is any number (positive or negative) that can be written as a fraction. All integers and all terminating and recurring decimals are rational numbers.

real number The real numbers are the all the rational and irrational numbers. So any integer, fraction or decimal is a real number.

reciprocal The reciprocal of a number is 1 divided by that number. So the reciprocal of 4 is $1 \div 4 = \frac{1}{4}$ = 0.25 and the reciprocal of $\frac{1}{5}$ is $1 \div \frac{1}{5} = 5$.

reciprocal function A reciprocal function is in the form $y = \frac{k}{x}$, where *k* is a constant.

rectangle A rectangle is a quadrilateral with the following properties:
- two pairs of parallel sides
- opposite sides are equal
- four equal angles (each 90°).

recurring decimal A recurring decimal has digits that repeat forever. For example, $\frac{2}{9} = 0.2222... = 0.\dot{2}$ and $\frac{415}{999} = 0.415415415415... = 0.\dot{4}1\dot{5}$

reflection A reflection is a 'flip' movement in a mirror line. The mirror line is the line of symmetry between the object and its image.

reflective symmetry A shape has reflective symmetry if it has one or more lines or planes of symmetry.

reflex angle A reflex angle lies between 180° and 360°.

region A region is a part of a graph, shape or Venn diagram.

regular polygon A regular polygon has all sides of equal length and all angles of equal size.

relative frequency

relative frequency = $\frac{\text{number of successful trials}}{\text{total number of trials}}$

rhombus A rhombus is a quadrilateral with the following properties:
- two pairs of parallel sides
- four equal sides
- opposite angles are equal
- diagonals which cross at right angles.

right angle A right angle is 90°.

right-angled triangle In a right-angled triangle, one angle is 90°.

roots (of an equation) The root(s) of an equation are the value(s) of *x* when $y = 0$. On a graph, these are the values of *x* where the curve crosses the *x*-axis.

rotation A rotation is a 'turning' movement about a specific point known as the centre of rotation.

rotational symmetry A shape has rotational symmetry if, when rotated about a central point, it fits its outline more than once in a complete turn.

round (or rounding) Rounding is a way of rewriting a number so it is simpler than the original number. A rounded number should be approximately equal to the unrounded (exact) number and be of the same order of magnitude (size). Rounded numbers are often given to 2 decimal places (2 d.p.) or 3 significant figures (3 s.f.), for example.

sample space diagram A sample space diagram shows all the possible outcomes of an experiment.

scalar A scalar is a quantity with magnitude (size) only.

scale A scale on a drawing shows the ratio of a length on the drawing to the length on the actual object.

scale factor of enlargement The scale factor of enlargement is the ratio between corresponding sides on an object and its image.

scalene triangle In a scalene triangle, none of the angles are of equal size and none of the sides are of equal length.

scatter diagram A scatter diagram is a graph of plotted points which shows the relationship between two variables.

sector A sector is the region of a circle enclosed by two radii and an arc.

segment A segment is an area of a circle formed by a line (chord) and an arc.

selling price The selling price is the total amount of money that an item is sold for.

semicircle A semicircle is made when a circle is cut into two congruent halves. A semicircle is half a circle.

sequence A sequence is a collection of terms arranged in a specific order, where each term is obtained according to a rule.

set A set is a well-defined group of objects or symbols.

significant figures The first significant figure of a number is the first non-zero digit in the number. The second significant figure is the next digit in the number, and so on. For example, in the numbers 78 046 and 0.0078 046 the first significant figure is 7, the second significant figure is 8 and the third significant figure is 0.

similar Two shapes are similar if the corresponding angles are equal and the corresponding sides are in proportion to each other.

simple interest Simple interest is calculated only on the principal (initial) amount deposited in an account. When simple interest is earned, the amount of interest paid is the same each year.

$$\text{simple interest} = \frac{\text{principal} \times \text{time in years} \times \text{rate per cent}}{100}$$

simplest form (or lowest terms) A fraction is in its simplest form when the highest common factor of the numerator and denominator is 1. In other words, the fraction cannot be cancelled down any further. For example, $\frac{30}{45} = \frac{6}{9} = \frac{2}{3}$, so $\frac{2}{3}$ is a fraction in its simplest form.

simultaneous equations Simultaneous equations are a pair of equations involving two unknowns.

sine The sine of an angle, sin x, in a right-angled triangle is the ratio of the side opposite the angle and the hypotenuse.

$$\sin x = \frac{\text{length of opposite side}}{\text{length of hypotenuse}}$$

sine rule The sine rule is: $\frac{a}{\sin A} = \frac{b}{\sin B} = \frac{c}{\sin C}$ or: $\frac{\sin A}{a} = \frac{\sin B}{b} = \frac{\sin C}{c}$

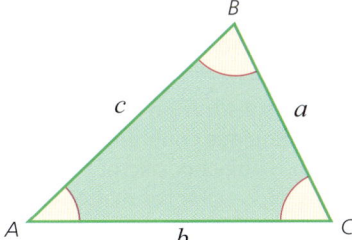

speed speed = $\frac{\text{distance}}{\text{time}}$ When the speed is not constant: average speed = $\frac{\text{total distance}}{\text{total time}}$

sphere A sphere is a three-dimensional shape which is a ball.

square A square is a quadrilateral with the following properties:
- two pairs of parallel sides
- four equal sides
- four equal angles (90°)
- diagonals which cross at right angles.

square number A square number is the result when an integer is multiplied by itself. The square numbers are 1, 4, 9, 16, 25, …

square root The square root of a number is the number which when multiplied by itself gives the original number. The inverse of squaring is square rooting. Every number has two square roots, for example, the square root of 9 is 3 (as $3 \times 3 = 9$) and -3 (as $-3 \times -3 = 9$). The symbol $\sqrt{\ }$ is used for the positive square root of a number, so $\sqrt{9} = 3$.

standard form Standard form is a way of writing very large or very small numbers. A number in standard form is written as $A \times 10^n$, where $1 \leq A < 10$ and n is a positive or negative integer. Examples of numbers in standard from are 5×10^3 and 2.7×10^{-18}.

stationary point A stationary point is a point on a curve where the gradient is zero.

stem-and-leaf diagram A stem-and-leaf diagram is a diagram where each date value is split into two parts – the 'stem' and the 'leaf' (usually the last digit). The data is then grouped so that data values with the same stem appear on the same line.

straight line A straight line is the shortest distance between two points.

subject The subject of a formula is the single variable (often on the left-hand side of a formula) that the rest of the formula is equal to. For example, in $C = 2\pi r$, C is the subject and in $a^2 + b^2 = c^2$, c2 is the subject.

subset When all the elements of set X are also elements of set Y, then X is a subset of Y. Every set has itself and the empty set as subsets.

substitute / substitution Substitution is replacing the variables (letter symbols) in an expression or formula with numbers.

substitution method The substitution method is a method for solving simultaneous equations, where one unknown is made the subject of one of the equations, and then this expression is substituted into the second equation.

subtraction Subtraction is one of the four operations: addition, subtraction, multiplication and division. It means to take one number away from another.

supplementary angle Two angles that add together to total 180° are called supplementary angles.

surd A surd is a square root or cube root of a number which cannot be simplified by removing the root. A surd is an irrational number. For example, $\sqrt{4}$ is not a surd as $\sqrt{4} = 2$. $\sqrt{5} = 2.2360679...$ is a surd.

surface area of a cuboid The surface area of a cuboid of length l, width w and height h is:
surface area $= 2(wl + lh + wh)$

surface area of a cylinder The surface area of a cylinder of radius r and height h is:
surface area $= 2\pi r(r + h)$

surface area of a sphere The surface area of a sphere is $4\pi r^2$.

tally table A tally table is table where the frequencies of each outcome are recorded using marks like ||| for 3 or |||| for 5.

tangent The tangent to a curve at a point is a straight line that just touches the curve at that point. The gradient of the tangent is the same as the gradient of the curve at that point.

tangent (x) The tangent of an angle, tan x, in a right-angled triangle is the ratio of the sides opposite and adjacent to the angle.

$\tan x = \dfrac{\text{length of opposite side}}{\text{length of adjacent side}}$

term Each number in a sequence is called a term.

terminating decimal A terminating decimal has digits after the decimal point that do not continue forever. For example, 0.123 and 0.987654321.

term-to-term rule A term-to-term rule describes how to use one term in a sequence to find the next term.

three-figure bearing A three-figure bearing is a measure of the direction in which an object is travelling. North is 000° and South is 180°.

time and a half Overtime is often paid at a higher rate. When overtime is paid at 1.5 × basic pay, it is called time and a half.

total number of possible outcomes The total number of possible outcomes refers to all the different types of outcomes one can get in a particular situation.

transformation A transformation changes either the position or size of an object, such as translation, rotation, reflection and enlargement.

translation A translation is a sliding movement. Each point on the object moves in the same way to its corresponding point on the image, as described by its translation vector.

translation vector A translation vector describes a translation in terms of its horizontal and vertical movement. For example, the translation vector $\begin{pmatrix} 2 \\ -3 \end{pmatrix}$ describes a translation of 2 units right and 3 units down.

trapezium A trapezium is a quadrilateral with one pair of parallel sides.

travel graph A travel graph is a diagram showing the journey of one or more objects on the same pair of axes. The vertical axis is distance and the horizontal axis is time.

triangle A triangle is a 3-sided polygon.

triangular numbers The triangular numbers are the numbers in the sequence 1, 3, 6, 10, 15, etc., where the difference between terms increases by 1 each time. The formula for the nth triangular number is: $\frac{1}{2}n(n + 1)$

turning point The turning point of a quadratic graph is its highest or lowest point. If the x^2 term is positive, the graph will have a lowest point. If the x^2 term is negative, it will have a highest point.

union The union of two sets is everything that belongs to EITHER or BOTH sets. It is represented by the symbol ∪.

unitary method The unitary method is used to solve problems involving direct proportion by first finding the value of a single unit. For example, 5 pens cost $8. Work out the cost of 7 pens.

5 pens cost $8

1 pen costs $8 ÷ 5 = $1.60 (the cost of 1 unit)

So 7 pens cost $1.60 × 7 = $11.20

universal set The universal set for any particular problem is the set which contains all the possible elements for that problem. It is represented by the symbol $\mathscr{E}$.

upper bound Measurement is only approximate; the actual value of a measurement could be half the rounded unit above or below the given value. The upper bound is the greatest value up to which the true measurement can be. For example, the length of a pencil is 15.5 cm to the nearest millimetre. So, the upper bound is the measurement up to, but not equalling, 15.5 cm + 0.5 mm = 15.55 cm. Therefore, the actual length, l, of the pencil is less than 15.55 cm, so $l < 15.55$.

upper quartile The upper quartile is the 75th percentile.

vector A vector is a quantity with both magnitude (size) and direction. A vector can be used to describe the position of one point in space relative to another.

Venn diagram A Venn diagram is a diagram comprising of overlapping circles, which is used to display sets.

vertex (plural: vertices) A vertex of a shape is a point where two sides meet.

vertically opposite angles Vertically opposite angles are formed when two lines cross. Vertically opposite angles are equal.

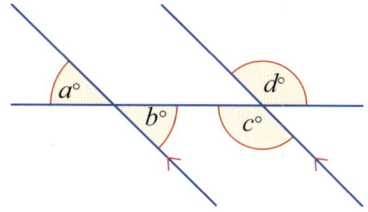

volume (capacity) The volume of a 3D solid is the amount of space the solid fills.

volume factor When shape A is an enlargement by scale factor k of shape B, the volume factor is k^3.

volume of a cylinder The volume of a cylinder of radius r and height h is given by: $volume = \pi r^2 h$

volume of a cone The volume of a cone with height h and a base of radius r is given by: $volume = \frac{1}{3}\pi r^2 h$

volume of a prism The volume of a prism is given by: volume = area of cross-section × length

volume of a pyramid The volume of pyramid is given by: $volume = \frac{1}{3} \times$ area of base × perpendicular height

volume of a sphere The volume of a sphere is given by: $volume = \frac{4}{3}\pi r^3$

x-axis The x-axis is the horizontal axis on a graph.

y-axis The y-axis is the vertical axis on a graph.

Index

Numbers
24-hour clock 89

A
acceleration 182, 185, 189
accuracy 15
acute-angled triangles 291
acute angles 290
addition of fractions 37–8
   algebraic fractions 121, 122
addition of vectors 446
adjacent side 394
algebra
   constructing equations 133–4, 141
   difference of two squares 114–15
   evaluation of expressions 115
   expanding a bracket 108
   expanding a pair of brackets 109, 112–13
   factorising quadratic expressions 116–17
   factorisation by grouping 114
   founders of 107
   linear equations 132
   linear inequalities 148–50
   quadratic equations 143–4, 146, 199–200
   quadratic formula 146
   rearranging formulas 111–12
   simple factorising 110
   simultaneous equations 136–40, 147
   substitution 110–11
algebraic fractions
   addition and subtraction 121–2
   simplifying 120, 123–4
algebraic indices 128–30
al-Karaji 107
al-Khwārizmī 107
Al-Kindi 505
alternate angles 323
alternate segment theorem 337–8
angles 290
   at the centre of a circle 335, 347
   of a circle, ICT investigation 347
   of elevation and depression 405
   in irregular polygons 333
   between a line and a plane 434–5
   in opposite segments of a circle 337
   within parallel lines 323–4
   at a point 322
   in a quadrilateral 327–8
   in the same segment of a circle 336
   in a semicircle 331
   sum of exterior angles of a polygon 330
   sum of interior angles of a polygon 329
   supplementary 337
   between tangent and radius of a circle 332
   in a triangle 325
Apollonius 289
approximation 13
   appropriate accuracy 15
   rounding 13–14
   significant figures 14
   upper and lower bounds 18–20
Archimedes 289
arcs 293
   angles subtended by 335, 347
   length of 369–70
area
   of a circle 360–1
   converting between units 353
   of a parallelogram 358
   of a rectangle 355
   of a sector of a circle 371–2
   of similar shapes (area factor) 301, 305
   of a trapezium 358–9
   of a triangle 356, 427
   see also surface area
area under a graph 184–5
arithmetic sequences see linear sequences
Aryabhata 3
asymptotes
   exponential functions 219
   reciprocal functions 217–18
averages 506, 507
   for grouped data 509
average speed 58, 59, 177
axes 258

B
back bearings 392–3
bar charts 513
base of a triangle 356
bearings 392–3
Bernoulli family 391
Bhascara 3
bounds, upper and lower 18–20
box-and-whisker plots 539–40
brackets 27, 35
   expanding a pair of brackets 112–13
   expanding a single bracket 108–9
Brahmagupta 3
Buffon's needle experiment 503

C
calculators, order of operations 27–8
capacity, units of 350, 352
Cardano, Girolamo 443
centre of enlargement 464–5
centre of rotation 459
chequered boards investigation 254
chords 293
   equal 318
circles 293
   alternate segment theorem 337–8
   angle at the centre of 335
   angle between tangent and radius 332
   angle in a semicircle 331
   angles in opposite segments 337
   angles in the same segment 336
   arc length 369–70
   area 360–1

area of a sector 371–2
circumference 360–1
equal chords and
    perpendicular bisectors
    318
ICT activity 347
tangents from an external
    point 319–20
circumference 293, 360–1
class intervals, histograms 530–1
co-interior angles 324
column vectors 444–5
    see also vectors
combined events 491
common difference, sequences
    157, 158
common ratio, sequences 164
compass points 392
complement of a set 94
completing the square 146
    and graphs of quadratic
        equations 201–2
composite functions 251
compound interest 84–6, 165–6
    formula for 86
compound measures 58–60
conditional probability 496–7
cones
    ICT activity 389
    surface area 384
    volume 380–1
congruent shapes 292, 309
constant of proportionality 170
constructing a triangle 296–7
constructing equations 133–4,
    141
conversion graphs 176
coordinates 258
    equation of a line through
        two points 278
    length of a line segment
        276–7
    midpoint of a line segment
        277
correlation 522, 524
corresponding angles 323
cosine (cos) 399–400
    graph of 411–12, 414
    ICT activity 441
    special angles 408–10
cosine rule 424–5
cube numbers 5, 167
cube roots 10
cubes, nets of 295
cubic functions 204
    sketching 215–16

cubic rules, sequences 162–3
cuboids
    surface area 363–4
    symmetry 314, 315
    trigonometry 430–1, 434–5
cumulative frequency 535–6
    ICT activity 546
    percentiles 537–8
currency conversions 79, 176
curves
    equation of a tangent at a
        given point 238–9
    stationary points 239–42
curves, gradient of 188–9, 191,
    207–8, 224–6
    calculating x for a given
        gradient 236–7
    gradient function 227–9
    at a point 234–5
    see also differentiation
cyclic quadrilaterals 337, 347
cylinders
    surface area 363–4
    volume 366

D

data displays
    bar charts 513
    box-and-whisker plots 539–40
    cumulative frequency
        diagrams 535–6
    grouped frequency tables
        514–15
    histograms 528–31
    pictograms 512
    pie charts 515–17
    scatter diagrams 522–3
    stem-and-leaf diagrams
        513–14
    tally charts 512
data interpretation
    correlation 524
    lines of best fit 523
decagons 295
deceleration 182, 189
decimal places (d.p.) 13–14
decimals 32–3
    conversion from fractions 40
    conversion to and from
        percentages 46
    conversion to fractions 40–1
    fraction and percentage
        equivalents 45
    recurring 41–3
denominator 31

rationalising 73–4
density 58
depreciation 82
depression, angles of 405
derivatives 229–33
    second derivative 233–4
Descartes, René 257
diameter 293
dice sum investigation 502
difference of two squares 114–15
differentiation 229–33
    second derivative 233–4
directed numbers 11
direct proportion 53–4, 170–2
discounts 49, 81
distance–time graphs 178, 180
    non-linear 188, 190
distance travelled 184–5
division
    of fractions 39
    long 35
    short 34
dodecagons 295
domain of a function 245, 246

E

earnings 79–80
Egyptian mathematicians 349
elements of a set 92
elevation, angles of 405
elimination method,
    simultaneous equations
    137
empty set 93
enlargement 464–5
    negative 467
equation of a straight line
    265–6, 270–1
    parallel lines 273
    perpendicular lines 280
    through two points 278
equations
    constructing 133–4, 141
    exponential 65
    linear 132
    quadratic 143–4, 146, 199–200
    simultaneous 136–40, 147
    solving by graphical methods
        208–10
    trigonometric 415–16
equilateral triangles 291
equivalent fractions 36
estimation 15–16
Euclid 289
Eudoxus 289

Euler, Leonhard 391
evaluation of expressions 115
evaluation of functions 247–8
exponential equations 65
    ICT activity 255
exponential functions 206, 218–19
exponential growth (and decay) 86
exponential sequences (geometric sequences) 86, 164
    compound interest 165–6
exterior angles of a polygon 330, 333

F

factorising 110
    difference of two squares 114–15
    by grouping 114
    quadratic expressions 116–17
    solving quadratic equations 143–4
factors 5
    highest common factor 7
    prime 5–6
favourable outcomes 480
Fermat, Pierre de 257, 479
Fibonacci (Leonardo Pisano) 443
football leagues investigation 104
formulas
    quadratic 146
    rearranging 111–12, 118
fountain borders investigation 345
fractional indices 69–70
    algebraic 129
fractions 31
    addition and subtraction 37–8, 121–2
    algebraic 120–4
    conversion from decimals 40–1
    conversion from recurring decimals 41–3
    conversion to and from percentages 46
    conversion to decimals 40
    decimal and percentage equivalents 45
    equivalent 36
    multiplication and division 38–9
    simplest form 36–7
French mathematicians 257, 479
frequency density 531
frequency tables 507, 512
    grouped 514–15
frustums 376
functions
    composite 251
    cubic 204
    domain and range of 246–7
    evaluation of 247–8
    exponential 206
    inverse 250
    linear 203, 212–13
    mappings 245–6
    quadratic 197–8, 201–2, 203, 205
    reciprocal 202, 203, 204, 205

G

Gauss, Carl Friedrich 479
geometric sequences *see* exponential sequences
gradient
    of a curve 188–9, 191, 207–8, 224–6
    of a curve at a point 234–5
    of distance–time graphs 178
    of parallel lines 273
    of perpendicular lines 279–80
    sign of 263
    of speed–time graphs 182
    of a straight line 224, 260–3, 270
gradient function 227–9
    see also differentiation
graphs
    conversion graphs 176
    cubic functions 215–16
    of direct proportion 170–2
    distance–time 178, 180
    exponential functions 206, 218–19
    gradient of a straight line 260–3
    gradients of curves 188–9, 191, 207–8, 224–6
    of inequalities 153–5
    linear functions 212–13
    non-linear travel graphs 187–91
    quadratic functions 197–8, 213–14
    reciprocal functions 202, 217–18
    sketching functions 212–19
    solving equations 199–200, 208–10
    solving simultaneous equations 275–6
    speed–time 181–2, 184–5
    stationary points 239–42
    straight-line 274
    trigonometric functions 411–14
    turning points 201–2
    types of 203–5
Greek mathematicians 289
gross earnings 79
grouped data, mean 509
grouped frequency tables 514–15

H

height data ICT activity 546
heights and percentiles investigation 544–5
hemispheres 375
heptagons 295
hexagons 295
hidden treasure investigation 286–7
highest common factor (HCF) 7
Hindu mathematicians 3
histograms 528–9
    with different class intervals 530–1
house of cards investigation 254
Huygens, Christiaan 479
hyperbolas 202, 203
    see also reciprocal functions
hypotenuse 394

I

ICT activities
    100m sprint 104
    angle properties of a circle 347
    Buffon's needle experiment 503
    cones 389
    exponential equations 255
    height data 546
    inequalities 287
    share prices 104

triangles 346–7
trigonometric ratios 441
vectors 477
images 456
improper fractions 31–2
indices (singular: index)
    algebraic 128–30
    exponential equations 65
    fractional 69–70, 129–30
    laws of 63
    negative 65, 129
    positive 63
    standard form 66–9
    zero index 64, 129
inequalities 25, 148–50
    graphical solution 153–5
    ICT activity 287
inflexion points 240
integers 4
interest
    compound 84–6, 165–6
    simple 82–3
interior angles of a polygon 329, 333
interquartile range 506, 537–8, 539–40
intersection of sets 94–5
inverse functions 250
inverse proportion 57, 172
investigations 102
    chequered boards 254
    dice sum 502
    football leagues 104
    fountain borders 345
    heights and percentiles 544–5
    hidden treasure 286–7
    house of cards 254
    metal trays 388
    mystic rose 102–4
    numbered balls 440
    painted cube 475
    plane trails 285–6
    primes and squares 104
    probability drop 501
    reading ages 545–6
    stretching a spring 255
    tennis balls 388–9
    tiled walls 346
    towers of Hanoi 440–1
    triangle count 476–7
irrational numbers 8
irregular polygons 294
    angle properties 333
isosceles triangles 291
Italian mathematicians 443

K
Khayyam, Omar 107
kites 294

L
Laghada 3
Laplace, Pierre-Simon 479
laws of indices 63
length, units of 350, 351
linear equations 132
linear functions 203
    sketching 212–13
linear inequalities 148–50
linear sequences (arithmetic sequences) 157–9
line segments
    length of 276–7
    midpoint of 277
lines of best fit 523
lines of symmetry 314
long division 35
long multiplication 34
loss 81–2
lower bounds 18–20
lower quartile 537, 538
lowest common multiple (LCM) 7
lowest terms 36–7

M
magnitude of a vector 448
many-to-one functions 245
mappings 245–6
mass, units of 350, 352
maximum points 201–2, 240–2
mean 506, 507
    for grouped data 509
median 506, 507, 514
    from cumulative frequency diagrams 536, 538
metal trays investigation 388
metric units 350
    converting between 351–3
midpoint of a line segment 277
minimum points 201–2, 240–2
mirror lines 456–7
mixed numbers 31–2
modal class 509
mode 506, 507
modelling, stretching a spring 255
money
    currency conversions 79, 176
    earnings 79–80
    interest 82–6

profit and loss 81–2
multiples 7
multiplication, long 34
multiplication of fractions 38–9
multiplication of vectors by a scalar 446–7
mystic rose investigation 102–4

N
natural numbers 4
negative correlation 524
negative enlargement 467
negative gradients 263
negative indices 65, 129
net pay 79
nets 295
Nightingale, Florence 505
nth term rules, sequences 158, 159
    exponential sequences 165
    quadratic and cubic rules 161–3
numbered balls investigation 440
number line 25, 149–50
numbers, types of 4–5, 8, 11
numerator 31

O
obtuse-angled triangles 291
obtuse angles 290
octagons 295
one-to-one functions 245
opposite side 394
order of operations 27–8, 35
order of rotational symmetry 314–15
origin 258
overtime 80

P
painted cube investigation 475
parabolas 197–8, 203, 204, 205
    see also quadratic functions
parallel lines 290
    angles formed within 323–4
    equations of 273
parallelograms 294
    area 358
Parimala, Raman 3
Pascal, Blaise 257, 479
pentagons 295
percentage increases and decreases 49
percentage interest 82

percentage profit and loss 82
percentages 33, 46
    fraction and decimal equivalents 45
    one quantity as percentage of another 48
    reverse 50
percentages of a quantity 47
percentiles 537–8
    heights and percentiles investigation 544–5
perimeter of a rectangle 355
perpendicular bisectors of equal chords 318
perpendicular lines 279–80, 290
pictograms 512
piece work 80
pie charts 515–17
planes of symmetry 314
plane trails investigation 285–6
points
    angles at 322
    gradients at 234–5
    shortest distance from a point to a line 428
polygons 294–5
    regular 329
    sum of exterior angles 330, 333
    sum of interior angles 329, 333
population density 58–60
position vectors 449
positive correlation 522, 524
positive gradients 263
positive indices 63
powers 10
    squares 5
    of two 167
    see also indices
prime factors 5–6
prime numbers 5
primes and squares investigation 104
prisms 365
    surface area 366–7
    volume 366
probability 479–81
    combined events 491
    conditional 496–7
    relative frequency 486–7
    tree diagrams 492, 494
    Venn diagrams 484, 496
probability drop investigation 501

probability scale 481
profit 81–2
proper fractions 31
proportion
    direct 53–4, 170–2
    inverse 57, 172
pyramids 376
    surface area 379
    volume 376–7
Pythagoras 289
Pythagoras' theorem 401

Q

quadratic equations
    completing the square 146, 201–2
    graphical solution 199–200, 209
    solving by factorising 143–4
quadratic expressions, factorising 116–17
quadratic formula 146
quadratic functions 197–8, 203, 205
    sketching 213–14
    turning points 201–2
quadratic rules, sequences 162
quadrilaterals 293–4, 295
    angles in 327–8
    cyclic 337
quartiles 537, 538

R

radius 293
    angle with tangent to a circle 332
range of a function 245
    calculation from the domain 246
range of data 506, 507
    interquartile range 537–8, 539–40
ratio method, direct proportion 53–4
rationalising the denominator 73–4
rational numbers 8
ratios, dividing a quantity in a given ratio 55–6
reading ages investigation 545–6
real numbers 8
rearranging formulas 111–12, 118
    equation of a straight line 271

reciprocal functions 202, 203, 204, 205
    sketching 217–18
reciprocals 4, 39
rectangles 293
    area 355
    perimeter 355
recurring decimals 41
    conversion to fractions 41–3
reflection 456–7
reflective symmetry 314
reflex angles 290
regular polygons 294, 329
relative frequency 486–7
reverse percentages 50
rhombuses 293
right-angled triangles 291, 394
    Pythagoras' theorem 401
    trigonometric ratios 394–400
right angles 290
roots 10–11
    cube 10
    fractional indices 69–70, 129–30
    square 9
    surds 71–4
rotation 459
rotational symmetry 314–15
rounding 13–14

S

scale drawings 297–8
scale factors 301
    of enlargement 464–5
scalene triangles 291
scales 259
scatter diagrams 522–3
second derivative 233–4
    and stationary points 241–2
segments of a circle 293
    alternate segment theorem 337–8
    angles in 336, 347
    angles in opposite segments 337
semicircle, angle in 331
sequences 157
    combinations of 167–8
    exponential 164–6
    key sequences 167
    linear 157–9
    nth term rules 158, 159, 161–3, 165

with quadratic and cubic rules 161–3
   term-to-term rules 158, 159, 164
sets 92
   complement of 94
   empty set 93
   problem solving 98
   subsets 93
   universal set 94
   Venn diagrams 94–5
share prices activity 104
short division 34
shortest distance from a point to a line 428
significant figures (s.f.) 14
similar shapes 292, 295, 301
   area 301, 305
   volume 305
simple interest 82–3
simplest form 36–7
   algebraic fractions 120, 123–4
simultaneous equations 136
   eliminating variables by multiplying 139–40
   elimination method 137
   graphical solution 275–6
   one linear and one non-linear equation 147
   substitution method 138
sine (sin) 397–8
   area of a triangle 427
   graph of 411, 413
   ICT activity 441
   special angles 408–10, 415
sine rule 422–3
   ambiguous case of 423
sketching graphs
   cubic functions 215–16
   exponential functions 218–19
   linear functions 212–13
   quadratic functions 213–14
   reciprocal functions 217–18
speed 58, 59, 177
   from a distance–time graph 178, 180, 188, 191
speed–time graphs 181–2
   area under 184–5
   non-linear 189
spheres
   surface area 375
   volume 374
spreadsheets
   Buffon's needle experiment 503
   cone volume activity 389

share prices activity 104
square numbers 5, 10, 167
square roots 8, 9, 10
squares 4, 293
standard form 66
   negative indices and small numbers 68–9
   positive indices and large numbers 66–7
stationary points 239–42
statistics
   averages 506, 507, 509
   correlation 524
   cumulative frequency 535–6
   data displays 512–18, 522, 528–31, 539–40
   grouped data 509
   in history 505
   percentiles 537–8
   spread of data 506, 507
stem-and-leaf diagrams 513–14
   back-to-back 514
straight-line graphs 274
straight lines
   angles at a point on 322
   equation of 265–6, 270–1, 273
   gradient of 260–3
   parallel 273, 290
   perpendicular 279–80, 290
   shortest distance from a point 428
   through two points, equation of 278
stretching a spring investigation 255
subsets 93
substitution 110–11
substitution method, simultaneous equations 138
subtraction of fractions 37–8
   algebraic fractions 122
subtraction of vectors 446
supplementary angles 337
surds 71
   leaving answers as 401
   rationalising the denominator 73–4
   rules of 72
   simplifying 72
surface area
   of a cone 384
   of a cuboid 363–4
   of a cylinder 363–4
   of a prism 366–7
   of a pyramid 379

   of a sphere 375
surveys 520
Swiss mathematicians 391
symmetry 314–15

T

tally charts 512
tangent (tan, trigonometric ratio) 394–5
   graph of 412–13
   ICT activity 441
   special angles 408–10, 416
tangents 293
   angle with radius of a circle 332
   equation of at a given point 238–9
   from an external point 319–20
tennis balls investigation 388–9
term-to-term rules, sequences
   exponential sequences 164
   linear sequences 157, 159
Thales 289
tiled walls investigation 346
time, 24-hour clock 89
time calculations 89
towers of Hanoi investigation 440–1
transformations 456
   combinations of 469–70
   enlargement 464–7
   reflection 456–7
   rotation 459
   translation 444–5, 462
translation 444–5, 462
trapeziums 294
   area 358–9
travel graphs
   distance–time 178, 180
   non-linear 187–91
   speed–time 181–2, 184–5
tree diagrams 492, 494
triangle count investigation 476–7
triangle numbers 167
triangles 291, 295
   angles in 325
   area 356, 427
   congruent 292, 309
   construction of 296–7
   height (altitude) 290, 356
   ICT activity 346–7
   similar 292, 301
trigonometric equations 415–16
trigonometry

cosine 399–400
cosine rule 424–5
    ICT activity 441
sine 397–8
sine rule 422–3
special angles 408–10, 415–16
tangent 394–5
in three dimensions 430–1, 434–5
truncated pyramids (frustums) 376
turning points 239–42
    quadratic functions 201–2, 214
two-way tables 491

U

union of sets 94–5
unitary method, direct proportion 53–4
units 350
    converting between 351–3
universal set 94
upper bounds 18–20
upper quartile 537, 538

V

vector geometry 450
vectors 444–5
    addition and subtraction 446
    ICT activity 477
    magnitude of 448
    multiplication by a scalar 446–7
    position vectors 449
Venn diagrams 94–5
    probability calculations 484, 496
    problem solving 98
vertex of a triangle 356
vertically opposite angles 323

Villani, Giovanni 505
volume
    of a cone 380–1
    converting between units 353
    of a prism 366
    of a pyramid 376
    of similar shapes (volume factor) 305
    of a sphere 374

X

x-axis 258
x-coordinates 258

Y

y-axis 258
y-coordinates 258
y-intercept 270

Z

zero index 64, 129